THE THEORY OF DETERMINANTS
IN THE HISTORICAL ORDER OF DEVELOPMENT

MACMILLAN AND CO., LIMITED
LONDON · BOMBAY · CALCUTTA
MELBOURNE

THE MACMILLAN COMPANY
NEW YORK · BOSTON · CHICAGO
ATLANTA · SAN FRANCISCO

THE MACMILLAN CO. OF CANADA, LTD.
TORONTO

THE THEORY OF DETERMINANTS

IN THE

HISTORICAL ORDER OF DEVELOPMENT

BY

THOMAS MUIR, C.M.G., LL.D., F.R.S.

SUPERINTENDENT-GENERAL OF EDUCATION IN CAPE COLONY

VOL. II.

THE PERIOD 1841 TO 1860

MACMILLAN AND CO., LIMITED

ST. MARTIN'S STREET, LONDON

1911

PREFACE.

THE story of the production of this volume is quite similar to that of the first. Its bibliographical basis was the three "Lists of Writings on Determinants" formerly referred to (see *History*, I., Introduction). These were followed at irregular intervals by a long series of papers in the *Proceedings of the Royal Society of Edinburgh*, containing historically arranged accounts of the writings in question up to 1860. A certain amount of outside attention having thus been attracted to the subject, neglected writings known to other mathematicians came gradually to be brought to my notice, and of course these were steadily supplemented by assiduous research on my own part. As a consequence, the "Fourth List of Writings," published in 1906,* was able to show a score or so of titles belonging to the period prior to 1860, and the "Fifth List," published about a fortnight ago,† a still greater number of similarly belated items. Considerable intercalations have thus been made in the Edinburgh series of papers, and much care has been bestowed on a complete revision of the whole. It is confidently hoped that now very little matter of real moment has escaped attention.

T. M.

CAPETOWN, SOUTH AFRICA,
19*th July*, 1911.

* *Quart. Journ. of Math.*, xxxvii. pp. 237–264.
† *Quart. Journ. of Math.*, xlii. pp. 343–378.

CONTENTS.

CHAPTER I.

CHAPTER II.

CHAPTER III.

CHAPTER IV.

CHAPTER V.

CHAPTER VI.

CHAPTER VII.

CHAPTER VIII.

CHAPTER XI.

CHAPTER XII.

CHAPTER XIII.

CHAPTER XIV.

CHAPTER XV.

CHAPTER I.

DETERMINANTS IN GENERAL, PRIOR TO 1841.

SINCE the publication of the first volume of this History, two other writings have been discovered which concern the general subject and belong to the period therein dealt with. The authors, Bianchi and Chelini, being very imperfectly acquainted with the relevant literature of the previous part of their own century, could scarcely be expected to make any advance: and, as the following notices will show, no advance was really made.

BIANCHI, G. (1839, January).

[Sopra l'analisi lineare per la risoluzione dei problemi di primo grado. *Mem. della Soc. ital. delle Sci.*, xxii. pp. 184–227.]

Bianchi's knowledge of previous work on simultaneous linear equations must have been slight—confined, probably, to an acquaintance with Cramer's rule and with Cauchy's so-called "symbolical" solution as given in the *Cours d'Analyse* of 1821: unless this were so, he would scarcely have referred to the methods given in such text-books as Ruffini's *Elementary Algebra* and Euler's *Elements*. One is thus prepared to find little new in his conscientiously laboured monograph, consisting of an introduction of five pages, a section of twenty-seven pages on the solution of a set of n equations with n unknowns, and a section of twelve pages on n equations with fewer unknowns. The main interest lies in the first fifteen pages (pp. 189–204) of the earlier section, these being devoted to establishing the validity of Cramer's rule. The procedure consists in eliminating one and the same unknown between the first equation and each of the other equations of the set, then in treating in the same way the set of $n-1$ equations thus derived, and so on until a single equation $Nx_n = D$ results. As negligible factors are not struck out in the course of the work, the discovery of the law of formation of the coefficients in the successive sets of equations is made unnecessarily difficult, and N and D are obtained in unwieldy forms. Thus, in the case of the six equations

$$\left.\begin{array}{l} a_1x_1+b_1x_2+\ldots\ldots+f_1x_6 = s_1 \\ a_2x_1+b_2x_2+\ldots\ldots+f_2x_6 = s_2 \\ \ldots\ldots\ldots\ldots\ldots\ldots \end{array}\right\}$$

the expression found for the last coefficient of x_6 is, in later notation,

$$|a_1b_2c_3d_4e_5f_6| \cdot |a_1b_2c_3d_4| \cdot |a_1b_2c_3|^2 \cdot |a_1b_2|^4 \cdot a_1^8,$$

and for the term independent of x_6

$$|a_1b_2c_3d_4e_5s_6| \cdot |a_1b_2c_3d_4| \cdot |a_1b_2c_3|^2 \cdot |a_1b_2|^4 \cdot a_1^8,$$

with the result, of course, that

$$x_6 = \frac{|a_1b_2c_3d_4e_5s_6|}{|a_1b_2c_3d_4e_5f_6|}.$$

It will readily be agreed that this procedure, though fresh, is not an improvement on others previously known.

CHELINI, D. (1840).

[Formazione e dimostrazione della formula che dà i valori delle incognite nelle equazioni di primo grado. *Giornale Arcadico di Sci. . . .* , lxxxv. pp. 3–12.]

The writings known to Chelini were Terquem's *Manuel d'Algèbre,* Bianchi's paper of 1839, and Molins' of the same year. The paper, however, which his short and clearly written exposition most readily calls to mind is Gergonne's of the year 1813. The "formazione" is essentially Bezout's, and the "dimostrazione" essentially Laplace's.

Attention must also be drawn to a certain neglect in regard to the theorem about the effect of increasing each element of a row (or column) of a determinant by a constant multiple of the corresponding element of another row (or column). It is unquestionably curious that this very elementary property should not have been formulated at a comparatively early date. The writers whose work came nearest to it were Scherk (1825) and Drinkwater (1831): indeed it is little short of marvellous that the latter author should have written his eighth and ninth propositions in the form

$$f(v+w, x, y, z, \dots) = f(v, x, y, z, \dots) + f(w, x, y, z, \dots),$$

$$f(mx, y, z, t, \dots) = m.f(x, y, z, t, \dots),$$

(see *History*, i. p. 199), and should not have added

$$f(v+mx, x, y, z, \dots) = f(v, x, y, z, \dots).$$

Though not formulated, the property in question, however, may actually have been used: and there is certainly evidence of such use in the last year of the period, as may be seen on turning to the closing lines of Jacobi's *De functionibus alternantibus* (see *History*, i. pp. 341–342).

CHAPTER II.

DETERMINANTS IN GENERAL, FROM 1841 TO 1860.

THE number of writings to be considered under this heading is sixty-six, and the number of writers thirty-eight, both numbers being slightly in excess of the corresponding numbers for the period of the whole previous history of the subject. For the first time a fair share of the advance is contributed by English mathematicians: indeed, from being practically of no account, England suddenly leaps to a position of predominating influence. As, therefore, we have already spoken of a *French* period (1693–1812) and a *German* period (1813–1841), it would not be inappropriate to style this the *English* period, or the period of Cayley-and-Sylvester.* Italy also became notably active, emerging from a still more unproductive condition: while Germany and France steadily continued their former labours. The period is also marked by the first appearance of text-books specially devoted to the subject.

CAYLEY, A. (1841).

[On a theorem in the geometry of position. *Cambridge Math. Journ.*, ii. pp. 267–271; or *Collected Math. Papers*, i. pp. 1–4.]

Of the two English mathematicians whose names are inseparably associated with the development of what has been called

* It must be remembered, of course, that Sylvester's work began in the latter part of the previous period (see *History*, i. pp. 227, etc.).

Modern Higher Algebra, Sylvester, as we have seen, was the first to direct public attention to the functions then partially known as determinants, but called by him in the heat of supposed discovery "zetaic products of differences." Cayley it was, however, who gave the great impetus to the study of them—an impetus due to two different causes, the choice of an exceedingly apt notation and the masterly manner in which he put the functions to use. How he obtained his knowledge we know not. It may be that Sylvester's two early papers had directed his attention to the matter, and that he had then read some of the authors who preceded Cauchy; but, whether this be true or not, it is certain that by his own independent research he had attained in 1841 a powerful and comprehensive grasp of the subject. The little paper to which we have now come is ample evidence of this. A peculiar interest attaches to it also, as being the first fruits of Cayley's genius, the earliest of that long and varied series of papers which has done so much to extend the bounds of pure mathematics.*

With characteristic directness and concision he opens as follows:—

"We propose to apply the following (new?) theorem to the solution of two problems in Analytical Geometry.

"Let the symbols

$$|\,\alpha\,|, \quad \begin{vmatrix} \alpha, & \beta \\ \alpha' & \beta' \end{vmatrix}, \quad \begin{vmatrix} \alpha, & \beta, & \gamma \\ \alpha', & \beta', & \gamma' \\ \alpha'', & \beta'', & \gamma'' \end{vmatrix}, \text{ \&c.}$$

denote the quantities

$$\alpha, \quad \alpha\beta' - \alpha'\beta, \quad \alpha\beta'\gamma'' - \alpha\beta''\gamma' + \alpha'\beta''\gamma - \alpha'\beta\gamma'' + \alpha''\beta\gamma' - \alpha''\beta'\gamma, \quad \text{\&c.}$$

the law of whose formation is tolerably well known, but may be thus expressed,

$$|\,\alpha\,| = \alpha, \quad \begin{vmatrix} \alpha, & \beta \\ \alpha', & \beta' \end{vmatrix} = \alpha\,|\,\beta'\,| - \alpha'\,|\,\beta\,|,$$

* In a strictly chronological arrangement Cayley's paper would not follow, but precede the papers of Craufurd, Cauchy, and Jacobi of the same year. It was published in February: Cauchy's note was presented to the Academy on 8th March, and Jacobi's memoir bears the date 17th March, though not published for more than two months afterwards. As Cayley's first appearance, however, marks the beginning of a new epoch, and as the other papers referred to belong by their character to the preceding epoch, a slight deviation from the chronological order seems warranted.

$$\begin{vmatrix} \alpha, & \beta, & \gamma \\ \alpha', & \beta', & \gamma' \\ \alpha'', & \beta'', & \gamma'' \end{vmatrix} = \alpha \begin{vmatrix} \beta', & \gamma' \\ \beta'', & \gamma'' \end{vmatrix} + \alpha' \begin{vmatrix} \beta'', & \gamma'' \\ \beta, & \gamma \end{vmatrix} + \alpha'' \begin{vmatrix} \beta, & \gamma \\ \beta', & \gamma' \end{vmatrix}, \quad \&c.$$

the signs + being used when the number of terms in the side of the square is odd, and + and − alternately when it is even. Then the theorem in question is

$$\begin{vmatrix} \rho\alpha + \sigma\beta + \tau\gamma \ldots, & \rho\alpha' + \sigma\beta' + \tau\gamma' \ldots, & \rho\alpha'' + \sigma\beta'' + \tau\gamma'' \ldots \\ \rho'\alpha + \sigma'\beta + \tau'\gamma \ldots, & \rho'\alpha' + \sigma'\beta' + \tau'\gamma' \ldots, & \rho'\alpha'' + \sigma'\beta'' + \tau'\gamma'' \ldots \\ \rho''\alpha + \sigma''\beta + \tau''\gamma \ldots, & \rho''\alpha' + \sigma''\beta' + \tau''\gamma' \ldots, & \rho''\alpha'' + \sigma''\beta'' + \tau''\gamma'' \ldots \\ \cdot \quad \cdot \quad \cdot & \cdot \quad \cdot \quad \cdot & \cdot \quad \cdot \quad \cdot \\ \cdot \quad \cdot \quad \cdot & \cdot \quad \cdot \quad \cdot & \cdot \quad \cdot \quad \cdot \end{vmatrix}$$

$$= \begin{vmatrix} \rho, & \sigma, & \tau \ldots \\ \rho', & \sigma', & \tau' \ldots \\ \rho'', & \sigma'', & \tau'' \ldots \\ \cdot & \cdot & \cdot \\ \cdot & \cdot & \cdot \end{vmatrix} \begin{vmatrix} \alpha, & \beta, & \gamma \ldots \\ \alpha', & \beta', & \gamma' \ldots \\ \alpha'', & \beta'', & \gamma'' \ldots \\ \cdot & \cdot & \cdot \\ \cdot & \cdot & \cdot \end{vmatrix}.$$

"This theorem admits of a generalisation which we shall not have occasion to make use of, and which therefore we may notice at another opportunity."

Here, then, we have for the first time in the notation of determinants the pair of upright lines so familiar in all the later work. The introduction of them marks an epoch in the history, so important to the mathematician is this apparently trivial matter of notation. By means of them every determinant became representable, no matter how heterogeneous or complicated its elements might be; and the most disguised member of the family could be exhibited in its true lineaments. While the common characteristic of previous notations is their ability to represent the determinant of such a system as

$$\begin{matrix} a_1 & a_2 & a_3 \\ b_1 & b_2 & b_3 \\ c_1 & c_2 & c_3 \end{matrix} \qquad \text{or} \qquad \begin{matrix} a_{1,1} & a_{1,2} & a_{1,3} \\ a_{2,1} & a_{2,2} & a_{2,3} \\ a_{3,1} & a_{3,2} & a_{3,3} \end{matrix}$$

and failure to represent in the case of systems like

$$\begin{matrix} a & b & c \\ c & a & b \\ b & c & a, \end{matrix} \qquad \begin{matrix} a & b & c \\ 1 & a & b \\ 0 & 1 & a, \end{matrix} \qquad \begin{matrix} 4 & 5 & 6 \\ 3 & 2 & 7 \\ 8 & 1 & 0: \end{matrix}$$

Cayley's notation is equally suitable for all. To illustrate by analogy,—the infinitesimal calculus supplied with Lagrange's notation for the differential coefficient of $\phi(x)$, but unable to symbolise the differential coefficients of such a special function as ax^3+bx^2, or $\log(1-x)/(1+x)$ would be in the exact predicament of the theory of determinants prior to Cayley.

Of less importance is the fact, which the quotation indicates, that Cayley had discovered for himself the multiplication-theorem, but characteristically hesitated to proclaim it *new*: also, that, probably following Vandermonde, he took the recurrent law of formation for his definition, making the signs all + in one case and + and − alternately in the next, exactly as Vandermonde did.

He then proceeds to the seemingly geometrical problem:—

"To find the relation that exists between the distances of five points in space."

"We have, in general, whatever x_1, y_1, z_1, w_1, &c., denote,

$$\begin{vmatrix} x_1^2+y_1^2+z_1^2+w_1^2, & -2x_1, & -2y_1, & -2z_1, & -2w_1, & 1 \\ x_2^2+y_2^2+z_2^2+w_2^2, & -2x_2, & -2y_2, & -2z_2, & -2w_2, & 1 \\ \cdot & \cdot & \cdot & \cdot & \cdot & \cdot \\ \cdot & \cdot & \cdot & \cdot & \cdot & \cdot \\ x_5^2+y_5^2+z_5^2+w_5^2, & -2x_5, & -2y_5, & -2z_5, & -2w_5, & 1 \\ 1 \quad , & 0\,, & 0\,, & 0\,, & 0\,, & 0 \end{vmatrix}$$

multiplied into

$$\begin{vmatrix} 1, & x_1, & y_1, & z_1, & w_1, & x_1^2+y_1^2+z_1^2+w_1^2 \\ 1, & x_2, & y_2, & z_2, & w_2, & x_2^2+y_2^2+z_2^2+w_2^2 \\ \cdot & \cdot & \cdot & \cdot & \cdot & \cdot \\ \cdot & \cdot & \cdot & \cdot & \cdot & \cdot \\ 1, & x_5, & y_5, & z_5, & w_5, & x_5^2+y_5^2+z_5^2+w_5^2 \\ 0, & 0, & 0, & 0, & 0, & 1 \end{vmatrix}$$

$$-\begin{vmatrix} \overline{x_1-x_1}^2+\overline{y_1-y_1}^2+\overline{z_1-z_1}^2+\overline{w_1-w_1}^2, & \overline{x_1-x_2}^2+\ldots, & \overline{x_1-x_3}^2+\ldots, & \overline{x_1-x_4}^2+\ldots, & \overline{x_1-x_5}^2+\ldots, & 1 \\ \overline{x_2-x_1}^2+\ldots\ldots\ldots\ldots, & \overline{x_2-x_2}^2+\ldots, & \overline{x_2-x_3}^2+\ldots, & \overline{x_2-x_4}^2+\ldots, & \overline{x_2-x_5}^2+\ldots, & 1 \\ \cdot\quad, & \cdot & \cdot & \cdot & \cdot & \cdot \\ \cdot\quad, & \cdot & \cdot & \cdot & \cdot & \cdot \\ \overline{x_5-x_1}^2+\ldots\ldots\ldots\ldots, & \overline{x_5-x_2}^2+\ldots, & \overline{x_5-x_3}^2+\ldots, & \overline{x_5-x_4}^2+\ldots, & \overline{x_5-x_5}^2+\ldots, & 1 \\ 1 & 1\,, & 1 & 1\,, & 1\,, & 0 \end{vmatrix}.$$

Putting the w's equal to 0, each factor of the first side of the equation vanishes, and therefore in this case the second side of the equation becomes equal to zero. Hence x_1, y_1, z_1, x_2, y_2, z_2, &c., being the coordinates of the points 1, 2, &c., situated arbitrarily in space, and $\overline{12}^2$, $\overline{13}^2$, &c., denoting the squares of the distances between these points, we have immediately the required relation

$$\left|\begin{array}{cccccc} 0, & \overline{12}^2, & \overline{13}^2, & \overline{14}^2, & \overline{15}^2, & 1 \\ \overline{21}^2, & 0, & \overline{23}^2, & \overline{24}^2, & \overline{25}^2, & 1 \\ \overline{31}^2, & \overline{32}^2, & 0, & \overline{34}^2, & \overline{35}^2, & 1 \\ \overline{41}^2, & \overline{42}^2, & \overline{43}^2, & 0, & \overline{45}^2, & 1 \\ \overline{51}^2, & \overline{52}^2, & \overline{53}^2, & \overline{54}^2, & 0, & 1 \\ 1, & 1, & 1, & 1, & 1, & 0 \end{array}\right| = 0,$$

which is easily expanded, though from the mere number of terms the process is somewhat long."

Than this no better example could have been chosen to illustrate what has just been said above regarding the great advantages of Cayley's notation. As is well known, the result arrived at had been given in forms, lengthy and forbidding, many years before by Lagrange and Carnot. What Cayley did was to rob it of all disguise, by expressing it as the vanishing of an elegantly formed determinant; and secondly, to show that the said determinant vanished because it was eight times the square* of another determinant whose zero character could not be overlooked. As has been implied, the result is purely algebraical, its geometrical character only appearing when x, y, z are taken to denote the coordinates of a point.

The corresponding identities for the cases of four points in a plane and three points in a straight line are given; and the latter of the two is most interestingly shown to be deducible also from the general theory of elimination. This is done as follows:—

"Let $\qquad x_{\prime\prime} - x_{\prime\prime\prime} = \alpha, \quad x_{\prime\prime\prime} - x_{\prime} = \beta, \quad x_{\prime} - x_{\prime\prime} = \gamma;$

then $\qquad \overline{12}^2 = \gamma^2, \quad \overline{23}^2 = \alpha^2, \quad \overline{31}^2 = \beta^2, \quad$ and $\quad \alpha + \beta + \gamma = 0;$

from which α, β, γ are to be eliminated. Multiplying the last equation by $\beta\gamma$, $\gamma\alpha$, $\alpha\beta$, and reducing by the three first,

* The first factor being 16 times the second, and the w's unnecessary.

$$\begin{aligned}
0.\alpha + \overline{12}^2.\beta + \overline{31}^2.\gamma + \quad \alpha\beta\gamma &= 0,\\
\overline{12}^2.\alpha + \quad 0.\beta + \overline{23}^2.\gamma + \quad \alpha\beta\gamma &= 0,\\
\overline{31}^2.\alpha + \overline{32}^2.\beta + \quad 0.\gamma + \quad \alpha\beta\gamma &= 0,\\
\alpha + \quad \beta + \quad \gamma + 0.\alpha\beta\gamma &= 0;
\end{aligned}$$

from which, eliminating α, β, γ, $\alpha\beta\gamma$ by the general theory of simple equations

$$\begin{vmatrix} 0, & \overline{12}^2, & \overline{13}^2, & 1 \\ \overline{21}^2, & 0, & \overline{23}^2, & 1 \\ \overline{31}^2, & \overline{32}^2, & 0, & 1 \\ 1, & 1, & 1, & 0 \end{vmatrix} = 0."$$

The conviction that the identity ought to come out as a result of elimination, and the ingenious fulfilment of it by using the identity $\alpha+\beta+\gamma=0$ after the manner of Sylvester's paper of 1840 are very noteworthy.

It is finally noticed that "the additional equation that exists between the distances of five points on a sphere" can be similarly obtained, and the process is given.

GRUNERT, J. A. (1842).

[Ueber die Theorie der Elimination. *Archiv der Math. u. Phys.*, ii. pp. 76–105, 345–377.]

This paper, extending to more than sixty pages, is little else than an amplified reproduction of work by Cauchy. Nine pages at the beginning concern simultaneous linear equations; the rest is entirely taken up with the various modes of eliminating x between two algebraical equations, $\phi(x)=0$, $\psi(x)=0$.

In the former part, which seems based on the third chapter of the *Cours d'Analyse*, the only fresh matter is a lengthy proof of the proposition that *the difference-product of any number of quantities changes sign when two of the quantities are transposed.* It will suffice to note in regard to it that the so-called inductive method is followed, and that two cases have to be considered, viz. (1) when the new quantity is not one of the two which are interchanged, (2) when it is.

The second part follows closely Cauchy's memoir of 1840.

TERQUEM, O. (1842).

[Notice sur l'élimination. Formules de Cramer. *Nouv. Annales de Math.*, i. pp. 125–131.*]

This is merely a simply written exposition of Cramer's rule, and of Bezout's rule of 1779, and contains nothing noteworthy. It is curious, however, to observe the reason given for directing attention to Cramer's rule,—"Comme ce procédé ne se trouve décrit, que je sache, que dans un seul ouvrage élémentaire français, peu répandu (*Manuel d'Algèbre*, p. 80, 2ᵉ édition, 1836)." This indicates a sad contrast to the state of matters attested to by Gergonne,† showing that there is a fashion which changeth even in things mathematical. The new favourite, it also appears, was Bezout's rule of 1764; for in passing this over, in order to give an account of the same author's rule of later date, Terquem says in regard to it, "Comme ce procédé est décrit dans tous les ouvrages à l'usage des classes, nous ne nous y arrêterons pas."

CAYLEY, A. (1843).

[Demonstration of Pascal's Theorem. *Cambridge Math. Journ.*, iv. pp. 18–20; or *Collected Math. Papers*, i. pp. 43–45.]

At the outset of this paper two lemmas are given, the second of which stands as follows:—

"Lemma 2. Representing the determinants

$$\begin{vmatrix} x_1, & y_1, & z_1 \\ x_2, & y_2, & z_2 \\ x_3, & y_3, & z_3 \end{vmatrix}, \text{ \&c.}$$

by the abbreviated notation $\overline{123}$, &c.; the following equation is identically true:

$$\overline{345}\,.\,\overline{126} - \overline{346}\,.\,\overline{125} + \overline{356}\,.\,\overline{124} - \overline{456}\,.\,\overline{123} = 0.$$

* The continuation intimated at the close (p. 131) was never made.

† The passage in question, which we quoted under Cramer, is to be found in the *Annales de Math.*, xx. p. 45.

This is an immediate consequence of the equations

$$\begin{vmatrix} \cdot & \cdot & x_3, & x_4, & x_5, & x_6 \\ \cdot & \cdot & y_3, & y_4, & y_5, & y_6 \\ \cdot & \cdot & z_3, & z_4, & z_5, & z_6 \\ x_1, & x_2, & x_3, & x_4, & x_5, & x_6 \\ y_1, & y_2, & y_3, & y_4, & y_5, & y_6 \\ z_1, & z_2, & z_3, & z_4, & z_5, & z_6 \end{vmatrix} = \begin{vmatrix} \cdot & \cdot & x_3, & x_4, & x_5, & x_6 \\ \cdot & \cdot & y_3, & y_4, & y_5, & y_6 \\ \cdot & \cdot & z_3, & z_4, & z_5, & z_6 \\ x_1, & x_2, & \cdot & \cdot & \cdot & \cdot \\ y_1, & y_2, & \cdot & \cdot & \cdot & \cdot \\ z_1, & z_2, & \cdot & \cdot & \cdot & \cdot \end{vmatrix} = 0."$$

The identity is readily recognisable as Bezout's (1779). The mode of arriving at it, however, is fresh, and worthy of every attention. The determinant of the sixth order on the left is shown to be equal to zero; and it is implied that the identity is got by transforming the said vanishing determinant into an aggregate of products of pairs of determinants by means of Laplace's expansion-theorem. The method is far-reaching in its application, and manifestly Cayley could have used it to produce a host of identities of similar kind.

The equatement of the two determinants of the sixth order deserves also to be noted, and may be taken as evidence that Cayley was familiar with the theorem that a determinant is not altered if each element of one row be diminished by the corresponding element of another row.

Lastly, it may be pointed out that we have here the first instance of a practice which afterwards became very general, viz., putting a dot instead of a zero element when writing a determinant.

The other lemma and the main body of the paper are geometrical; but as an important determinant identity is implicitly established in the course of the investigation, and as it is of the greatest historical importance to make evident the wonderful command which Cayley with his new notation had suddenly obtained over determinants, we shall give the full text of these portions also, at least up to a certain point.

"Lemma 1. Let $U = Ax + By + Cz = 0$ be the equation of a plane passing through a given point taken for the origin, and consider the planes

$$U_1 = 0, \quad U_2 = 0, \quad U_3 = 0, \quad U_4 = 0, \quad U_5 = 0, \quad U_6 = 0;$$

the condition which expresses that the intersections of the planes (1) and (2), (3) and (4), (5) and (6), lie in the same plane, may be written down under the form.*

$$\begin{vmatrix} A_1 & A_2 & A_3 & A_4 & . & . \\ B_1 & B_2 & B_3 & B_4 & . & . \\ C_1 & C_2 & C_3 & C_4 & . & . \\ . & . & A_3 & A_4 & A_5 & A_6 \\ . & . & B_3 & B_4 & B_5 & B_6 \\ . & . & C_3 & C_4 & C_5 & C_6 \end{vmatrix} = 0.$$

"Consider now the points 1, 2, 3, 4, 5, 6, the coordinates of these being respectively $x_1, y_1, z_1, \ldots\ldots, x_6, y_6, z_6$. I represent, for shortness, the equation to the plane passing through the origin, and the points 1, 2, which may be called the plane $\overline{12}$, in the form

$$x\,\overline{12}_x + y\,\overline{12}_y + z\,\overline{12}_z = 0\,;$$

consequently the symbols $\overline{12}_x$, $\overline{12}_y$, $\overline{12}_z$ denote respectively $y_1z_2 - y_2z_1$, $z_1x_2 - z_2x_1$, $x_1y_2 - x_2y_1$, and similarly for the planes $\overline{13}$, &c. If now the intersections of $\overline{12}$ and $\overline{45}$, $\overline{23}$ and $\overline{56}$, $\overline{34}$ and $\overline{61}$ lie in the same plane, we must have by lemma (1) the equation

$$\begin{vmatrix} 12_x & 45_x & 23_x & 56_x & . & . \\ 12_y & 45_y & 23_y & 56_y & . & . \\ 12_z & 45_z & 23_z & 56_z & . & . \\ . & . & 23_x & 56_x & 34_x & 61_x \\ . & . & 23_y & 56_y & 34_y & 61_y \\ . & . & 23_z & 56_z & 34_z & 61_z \end{vmatrix} = 0.$$

Multiplying the two sides of this equation by the two sides respectively of the equation

$$\begin{vmatrix} x_6 & x_1 & x_2 & . & . & . \\ y_6 & y_1 & y_2 & . & . & . \\ z_6 & z_1 & z_2 & . & . & . \\ . & . & . & x_3 & x_4 & x_5 \\ . & . & . & y_3 & y_4 & y_5 \\ . & . & . & z_3 & z_4 & z_5 \end{vmatrix} = \overline{612}\,.\,\overline{345},$$

and observing the equations

$$x_6\overline{12}_x + y_6\overline{12}_y + z_6\overline{12}_z = \overline{612}, \quad \overline{112} = 0, \quad \text{\&c.}$$

* The commas which Cayley prints after the elements in a determinant we omit here and henceforth.

this becomes

$$\begin{vmatrix} \overline{612} & . & . & . & . & . \\ \overline{645} & \overline{145} & \overline{245} & . & . & . \\ \overline{623} & \overline{123} & . & . & \overline{423} & \overline{523} \\ . & \overline{156} & \overline{256} & \overline{356} & \overline{456} & . \\ . & . & . & . & . & \overline{534} \\ . & . & . & \overline{361} & \overline{461} & \overline{561} \end{vmatrix} = 0,$$

reducible to

$$\overline{612}\,.\,\overline{534} \begin{vmatrix} \overline{145} & \overline{245} & . & . \\ \overline{123} & . & . & \overline{423} \\ \overline{156} & \overline{256} & \overline{356} & \overline{456} \\ . & . & \overline{361} & \overline{461} \end{vmatrix} = 0:$$

or, omitting the factor $\overline{612}\,.\,\overline{534}$, and expanding

$$\overline{145}\,.\,\overline{256}\,.\,\overline{423}\,.\,\overline{361} + \overline{245}\,.\,\overline{123}\,.\,\overline{456}\,.\,\overline{361}$$
$$- \overline{245}\,.\,\overline{123}\,.\,\overline{356}\,.\,\overline{461} - \overline{245}\,.\,\overline{156}\,.\,\overline{423}\,.\,\overline{361} = 0."$$

The purely algebraical identity involved in this is in later notation

$$\left\| \begin{matrix} |y_1z_2| & |y_4z_5| & |y_2z_3| & |y_5z_6| & . & . \\ |z_1x_2| & |z_4x_5| & |z_2x_3| & |z_5x_6| & . & . \\ |x_1y_2| & |x_4y_5| & |x_2y_3| & |x_5y_6| & . & . \\ . & . & |y_2z_3| & |y_5z_6| & |y_3z_4| & |y_6z_1| \\ . & . & |z_2x_3| & |z_5x_6| & |z_3x_4| & |z_6x_1| \\ . & . & |x_2y_3| & |x_5y_6| & |x_3y_4| & |x_6y_1| \end{matrix} \right\| = \left\| \begin{matrix} |x_1y_2z_3| & . & . & |x_4y_2z_3| \\ |x_1y_4z_5| & |x_2y_4z_5| & . & . \\ |x_1y_5z_6| & |x_2y_5z_6| & |x_3y_5z_6| & |x_4y_5z_6| \\ . & . & |x_3y_6z_1| & |x_4y_6z_1| \end{matrix} \right\|.$$

BOOLE, G. (1843).

[On the transformation of multiple integrals. *Cambridge Math. Journ.*, iv. pp. 20–28.]

Boole had to use in his paper the resultant of a system of n linear homogeneous equations, and he therefore thought proper, by way of introduction, to state a mode of forming the resultant, and to prove that the result was correct. As the mode is that in which the rule of signs is dependent on the number of interchanges,* or, as Boole calls them, "binary permutations," any

* See Rothe's paper of the year 1800.

interest attaching to the little exposition is connected with the "proof." The first essential paragraph is:—

"The result of the elimination of the variables from the equations

$$
\begin{aligned}
a_1x_1 + a_2x_2 + \ldots + a_nx_n &= 0,\\
b_1x_1 + b_2x_2 + \ldots + b_nx_n &= 0,\\
\ldots\ldots\ldots\ldots\ldots\ldots\ldots&\ldots\\
r_1x_1 + r_2x_2 + \ldots + r_nx_n &= 0,
\end{aligned}
$$

is an equation of which the second member is 0, and of which the first member is formed from the coefficient of $x_1x_2 \ldots x_n$ in the product of the given equations, by assuming a particular term, as $a_1b_2 \ldots r_n$, positive, and applying to every other term a change of sign for every binary permutation which it may exhibit, when compared with the proposed term $a_1b_2 \ldots r_n$."

The curious point worth noting here is that we are directed first to form the terms of the expression afterwards denoted by $\overset{+}{|} a_1b_2 \ldots r_n \overset{+}{|}$ and called a "permanent," and then to alter the signs of certain terms of it. Boole then proceeds:—

"The truth of the above theorem is shown by the following considerations. The elimination of x_1 from the first and second equation of the system introduces terms of the form $a_1b_2 - a_2b_1$, $a_1b_3 - a_3b_1$, etc., in which the law of binary permutation is apparent, and as we may begin the process of elimination with any variable and with any pair of equations, the law is universal. From the same instance it is evident that no proposed suffix can occur twice in a given term, which condition is also characteristic of the coefficient of $x_1x_2 \ldots x_n$ in the product of the equations of the system, whence the theorem is manifest."

It will be observed that neither the word "determinant" nor the word "resultant" occurs: indeed, throughout the paper, instead of resultant he uses "final derivative," a term which probably may be traced to Sylvester.*

CAYLEY, A. (1843).

[Chapters in the analytical geometry of n dimensions. *Cambridge Math. Journ.*, iv. pp. 119–127; or *Collected Math. Papers*, i. pp. 55–62.]

Of the four short chapters which compose this paper, the only one which concerns us is the first, although in the others deter-

* See Sylvester's paper of 1840.

minants are constantly made use of. At the outset an important notation is introduced which afterwards came to be generally adopted. The passage in regard to it is:—

"Consider the series of terms—

$$\begin{matrix} x_1 & x_2 & \cdots & x_n \\ A_1 & A_2 & \cdots & A_n \\ \cdot & \cdot & \cdots & \cdot \\ K_1 & K_2 & \cdots & K_n, \end{matrix}$$

the number of quantities A, . . . , K being equal to q $(q < n)$. Suppose $q+1$ vertical rows selected, and the quantities contained in them formed into a determinant, this may be done in

$$\frac{n(n-1)\ .\ .\ .\ (q+2)}{1\,.\,2\ .\ .\ .\ (n-q-1)}$$

different ways. The system of determinants so obtained will be represented by the notation

$$\left\| \begin{matrix} x_1 & x_2 & \cdots & x_n \\ A_1 & A_2 & \cdots & A_n \\ \cdot & \cdot & \cdots & \cdot \\ K_1 & K_2 & \cdots & K_n \end{matrix} \right\| ;$$

and the system of equations, obtained by equating each of these determinants to zero, by the notation

$$(3) \qquad \left\| \begin{matrix} x_1 & x_2 & \cdots & x_n \\ A_1 & A_2 & \cdots & A_n \\ \cdot & \cdot & \cdots & \cdot \\ K_1 & K_2 & \cdots & K_n \end{matrix} \right\| = 0."$$

A theorem is next enunciated in regard to the expression of any one of the determinants in terms of $n-q$ of them.

"The $\frac{n(n-1)\ .\ .\ .\ (q+2)}{1\,.\,2\ .\ .\ .\ (n-q-1)}$ equations represented by this formula reduce themselves to $n-q$ independent equations. Imagine these expressed by

$$(1)=0, \quad (2)=0, \quad \ldots, \quad (n-q)=0,$$

any one of the determinants is reducible to the form

$$\Theta_1(1) + \Theta_2(2) + \ .\ .\ .\ + \Theta_{n-q}(n-q)$$

where $\Theta_1, \Theta_2, \ldots, \Theta_{n-q}$ are coefficients independent of $x_1, x_2, \ldots, x_n$."

No proof is given.

The introduction of the notation is fully justified by two theorems which follow. The first is virtually to the effect that we may multiply both sides of (3) by the determinant

$$(5) \qquad \begin{vmatrix} \lambda_1 & \lambda_2 & \dots & \lambda_n \\ \mu_1 & \mu_2 & \dots & \mu_n \\ \cdot & \cdot & \cdot & \cdot \\ \tau_1 & \tau_2 & \dots & \tau_n \end{vmatrix}$$

just as if (3) were a single equation instead of $C_{n,q+1}$ equations, and as if the left-hand side were a determinant; and the result, written in the form

$$(6) \qquad \left\Vert \begin{matrix} \lambda_1 x_1 + \dots + \lambda_n x_n & \mu_1 x_1 + \dots + \mu_n x_n & \dots & \tau_1 x_1 + \dots + \tau_n x_n \\ \lambda_1 A_1 + \dots + \lambda_n A_n & \mu_1 A_1 + \dots + \mu_n A_n & \dots & \tau_1 A_1 + \dots + \tau_n A_n \\ \cdot\ \cdot\ \cdot & \cdot\ \cdot\ \cdot & & \cdot\ \cdot\ \cdot \\ \lambda_1 K_1 + \dots + \lambda_n K_n & \mu_1 K_1 + \dots + \mu_n K_n & \dots & \tau_1 K_1 + \dots + \tau_n K_n \end{matrix} \right\Vert = 0$$

will be true; that is to say, we shall have a new set of $C_{n,q+1}$ equations, which follows logically from the original set. Further. and conversely, if the set (6) hold, we can deduce the set (3) provided that the determinant (5) be not zero. The other theorem is quite similar, being to the effect that the equations (3) may be replaced by the set

$$(8) \qquad \left\Vert \begin{matrix} x_1 & x_2 & \dots & x_n \\ \lambda_1 A_1 + \dots + \omega_1 K_1 & \lambda_1 A_2 + \dots + \omega_1 K_2 & \dots & \lambda_1 A_n + \dots + \omega_1 K_n \\ \cdot\ \cdot\ \cdot & \cdot\ \cdot\ \cdot & \dots & \cdot\ \cdot\ \cdot \\ \lambda_q A_1 + \dots + \omega_q K_1 & \lambda_q A_2 + \dots + \omega_q K_2 & \dots & \lambda_q A_n + \dots + \omega_q K_n \end{matrix} \right\Vert = 0,$$

and that conversely from the set (8) the set (3) is deducible provided the determinant

$$\begin{vmatrix} \lambda_1 & \mu_1 & \dots & \omega_1 \\ \lambda_2 & \mu_2 & \dots & \omega_2 \\ \cdot & \cdot & \cdot & \cdot \\ \lambda_q & \mu_q & \dots & \omega_q \end{vmatrix}$$

be not zero.

As the "derivation of coexistence" came prominently before us in examining Sylvester's early work, it may be noted here in passing that Cayley's second chapter, extending to about a page, consists of the enunciation of a theorem on this subject.

HESSE, O. (1843).

[Ueber die Bildung der Endgleichung, welche durch Elimination einer Variabeln aus zwei algebraischen Gleichungen hervorgeht, und die Bestimmung ihres Grades. *Crelle's Journ.*, xxvii. pp. 1–5; or *Werke*, pp. 83–88.]

Hesse, at this time, must have been unaware of Richelot's paper (dated from the same University), and Grunert's paper, not to speak of writings published outside Germany, for the method which he gives of finding the final equation is nothing more nor less than Sylvester's dialytic method. His exposition, to say the least, is not preferable to Grunert's, and the determinant of the $(m+n)^{\text{th}}$ order which he prints is misleading in points of detail.

CAYLEY, A. (1843).

[On the theory of determinants. *Transac. Cambridge Philos. Soc.*, viii. pp. 1–16; or *Collected Math. Papers*, i. pp. 63–79.]

Up to this point Cayley had dealt with determinants, only, as it were, incidentally. Now, however, he devotes a memoir of sixteen quarto pages to the study of them.

The introductory page shows a pretty wide acquaintance with previous writings on the subject, the authors mentioned being Cramer, Bezout (1764), Laplace, Vandermonde, Lagrange,* Bezout (1779), Gauss, Binet, Cauchy (1812), Lebesgue, Jacobi (1841), and Cauchy (1841).

The first section of the paper is said to deal with "the properties of determinants considered as *derivational functions*."

* As the memoir of Lagrange which Cayley refers to is not one of those brought into notice in the early part of our history, but is one bearing the title "Sur le problème de la détermination des orbites des comètes d'après trois observations," it may be well to mention that the substance of the only sentence in it which concerns us had already appeared in the memoir of 1773. The sentence is

"De là il s'ensuit aussi qu'on aura

$$\begin{aligned}(t''u'-t'u'')^2 &= (x''z'-x'z'')^2+(y''z'-y'z'')^2+(x''y'-x'y'')^2,\\ &= (x'^2+y'^2+z'^2)(x''^2+y''^2+z''^2)-(x'x''+y'y''+z'z'')^2."\end{aligned}$$

—*Nouv. Mém. de l'Acad. Roy.* . . . (Berlin), ann. 1778, p. 160.

As a matter of fact, however, a close examination shows that the functions whose properties are investigated are not strictly determinants, but belong to a class afterwards named *bipartites* by Cayley himself. It is true that it is the determinant notation which is employed in specifying the functions, but this is due to the fact that the bipartite under discussion is of a very special type, and so happens to be expressible as a determinant.

The function U from which he considers his three determinants to be "derived" is

$$\begin{aligned} &x(\alpha\xi + \beta\eta + \ldots\ldots) \\ +\ &y(\alpha'\xi + \beta'\eta + \ldots\ldots) \\ +\ &\ldots\ldots\ldots\ldots \end{aligned}$$

there being n lines and n terms in each line. This at a somewhat later date (1855) he would have denoted by

$$\left(\begin{array}{|cccc|} \alpha & \beta & \ldots\ldots \\ \alpha' & \beta' & \ldots\ldots \\ \ldots & \ldots & \ldots \end{array} \between \xi, \eta, \ldots \between x, y, \ldots\right)$$

and called a *bipartite*. A still later notation is

$$\begin{array}{ccc|c} \xi & \eta & \ldots\ldots & \\ \hline \alpha & \beta & \ldots\ldots & x \\ \alpha' & \beta' & \ldots\ldots & y \\ \ldots & \ldots & \ldots & . \end{array}$$

from which each term of the final expansion is very readily obtained by multiplying an element, β' say, of the square array by the two elements (y, η) which lie in the same row and column with it but outside the array. The three determinants which are viewed as "derivational functions" of this function U are

$$\left|\begin{array}{ccc} \alpha & \beta & \ldots\ldots \\ \alpha' & \beta' & \ldots\ldots \\ \ldots & \ldots & \ldots \end{array}\right|,$$

$$-\left|\begin{array}{cccc} & Ax + A'y + \ldots & Bx + B'y + \ldots & \ldots\ldots \\ R\xi + S\eta + \ldots & \alpha & \beta & \ldots\ldots \\ R'\xi + S'\eta + \ldots & \alpha' & \beta' & \ldots\ldots \\ \ldots & \ldots & \ldots & \ldots \end{array}\right|,$$

and

$$-\left|\begin{array}{cccc} & \mathrm{R}x+\mathrm{R}'y+\dots & \mathrm{S}x+\mathrm{S}'y+\dots & \dots\dots \\ \mathrm{A}\xi+\mathrm{B}\eta+\dots & \alpha & \beta & \dots\dots \\ \mathrm{A}'\xi+\mathrm{B}'\eta+\dots & \alpha' & \beta' & \dots\dots \\ \dots & \dots & \dots & \dots \end{array}\right|.$$

These are denoted by KU, FU, ℲU; and the closing sentence of the introduction is, "The symbols K, F, Ⅎ possess properties which it is the object of this section to investigate."

KU, it will be observed, is what afterwards came to be called the *discriminant* of U; and FU, ℲU are the results of making certain linear substitutions for the elements of the first row and of the first column of the determinant

$$\left|\begin{array}{ccccc} & x & y & z & \dots\dots \\ \xi & \alpha & \beta & \gamma & \dots\dots \\ \eta & \alpha' & \beta' & \gamma' & \dots\dots \\ \zeta & \alpha'' & \beta'' & \gamma'' & \dots\dots \\ \dots & \dots & \dots & \dots & \dots \end{array}\right|.$$

It is this determinant, therefore, which is under investigation and under comparison with U. That it is a bipartite function of $x, y, z, \dots$ and $\xi, \eta, \zeta, \dots$ is manifest when we think of expanding it according to binary products of the elements of the first row and of the first column, the expression for it in the notation of bipartites being thus seen to be

$$\begin{array}{cccc|c} x & y & z & \dots\dots & \\ \hline -|\beta'\gamma''\dots| & |\alpha'\gamma''\dots| & -|\alpha'\beta''\dots| & \dots\dots & \xi \\ |\beta\gamma''\dots| & -|\alpha\gamma''\dots| & |\alpha\beta''\dots| & \dots\dots & \eta \\ -|\beta\gamma'\dots| & |\alpha\gamma'\dots| & -|\alpha\beta'\dots| & \dots\dots & \zeta \\ \dots & \dots & \dots & \dots & . \end{array}$$

Now the properties of this which are investigated by Cayley are properties possessed by the more general bipartite

$$\begin{array}{cccc|c} x & y & z & \dots\dots & \\ \hline a_1 & a_2 & a_3 & \dots\dots & \xi \\ b_1 & b_2 & b_3 & \dots\dots & \eta \\ c_1 & c_2 & c_3 & \dots\dots & \zeta \\ \dots & \dots & \dots & \dots & . \end{array}$$

which is not expressible in the form of a determinant. So far, therefore, as this section of the memoir is concerned, it is evident that the title is somewhat misleading, and it is unnecessary to enter into detail regarding the properties in question.

In the course of the section, however, having occasion to use Jacobi's theorem regarding a coaxial minor of the adjugate, Cayley gives at the outset a formal proof which it is most important to note, as it is the natural generalisation of Cauchy's proof for the ultimate case, and consequently has since become the standard proof given in text-books. The passage is

"Let A, B, ..., A′, B′, ... be given by the equations

$$A = \begin{vmatrix} \beta' & \gamma' & \dots \\ \beta'' & \gamma'' & \dots \\ . & . & . \end{vmatrix}, \qquad B = \pm\begin{vmatrix} \gamma' & \delta' & \dots \\ \gamma'' & \delta'' & \dots \\ . & . & . \end{vmatrix}, \quad \dots$$

$$A' = \pm\begin{vmatrix} \beta'' & \gamma'' & \dots \\ \beta''' & \gamma''' & \dots \\ . & . & . \end{vmatrix}, \qquad B' = \begin{vmatrix} \gamma'' & \delta'' & \dots \\ \gamma''' & \delta''' & \dots \\ . & . & . \end{vmatrix}, \quad \dots$$

the upper or lower signs being taken according as n is odd or even.

"These quantities satisfy the double series of equations

$$\left.\begin{array}{l} A\alpha + B\beta + \dots\dots = \kappa \\ A\alpha' + B\beta' + \dots\dots = 0 \\ \dots\dots\dots\dots\dots \\ A'\alpha + B'\beta + \dots\dots = 0 \\ A'\alpha' + B'\beta' + \dots\dots = \kappa \\ \dots\dots\dots\dots\dots \\ \\ A\alpha + A'\alpha' + \dots\dots = \kappa \\ A\beta + A'\beta' + \dots\dots = 0 \\ \dots\dots\dots\dots\dots \\ B\alpha + B'\alpha' + \dots\dots = 0 \\ B\beta + B'\beta' + \dots\dots = \kappa \\ \dots\dots\dots\dots\dots \end{array}\right\} \qquad (6)$$

the second side of each equation being 0, except for the rth equation of the rth set of equations in the systems.

"Let λ, μ, ... represent the rth, $(r+1)^{\text{th}}$, ... terms of the series α, β, ...; L, M, ... the corresponding terms of the series A, B, ..., where r is any number less than n, and consider the determinant

$$\left|\begin{array}{ccc} A & \ldots\ldots & L \\ \cdot & \cdot\;\cdot\;\cdot\;\cdot & \cdot \\ A^{(r-1)} & \ldots\ldots & L^{(r-1)} \end{array}\right|$$

which may be expressed as a determinant of the nth order, in the form

$$\left|\begin{array}{cccccc} A & \ldots\ldots & L & 0 & 0 & \ldots \\ \cdot & \cdot\;\cdot\;\cdot\;\cdot & \cdot & \cdot & \cdot & \cdot\;\cdot \\ A^{(r-1)} & \ldots\ldots & L^{(r-1)} & 0 & 0 & \ldots \\ 0 & \ldots\ldots & 0 & 1 & 0 & \ldots \\ 0 & \ldots\ldots & 0 & 0 & 1 & \ldots \\ \cdot & \cdot\;\cdot\;\cdot\;\cdot & \cdot & \cdot & \cdot & \cdot\;\cdot \end{array}\right|.$$

Multiplying this by the two sides of the equation

$$\kappa = \left|\begin{array}{ccc} \alpha & \beta & \ldots \\ \alpha' & \beta' & \ldots \\ \cdot & \cdot & \cdot\;\cdot\;\cdot \end{array}\right|$$

and reducing the result by the equation ($\odot$) [*i.e.* the multiplication-theorem] and the equations (6), the second side becomes

$$\left|\begin{array}{cccccc} \kappa & 0 & \ldots & & & \\ 0 & \kappa & \ldots & & & \\ \cdot & \cdot & \cdot\;\cdot\;\cdot & & & \\ \cdot\;\cdot & \cdot\;\cdot & \kappa & 0 & 0 & \ldots \\ & & 0 & \mu^{(r)} & \nu^{(r)} & \ldots \\ & & 0 & \mu^{(r+1)} & \nu^{(r+1)} & \ldots \\ & & \cdot & \cdot & \cdot & \cdot\;\cdot\;\cdot \end{array}\right|$$

which is equivalent to

$$\kappa^r \left|\begin{array}{ccc} \mu^{(r)} & \nu^{(r)} & \ldots\ldots \\ \mu^{(r+1)} & \nu^{(r+1)} & \ldots\ldots \\ \cdot & \cdot & \cdot\;\cdot\;\cdot\;\cdot\;\cdot \end{array}\right|,$$

or we have the equation

$$\left|\begin{array}{ccc} A & \ldots\ldots & L \\ \cdot & \cdot\;\cdot\;\cdot\;\cdot & \cdot \\ A^{(r-1)} & \ldots\ldots & L^{(r-1)} \end{array}\right| = \kappa^{r-1} \left|\begin{array}{ccc} \mu^{(r)} & \nu^{(r)} & \ldots \\ \mu^{(r+1)} & \nu^{(r+1)} & \ldots \\ \cdot & \cdot & \cdot\;\cdot\;\cdot\;\cdot \end{array}\right|,$$

which in the particular case of $r = n$ becomes

$$\left|\begin{array}{ccc} A & B & \ldots \\ A' & B' & \ldots \\ \cdot & \cdot & \cdot\;\cdot\;\cdot \end{array}\right| = \kappa^{n-1}."$$

The Second Section is said to concern "the notation and properties of certain functions resolvable into a series of determinants," and it is at once seen that the functions in question are obtainable from the use of m sets of n indices in the way in which a determinant is obtainable from only two sets. Sylvester spoke of them later (1851) as *commutants.**

CAUCHY, A. L. (1844).

[Mémoire sur les arrangements que l'on peut former avec des lettres données, et sur les permutations ou substitutions à l'aide desquelles on passe d'un arrangement à un autre. *Exercices d'Analyse et de Phys. Math.*, iii. pp. 151–252; or *Œuvres complètes*, (2) xiii.]

The nature of the connection of this with the theory of determinants is evident from the title. Some of the elementary portions of the memoir had in fact already appeared in Cauchy's determinant papers of the years 1812, 1840, 1841, and have been noted in our accounts of the latter. In these papers, as was natural, only such isolated properties were given as might be of immediate application to the main subject: here we have a methodically arranged and lucidly written *treatise.* As, however, in dealing with permutations the question of signature is not taken up, there is no explicit reference to determinants: and all that is therefore necessary is to direct attention to a storehouse of information regarding a subject closely connected with them.

CAUCHY, A. L. (1844).

[Mémoire sur quelques propriétés des résultantes à deux termes. *Exercices d'Analyse et de Phys. Math.*, iii. pp. 274–304; or *Œuvres complètes*, (2) xiii.]

By "résultantes à deux termes" are meant *determinants of the second order.* The expression recalls "résultantes à deux lettres," used by Binet in his memoir of November 1812; and as

* See Postscript to Cayley's paper "On the Theory of Permutants," *Camb. and Dub. Math. Journ.*, vii. pp. 40–51; or *Collected Math. Papers*, ii. pp. 16–26.

the said memoir is here referred to by Cauchy and contains the foundation of the latter's results, it is not improbable that the one expression suggested the other.

Five theorems with attendant corollaries are carefully formulated and proved, extreme simplicity and fulness of exposition being in evidence throughout. The first three theorems are mere variants of the first case of Binet's multiplication-theorem for two non-quadrate matrices, namely, in later notation—

$$\begin{vmatrix} a_1x_1 + a_2x_2 + a_3x_3 + \dots & a_1y_1 + a_2y_2 + a_3y_3 + \dots \\ \beta_1x_1 + \beta_2x_2 + \beta_3x_3 + \dots & \beta_1y_1 + \beta_2y_2 + \beta_3y_3 + \dots \end{vmatrix}$$

$$\equiv \begin{vmatrix} a_1 & a_2 & a_3 & \dots \\ \beta_1 & \beta_2 & \beta_3 & \dots \end{vmatrix} \cdot \begin{vmatrix} x_1 & x_2 & x_3 & \dots \\ y_1 & y_2 & y_3 & \dots \end{vmatrix}$$

$$= \begin{vmatrix} a_1 & a_2 \\ \beta_1 & \beta_2 \end{vmatrix} \cdot \begin{vmatrix} x_1 & x_2 \\ y_1 & y_2 \end{vmatrix} + \begin{vmatrix} a_1 & a_3 \\ \beta_1 & \beta_3 \end{vmatrix} \cdot \begin{vmatrix} x_1 & x_3 \\ y_1 & y_3 \end{vmatrix} + \dots + \begin{vmatrix} a_2 & a_3 \\ \beta_2 & \beta_3 \end{vmatrix} \cdot \begin{vmatrix} x_2 & x_3 \\ y_2 & y_3 \end{vmatrix} + \dots$$

The real interest arises when the a's and β's of this are so taken that the first determinant of every pair on the right is of the same form as the determinant on the left, and can therefore be expanded in exactly the same way as the latter. The outcome is

"4ᵉ *Théorème.* Soient

$$P_x, \quad P_y, \quad P_z, \quad \dots$$

n fonctions homogènes et linéaires de n variables

$$x, \quad y, \quad z, \quad \dots$$

[viz., $P_x = xP_{x,x} + yP_{x,y} + zP_{x,z} + \dots, \ P_y = \dots$] et nommons

$$\mathrm{P_x}, \quad \mathrm{P_y}, \quad \mathrm{P_z}, \quad \dots$$

ce que deviennent les fonctions P_x, P_y, P_z, ... quand on remplace les n variables x, y, z, . . . par n autres variables

$$\mathrm{x}, \quad \mathrm{y}, \quad \mathrm{z}, \quad \dots$$

[viz., $\mathrm{P_x} = \mathrm{x}P_{x,x} + \mathrm{y}P_{x,y} + \mathrm{z}P_{x,z} + \dots, \ \mathrm{P_y} = \dots$]. Concevons d'ailleurs, que l'on ajoute entre eux les termes de la suite

$$P_x, \quad P_y, \quad P_z, \quad \dots,$$

ou de la suite

$$P_\mathrm{x}, \quad P_\mathrm{y}, \quad P_\mathrm{z}, \quad \dots,$$

respectivement multipliés par les variables

$$x,\ \ y,\ \ z,\ \ \dots\dots,$$

ou par les variables

$$\mathrm{x},\ \ \mathrm{y},\ \ \mathrm{z},\ \ \dots\dots;$$

et nommons

$$\begin{matrix} P & Q \\ \mathrm{Q} & \mathrm{P} \end{matrix}$$

les quatre sommes ainsi obtenus, P étant celle qui renferme les seules variables $x, y, z, \dots$, et P celle qui renferme les seules variables $\mathrm{x}, \mathrm{y}, \mathrm{z}, \dots$, en sorte qu'on ait

$$P = xP_x + yP_y + zP_z + \dots,\qquad Q = \mathrm{x}P_x + \mathrm{y}P_y + \mathrm{z}P_z + \dots$$
$$\mathrm{Q} = x\mathrm{P_x} + y\mathrm{P_y} + z\mathrm{P_z} + \dots,\qquad \mathrm{P} = \mathrm{x}\mathrm{P_x} + \mathrm{y}\mathrm{P_y} + \mathrm{z}\mathrm{P_z} + \dots$$

La résultante

$$P\mathrm{P} - Q\mathrm{Q},$$

formée avec ces quatres sommes, dépendra uniquement des binômes qui représentent les divers termes de la série

$$x\mathrm{y} - \mathrm{x}y,\quad x\mathrm{z} - \mathrm{x}z,\quad \dots\dots,\quad y\mathrm{z} - \mathrm{y}z,\quad \dots\dots$$

et sera une fonction de ces binômes, non-seulement entière, mais encore homogène et du second degré."

The full meaning of the theorem and the mode of establishing it will be readily understood from working out the case where the number of terms in each element of the initial determinant is *three*. The process is—

$$\begin{vmatrix} P & Q \\ \mathrm{Q} & \mathrm{P} \end{vmatrix} \equiv \begin{vmatrix} xP_x+yP_y+zP_z & \mathrm{x}P_x+\mathrm{y}P_y+\mathrm{z}P_z \\ x\mathrm{P_x}+y\mathrm{P_y}+z\mathrm{P_z} & \mathrm{x}\mathrm{P_x}+\mathrm{y}\mathrm{P_y}+\mathrm{z}\mathrm{P_z} \end{vmatrix},$$

$$\equiv \begin{vmatrix} P_x & P_y & P_z \\ \mathrm{P_x} & \mathrm{P_y} & \mathrm{P_z} \end{vmatrix} \cdot \begin{vmatrix} x & y & z \\ \mathrm{x} & \mathrm{y} & \mathrm{z} \end{vmatrix}$$

$$= \begin{vmatrix} P_x & P_y \\ \mathrm{P_x} & \mathrm{P_y} \end{vmatrix} \cdot \begin{vmatrix} x & y \\ \mathrm{x} & \mathrm{y} \end{vmatrix} + \begin{vmatrix} P_x & P_z \\ \mathrm{P_x} & \mathrm{P_z} \end{vmatrix} \cdot \begin{vmatrix} x & z \\ \mathrm{x} & \mathrm{z} \end{vmatrix} + \begin{vmatrix} P_y & P_z \\ \mathrm{P_y} & \mathrm{P_z} \end{vmatrix} \cdot \begin{vmatrix} y & z \\ \mathrm{y} & \mathrm{z} \end{vmatrix},$$

$$= \begin{vmatrix} xP_{x,x}+yP_{x,y}+zP_{x,z} & xP_{y,x}+yP_{y,y}+zP_{y,z} \\ \mathrm{x}P_{x,x}+\mathrm{y}P_{x,y}+\mathrm{z}P_{x,z} & \mathrm{x}P_{y,x}+\mathrm{y}P_{y,y}+\mathrm{z}P_{y,z} \end{vmatrix} \cdot \begin{vmatrix} x & y \\ \mathrm{x} & \mathrm{y} \end{vmatrix}$$
$$+ \begin{vmatrix} xP_{x,x}+yP_{x,y}+zP_{x,z} & xP_{z,x}+yP_{z,y}+zP_{z,z} \\ \mathrm{x}P_{x,x}+\mathrm{y}P_{x,y}+\mathrm{z}P_{x,z} & \mathrm{x}P_{z,x}+\mathrm{y}P_{z,y}+\mathrm{z}P_{z,z} \end{vmatrix} \cdot \begin{vmatrix} x & z \\ \mathrm{x} & \mathrm{z} \end{vmatrix}$$
$$+ \begin{vmatrix} xP_{y,x}+yP_{y,y}+zP_{y,z} & xP_{z,x}+yP_{z,y}+zP_{z,z} \\ \mathrm{x}P_{y,x}+\mathrm{y}P_{y,y}+\mathrm{z}P_{y,z} & \mathrm{x}P_{z,x}+\mathrm{y}P_{z,y}+\mathrm{z}P_{z,z} \end{vmatrix} \cdot \begin{vmatrix} y & z \\ \mathrm{y} & \mathrm{z} \end{vmatrix},$$

$$
\begin{aligned}
= \begin{vmatrix} x & y & z \\ \mathrm{x} & \mathrm{y} & \mathrm{z} \end{vmatrix} \cdot \begin{vmatrix} P_{x,x} & P_{x,y} & P_{x,z} \\ P_{y,x} & P_{y,y} & P_{y,z} \end{vmatrix} \cdot \begin{vmatrix} x & y \\ \mathrm{x} & \mathrm{y} \end{vmatrix} &+ \begin{vmatrix} x & y & z \\ \mathrm{x} & \mathrm{y} & \mathrm{z} \end{vmatrix} \cdot \begin{vmatrix} P_{x,x} & P_{x,y} & P_{x,z} \\ P_{z,x} & P_{z,y} & P_{z,z} \end{vmatrix} \cdot \begin{vmatrix} x & z \\ \mathrm{x} & \mathrm{z} \end{vmatrix} \\
&+ \begin{vmatrix} x & y & z \\ \mathrm{x} & \mathrm{y} & \mathrm{z} \end{vmatrix} \cdot \begin{vmatrix} P_{y,x} & P_{y,y} & P_{y,z} \\ P_{z,x} & P_{z,y} & P_{z,z} \end{vmatrix} \cdot \begin{vmatrix} y & z \\ \mathrm{y} & \mathrm{z} \end{vmatrix},
\end{aligned}
$$

$$
\begin{aligned}
= \ & \left\{ \begin{vmatrix} x & y \\ \mathrm{x} & \mathrm{y} \end{vmatrix} \cdot \begin{vmatrix} P_{x,x} & P_{x,y} \\ P_{y,x} & P_{y,y} \end{vmatrix} + \begin{vmatrix} x & z \\ \mathrm{x} & \mathrm{z} \end{vmatrix} \cdot \begin{vmatrix} P_{x,x} & P_{x,z} \\ P_{y,x} & P_{y,z} \end{vmatrix} + \begin{vmatrix} y & z \\ \mathrm{y} & \mathrm{z} \end{vmatrix} \cdot \begin{vmatrix} P_{x,y} & P_{x,z} \\ P_{y,y} & P_{y,z} \end{vmatrix} \right\} \cdot \begin{vmatrix} x & y \\ \mathrm{x} & \mathrm{y} \end{vmatrix} \\
+ & \left\{ \begin{vmatrix} x & y \\ \mathrm{x} & \mathrm{y} \end{vmatrix} \cdot \begin{vmatrix} P_{x,x} & P_{x,y} \\ P_{z,x} & P_{z,y} \end{vmatrix} + \begin{vmatrix} x & z \\ \mathrm{x} & \mathrm{z} \end{vmatrix} \cdot \begin{vmatrix} P_{x,x} & P_{x,z} \\ P_{z,x} & P_{z,z} \end{vmatrix} + \begin{vmatrix} y & z \\ \mathrm{y} & \mathrm{z} \end{vmatrix} \cdot \begin{vmatrix} P_{x,y} & P_{x,z} \\ P_{z,y} & P_{z,z} \end{vmatrix} \right\} \cdot \begin{vmatrix} x & z \\ \mathrm{x} & \mathrm{z} \end{vmatrix} \\
+ & \left\{ \begin{vmatrix} x & y \\ \mathrm{x} & \mathrm{y} \end{vmatrix} \cdot \begin{vmatrix} P_{y,x} & P_{y,y} \\ P_{z,x} & P_{z,y} \end{vmatrix} + \begin{vmatrix} x & z \\ \mathrm{x} & \mathrm{z} \end{vmatrix} \cdot \begin{vmatrix} P_{y,x} & P_{y,z} \\ P_{z,x} & P_{z,z} \end{vmatrix} + \begin{vmatrix} y & z \\ \mathrm{y} & \mathrm{z} \end{vmatrix} \cdot \begin{vmatrix} P_{y,y} & P_{y,z} \\ P_{z,y} & P_{z,z} \end{vmatrix} \right\} \cdot \begin{vmatrix} y & z \\ \mathrm{y} & \mathrm{z} \end{vmatrix},
\end{aligned}
$$

or, in still later notation,

$$
\begin{array}{ccc|c}
|x\,\mathrm{y}| & |x\,\mathrm{z}| & |y\,\mathrm{z}| & \\
\hline
|P_{x,x}\ P_{y,y}| & |P_{x,x}\ P_{y,z}| & |P_{x,y}\ P_{y,z}| & |x\ \mathrm{y}| \\
|P_{x,x}\ P_{z,y}| & |P_{x,x}\ P_{z,z}| & |P_{x,y}\ P_{z,z}| & |x\ \mathrm{z}| \\
|P_{y,x}\ P_{z,y}| & |P_{y,x}\ P_{z,z}| & |P_{y,y}\ P_{z,z}| & |y\ \mathrm{z}|,
\end{array}
$$

—a result which loses half its interest if we do not note that each element of the initial determinant is presentable in the same form, viz.,

$$
P = \begin{array}{ccc|c}
x & y & z & \\
\hline
P_{x,x} & P_{x,y} & P_{x,z} & x \\
P_{y,x} & P_{y,y} & P_{y,z} & y \\
P_{z,x} & P_{z,y} & P_{z,z} & z
\end{array},
\qquad
Q = \begin{array}{ccc|c}
\mathrm{x} & \mathrm{y} & \mathrm{z} & \\
\hline
 & & & x \\
 & & & y \\
 & & & z
\end{array},
$$

$$
\mathrm{Q} = \begin{array}{ccc|c}
x & y & z & \\
\hline
 & & & \mathrm{x} \\
 & & & \mathrm{y} \\
 & & & \mathrm{z}
\end{array},
\qquad
\mathrm{P} = \begin{array}{ccc|c}
\mathrm{x} & \mathrm{y} & \mathrm{z} & \\
\hline
 & & & \mathrm{x} \\
 & & & \mathrm{y} \\
 & & & \mathrm{z}
\end{array}.
$$

The fifth theorem, which is obtained from the fourth by further specialisation, viz., by putting in every instance $P_{r,s} = P_{s,r}$, is enunciated at equal length; and then, evidently for the sake of historical connection, it is illustrated by the two simplest cases, that is to say, the case where the number of variables is two and where the number is three. In the former case "on obtiendra l'équation identique

$$(ax^2+by^2+2cxy)(a\mathrm{x}^2+b\mathrm{y}^2+2c\mathrm{xy}) - \{ax\mathrm{x}+by\mathrm{y}+c(x\mathrm{y}+\mathrm{x}y)\}^2$$
$$= (ab-c^2)(x\mathrm{y}-\mathrm{x}y)^2,$$

qui a été donnée par Lagrange dans les *Mémoires de Berlin* de 1773";

in the latter case, it being explained that

$$\mathrm{A}=bc-d^2,\quad \mathrm{B}=ca-e^2,\quad \mathrm{C}=ab-f^2,$$
$$\mathrm{D}=ef-ad,\quad \mathrm{E}=fd-be,\quad \mathrm{F}=de-cf,$$

and

$$\mathrm{X}=y\mathrm{z}-\mathrm{y}z,\quad \mathrm{Y}=z\mathrm{x}-\mathrm{z}x,\quad \mathrm{Z}=x\mathrm{y}-\mathrm{x}y,$$

"on obtiendra l'équation identique

$$(ax^2+by^2+cz^2+2dyz+2ezx+2fxy)(a\mathrm{x}^2+b\mathrm{y}^2+c\mathrm{z}^2+2d\mathrm{yz}+2e\mathrm{zx}+2f\mathrm{xy})$$
$$-\Big\{ax\mathrm{x}+by\mathrm{y}+cz\mathrm{z}+d(y\mathrm{z}+\mathrm{y}z)+e(z\mathrm{x}+\mathrm{z}x)+f(x\mathrm{y}+\mathrm{x}y)\Big\}^2$$
$$=\mathrm{AX}^2+\mathrm{BY}^2+\mathrm{CZ}^2+2\mathrm{DYZ}+2\mathrm{EZX}+2\mathrm{FXY},$$

que l'on pourrait déduire de l'une des formules données par M. Binet dans le xvie cahier du *Journal de l'École Polytechnique.*"

This latter, for the sake of future reference, it is well to restate in the form

$$\left|\begin{array}{cc} \begin{array}{ccc|c} x & y & z & \\ \hline a & f & e & x \\ f & b & d & y \\ e & d & c & z \end{array} & \begin{array}{ccc|c} \mathrm{x} & \mathrm{y} & \mathrm{z} & \\ \hline a & f & e & x \\ f & b & d & y \\ e & d & c & z \end{array} \\ \\ \begin{array}{ccc|c} x & y & z & \\ \hline a & f & e & \mathrm{x} \\ f & b & d & \mathrm{y} \\ e & d & c & \mathrm{z} \end{array} & \begin{array}{ccc|c} \mathrm{x} & \mathrm{y} & \mathrm{z} & \\ \hline a & f & e & \mathrm{x} \\ f & b & d & \mathrm{y} \\ e & d & c & \mathrm{z} \end{array} \end{array}\right| = \begin{array}{ccc|c} \mathrm{X} & \mathrm{Y} & \mathrm{Z} & \\ \hline \mathrm{A} & \mathrm{F} & \mathrm{E} & \mathrm{X} \\ \mathrm{F} & \mathrm{B} & \mathrm{D} & \mathrm{Y} \\ \mathrm{E} & \mathrm{D} & \mathrm{C} & \mathrm{Z}. \end{array}$$

A concluding paragraph is devoted to noting that $P_{x,x}, P_{x,y}, \ldots$ in the fourth theorem are expressible as halved differential-quotients of P, viz.,

$$P_{x,x}=\tfrac{1}{2}\mathrm{D}_x^2P,\qquad P_{y,y}=\tfrac{1}{2}\mathrm{D}_y^2P,\qquad P_{z,z}=\tfrac{1}{2}\mathrm{D}_z^2P,$$
$$P_{x,y}=\tfrac{1}{2}\mathrm{D}_x\mathrm{D}_yP,\quad P_{x,z}=\tfrac{1}{2}\mathrm{D}_x\mathrm{D}_zP,\quad P_{y,z}=\tfrac{1}{2}\mathrm{D}_y\mathrm{D}_zP;$$

that therefore

$$2P = x^2 D_x^2 P + y^2 D_y^2 P + z^2 D_z^2 P + \ldots$$
$$+ 2xy D_x D_y P + 2xz D_x D_z P + \ldots .$$

and

$$P_x = \tfrac{1}{2} D_x P, \quad P_y = \tfrac{1}{2} D_y P, \quad P_z = \tfrac{1}{2} D_z P, \quad \ldots .$$

After this the second part of the memoir, consisting of geometrical applications, is entered upon.

GRASSMANN, H. (June 1844).

[DIE WISSENSCHAFT DER EXTENSIVEN GRÖSSE, oder die Ausdehnungslehre, eine neue mathematische Disciplin dargestellt und durch Anwendungen erläutert. Erster Theil, die lineale Ausdehnungslehre enthaltend. xxxii+279 pp. Leipzig, 1844. Abstract in *Archiv d. Math. u. Phys.* vi. (1845), pp. 337–369.]

A quite peculiar form of the law of formation of a determinant had its origin with Grassmann. Grassmann, it will be remembered, was one of the most distinguished of the mathematicians who occupied themselves with the search for an *Algebra of directed quantities*, or with the allied problem of the geometrical interpretation of the so-called imaginary expressions of ordinary algebra. By the beginning of the third decade of the century, the way had been gradually, though intermittingly, prepared for important discoveries on the subject by the writings of Wallis (1685), Buée (1805), Argand (1806), Servois (1813), Mourey (1828), Warren (1828), and Gauss (1831).* With Hamilton and Grassmann important discoveries came. Hamilton, whose writings of 1833 and 1835 show that even then he had meditated to some purpose on the matter, announced in 1843 his great invention of Quaternions. In 1844 Grassmann followed with the first part of the Ausdehnungslehre.

In his preface Grassmann explains the steps by which he had been led to his theory. First, there was the question of the addition of directed straight lines (Strecken), or *vectors*, to use

* See art. "Quaternions," by Professor Tait, in *Encyclopædia Britannica*; or Hamilton's *Lectures on Quaternions*.

Hamilton's widely accepted term. This it was unnecessary to linger over, as his predecessors had already dealt satisfactorily with it. Then came the question of multiplication of vectors. Seeing that when a and b represent two lines in magnitude only, in other words, are scalars and not vectors, the product ab represents the rectangle of which a and b are adjacent sides, Grassmann ventured to denote by the product ab, when a and b are vectors, a parallelogram having the vectors for adjacent sides. This definition of multiplication manifestly entailed the result.

$$a^2 = 0;$$

and along with the definition of addition required further that

$$a(b+c) = ab+ac.$$

These two again involved a third, viz.,

$$ab = -ba;$$

for from the two we have

$$\begin{aligned} 0 &= (a+b)^2, \\ &= (a+b)a+(a+b)b, \\ &= a^2+ba+ab+b^2, \\ &= ba+ab. \end{aligned}$$

The remaining steps of the building up of the theory need not be told, as these laws of *outer* multiplication (“*äussere Multiplication*”) suffice for the purpose we have in view.

The exposition of the theory itself is broken up into an introduction and nine chapters, all of them marked by ability and much originality. It is the second chapter which deals specially with outer multiplication, and at the end of it (pp. 70–73) occurs the application which concerns determinants. The matter is introduced by a sentence or two pointing out that it is scarcely to be expected that outer multiplication can be so directly applied to ordinary algebra as to geometry and dynamics, because in ordinary algebra the quantities are essentially alike (*gleichartige*, in the sense of the Ausdehnungslehre), and outer multiplication presupposes the idea of unlikeness. In certain circumstances, however, we are told that we may impose distinctions upon the quantities, and then outer multiplication may be applied with notable results.

"Um hiervon eine Idee zu geben, will ich n Gleichungen ersten Grades mit n Unbekannten setzen, von der Form

$$\begin{aligned} a_1x_1 + a_2x_2 + \ldots\ldots + a_nx_n &= a_0, \\ b_1x_1 + b_2x_2 + \ldots\ldots + b_nx_n &= b_0, \\ \ldots\ldots\ldots\ldots\ldots\ldots \\ s_1x_1 + s_2x_2 + \ldots\ldots + s_nx_n &= s_0, \end{aligned}$$

wo $x_1, \ldots, x_n$ die Unbekannten seien. Hier können wir die Zahlencoefficienten, welche verschiedenen Gleichungen angehören, sofern wir diese Verschiedenheit an ihrem Begriff noch festhalten, als verschiedenartig ansehen, und zwar alle als an sich verschiedenartig, d. h. als unabhängig in dem Sinne unserer Wissenschaft, die einer und derselben Gleichung als unter sich in derselben Beziehung gleichartig. Addiren wir nun in diesem Sinne alle n Gleichungen und bezeichnen die Summe des Verschiedenartigen in dem Sinne unserer Wissenschaft mit dem Verknüpfungszeichen $\dotplus$, indem die gleichen Stellen in den so gebildeten Summenausdrücken immer dem Gleichartigen zukommen sollen, so erhalten wir

$$\begin{aligned} &(a_1 \dotplus b_1 \dotplus \ldots \dotplus s_1)x_1 \dotplus (a_2 \dotplus b_2 \dotplus \ldots \dotplus s_2)x_2 \\ &\qquad \ldots\ldots \dotplus (a_n \dotplus b_n \dotplus \ldots \dotplus s_n)x_n = a_0 \dotplus b_0 \dotplus \ldots \dotplus s_0, \end{aligned}$$

oder bezeichnen wir $(a_1 \dotplus b_1 \dotplus \ldots \dotplus s_1)$ mit p_1, und entsprechend die übrigen Summen, so haben wir

$$p_1x_1 \dotplus p_2x_2 \dotplus \ldots \dotplus p_nx_n = p_0.$$

Aus dieser Gleichung, welche die Stelle jener n Gleichungen vertritt, lässt sich nun auf der Stelle jede der Unbekannten, z. B. x_1 finden, wenn wir die beiden Seiten mit dem äusseren Produkte aus den Coefficienten der übrigen Unbekannten äusserlich multipliciren, also hier mit $p_2p_3\ldots p_n$. Da nämlich, wenn man die Glieder der linken Seite einzeln multiplicirt, nach dem Begriff des äusseren Produktes, alle Produkte wegfallen, welche zwei gleiche Factoren enthalten, so erhält man

$$p_1p_2p_3\ldots p_nx_1 = p_0p_2p_3\ldots p_n.$$

Also da beide Produkte, als demselben System n-ter Stufe angehörig einander gleichartig sind, so hat man

$$x_1 = \frac{p_0p_2p_3\ldots p_n}{p_1p_2p_3\ldots p_n}."$$

The method is thus seen to consist in the deduction of a new equation by addition, and in the elimination of all the unknowns, except one, from the equation, by multiplying both sides by the product of the coefficients of the other unknowns,—the multiplication in question being "outer," and for the purposes of the multiplication, any two coefficients of one and the same equation being considered as "like," and any two belonging to different equations as "unlike." For example, in the case of $n=3$, we have

$$x_1 = \frac{(a_0+b_0+c_0).(a_2+b_2+c_2).(a_3+b_3+c_3)}{(a_1+b_1+c_1).(a_2+b_2+c_2).(a_3+b_3+c_3)},$$
$$= \frac{(a_0a_2+a_0b_2+a_0c_2+b_0a_2+b_0b_2+\ldots).(a_3+b_3+c_3)}{(a_1a_2+a_1b_2+a_1c_2+b_1a_2+b_1b_2+\ldots).(a_3+b_3+c_3)},$$
$$= \frac{(a_0b_2+a_0c_2+b_0a_2+b_0c_2+c_0a_2+c_0b_2).(a_3+b_3+c_3)}{(a_1b_2+a_1c_2+b_1a_2+b_1c_2+c_1a_2+c_1b_2).(a_3+b_3+c_3)},$$

since $a_0a_2 = b_0b_2 = \ldots = c_1c_2 = 0$; and finally

$$x_1 = \frac{a_0b_2c_3-a_0b_3c_2+a_2b_3c_0-a_2b_0c_3+a_3b_0c_2-a_3b_2c_0}{a_1b_2c_3-a_1b_3c_2+a_2b_3c_1-a_2b_1c_3+a_3b_1c_2-a_3b_2c_1},$$

"worin wir, da alles entsprechend geordnet ist, wieder die gewöhnliche Multiplicationsbezeichnung einführen konnten."

All this semblance of demonstration—for it is nothing else—is of little moment compared with the fact sought to be demonstrated, viz., that a determinant is expressible as the outer product of the sums of the elements of its columns. Grassman, however, makes no reference to determinants.

In a paragraph of a subsequent chapter (p. 129), he takes up the problem of elimination between two equations of the mth and n^{th} degrees. What it contains is a reproduction of Sylvester's dialytic method, without any reference to the author of the method.

CAUCHY, A. L. (1845 Augt.).

[Mémoire sur divers théorèmes d'analyse et de calcul intégral. *Comptes rendus .... Acad. des Sci.* (Paris), xxi. pp. 407–415; or *Œuvres complètes* (1), ix. pp. 266–275.]

Cauchy's results are arrived at by taking two different ways of eliminating α, β, γ from the equations

$$\left.\begin{aligned} u_1\alpha + u_2\beta + u_3\gamma &= \alpha s \\ v_1\alpha + v_2\beta + v_3\gamma &= \beta s \\ w_1\alpha + w_2\beta + w_3\gamma &= \gamma s \end{aligned}\right\}$$

where the u's, v's, w's are definable in Cayley's notation by

$$\left(\begin{array}{|ccc|} u_1 & v_1 & w_1 \\ u_2 & v_2 & w_2 \\ u_3 & v_3 & w_3 \end{array}\right) \equiv \left(\begin{array}{|ccc|} a_1x_1+b_1y_1+c_1z_1 & a_2x_1+b_2y_1+c_2z_1 & a_3x_1+b_3y_1+c_3z_1 \\ a_1x_2+b_1y_2+c_1z_2 & a_2x_2+b_2y_2+c_2z_2 & a_3x_2+b_3y_2+c_3z_2 \\ a_1x_3+b_1y_3+c_1z_3 & a_2x_3+b_2y_3+c_2z_3 & a_3x_3+b_3y_3+c_3z_3 \end{array}\right).$$

The most direct mode of elimination gives of course

$$\begin{vmatrix} u_1-s & u_2 & u_3 \\ v_1 & v_2-s & v_3 \\ w_1 & w_2 & w_3-s \end{vmatrix} = 0. \qquad (1)$$

But by writing X, Y, Z for

$$\alpha x_1+\beta x_2+\gamma x_3, \quad \alpha y_1+\beta y_2+\gamma y_3, \quad \alpha z_1+\beta y_2+\gamma z_3,$$

the given set of equations is changed into

$$\left.\begin{aligned} a_1\mathrm{X} + b_1\mathrm{Y} + c_1\mathrm{Z} &= \alpha s \\ a_2\mathrm{X} + b_2\mathrm{Y} + c_2\mathrm{Z} &= \beta s \\ a_3\mathrm{X} + b_3\mathrm{Y} + c_3\mathrm{Z} &= \gamma s \end{aligned}\right\},$$

whence by solution we obtain

$$\left.\begin{aligned} \mathrm{X}\Delta &= (\alpha\mathrm{A}_1+\beta\mathrm{A}_2+\gamma\mathrm{A}_3)s \\ \mathrm{Y}\Delta &= (\alpha\mathrm{B}_1+\beta\mathrm{B}_2+\gamma\mathrm{B}_3)s \\ \mathrm{Z}\Delta &= (\alpha\mathrm{C}_1+\beta\mathrm{C}_2+\gamma\mathrm{C}_3)s \end{aligned}\right\},$$

where Δ is put for $|a_1b_2c_3|$ and $\mathrm{A}_1, \mathrm{A}_2, \ldots, \mathrm{C}_3$ are respectively the cofactors of $a_1, a_2, \ldots, c_3$ in Δ. This, however, if we dispense with the use of X, Y, Z is again a set of linear homogeneous equations in α, β, γ, and elimination gives

$$\begin{vmatrix} x_1\Delta - s\mathrm{A}_1 & x_2\Delta - s\mathrm{A}_2 & x_3\Delta - s\mathrm{A}_3 \\ y_1\Delta - s\mathrm{B}_1 & y_2\Delta - s\mathrm{B}_2 & y_3\Delta - s\mathrm{B}_3 \\ z_1\Delta - s\mathrm{C}_1 & z_2\Delta - s\mathrm{C}_2 & z_3\Delta - s\mathrm{C}_3 \end{vmatrix} = 0. \qquad (2)$$

Now when (1) and (2) have their left-hand members arranged according to ascending powers of s, they become, in later notation,

$$|u_1v_2w_3| - s\Big\{|v_2w_3| + |u_1w_3| + |u_1v_2|\Big\} + s^2(u_1+v_2+w_3) - s^3 = 0,$$

and

$$\Delta^3|x_1y_2z_3| - \Delta^2 s\left\{\begin{vmatrix} x_1 & x_2 & \mathrm{A}_3 \\ y_1 & y_2 & \mathrm{B}_3 \\ z_1 & z_2 & \mathrm{C}_3 \end{vmatrix} + \begin{vmatrix} x_1 & \mathrm{A}_2 & x_3 \\ y_1 & \mathrm{B}_2 & y_3 \\ z_1 & \mathrm{C}_2 & z_3 \end{vmatrix} + \begin{vmatrix} \mathrm{A}_1 & x_2 & x_3 \\ \mathrm{B}_1 & y_2 & y_3 \\ \mathrm{C}_1 & z_2 & z_3 \end{vmatrix}\right\}$$

$$+ \Delta s^2 \left\{ \begin{vmatrix} x_1 & A_2 & A_3 \\ y_1 & B_2 & B_3 \\ z_1 & C_2 & C_3 \end{vmatrix} + \begin{vmatrix} A_1 & x_2 & A_3 \\ B_1 & y_2 & B_3 \\ C_1 & z_2 & C_3 \end{vmatrix} + \begin{vmatrix} A_1 & A_2 & x_3 \\ B_1 & B_2 & y_3 \\ C_1 & C_2 & z_3 \end{vmatrix} \right\}$$

$$- s^3 | A_1B_2C_3 | = 0;$$

and, by equating like powers of s in the two, three results are reached, namely,

$$| u_1v_2w_3 | = \frac{\Delta^3}{| A_1B_2C_3 |} | x_1y_2z_3 |,$$

$$| v_2w_3 | + | u_1w_3 | + | u_1v_2 | = \frac{\Delta^2}{| A_1B_2C_3 |} \left\{ \begin{vmatrix} x_1 & x_2 & A_3 \\ y_1 & y_2 & B_3 \\ z_1 & z_2 & C_3 \end{vmatrix} + \begin{vmatrix} x_1 & A_2 & x_3 \\ y_1 & B_2 & y_3 \\ z_1 & C_2 & z_3 \end{vmatrix} + \begin{vmatrix} A_1 & x_2 & x_3 \\ B_1 & y_2 & y_3 \\ C_1 & z_2 & z_3 \end{vmatrix} \right\};$$

$$u_1 + v_2 + w_3 = \frac{\Delta}{| A_1B_2C_3 |} \left\{ \begin{vmatrix} x_1 & A_2 & A_3 \\ y_1 & B_2 & B_3 \\ z_1 & C_2 & C_3 \end{vmatrix} + \begin{vmatrix} A_1 & x_2 & A_3 \\ B_1 & y_2 & B_3 \\ C_1 & z_2 & C_3 \end{vmatrix} + \begin{vmatrix} A_1 & A_2 & x_3 \\ B_1 & B_2 & y_3 \\ C_1 & C_2 & z_3 \end{vmatrix} \right\}.$$

In the first of these the cofactor of $| x_1y_2z_3 |$ is independent of the x's, y's, z's, so that it does not alter on putting

$$\begin{pmatrix} x_1 & y_1 & z_1 \\ x_2 & y_2 & z_2 \\ x_3 & y_3 & z_3 \end{pmatrix} = \begin{pmatrix} 1 & . & . \\ . & 1 & . \\ . & . & 1 \end{pmatrix};$$

and as on doing so we have

$$\begin{vmatrix} a_1x_1 & a_2x_1 & a_3x_1 \\ b_1y_2 & b_2y_2 & b_3y_2 \\ c_1z_3 & c_2z_3 & c_3z_3 \end{vmatrix} = \frac{\Delta^3}{| A_1B_2C_3 |} \cdot x_1y_2z_3,$$

it is seen that the said cofactor is equal to Δ. This is Cauchy's old theorem regarding the adjugate determinant, and substituting Δ for the cofactor, we have another equally old, namely, the multiplication-theorem.

The reader should note, however, that it is better not to view these old results as the goal of the memoir; but, viewing them as already known, to note the theorem then reached regarding the sum of the m-line coaxial minors of the product-determinant.

CAYLEY, A. (1845).

[On the theory of linear transformations. *Camb. Math. Journ.*, iv. pp. 193–209; or *Collected Math. Papers*, i. pp. 80–94.]

[Mémoire sur les hyperdéterminants. *Crelle's Journ.*, xxx. pp. 1–37.]*

[On linear transformations. *Camb. and Dub. Math. Journ.*, i. pp. 104–122; or *Collected Math. Papers*, i. pp. 95–112.]

These memoirs, afterwards so famous in the history of what is now known as the algebra of quantics, contain exceedingly little on determinants. It is important, however, to direct attention to them, because the basis of them is a generalisation of determinants. Using language which came into vogue two or three years later, we may say that just as the idea and notation of determinants provided the means of expressing *one* of the invariants (viz., the discriminant) of a function, the idea and notation of hyperdeterminants were brought forward for the purpose of expressing *all* the invariants.† The generalisation is of great width, hyperdeterminants including as a very special case the generalisation previously made, viz., *commutants.*

The first memoir gives incidentally a more general mode of using what we may call the *notation of multiple determinants* than that specified in his paper of 1843. The first usage, it will be remembered, is exemplified by

$$\left\| \begin{matrix} a_1 & a_2 & a_3 & a_4 \\ b_1 & b_2 & b_3 & b_4 \end{matrix} \right\| = 0$$

which is meant to signify that

$$\begin{vmatrix} a_1 & a_2 \\ b_1 & b_2 \end{vmatrix} = \begin{vmatrix} a_1 & a_3 \\ b_1 & b_3 \end{vmatrix} = \begin{vmatrix} a_1 & a_4 \\ b_1 & b_4 \end{vmatrix} = \begin{vmatrix} a_2 & a_3 \\ b_2 & b_3 \end{vmatrix} = \begin{vmatrix} a_2 & a_4 \\ b_2 & b_4 \end{vmatrix} = \begin{vmatrix} a_3 & a_4 \\ b_3 & b_4 \end{vmatrix} = 0.$$

A corresponding example of the new usage is

$$\left\| \begin{matrix} a_1 & a_2 & a_3 & a_4 \\ b_1 & b_2 & b_3 & b_4 \end{matrix} \right\| = \left\| \begin{matrix} x_1 & x_2 & x_3 & x_4 \\ y_1 & y_2 & y_3 & y_4 \end{matrix} \right\|,$$

*This is stated to be a translation of the preceding paper, with certain additions by the author; and as such it is not reprinted in *Collected Math. Papers*. It also contains the substance of the paper which follows, the latter having been delayed in publication.

†And indeed the covariants also.

where six equations are again intended to be specified, viz.,

$$\begin{vmatrix} a_1 & a_2 \\ b_1 & b_2 \end{vmatrix} = \begin{vmatrix} x_1 & x_2 \\ y_1 & y_2 \end{vmatrix}, \quad \begin{vmatrix} a_1 & a_3 \\ b_1 & b_3 \end{vmatrix} = \begin{vmatrix} x_1 & x_3 \\ y_1 & y_3 \end{vmatrix}, \quad \cdot \; \cdot \; \cdot \; \cdot$$

each determinant of the one group of six being meant to be equal to the corresponding determinant of the other group.

The example actually employed by Cayley is a result of the multiplication-theorem, and fully justifies the usage. It is

$$\begin{Vmatrix} \lambda\alpha+\lambda'\alpha'+\ldots, & \lambda\beta+\lambda'\beta'+\ldots, & \ldots \\ \mu\alpha+\mu'\alpha'+\ldots, & \mu\beta+\mu'\beta'+\ldots, & \ldots \\ \cdot\;\cdot\;\cdot & \cdot\;\cdot\;\cdot & \cdot \end{Vmatrix} = \begin{vmatrix} \lambda & \mu & \ldots \\ \lambda' & \mu' & \ldots \\ \cdot & \cdot & \cdot \end{vmatrix} \cdot \begin{Vmatrix} \alpha & \beta & \ldots \\ \alpha' & \beta' & \ldots \\ \cdot & \cdot & \cdot \end{Vmatrix}$$

where, of course, the number of columns in the multiplier must be greater than the number in the determinant which is its cofactor.

It may be worth adding that the *Mémoire sur les hyper-déterminants* affords the first instance of the occurrence of Cayley's vertical-line notation in *Crelle's Journal.**

FÉRUSSAC, DE (1845).

[Sur la résolution d'un système général de m équations du premier degré entre m inconnues. *Nouv. Annales de Math.*, iv. pp. 28–32.]

This is a belated contribution, having no connection with any of those immediately preceding it. The author in all probability knew nothing of the subject, with the exception of Cramer's rule, which by this time was almost a century old.

The theorem which he seeks to establish is:—

"Connaissant les valeurs des inconnues d'un système de n équations à n inconnues, pour avoir le dénominateur commun des valeurs d'un système de $n+1$ équations à $n+1$ inconnues, on multiplie le dénominateur du valeur du premier système, par le coefficient de la nouvelle inconnue dans la nouvelle équation. Puis on en retranche les produits respectifs des numérateurs des n inconnues du premier système par leurs coefficients dans la dernière du nouveau système. Quant au numérateur il se forme toujours du dénominateur en remplaçant le coefficient de l'inconnue que l'on considère par le terme tout connu."

* In *Liouville's Journal* brackets, [ ] or { }, were used in Cayley's own papers of the year 1845. See vol. x.

The method of proof is that known as "mathematical induction." The details of it need not be given, as they correspond closely with what are to be found in Scherk's paper of the year 1825, the main differences being that Férussac uses no special determinant notation, and, while clear and simple, is not nearly so lengthy nor so laboriously logical.

TERQUEM, O. (1846).

[Notice sur l'élimination. *Nouv. Annales de Math.*, v. pp. 153–162.]

This is a continuation of Terquem's paper of the year 1842. Just as the previous portion dealt with Cramer and Bezout, this deals with Fontaine (des Bertins), Vandermonde, and Laplace, explaining concisely and clearly their main contributions to the subject.

The only portion of it calling for notice is that in which attention is drawn to the curious fact that Laplace makes no reference to Vandermonde's paper read to the Academy in the preceding year. In regard to this Terquem's remark is—

> "Il est extrêmement probable que Laplace n'a pas pris connaissance du mémoire de son confrère: on sait, d'ailleurs, que les analystes français lisent peu les ouvrages les uns des autres. Ceci nous explique également comment la résolution de l'équation du onzième degré à deux termes, la plus importante découverte de Vandermonde, soit restée ignorée jusqu'à ce qu'elle ait attiré l'attention de Lagrange, après la découverte similaire de M. Gauss."

Not only, however, does this explanation not carry us far, but the question arises whether the point sought to be explained is really the point which stands most in need of explanation. Vandermonde's paper was read at the very beginning of 1771 and Laplace's in 1772: yet in the History of the Academy for the latter year Laplace's occupies pp. 267–376 and Vandermonde's pp. 516-532, and neither refers to the other's work.

It may be noted here that, notwithstanding Terquem's knowledge of the early history of determinants and his manifest desire to induce his readers to take up the subject, he does not himself hold the new weapon with a very firm grasp. For example, in giving in this volume an account of a paper of Grunert's in *Crelle's Journal*, viii. pp. 153–159, in which the author says—

"Entwickeln wir nemlich x', y', z', durch Elimination aus den Gleichungen:

$$x = Ax' + By' + Cz',$$
$$y = A'x' + B'y' + C'z',$$
$$z = A''x' + B''y' + C''z',$$

so erhalten wir:

$$x' = \frac{(B'C'' - B''C')x + (B''C - BC'')y + (BC' - B'C)z}{L},$$
$$y' = \dots\dots$$
$$z' = \dots\dots$$

wenn wir

$$L = AB'C'' - A'BC'' + A''BC' - AB''C' + A'B''C - A''B'C$$

setzen"—

he paraphrases the passage as follows:

"Les équations donnent

$$x' = \frac{x[B'C''] + y[B''C] + z[BC']}{L},$$
$$y' = \dots\dots$$
$$z' = \dots\dots$$

où les crochets représentent des *binômes alternés*;

$$[B'C''] = B'C'' - B''C',$$

et ainsi des autres: L est la *résultante*, dénominateur commun."

The simultaneous use of *binôme alterné* and *résultante* is far from happy.*

*Two years later we find him, in referring to a paper of Cayley's where the determinant

$$\begin{vmatrix} L & T & S & \xi \\ T & M & R & \eta \\ S & R & N & \zeta \\ \xi & \eta & \zeta & \end{vmatrix}$$

occurs, calling it, as he did in 1845, a "fonction cramérienne," and writing it

$$\left\{\begin{matrix} L & T & S & \xi \\ T & M & R & \eta \\ S & R & N & \zeta \\ \xi & \eta & \zeta & \end{matrix}\right\}.$$

See *Nouv. Annales de Math.*, iv. (1845), p. 535; vii. (1848), p. 420.

CATALAN, E. (1846).

[Recherches sur les déterminants. *Bull. de l'Acad. roy. ... de Belgique*, xiii. pp. 534–555.]

As is known, Catalan had already dealt with determinants in the year 1839 in a memoir regarding the change of variables in a multiple integral. In the paper which we have now come to he leads up to examples of the same kind of transformation; but the greater part of it—seventeen out of the total twenty-two pages—is occupied with determinants pure and simple. Half of this amount consists of an elementary exposition of known properties, and calls for no remark save that what Cauchy called "produit principal" or "terme indicatif" is here called "terme caractéristique," and that he makes constant use of the symbolism

$$\text{dét.}(A, B, C, \ldots)$$

to stand for the determinant whose first row consists of a's, second row of b's, and so on: for example,

$$\begin{aligned} \text{dét.}(B, A, C, \ldots) &= -\,\text{dét.}(A, B, C, \ldots), \\ \text{dét.}(A, A, C, \ldots) &= 0, \\ \text{dét.}(A + M, B) &= \text{dét.}(A, B) + \text{dét.}(M, B), \\ \ldots & \ldots \end{aligned}$$

When we come to § 13, however, we find fresh ground struck. The exact words are:—

"Supposons maintenant qu'étant donné la système—

$$\left.\begin{matrix} A_1, \\ A_2, \\ \cdot \\ \cdot \\ \cdot \\ A_n \end{matrix}\right\} \quad \ldots\ldots (A)$$

dont le déterminant est Δ, on ait combiné par voie d'addition et de soustraction les équations dont les premiers membres sont représentés par $A_1, A_2, \ldots, A_n$; et, par exemple, qu'on ait déduit du système (A) le système suivant

$$\left.\begin{matrix} A_1 + A_2 + \ldots + A_n, \\ A_1 - A_2, \\ A_2 - A_3, \\ \cdot\ \cdot\ \cdot\ \cdot \\ A_{n-1} - A_n \end{matrix}\right\} \quad \ldots\ldots (B)$$

dont la considération nous sera utile plus loin. Soit Δ' le déterminant de ce nouveau système : d'après les n^{os} (3) et (4), nous aurons

$$\begin{aligned}\Delta' = &\ \text{dét.}(A_1, -A_2, -A_3, \ldots, -A_n)\\ &+ \text{dét.}(A_1, A_2, -A_3, -A_4, \ldots, -A_n)\\ &+ \text{dét.}(A_1, A_2, A_3, -A_4, \ldots, -A_n)\\ &+ \ldots\ldots\ldots\ldots\\ &+ \text{dét.}(A_n, A_1, A_2, \ldots, A_{n-1}).\end{aligned}$$

On sait que si l'on change les signes des termes d'une colonne horizontale, le déterminant change de signe ; donc

$$\begin{aligned}\Delta' = &(-1)^{n-1}\text{dét.}(A_1, A_2, \ldots, A_n) + (-1)^{n-2}\text{dét.}(A_2, A_1, A_3, \ldots, A_n)\\ &+ (-1)^{n-3}\text{dét.}(A_3, A_1, A_2, A_4, \ldots, A_n) + \ldots\ldots\ldots\\ &+ (-1)\text{dét.}(A_{n-1}, A_1, A_2, \ldots, A_{n-2}, A_n) + \text{dét.}(A_n, A_1, A_2, \ldots A_{n-1}).\end{aligned}$$

Dans la première parenthèse, il n'y a pas d'inversion ; dans la seconde, il y a une inversion, etc. ; donc

$$\Delta' = (-1)^{n-1} n\, \Delta."$$

The theorem thus reached may be enunciated as follows:—*If from a determined Δ of the* nth *order, we form another Δ' such that the first row of Δ' is the sum of all the rows of Δ and every other row of Δ' is got by subtracting the corresponding row of Δ from the row preceding it in Δ, then*

$$\Delta' = (-1)^{n-1} n\, \Delta.$$

In Catalan's notation it is

$$\begin{aligned}&\text{dét.}(A_1 + A_2 + \ldots + A_n, A_1 - A_2, A_2 - A_3, \ldots, A_{n-1} - A_n)\\ &\quad = (-1)^{n-1} n \,.\, \text{dét.}(A_1, A_2, \ldots, A_n),\end{aligned}$$

although, strange to say, it is never so formulated by him.

A generalisation of it is next given by saying :—

"Si la première ligne du système (B) avait renfermé seulement p des quantites $A_1, A_2, \ldots, A_n$, nous aurions trouvé, pour la déterminant de ce système,

$$\Delta' = (-1)^{n-1} p \Delta,"$$

and then there follow a number of applications to the evaluation of certain special determinants. Thus, to take the simplest example, having

$$\Delta = \begin{vmatrix} 1 & . & . & . \\ . & 1 & . & . \\ . & . & 1 & . \\ . & . & . & 1 \end{vmatrix} = 1$$

the theorem gives

$$\begin{vmatrix} 1 & 1 & 1 & 1 \\ 1 & -1 & . & . \\ . & 1 & -1 & . \\ . & . & -1 & -1 \end{vmatrix} = (-1)^3 4\Delta = -4.$$

The other illustrations all concern determinants of the special form afterwards known as "circulants"; for example, C(-1, 1, 1, . . . , 1), C(-1, -1, 1, 1, . . . , 1), etc., C(1, 1, . . . , 1, 0), C(1, 1, . . . , 1, 0, 0), etc. They therefore fall to be dealt with in a different place.

SARRUS, P. F. (1846).

[Finck, P. J. E. Éléments d'Algèbre. Seconde édition. iv+544 pages. Strasbourg.]

In the course of his discussion of the solution of a set of linear equations with three unknowns, the author interjects the following paragraph (No. 52, p. 95):—

"Pour calculer, dans un exemple donné, les valeurs de x, y et z, M. Sarrus a imaginé la méthode pratique suivante, qui est fort ingénieuse. D'abord on peut calculer le dénominateur, et à cet effet on écrit les coefficients des inconnues ainsi

$$\begin{array}{lccc} & a & b & c \\ & a' & b' & c' \\ & a'' & b'' & c'' \\ \text{On répète les trois premiers} & a & b & c \\ \text{et les trois suivants} & a' & b' & c'. \end{array}$$

Actuellement partant de a, on prend diagonalement du haut en bas, en descendant à la fois d'un rang, et reculant d'autant à droite, $ab'c''$: on part de a' de même, et on a $a'b''c$; de a'', et on trouve $a''bc'$; on a ainsi les trois termes positifs (c'est-à-dire à prendre avec leur signes) du dénominateur. On commence ensuite par c et descendant de même vers la gauche on a $cb'a''$, $c'b''a$, $c''ba'$, ou les trois termes négatifs (ou plutôt les termes qu'il faut changer de signe)."

This "méthode pratique" or mnemonic is the original form of the so-called "règle de Sarrus" which came later to have unnecessary prominence given to it by writers on determinants when dealing with those of the third order.*

* The date 1833 has been assigned to this "rule" in a recent German text-book on determinants (Weichold's), but without adducing any evidence.

HANSEN, P. A. (1846, Aug.).

[Ueber eine allgemeine Auflösung eines beliebigen Systems von linearischen Gleichungen. *Berichte . . . k. sächs. Ges. d. Wiss.* (Leipzig), pp. 333–339.]

The solution of a set of linear equations is here looked at from the point of view of an astronomical computer, and though determinants are not used—are, in fact, eschewed—it is interesting to note that the hidden identities on which the new process of solution depends are

$$\frac{|a_3b_2|}{|a_1b_2|} = \frac{a_3}{a_1} + \frac{|a_3b_1|}{|a_1b_2|}\cdot\frac{a_2}{a_1},$$

$$\frac{|a_4b_2c_3|}{|a_1b_2c_3|} = \frac{a_4}{a_1} + \frac{|a_4b_1|}{|a_1b_2|}\cdot\frac{a_2}{a_1} + \frac{|a_4b_1c_2|}{|a_1b_2c_3|}\cdot\frac{|a_2b_3|}{|a_1b_2|},$$

$$\frac{|a_5b_2c_3d_4|}{|a_1b_2c_3d_4|} = \frac{a_5}{a_1} + \frac{|a_5b_1|}{|a_1b_2|}\cdot\frac{a_2}{a_1} + \frac{|a_5b_1c_2|}{|a_1b_2c_3|}\cdot\frac{|a_2b_3|}{|a_1b_2|} + \frac{|a_5b_1c_2d_3|}{|a_1b_2c_3d_4|}\cdot\frac{|a_2b_3c_4|}{|a_1b_2c_3|},$$

$$\cdots\cdots\cdots\cdots\cdots$$

and that these are to be found explicitly stated in Schweins' memoir of the year 1825.

TERQUEM, O. (1846).

[Note sur les équations du premier degré en nombre plus grand que celui des inconnues. . . . *Nouv. Annales de Math.*, v. pp. 551–556.]

Knowing from the four equations (Terquem uses n)

$$\left.\begin{array}{l} a_1x + a_2y + a_3z + a_4w = a_5 \\ b_1x + b_2y + b_3z + b_4w = b_5 \\ \cdots\cdots\cdots\cdots \end{array}\right\}$$

the usual expressions for $w, z, \ldots$ Terquem affirms that if w is to be equal to 0 we must have

$$|a_1\,b_2\,c_3\,d_5| = 0,$$

and that therefore this last equation is the equation of condition for the simultaneous existence of four equations between three unknowns. Continuing, he says that if we are to have $z = w = 0$, we must have

$$| a_1 b_2 c_3 d_5 | = | a_1 b_2 c_4 d_5 | = 0,^* \qquad (\beta)$$

and that therefore these two equations are "les deux équations de condition pour que 4 équations entre 2 inconnues puissent être satisfaites par les mêmes valeurs." The words "et ainsi de suite" are added to draw attention to the general theorem.

On this we can only remark that the giving of the equations of condition in the form (β) in the second case, even although the real equations of condition

$$\left\| \begin{matrix} a_1 & b_1 & c_1 & d_1 \\ a_2 & b_2 & c_2 & d_2 \\ a_5 & b_5 & c_5 & d_5 \end{matrix} \right\| = 0$$

are thence deducible, seems quite inexcusable, especially in an exposition meant to be elementary.

CAYLEY, A. (1847).

[Sur les déterminants gauches. *Crelle's Journ.*, xxxviii. pp. 93–96; or *Collected Math. Papers*, i. pp. 410–413.]

As the title implies, the subject of this paper is not *general* determinants. Part of the purpose, however, which the author had in view necessitated reflection on the definition of such determinants, and the outcome was a suggestion which it would be a serious mistake to pass over. In explaining the character of the functions known afterwards as Pfaffians, and which he was about to show were closely connected with skew determinants, it was natural that he should be struck with certain points of resemblance between them and general determinants, and that in consequence he should seek a general definition which would include both. The new definition given is

"$\ldots$ en exprimant par ($1\ 2\ \ldots\ n$) une fonction quelconque dans laquelle entrent les nombres symboliques $1, 2, \ldots, n$, et par $\pm$ le signe

* The second determinant is incorrectly printed in the original.

correspondant à une permutation quelconque de ces nombres, la fonction

$$\Sigma \pm (1\ 2\ \ldots\ n)$$

(où Σ désigne la somme de tous les termes qu'on obtient en permutant ces nombres d'une manière quelconque) est ce qu'on nomme *Déterminant.*"

It is readily seen that this is much more general than any definition in use up to that time, and that it agrees with the ordinary definition only when the function ($1\ 2 \ldots n$) takes the particular form $\lambda_{\alpha 1}\, \lambda_{\beta 2} \ldots \lambda_{\kappa n}$ or $\lambda_{1\alpha}\, \lambda_{2\beta} \ldots \lambda_{n\kappa}$. Further, it is not the same generalisation as we are familiar with from Cauchy's great memoir of 1812, where determinants are viewed as a special class of *alternating symmetric functions.* This is shown quite clearly by the only other case brought forward by Cayley, viz., the case where the function ($1\ 2 \ldots n$) is given the form $\lambda_{12}\, \lambda_{34} \ldots \lambda_{n-1,n}$. Other examples, not given by Cayley, are—

$$\sum \pm a_{123} \quad i.e. \quad a_{123} - a_{132} - a_{213} + a_{231} + a_{312} - a_{321},$$

$$\sum \pm \frac{a_{12}}{a_{23}} \quad i.e. \quad \frac{a_{12}}{a_{23}} - \frac{a_{13}}{a_{32}} - \frac{a_{21}}{a_{13}} + \frac{a_{23}}{a_{31}} + \frac{a_{31}}{a_{12}} - \frac{a_{32}}{a_{21}},$$

. . . .

There is also in the same paper a more direct contribution to the theory of general determinants, viz., the theorem afterwards associated with Cayley's name, and which,—to use later phraseology,—gives the expression for a determinant in terms of its own devertebrated coaxial minors and its primary diagonal elements. In the actual wording of the description of the theorem it is, not unnaturally, applied to a skew determinant only; but there is clearly nothing in the nature of the case to confine it to this special form. The description is—

"En effet, soit Ω le déterminant gauche dont il s'agit, cette fonction peut être présentée sous la forme

$$\Omega = \Omega_0 + \Omega_1\lambda_{11} + \Omega_2\lambda_{22} + \ldots + \Omega_{12}\lambda_{11}\lambda_{22} + \ldots$$

où Ω_0 est ce que devient Ω si $\lambda_{11}, \lambda_{22}, \ldots$ sont réduits à zéro. Ω_1 est ce que devient le coëfficient de λ_{11} sous la même condition, et ainsi de suite; c'est-à-dire: Ω_0 est le déterminant formé par les quantités λ_{rs} en supposant que ces quantités satisfassent aux condition (2), et en donnant à r les valeurs $1, 2, \ldots, n$. Ω_1 est le déterminant formé pareillement en donnant à r, s les valeurs $2, 3, \ldots, n$; Ω_2 s'obtient en donnant à r, s les valeurs $1, 3, \ldots, n$, et ainsi de suite: cela est aisé de voir si l'on range les quantités λ_{rs} en forme de carré."

CAUCHY, A. L. (1847).

[Mémoire sur les clefs algébriques. *Exercices d'Analyse et de Phys. Math.*, iv. pp. 356–400, § 11; or *Œuvres complètes* (2), xiv.]

In this longish memoir it is the second section (§ ii.) that is of interest to us, its title being "Décomposition des sommes alternées, connues sous le nom de résultantes, en facteurs symboliques." The only previous writing with which it is clearly connected in subject is Grassmann's of the year 1844. To ensure the possibility of proper comparison between the two, it is necessary to do now as was done in the previous case, viz., to give the opening paragraph verbatim. No general explanation of 'algebraic keys' need be offered, all that is requisite for our present purpose being obtainable from the paragraph itself. It runs as follows:—

"En effet, considérons d'abord la somme alternée s, formée avec les quatre termes du tableau

(1) $$\begin{cases} a_1 & b_1 \\ a_2 & b_2 \end{cases}$$

et fournie par l'équation

(2) $$s = S(\pm a_1 b_2) = a_1 b_2 - a_2 b_1.$$

Si l'on donne pour coefficients à deux clefs algébriques α, β dans deux fonctions linéaires λ, μ les termes qui renferment la première et la seconde ligne horizontale du tableau (1) on aura non-seulement

(3) $$\lambda = a_1\alpha + b_1\beta, \quad \mu = a_2\alpha + b_2\beta,$$

mais encore

(4) $$|\lambda\mu| = a_1a_2|\alpha^2| + a_1b_2|\alpha\beta| + b_1a_2|\beta\alpha| + b_1b_2|\beta^2|;$$

et, pour que le produit symbolique $|\lambda\mu|$ se réduise à la résultante s, il suffira évidemment de poser

(5) $$|\alpha^2| = 0, \quad |\alpha\beta| = 1, \quad |\beta\alpha| = -1, \quad |\beta^2| = 0.$$

Sous cette condition, l'on aura

(6) $$s = |\lambda\mu|;$$

et λ, μ seront les *facteurs symboliques* de la résultante s.

Si aux formules (5) on substituait les suivantes:

(7) $$|\alpha^2| = 0, \quad |\beta\alpha| = -|\alpha\beta|, \quad |\beta^2| = 0,$$

alors, à la place de l'équation (6) on obtiendrait la formule

$$| \lambda\mu | = s \, | \alpha\beta |,$$

ou

(8) $$s = \frac{| \lambda\mu |}{| \alpha\beta |},$$

dans laquelle il suffirait de poser $| \alpha\beta | = 1$ pour retrouver l'équation (6). Remarquons d'ailleurs que la seconde des formules (7) peut s'écrire comme il suit :

(9) $$| \alpha\beta | + | \beta\alpha | = 0,$$

et que la formule (9) donne $| \alpha^2 | = 0$ ou $| \beta^2 | = 0$, quand on y suppose $\beta = \alpha$. Donc, et définitive, les transmutations (7) sont toutes trois comprises dans la formule (9). Donc il suffit de recourir à cette formule et à celles qui s'en déduisent, pour obtenir l'équation (8), et, par suite, pour décomposer en facteurs symboliques la somme alternée s, c'est-à-dire la rèsultante algébrique formée avec les quatre termes du tableau (1)."

From the case of the second order he proceeds at once to the case of the nth order, inquiring as before under what conditions the symbolic product

$$\begin{aligned} &(a_1\alpha + b_1\beta + c_1\gamma + \ \ldots \ + h_1\eta) \\ &.\,(a_2\alpha + b_2\beta + c_2\gamma + \ \ldots \ + h_2\eta) \\ &.\,(a_3\alpha + b_3\beta + c_3\gamma + \ \ldots \ + h_3\eta) \\ &\ldots\ldots\ldots\ldots\ldots\ldots \\ &.\,(a_n\alpha + b_n\beta + c_n\gamma + \ \ldots \ + h_n\eta), \end{aligned}$$

where $\alpha, \beta, \gamma, \ldots, \eta$ are n distinct 'algebraic keys,' reduces to

$$S(\pm a_1 b_2 c_3 \ \ldots \ h_n).$$

Denoting by $| \kappa |$ the symbolic portion of any term of the final product, he finds that

"on devra poser

(14) $$| \kappa | = 0$$

quand l'une quelconque des lettres

$$\alpha, \ \beta, \ \gamma, \ \ldots, \ \eta$$

entrera deux ou plusieurs fois comme facteur dans le produit κ ; et poser, au contraire,

(15) $$| \kappa | = 1,$$

ou

(16) $$| \kappa | = -1$$

quand le produit κ renfermera une seule fois chacune des lettres

$$\alpha,\ \beta,\ \gamma,\ \ldots,\ \eta,$$

la formule (15) étant relative au cas où l'on sera obligé d'opérer entre ces lettres prises deux à deux un nombre pair d'échanges, pour passer du produit $|\ \alpha\beta\gamma\ \ldots\ \eta\ |$ au produit $|\ \kappa\ |$."

The obtaining of the results

$$\left.\begin{array}{rrr} |\ \alpha\beta\gamma\ | = & |\ \beta\gamma\alpha\ | = & |\ \gamma\alpha\beta\ | \\ = -\ |\ \alpha\gamma\beta\ | = & -\ |\ \beta\alpha\gamma\ | = & -\ |\ \gamma\beta\alpha\ | \end{array}\right\}$$

from transformations of the form

$$|\ \alpha\beta\ | = -\ |\ \beta\alpha\ |$$

is then shortly considered; and this is followed by a concluding paragraph, in which the statement occurs that "cette décomposition (des sommes alternées en facteurs symboliques) une fois operée on peut s'en servir avec avantage pour découvrir ou pour démontrer les principales propriétés des sommes alternées."

Cauchy's position is thus seen to be very different from Grassmann's. Grassmann was not concerned with determinants: his problem was to solve the set of equations

$$\left.\begin{array}{l} a_1x + a_2y + a_3z = a_0 \\ b_1x + b_2y + b_3z = b_0 \\ c_1x + c_2y + c_3z = c_0 \end{array}\right\}$$

and he satisfied himself by a curious process of reasoning that

$$x = \frac{(a_0+b_0+c_0)(a_2+b_2+c_2)(a_3+b_3+c_3)}{(a_1+b_1+c_1)(a_2+b_2+c_2)(a_3+b_3+c_3)},$$

provided that the multiplications indicated be performed according to the laws of "outer multiplication." Cauchy, on the other hand, starts with the determinant formed from

$$\begin{array}{ccc} a_1 & a_2 & a_3 \\ b_1 & b_2 & b_3 \\ c_1 & c_2 & c_3, \end{array}$$

his problem being to find under what conditions, as regards the arbitrarily introduced symbols α, β, γ, the product

$$(a_1\alpha+a_2\beta+a_3\gamma)(b_1\alpha+b_2\beta+b_3\gamma)(c_1\alpha+c_2\beta+c_3\gamma)$$

will be identical with the determinant; and he is led to the need for imposing laws essentially the same as those of "outer multiplication." Grassmann's factors are each the sum of the elements of a *column* of the determinant, and, according to his quasi-demonstration, could not be anything else: Cauchy's factors are each formed from the elements of a *row*; but had they been formed from the elements of a column, the result would not have been different. The root idea, viz., *the expression of a determinant as a product of factors subject to multiplication of a special kind,* was certainly first reached by Grassmann: Cauchy attained the same result, adding somewhat to its width, and presenting it in a fresh and more reasonable form.

HERMITE, C. (1849, January).

[Sur une question relative à la théorie des nombres. *Journ.* (*de Liouville*) *de Math.*, xiv. pp. 21–30; or *Œuvres*, i. pp. 265–273.]

As a lemma in the process of attaining the main purpose of his paper Hermite gives an identity which for the 4th order we should nowadays write in the form

$$\begin{vmatrix} a_1 & a_2 & a_3 & a_4 \\ b_1 & b_2 & b_3 & b_4 \\ c_1 & c_2 & c_3 & c_4 \\ d_1 & d_2 & d_3 & d_4 \end{vmatrix} = -\frac{1}{b_1c_1} \begin{Vmatrix} |a_2b_1| & |a_3b_1| & |a_4b_1| \\ |b_2c_1| & |b_3c_1| & |b_4c_1| \\ |c_2d_1| & |c_3d_1| & |c_4d_1| \end{Vmatrix}.$$

This he establishes rather circuitously by taking four quantities $\xi_1, \xi_2, \xi_3, \xi_4$ which satisfy the equations

$$\left.\begin{aligned} a_1\xi_1 + b_1\xi_2 + c_1\xi_3 + d_1\xi_4 &= 1 \\ a_2\xi_1 + b_2\xi_2 + c_2\xi_3 + d_2\xi_4 &= 0 \\ a_3\xi_1 + b_3\xi_2 + c_3\xi_3 + d_3\xi_4 &= 0 \\ a_4\xi_1 + b_4\xi_2 + c_4\xi_3 + d_4\xi_4 &= 0 \end{aligned}\right\},$$

and then multiplying the original determinant columnwise by $(-1)^3b_1c_1$ in the form

$$\begin{vmatrix} \xi_1 & b_1 & \cdot & \cdot \\ \xi_2 & -a_1 & c_1 & \cdot \\ \xi_3 & \cdot & -b_1 & d_1 \\ \xi_4 & \cdot & \cdot & -c_1 \end{vmatrix}.$$

JOACHIMSTHAL, F. (1849, Nov.).

[Sur quelques applications des déterminants à la géométrie. *Crelle's Journ.*, xi. pp. 21–47.]

Joachimsthal's interesting series of 'applications' being mainly connected with the multiplication of determinants, he introduces them by enunciating the multiplication-theorem, and indicating a mode of proof "pour éviter aux lecteurs la peine de la chercher ailleurs."

The enunciation is

$$\text{dét.}\left\{\begin{matrix} x & y & z \\ x_1 & y_1 & z_1 \\ x_2 & y_2 & z_2 \end{matrix}\right\} \times \text{dét.}\left\{\begin{matrix} \xi & \eta & \zeta \\ \xi_1 & \eta_1 & \zeta_1 \\ \xi_2 & \eta_2 & \zeta_2 \end{matrix}\right\}$$

$$= \text{dét.}\left\{\begin{matrix} x\xi + y\eta + z\zeta & x\xi_1 + y\eta_1 + z\zeta_1 & x\xi_2 + y\eta_2 + z\zeta_2 \\ x_1\xi + y_1\eta + z_1\zeta & x_1\xi_1 + y_1\eta_1 + z_1\zeta_1 & x_1\xi_2 + y_1\eta_2 + z_1\zeta_2 \\ x_2\xi + y_2\eta + z_2\zeta & x_2\xi_1 + y_2\eta_1 + z_2\zeta_1 & x_2\xi_2 + y_2\eta_2 + z_2\zeta_2 \end{matrix}\right\}$$

and the proof, it is stated, consists in taking the equations

$$\begin{aligned} h &= x\text{U} + y\text{V} + z\text{W}, & \text{U} &= \xi u + \xi_1 v + \xi_2 w, \\ h_1 &= x_1\text{U} + y_1\text{V} + z_1\text{W}, & \text{V} &= \eta u + \eta_1 v + \eta_2 w, \\ h_2 &= x_2\text{U} + y_2\text{V} + z_2\text{W}, & \text{W} &= \zeta u + \zeta_1 v + \zeta_2 w, \end{aligned}$$

deducing from them two different triads of equations independent of U, V, W, solving these triads for u, v, w, and then comparing the results.*

In connection with this two points have to be noted. The first is the use of the notation

$$\text{dét.}\left\{\qquad\qquad\right\},$$

which may be compared with Catalan's of the year 1846, and

* The details of the proof not being given, one cannot guess how it was that a second theorem was not obtained, viz., the theorem

$$\begin{vmatrix} h & x\xi_1 + y\eta_1 + z\zeta_1 & x\xi_2 + y\eta_2 + z\zeta_2 \\ h_1 & x_1\xi_1 + y_1\eta_1 + z_1\zeta_1 & x_1\xi_2 + y_1\eta_2 + z_1\zeta_2 \\ h_2 & x_2\xi_1 + y_2\eta_1 + z_2\zeta_1 & x_2\xi_2 + y_2\eta_2 + z_2\zeta_2 \end{vmatrix} = \begin{Vmatrix} |hy_1z_2| & \xi_1 & \xi_2 \\ |xh_1z_2| & \eta_1 & \eta_2 \\ |xy_1h_2| & \zeta_1 & \zeta_2 \end{Vmatrix}.$$

with the modification of the vertical-line notation which the printers of Crelle's and Liouville's journals employed for two or three years in setting up Cayley's papers. That Joachimsthal was familiar with Catalan's paper of 1846 is made more probable by the occurrence of a footnote (p. 28) giving

$$\text{dét.}\left\{\begin{matrix} x+l & y & z \\ x'+l' & y' & z' \\ x''+l'' & y'' & z'' \end{matrix}\right\} = \text{dét.}\left\{\begin{matrix} x & y & z \\ x' & y' & z' \\ x'' & y'' & z'' \end{matrix}\right\} + \text{dét.}\left\{\begin{matrix} l & y & z \\ l' & y' & z' \\ l'' & y'' & z'' \end{matrix}\right\};$$

the identity which Catalan was the first to formulate in a similar way.

The second point is the unnaturalness, in view of the mode of proof, of not writing the second determinant of the theorem in the form

$$\text{dét.}\left\{\begin{matrix} \xi & \xi_1 & \xi_2 \\ \eta & \eta_1 & \eta_2 \\ \zeta & \zeta_1 & \zeta_2 \end{matrix}\right\}.$$

Towards the close of the paper (p. 44), his geometrical work having led him to use the identity

$$\begin{aligned}(\alpha'\beta' + \alpha''\beta'' + \alpha'''\beta''')(\gamma'\delta' + \gamma''\delta'' + \gamma'''\delta''') &- (\alpha'\delta' + \alpha''\delta'' + \alpha'''\delta''')(\beta'\gamma' + \beta''\gamma'' + \beta'''\gamma''') \\ = (\alpha''\gamma''' - \alpha'''\gamma'')(\beta''\delta''' - \beta'''\delta'') &+ (\alpha'\gamma''' - \alpha'''\gamma')(\beta'\delta''' - \beta'''\delta') \\ &+ (\alpha'\gamma'' - \alpha''\gamma')(\beta'\delta'' - \beta''\delta'),\end{aligned}$$

he devotes the last two pages to proving a generalisation of it,—a generalisation not so wide as that of Binet and Cauchy, but interesting because of the way in which it is arrived at. Putting

$$\text{D} \equiv \text{dét.}\left\{\begin{matrix} x_0 & y_0 & z_0 & \dots & t_0 \\ x_1 & y_1 & z_1 & \dots & t_1 \\ x_2 & y_2 & z_2 & \dots & t_2 \\ \cdot & \cdot & \cdot & \cdot & \cdot \\ x_n & y_n & z_n & \dots & t_n \end{matrix}\right\}, \quad \Delta \equiv \text{dét.}\left\{\begin{matrix} \xi_0 & \eta_0 & \zeta_0 & \dots & \tau_0 \\ \xi_1 & \eta_1 & \zeta_1 & \dots & \tau_1 \\ \xi_2 & \eta_2 & \zeta_2 & \dots & \tau_2 \\ \cdot & \cdot & \cdot & \cdot & \cdot \\ \xi_n & \eta_n & \zeta_n & \dots & \tau_n \end{matrix}\right\}$$

$$\text{and} \quad \text{N} \equiv \text{dét.}\left\{\begin{matrix} l_{0,0} & l_{0,1} & l_{0,2} & \dots & l_{0,n} \\ l_{1,0} & l_{1,1} & l_{1,2} & \dots & l_{1,n} \\ l_{2,0} & l_{2,1} & l_{2,2} & \dots & l_{2,n} \\ \cdot & \cdot & \cdot & \cdot & \cdot \\ l_{n,0} & l_{n,1} & l_{n,} & \dots & l_{n,n} \end{matrix}\right\}$$

where

$$\begin{aligned}
l_{0,0} &= x_0\xi_0 + y_0\eta_0 + z_0\zeta_0 + \ldots + t_0\tau_0,\\
l_{0,1} &= x_0\xi_1 + y_0\eta_1 + z_0\zeta_1 + \ldots + t_0\tau_1,\\
&\ldots\ldots\ldots\ldots\ldots\ldots\\
l_{n,0} &= x_n\xi_0 + y_n\eta_0 + z_n\zeta_0 + \ldots + t_n\tau_0,\\
l_{n,1} &= x_n\xi_1 + y_n\eta_1 + z_n\zeta_1 + \ldots + t_n\tau_1,\\
&\ldots\ldots\ldots\ldots\ldots\ldots\\
l_{n,n} &= x_n\xi_n + y_n\eta_n + z_n\zeta_n + \ldots + t_n\tau_n;
\end{aligned}$$

and where therefore

$$\mathrm{D}\Delta = \mathrm{N},$$

he differentiates both members of this identity with respect to $x_n, y_n, z_n, \ldots t_n$, obtaining

$$\begin{aligned}
\Delta\frac{\partial \mathrm{D}}{\partial x_n} &= \xi_0\frac{\partial \mathrm{N}}{\partial l_{n,0}} + \xi_1\frac{\partial \mathrm{N}}{\partial l_{n,1}} + \ldots\ldots + \xi_n\frac{\partial \mathrm{N}}{\partial l_{n,n}},\\
\Delta\frac{\partial \mathrm{D}}{\partial y_n} &= \eta_0\frac{\partial \mathrm{N}}{\partial l_{n,0}} + \eta_1\frac{\partial \mathrm{N}}{\partial l_{n,1}} + \ldots\ldots + \eta_n\frac{\partial \mathrm{N}}{\partial l_{n,n}},\\
&\ldots\ldots\ldots\ldots\ldots\ldots\\
\Delta\frac{\partial \mathrm{D}}{\partial t_n} &= \tau_0\frac{\partial \mathrm{N}}{\partial l_{n,0}} + \tau_1\frac{\partial \mathrm{N}}{\partial l_{n,1}} + \ldots\ldots + \tau_n\frac{\partial \mathrm{N}}{\partial l_{n,n}}.
\end{aligned}$$

From these last equations, on multiplying by

$$\frac{\partial\Delta}{\partial\xi_n},\ \frac{\partial\Delta}{\partial\eta_n},\ \ldots\ldots,\ \frac{\partial\Delta}{\partial\tau_n}$$

respectively, and adding, there is obtained with the help of the known identities

$$\begin{aligned}
\xi_0\frac{\partial\Delta}{\partial\xi_n} + \eta_0\frac{\partial\Delta}{\partial\eta_n} + \ldots\ldots + \tau_0\frac{\partial\Delta}{\partial\tau_n} &= 0,\\
\xi_1\frac{\partial\Delta}{\partial\xi_n} + \eta_1\frac{\partial\Delta}{\partial\eta_n} + \ldots\ldots + \tau_1\frac{\partial\Delta}{\partial\tau_n} &= 0,\\
\ldots\ldots\ldots\ldots\ldots\ldots&\\
\xi_n\frac{\partial\Delta}{\partial\xi_n} + \eta_n\frac{\partial\Delta}{\partial\eta_n} + \ldots\ldots + \tau_n\frac{\partial\Delta}{\partial\tau_n} &= \Delta,
\end{aligned}$$

the result

$$\frac{\partial \mathrm{D}}{\partial x_n}\cdot\frac{\partial\Delta}{\partial\xi_n} + \frac{\partial \mathrm{D}}{\partial y_n}\cdot\frac{\partial\Delta}{\partial\eta_n} + \ldots\ldots + \frac{\partial \mathrm{D}}{\partial t_n}\cdot\frac{\partial\Delta}{\partial\tau_n} = \frac{\partial \mathrm{N}}{\partial l_{n,n}},$$

i.e.

$$\text{dét.}\begin{Bmatrix} y_0 & z_0 & \cdots & t_0 \\ y_1 & z_1 & \cdots & t_1 \\ \cdot & \cdot & \cdot & \cdot \\ y_{n-1} & z_{n-1} & \cdots & t_{n-1} \end{Bmatrix}.\,\text{dét.}\begin{Bmatrix} \eta_0 & \zeta_0 & \cdots & \tau_0 \\ \eta_1 & \zeta_1 & \cdots & \tau_1 \\ \cdot & \cdot & \cdot & \cdot \\ \eta_{n-1} & \zeta_{n-1} & \cdots & \tau_{n-1} \end{Bmatrix}$$

$$+\,\text{dét.}\begin{Bmatrix} x_0 & z_0 & \cdots & t_0 \\ x_1 & z_1 & \cdots & t_1 \\ \cdot & \cdot & \cdot & \cdot \\ x_{n-1} & z_{n-1} & \cdots & t_{n-1} \end{Bmatrix}.\,\text{dét.}\begin{Bmatrix} \xi_0 & \zeta_0 & \cdots & \tau_0 \\ \xi_1 & \zeta_1 & \cdots & \tau_1 \\ \cdot & \cdot & \cdot & \cdot \\ \xi_{n-1} & \zeta_{n-1} & \cdots & \tau_{n-1} \end{Bmatrix} + \cdots$$

$$= \text{dét.}\begin{Bmatrix} l_{0,0} & l_{0,1} & \cdots & l_{0,n-1} \\ l_{1,0} & l_{1,1} & \cdots & l_{1,n-1} \\ \cdot & \cdot & \cdot & \cdot \\ l_{n-1,0} & l_{n-1,1} & \cdots & l_{n-1,n-1} \end{Bmatrix}.$$

Using later phraseology we may say in describing this that from the theorem regarding the product of two square matrices of order $n+1$ there is obtained the theorem regarding the product of two rectangular but non-quadrate matrices, the latter product appearing as a principal coaxial minor of the former. Cauchy's generalisation concerned *any* minor of the former product, but even this further extension was not beyond Joachimsthal's reach, for he ends with the remark "En différentiant de nouveau par rapport à x_{n-1}, y_{n-1}, ... on obtiendra d'autres formules; et ainsi de suite."

SYLVESTER, J. J. (1850, Sep.).

[Additions to the articles in the September number of this journal "On a new class of theorems . . . ." and "On Pascal's theorem." *Philos. Magazine* (3), xxxvii. pp. 363–370; *Collected Math. Papers*, i. pp. 145–151.]

Of the three additions referred to in the title it is the last which concerns us, viz., that in which Sylvester introduces and explains his use of the term *minor* as applied to a determinant. Starting with the 'square array' of a given determinant, and leaving out one 'line' and one 'column,' he calls the determinant of the minor array which remains a 'First Minor Determinant';

similarly 'Second Minor Determinant' is explained; and then he adds,

"and so in general we can form a system of rth minor determinants by the exclusion of r lines and r columns, and such system in general will contain

$$\left\{\frac{n(n-1)\ \ldots\ (n-r+1)}{1\cdot 2\ \ldots\ r}\right\}^2$$

distinct determinants."

It is thus seen that 'minor determinant is used as 'partial determinant' had already been used by Lebesgue (1837), and as 'determinant of a derived system' had been used by Cauchy (1812), but that, whereas Cauchy added a distinguishing epithet to specify the order of the determinant, Sylvester did so to indicate how many lines or columns fewer it had than the 'principal' or 'complete' determinant originally started with.

The following proposition or 'law' is next given, viz.: *The whole of a system of* r^{th} *minors being zero implies only* $(\mathrm{r}+1)^2$ *equations, that is, by making* $(\mathrm{r}+1)^2$ *of these minors zero, all will become zero: and this is true, no matter what may be the dimensions or form of the complete determinant.* Then, after some geometrical applications concerned with *first* minors of a symmetrical determinant, there follows the explanation—

"The law which I have stated for assigning the number of independent or, to speak more accurately, non-coevanescent determinants belonging to a given system of minors, I call the Homaloidal law, because it is a corollary to a proposition which represents analytically the indefinite extension of a property, common to lines and surfaces, to all loci (whether in ordinary or transcendental space) of the first order, all of which loci may, by an abstraction derived from the idea of levelness common to straight lines and planes, be called Homaloids."

A further advance is made just before the close of his paper. Leaving the *square* array and taking m lines and n columns, he says

"This will not represent a determinant, but is, as it were, a Matrix out of which we may form various systems of determinants by fixing upon a number p and selecting at will p lines and p columns, the squares corresponding to which may be termed determinants of the pth order."

Here there is to be noticed the first use of the word *matrix* in connection with determinants, as well as the change back to Cauchy's mode of particularising the minors. The corresponding more general 'law' is said to be—*The number of uncoevanescent determinants constituting a system of the* p*th* *order derived from a given matrix,* n *terms broad and* m *terms deep, may equal but can never exceed in number*

$$(n-p+1)(m-p+1).$$

No proof of this is given, nor does he refer to Cayley's related theorem of 1843. (See p. 15 above.)

SCHLÄFLI, L. (1851, Jan.).

[Ueber die Resultante eines Systems mehrerer algebraischen Gleichungen: ein Beitrag zur Theorie der Elimination. *Denkschr. d. k. Akad. d. Wiss.* (Wien): *math.-naturw. Cl.*, iv. (2), pp. 1–54.]

The last section (§ 29, pp. 52–54) of this long memoir bears the heading "Ein zur Theorie der Determinanten gehörender Satz," being introduced as an aid to the reading of a geometrical memoir (pp. 54–74) on the ternary cubic.

The 'Satz' is not formally enunciated, but we may express it for ourselves as follows: *If a set of* n *homogeneous equations of the first degree in* n *unknowns be given, the determinant of the set being* Δ, *and there be formed another set consisting of all equations of the* r*th* *degree derivable from the equations of the given set by multiplication among themselves, the determinant of the latter set is equal to*

$$\Delta^{C_{n+r-1,\, n}}.$$

For example, the initial set of equations being

$$\left.\begin{aligned} a_1x + a_2y &= 0 \\ b_1x + b_2y &= 0 \end{aligned}\right\}$$

and $r=3$, the derived set of equations is

$$\left.\begin{aligned} (a_1x+a_2y)^3 &= 0 \\ (a_1x+a_2y)^2(b_1x+b_2y) &= 0 \\ (a_1x+a_2y)\,(b_1x+b_2y)^2 &= 0 \\ (b_1x+b_2y)^3 &= 0 \end{aligned}\right\}$$

and it has to be shown that

$$\begin{vmatrix} a_1^3 & 3a_1^2a_2 & 3a_1a_2^2 & a_2^3 \\ a_1^2b_1 & 2a_1a_2b_1 + a_1^2b_2 & 2a_1a_2b_2 + a_2^2b_1 & a_2^2b_2 \\ a_1b_1^2 & 2b_1b_2a_1 + b_1^2a_2 & 2b_1b_2a_2 + b_2^2a_1 & a_2b_2^2 \\ b_1^3 & 3b_1^2b_2 & 3b_1b_2^2 & b_2^3 \end{vmatrix} = |\,a_1b_2\,|^6.$$

To do this Schläfli takes the conjugate of the adjugate of $|\,a_1b_2\,|$, namely,

$$\begin{vmatrix} A_1 & B_1 \\ A_2 & B_2 \end{vmatrix};$$

forms with its elements the like determinant

$$\begin{vmatrix} A_1^3 & 3A_1^2B_1 & 3A_1B_1^2 & B_1^3 \\ A_1^2A_2 & 2A_1B_1A_2 + A_1^2B_2 & 2A_1B_1B_2 + B_1^2A_2 & B_1^2B_2 \\ A_1A_2^2 & 2A_2B_2A_1 + A_2^2B_1 & 2A_2B_2B_1 + B_2^2A_1 & B_1B_2^2 \\ A_2^3 & 3A_2^2B_2 & 3A_2B_2^2 & B_2^3 \end{vmatrix};$$

and then in row-by-column fashion multiplies the former four-line determinant by the latter, obtaining for the product

$$\begin{vmatrix} |\,a_1b_2\,|^3 & \cdot & \cdot & \cdot \\ \cdot & |\,a_1b_2\,|^3 & \cdot & \cdot \\ \cdot & \cdot & |\,a_1b_2\,|^3 & \cdot \\ \cdot & \cdot & \cdot & |\,a_1b_2\,|^3 \end{vmatrix}$$

whence the desired result follows.

It is the penultimate step of the demonstration, namely, the obtaining of the final form of the elements of the product, that is attended with difficulty: and the difficulty is not lessened to Schläfli by his taking n and r in all their generality. As soon, however, as the determinant under investigation is known in this or any other way to be a power of Δ, the index of the power is readily found to be

$$\frac{r}{n}\cdot C_{n+r-1,\,r}$$

by noting that the determinant is of the order $C_{n+r-1,\,r}$ and each of its elements of the rth degree.

SPOTTISWOODE, W. (1851).

[ELEMENTARY THEOREMS RELATING TO DETERMINANTS. viii+63 pp. London.]

This is noteworthy as being the first separately-published elementary work on the subject, the author explaining that he had been led to write it because determinants had come to be in frequent use, and there was no accessible text-book to which students could be referred. It consists of a preface and ten short chapters or sections, the mode of partitioning and arranging the matter being such as has often subsequently been followed.

The preface contains a short sketch of the history of the theory, the first of the kind that had appeared. In the first section (§ 1) the reader is introduced to determinants of the second and third orders as they actually occur in the solution of geometrical problems, and certain of the simpler properties are incidentally pointed out; the next seven sections (§§ 2-8) deal with determinants in general; and the rest of the book is occupied with determinants of special form, viz., § 9 with skew determinants and § 10 with functional determinants. The concluding portions of most of the sections consist of worked examples illustrative of applications of the theory to co-ordinate geometry.

The definition employed is that of Vandermonde * as expressed in Cayley's vertical-line notation, viz.—

$$|(1,1)| = (1,1),$$

$$\begin{vmatrix}(1,1) & (1,2)\\(2,1) & (2,2)\end{vmatrix} = (1,1)\,|(2,2)| - (1,2)\,|(2,1)|,$$

* It may be worth noting that while both Vandermonde and Schweins used the recurrent law of formation as a definition, they did not write it in exactly the same form. Schweins followed closely the form used by the original discoverer, Bezout, putting for example

$$|a_1b_2c_3d_4| = d_4\,|a_1b_2c_3| - d_3\,|a_1b_2c_4| + d_2\,|a_1b_3c_4| - d_1\,|a_2b_3c_4|,$$

the connecting signs being in all cases alternately positive and negative; whereas Vandermonde wrote

$$|a_1b_2c_3d_4| = a_1\,|b_2c_3d_4| - a_2\,|b_3c_4d_1| + a_3\,|b_4c_1d_2| - a_4\,|b_1c_2d_3|,$$

where the cyclical change of suffixes causes the connecting signs to be alternately positive and negative when the order of the determinant is even, and to be uniformly positive when the order is odd.

$$\begin{vmatrix} (1,1) & (1,2) & (1,3) \\ (2,1) & (2,2) & (2,3) \\ (3,1) & (3,2) & (3,3) \end{vmatrix} = (1,1)\begin{vmatrix} (2,2) & (2,3) \\ (3,2) & (3,3) \end{vmatrix} + (1,2)\begin{vmatrix} (2,3) & (2,1) \\ (3,3) & (3,1) \end{vmatrix} + (1,3)\begin{vmatrix} (2,1) & (2,2) \\ (3,1) & (3,2) \end{vmatrix},$$

. . . .

The quantities (1, 1), (1, 2), . . . which Cauchy called 'terms' and Jacobi 'elements,' are named '*constituents*'; and the determinant of the nth order having these constituents is denoted shortly by

$$\sum \pm (1,1)(2,2)\ldots(n,n)$$

The first result, deduced in somewhat loose fashion from the definition, is "Cramer's rule"; but the first that is actually formulated and numbered is one of much later date than Cramer, viz.—

"Theorem I.—*If the whole of a vertical or horizontal row be multiplied by the same quantity, the determinant is multiplied by that quantity.*"

In this form, as is well known, it afterwards became almost stereotyped. The second result, which is of about the same age, is that regarding a determinant whose vertical row consists of p-termed expressions, second vertical row of q-termed expressions third vertical row of r-termed expressions, and so on, being to the effect that such a determinant is expressible as a sum of pqr . . . determinants with monomial constituents. The next seven results are, like the first, new only in form, the wording being, as in Theorem I., more topographical in character than formerly, on account of the determinant being now more consciously viewed as connected with a square holding $n \cdot n$ quantities situate in n vertical rows and at the same time in n horizontal rows. The tenth result, which is a converse of the ninth, is new but unimportant, viz.—

"Theorem X.—*If a determinant of the* nth *order vanishes, a system of* n *homogeneous linear equations, the coefficients of which are the constituents of the given determinant, may always be established.*"

The ninth he establishes by the method of so-called 'mathematical induction,' deducing from it the solution of

$$(r1)x_1 + (r2)x_2 + \ldots + (rn)x_n = u_r \Big\}_{r=1}^{r=n}$$

and thence the ratios of $x_0, x_1, \ldots, x_n$ in the case of

$$(r0)x_0 + (r1)x_1 + (r2)x_2 + \ldots + (rn)x_n = 0 \quad \Big\}_{r=1}^{r=n}$$

as originally noted by Jacobi.*

The eleventh result is the multiplication-theorem; and here anything that is noteworthy is not in the enunciation but in the proof. Beginning with two sets of equations, exactly after the manner of Joachimsthal, viz.,

$$\left.\begin{aligned} a_1x + a_2y + a_3z &= u_1 \\ b_1x + b_2y + b_3z &= u_2 \\ c_1x + c_2y + c_3z &= u_3 \end{aligned}\right\}, \quad \left.\begin{aligned} l_1u_1 + l_2u_2 + l_3u_3 &= v_1 \\ m_1u_1 + m_2u_2 + m_3u_3 &= v_2 \\ n_1u_1 + n_2u_2 + n_3u_3 &= v_3 \end{aligned}\right\},$$

he views them as six linear equations in x, y, z, u_1, u_2, u_3, and seeks to find the value of one of the first three unknowns, say x, in two different ways. Firstly, by using the first set to eliminate u_1, u_2, u_3 from the second set, he obtains

$$\left.\begin{aligned} (l_1a_1 + l_2b_1 + l_3c_1)x + (l_1a_2 + l_2b_2 + l_3c_2)y + (l_1a_3 + l_2b_3 + l_3c_3)z &= v_1 \\ (m_1a_1 + m_2b_1 + m_3c_1)x + (m_1a_2 + m_2b_2 + m_3c_2)y + (m_1a_3 + m_2b_3 + m_3c_3)z &= v_2 \\ (n_1a_1 + n_2b_1 + n_3c_1)x + (n_1a_2 + n_2b_2 + n_3c_2)y + (n_1a_3 + n_2b_3 + n_3c_3)z &= v_3 \end{aligned}\right\},$$

and thence

$$x = \frac{\begin{vmatrix} v_1 & l_1a_2 + l_2b_2 + l_3c_2 & l_1a_3 + l_2b_3 + l_3c_3 \\ v_2 & m_1a_2 + m_2b_2 + m_3c_2 & m_1a_3 + m_2b_3 + m_3c_3 \\ v_3 & n_1a_2 + n_2b_2 + n_3c_2 & n_1a_3 + n_2b_3 + n_3c_3 \end{vmatrix}}{\begin{vmatrix} l_1a_1 + l_2b_1 + l_3c_1 & l_1a_2 + l_2b_2 + l_3c_2 & l_1a_3 + l_2b_3 + l_3c_3 \\ m_1a_1 + m_2b_1 + m_3c_1 & m_1a_2 + m_2b_2 + m_3c_2 & m_1a_3 + m_2b_3 + m_3c_3 \\ n_1a_1 + n_2b_1 + n_3c_1 & n_1a_2 + n_2b_2 + n_3c_2 & n_1a_3 + n_2b_3 + n_3c_3 \end{vmatrix}},$$

Secondly,—and here he differs from Joachimsthal,—by writing the six equations as one set in the form

$$\left.\begin{array}{llllllll} a_1x + a_2y + a_3z & - u_1 & & & = 0 \\ b_1x + b_2y + b_3z & & - u_2 & & = 0 \\ c_1x + c_2y + c_3z & & & - u_3 & = 0 \\ & l_1u_1 & + l_2u_2 & + l_3u_3 & = v_1 \\ & m_1u_1 & + m_2u_2 & + m_3u_3 & = v_2 \\ & n_1u_1 & + n_2u_2 & + n_3u_3 & = v_3 \end{array}\right\}$$

* *Crelle's Journ.*, xv. (1835), p. 104, and xxii. (1841), p. 296.

he obtains directly for x the value

$$\begin{vmatrix} \cdot & a_2 & a_3 & -1 & \cdot & \cdot \\ \cdot & b_2 & b_3 & \cdot & -1 & \cdot \\ \cdot & c_2 & c_3 & \cdot & \cdot & -1 \\ v_1 & \cdot & \cdot & l_1 & l_2 & l_3 \\ v_2 & \cdot & \cdot & m_1 & m_2 & m_3 \\ v_3 & \cdot & \cdot & n_1 & n_2 & n_3 \end{vmatrix} \div \begin{vmatrix} a_1 & a_2 & a_3 & -1 & \cdot & \cdot \\ b_1 & b_2 & b_3 & \cdot & -1 & \cdot \\ c_1 & c_2 & c_3 & \cdot & \cdot & -1 \\ \cdot & \cdot & \cdot & l_1 & l_2 & l_3 \\ \cdot & \cdot & \cdot & m_1 & m_2 & m_3 \\ \cdot & \cdot & \cdot & n_1 & n_2 & n_3 \end{vmatrix}$$

A comparison of the denominators in the two values of x then gives the desired result.*

The theorems of § 6, though, for some unexplained reason not formulated and numbered like the others, are of the highest importance, the subject-matter being the determinant of what is called the "inverse" system, that is to say, the determinant

$$\begin{vmatrix} [1,1] & [1,2] & \dots & [1,n] \\ [2,1] & [2,2] & \dots & [2,n] \\ \cdot & \cdot & \cdot & \cdot \\ [n,1] & [n,2] & \dots & [n,n] \end{vmatrix},$$

where $[r,s]$ is the cofactor of $(1,1)$ in $|1,2,\dots,n|$. Cauchy's theorem regarding the whole determinant is first proved, and then, instead of Jacobi's more general theorem, there is established a theorem said to include Jacobi's, viz.—

$$\begin{vmatrix} [i+1,\,i+1] & [i+1,\,i+2] & \dots & [i+1,\,n] \\ [i+2,\,i+1] & [i+2,\,i+2] & \dots & [i+2,\,n] \\ \cdot & \cdot & \cdot & \cdot \\ [n,\,i+1] & [n,\,i+2] & \dots & [n,\,n] \end{vmatrix}$$

$$= \begin{vmatrix} [i+j+1,\,i+j+1] & [i+j+1,\,i+j+2] \dots & [i+j+1,\,n] \\ [i+j+2,\,i+j+1] & [i+j+2,\,i+j+2] \dots & [i+j+2,\,n] \\ \cdot & \cdot & \cdot \\ [n,\,i+j+1] & [n,\,i+j+2] & \dots [n,\,n] \end{vmatrix} \cdot |1,2,\dots,n|^j \cdot \frac{|1,2,\dots,i|}{|1,2,\dots,i+j|}.$$

*Spottiswoode, like Joachimsthal, it will be observed, deduces nothing from a comparison of the *numerators*. Thus, by equating the two cofactors of v_1, he might have obtained

$$\begin{vmatrix} m_1 & m_2 & m_3 \\ n_1 & n_2 & n_3 \end{vmatrix} \cdot \begin{vmatrix} a_2 & b_2 & c_2 \\ a_3 & b_3 & c_3 \end{vmatrix} = - \begin{vmatrix} a_2 & a_3 & -1 & \cdot & \cdot \\ b_2 & b_3 & \cdot & -1 & \cdot \\ c_2 & c_3 & \cdot & \cdot & -1 \\ \cdot & \cdot & m_1 & m_2 & m_3 \\ \cdot & \cdot & n_1 & n_2 & n_3 \end{vmatrix}.$$

That it does include Jacobi's is at once seen on putting $i+j+1=n$, when we have

$$\begin{vmatrix} [i+1, i+1] & [i+1, i+2] & \dots & [i+1, n] \\ [i+2\ i+1] & [i+2, i+2] & \dots & [i+2, n] \\ \cdot & \cdot & \cdot & \cdot \\ [n, i+1] & [n, i+2] & \dots & [n, n] \end{vmatrix} = [n, n] \cdot |1, 2, \dots, n|^{n-i-1} \cdot \frac{|1, 2, \dots, i|}{|1, 2, \dots, n-1|}, \\ = |1, 2, \dots, n|^{n-i-1} \cdot |1, 2, \dots, i|.$$

It is equally true, however, that by a double use of Jacobi's theorem Spottiswoode's follows immediately.

The next section (§ 7) deals with the *differentiation* of a determinant, and with an application of the same by Malmsten to find the nth particular integral of a certain differential equation when $n-1$ particular integrals are already known.

The eighth section concerns the solution of what is called a "redundant system" of linear equations, that is to say, a system of m equations in n unknowns where $m > n$. Theorem XII. is the result obtained therefrom, and this being applied to the method of least squares, the last formulated result, Theorem XIII., is reached.

SYLVESTER, J. J. (1851, March).

[On the relation between the minor determinants of linearly equivalent quadratic functions. *Philos. Magazine* (4), i. pp. 295–305, 415; *Collected Math. Papers*, i. pp. 241–250, 251.]

In order to formulate the relation referred to in the title, that is to say, the relation between any minor of the determinant (later, discriminant) of a quadratic form and the 'pari-ordinal' minors of the determinant of the new form obtained from the old by means of a linear substitution, Sylvester found it necessary to introduce "a most powerful, because natural, method of notation" for determinants. He says—

"My method consists in expressing the same quantities biliterally as below:

$$\begin{matrix} a_1\alpha_1 & a_1\alpha_2 & \dots & a_1\alpha_n \\ a_2\alpha_1 & a_2\alpha_2 & \dots & a_2\alpha_n \\ \cdot & \cdot & \cdot & \cdot \\ a_n\alpha_1 & a_n\alpha_2 & \dots & a_n\alpha_n \end{matrix}$$

where, of course, whenever desirable, instead of $a_1, a_2, \ldots, a_n$ and $\alpha_1, \alpha_2, \ldots, \alpha_n$, we may write simply $a, b, \ldots, l$ and $\alpha, \beta, \ldots, \lambda$ respectively. Each quantity is now represented by two letters; the letters themselves, taken separately, being symbols neither of quantity nor of operation, but mere umbræ or ideal elements of quantitative symbols. We have now a means of representing the determinant above given in a compact form: for this purpose we need but to write one set of umbræ over the other as follows: $\begin{pmatrix} a_1 & a_2 & \ldots & a_n \\ \alpha_1 & \alpha_2 & \ldots & \alpha_n \end{pmatrix}$. If we now wish to obtain the algebraic value of this determinant, it is only necessary to take $\alpha_1, \alpha_2, \ldots, \alpha_n$ in all its $1 \cdot 2 \cdot 3 \ldots n$ different positions, and we shall have

$$\begin{Bmatrix} a_1 & a_2 & \ldots & a_n \\ \alpha_1 & \alpha_2 & \ldots & \alpha_n \end{Bmatrix} = \sum \pm \{a_1\alpha_{\theta_1} \times a_2\alpha_{\theta_2} \times \ldots \times a_n\alpha_{\theta_n}\},$$

in which expression $\theta_1, \theta_2, \ldots, \theta_n$ represents some order of the numbers $1, 2, \ldots, n$, and the positive or negative sign is to be taken according to the well-known dichotomous law."

An obvious extension of the notation is also indicated, whereby what he calls "compound" determinants may be appropriately represented. Since, in accordance with the foregoing,

$$\begin{Bmatrix} a & b \\ \alpha & \beta \end{Bmatrix}$$

is used to denote

$$a\alpha \,.\, b\beta - a\beta \,.\, b\alpha,$$

he considers that

$$\begin{Bmatrix} \overline{ab} & \overline{cd} \\ \alpha\beta & \gamma\delta \end{Bmatrix}$$

"will naturally denote

$$\begin{matrix} ab \\ \alpha\beta \end{matrix} \times \begin{matrix} cd \\ \gamma\delta \end{matrix} - \begin{matrix} ab \\ \gamma\delta \end{matrix} \times \begin{matrix} cd \\ \alpha\beta \end{matrix},$$

that is

$$\begin{Bmatrix} (a\alpha \times b\beta) \\ -(a\beta \times b\alpha) \end{Bmatrix} \times \begin{Bmatrix} (c\gamma \times d\delta) \\ -(c\delta \times d\gamma) \end{Bmatrix} - \begin{Bmatrix} (a\gamma \times b\delta) \\ -(a\delta \times b\gamma) \end{Bmatrix} \times \begin{Bmatrix} (c\alpha \times d\beta) \\ -(c\beta \times d\alpha) \end{Bmatrix}.$$

And in general the compound determinant

$$\begin{Bmatrix} \overline{a_1 \quad b_1 \quad \ldots \quad l_1} & \overline{a_2 \quad b_2 \quad \ldots \quad l_2} & \ldots\ldots & \overline{a_r \quad b_r \quad \ldots \quad l_r} \\ \alpha_1 \quad \beta_1 \quad \ldots \quad \lambda_1 & \alpha_2 \quad \beta_2 \quad \ldots \quad \lambda_2 & \ldots\ldots & \alpha_r \quad \beta_r \quad \ldots \quad \lambda_r \end{Bmatrix}$$

will denote

$$\sum \pm \begin{Bmatrix} a_1 & b_1 & \ldots & l_1 \\ \alpha_{\theta_1} & \beta_{\theta_1} & \ldots & \lambda_{\theta_1} \end{Bmatrix} \times \begin{Bmatrix} a_2 & b_2 & \ldots & l_2 \\ \alpha_{\theta_2} & \beta_{\theta_2} & \ldots & \lambda_{\theta_2} \end{Bmatrix} \times \ldots \times \begin{Bmatrix} a_r & b_r & \ldots & l_r \\ \alpha_{\theta_r} & \beta_{\theta_r} & \ldots & \lambda_{\theta_r} \end{Bmatrix},$$

where, as before, we have the disjunctive equation

$$\theta_1, \theta_2, \ldots, \theta_r = 1, 2, \ldots, r."$$

As an example of the power of this notation he gives the theorem

$$\left\{\begin{array}{ccccccc} \overline{a_1\ a_2\ \ldots\ a_r\ a_{r+1}} & \overline{a_1\ a_2\ \ldots\ a_r\ a_{r+2}} & \ldots\ldots & \overline{a_1\ a_2\ \ldots\ a_r\ a_{r+s}} \\ \alpha_1\ \alpha_2\ \ldots\ \alpha_r\ \alpha_{r+1} & \alpha_1\ \alpha_2\ \ldots\ \alpha_r\ \alpha_{r+2} & \ldots\ldots & \alpha_1\ \alpha_2\ \ldots\ \alpha_r\ \alpha_{r+s} \end{array}\right\}$$

$$= \left\{\begin{array}{c} a_1\ a_2\ \ldots\ a_r \\ \alpha_1\ \alpha_2\ \ldots\ \alpha_r \end{array}\right\}^{s-1} \times \left\{\begin{array}{c} a_1\ a_2\ \ldots\ a_r\ a_{r+1}\ a_{r+2}\ \ldots\ a_{r+s} \\ \alpha_1\ \alpha_2\ \ldots\ \alpha_r\ \alpha_{r+1}\ \alpha_{r+2}\ \ldots\ \alpha_{r+s} \end{array}\right\},$$

adding, in his enthusiasm, that without the aid of his "system of umbral or biliteral notation, this important theorem could not be made the subject of statement without an enormous periphrasis, and could never have been made the object of distinct contemplation or proof."

The main object of his paper is then taken up, but the subject of the notation is twice recurred to,—once to say that it is "very similiar to that of Vandermonde" as he had learned from "Mr. Spottiswoode's valuable treatise," and the second time to point out that on second thoughts it is better to unite two umbral elements in the form

$$\begin{array}{c} a \\ \alpha \end{array}$$

rather than in the form

$$a\alpha,$$

because then the analogy upon which the extension of the notation from simple to compound determinants is grounded will be better apprehended.

The theorem used above to illustrate the power of the 'umbral' notation should not be lightly passed over, being far more deserving of the epithet 'new' than the notation employed in formulating it. Although at a later date it came to be of less moment because of its inclusion as merely one of a *class* of theorems, viz., the class known as 'Extensionals,' there can be no doubt that at the time of its discovery it was, as Sylvester himself styled it, "a remarkable theorem." Taking, for the sake of illustration, the case of it where $r=3$ and $s=4$, viz.,

$$\begin{Vmatrix} |a_1b_2c_3d_4| & |a_1b_2c_3d_5| & |a_1b_2c_3d_6| & |a_1b_2c_3d_7| \\ |a_1b_2c_3e_4| & |a_1b_2c_3e_5| & |a_1b_2c_3e_6| & |a_1b_2c_3e_7| \\ |a_1b_2c_3f_4| & |a_1b_2c_3f_5| & |a_1b_2c_3f_6| & |a_1b_2c_3f_7| \\ |a_1b_2c_3g_4| & |a_1b_2c_3g_5| & |a_1b_2c_3g_6| & |a_1b_2c_3g_7| \end{Vmatrix}$$
$$= |a_1b_2c_3|^3 \cdot |a_1b_2c_3d_4e_5f_6g_7|,$$

we see that if we delete $a_1b_2c_3$ everywhere on both sides we are left with

$$\begin{vmatrix} d_4 & d_5 & d_6 & d_7 \\ e_4 & e_5 & e_6 & e_7 \\ f_4 & f_5 & f_6 & f_7 \\ g_4 & g_5 & g_6 & g_7 \end{vmatrix} = |d_4e_5f_6g_7|;$$

so that the theorem is seen to be the Extensional of a manifest identity.*

SYLVESTER, J. J. (1851, Aug.).

[On a certain fundamental theorem of determinants. *Philos. Magazine* (4), ii. pp. 142–145; *Collected Math. Papers*, i. pp. 252–255.]

After a characteristic introductory paragraph about the importance of the new theorem and his reasons for publishing it, Sylvester proceeds—

"The theorem is as follows:—Suppose that there are two determinants of the ordinary kind, each expressed by a square array of terms made up of n lines and n columns, so that in each square there are n^2 terms. Now let n be broken up in any given manner into two parts p and q, so that $p+q=n$. Let, firstly, one of the two given squares be divided in a given *definite* manner into two parts, one containing p of the n given lines, and the other part q of the same; and secondly, let the other of the two given squares be divided *in every possible way* into two parts, consisting of q and p lines respectively, so that on tacking on the part containing q lines of the second square to the part containing p lines of the first square, and the part containing p lines of the second square to the part containing q of the first, we get back a new couple of squares, each denoting a determinant different from the two given determinants: the number of such new couples will evidently be

$$\frac{n(n-1)\ \ldots\ (n-p+1)}{1\cdot2\ \ldots\ p};$$

* See *Trans. R. Soc. Edinburgh*, xxx. p. 4.

and my theorem is that *the product of the given couple of determinants is equal to the sum of the products (affected with the proper algebraical sign) of each of the new couples formed as above described.*"

The same is then stated in symbols, namely,

$$\begin{Bmatrix} a_1 & a_2 & \ldots & a_n \\ b_1 & b_2 & \ldots & b_n \end{Bmatrix} \times \begin{Bmatrix} a_1 & a_2 & \ldots & a_n \\ \beta_1 & \beta_2 & \ldots & \beta_n \end{Bmatrix}$$

$$= \sum \pm \begin{Bmatrix} a_1 \; a_2 \; \ldots \ldots \ldots \ldots \ldots \; a_n \\ b_1 \; b_2 \ldots b_p \beta_{\theta_{p+1}} \beta_{\theta_{p+2}} \ldots \beta_{\theta_n} \end{Bmatrix} \begin{Bmatrix} a_1 \; a_2 \; \ldots \ldots \ldots \ldots \ldots \; a_n \\ \beta_{\theta_1} \beta_{\theta_2} \ldots \beta_{\theta_p} b_{p+1} b_{p+2} \ldots b_n \end{Bmatrix}$$

where $\theta_1, \theta_2, \ldots, \theta_p$ are any p different integers taken from 1, 2, ..., n, or where, as Sylvester puts it, $\theta_1, \theta_2, \ldots, \theta_n$ are 'disjunctively' equal to 1, 2, ..., n.

A proof of a verificatory character is given at some length, the aim being to show that the terms arising from the expansion of the products occurring on the right-hand side are classifiable under two heads: (1) those which appear twice, namely, once with a positive sign and once with a negative sign; (2) those which appear only once, and which in their aggregate agree with those arising from the expansion of the single product on the left. An improvement of it by Faà di Bruno will be given later.

By removing to the right the solitary product at present on the left, the theorem is seen to belong to the class of *vanishing aggregates of products of pairs of determinants*, and therefore to be not entirely new, the first instances of it, viz.,

$$|ab'| \cdot |cd'| - |ac'| \cdot |bd'| + |ad'| \cdot |bc'| = 0,$$

$$|ab'c''| \cdot |de'f''| - |ab'd''| \cdot |ce'f''| + |ac'd''| \cdot |be'f''| - |bc'd''| \cdot |ae'f''| = 0,$$

that is to say, the cases where $n=2, p=1$, and $n=3, p=1$, being found in Bezout (1779). Nothing, however, done by Bezout, Cauchy, or Schweins ought to dissociate Sylvester's name from the theorem. His claim, too, is all the stronger from the fact that he it was who in 1839 first formulated the case $n=n$, $p=1$,—a fact which seems to have dropped entirely out of remembrance, writers giving the date 1851 when referring to the case enunciated a dozen years earlier.

SYLVESTER, J. J. (1851, Oct.).

[On a remarkable discovery in the theory of canonical forms and of hyperdeterminants. *Philos. Magazine* (4), ii. pp. 391–410; *Collected Math. Papers*, i. pp. 265–283.]

Here, amid matter of great algebraical importance, there is incidentally suggested the discarding of the use of the term 'determinant' as connected with a single function,—that is to say, Gauss' original use of the term,—and the substitution of the term 'discriminant' in its place. The introduction of a new word, it is explained, is for the purpose of avoiding the obscurity and confusion which arises from employing the same word in two different senses, and "'discriminant' because it affords the *discrimen* or test for ascertaining whether or not equal factors enter into a function of two variables, or more generally of the existence or otherwise of multiple points in the locus represented or characterised by any algebraical function." *

CAYLEY, A. (1851, end).

[On the theory of permutants. *Cambridge and Dubl. Math. Journ.*, vii. pp. 40–51; *Collected Math. Papers*, ii. pp. 16–26.]

The second part of his paper of 1843, as we have seen, Cayley devoted to the consideration of a class of functions obtainable from the use of m sets of n indices in the way in which a determinant is obtainable from only two sets. The general symbol used for such a function was

---

* Apropos of this happy coinage, Sylvester adds in a footnote the general remark:—"Progress in these researches is impossible without the aid of clear expression; and the first condition of a good nomenclature is that different things shall be called by different names. The innovations in mathematical language here and elsewhere (not without high sanction) introduced by the author, have been never adopted except under actual experience of the embarrassment arising from the want of them, and will require no vindication to those who have reached that point where the necessity of some such additions becomes felt." The truth of the remark is not appreciably diminished by the occurrence of the word 'meso-catalecticism' in another footnote two pages further on. The year of the paper (1851) was for Cayley and Sylvester a year teeming with fresh ideas as well as with fresh words.

$$\left\{ A \begin{matrix} \rho_1 & \sigma_1 & \tau_1 & \cdots \\ \rho_2 & \sigma_2 & \tau_2 & \cdots \\ \cdot & \cdot & \cdot & \cdot \\ \rho_n & \sigma_n & \tau_n & \cdots \end{matrix} \right\},$$

this standing for the sum of all the different terms of the form

$$\pm_r \cdot \pm_s \cdot \pm_t \ldots A_{\rho_{r_1}\,\sigma_{s_1}\,\tau_{t_1}\,\ldots} \times \ldots\ldots \times A_{\rho_{r_n}\,\sigma_{s_n}\,\tau_{t_n}\,\ldots}$$

where

$$r_1, r_2, \ldots, r_n; \quad s_1, s_2, \ldots, s_n; \quad t_1, t_2, \ldots, t_n; \quad \ldots\ldots$$

denote any permutation whatever, the same or different, of the series of integers

$$1, 2, \ldots, n,$$

and where $\pm_r$ denotes the sign + or − according as the number of inversions in $r_1, r_2, \ldots, r_n$ is even or odd. Using a † over the column of ρ's to indicate that these are unpermutable, he shows that

$$\left\{ A \begin{matrix} \rho_1 & \sigma_1 & \cdots \\ \cdot & \cdot & \cdot \\ \rho_n & \sigma_n & \cdots \end{matrix} \right\} = 1 \cdot 2 \ldots . m \left\{ A \begin{matrix} \overset{\dagger}{\rho_1} & \sigma_1 & \cdots \\ \cdot & \cdot & \cdot \\ \rho_n & \sigma_n & \cdots \end{matrix} \right\}$$

when the number of columns, m, is even; and vanishes when m is odd. In the former case it is clear that the placing of the † over any other column would have had the same result, and therefore that it is better to mark this indifference by placing it over the A. The other theorems, including a multiplication theorem, need not be given, our object being simply to show the position occupied by determinants among the new functions; and this we can now do by quoting one sentence, viz., "an ordinary determinant is represented by

$$\left\{ \overset{\dagger}{A} \begin{matrix} \alpha_1 & \beta_1 \\ \cdot & \cdot \\ \alpha_n & \beta_n \end{matrix} \right\} \quad \text{or} \quad \left\{ \overset{\dagger}{A} \begin{matrix} 1 & 1 \\ \cdot & \cdot \\ n & n \end{matrix} \right\},$$

the latter form being obviously equally general with the former one."

In his paper of 1845 a further generalisation was made, the functions then reached being called 'hyperdeterminants,' and a hyperdeterminant defined as an expression representable as a homogeneous p^{th}-degree function, H_p, of certain of the determinants of a rectangular array, each of whose elements is denoted by n umbræ, and each umbra one of the integers $1, 2, \ldots, m$. Thus when $n=3$ and $m=2$, the array is

$$\begin{matrix} 111 & 112 & 121 & 122 \\ 211 & 212 & 221 & 222, \end{matrix}$$

or

$$\begin{matrix} 111 & 112 & 211 & 212 \\ 121 & 122 & 221 & 222, \end{matrix}$$

or

$$\begin{matrix} 111 & 121 & 211 & 221 \\ 112 & 122 & 212 & 222, \end{matrix}$$

according as the first, second, or third umbra is made invariable throughout the two rows; and if $p=1$ we have the 'incomplete' hyperdeterminants

$$\begin{vmatrix} 111 & 122 \\ 211 & 222 \end{vmatrix} + \begin{vmatrix} 112 & 121 \\ 212 & 221 \end{vmatrix},$$

$$\begin{vmatrix} 111 & 212 \\ 121 & 222 \end{vmatrix} + \begin{vmatrix} 112 & 211 \\ 122 & 221 \end{vmatrix},$$

$$\begin{vmatrix} 111 & 221 \\ 112 & 222 \end{vmatrix} + \begin{vmatrix} 121 & 211 \\ 122 & 212 \end{vmatrix};$$

whereas if $p=2$ we have the hyperdeterminants

$$\left\{\begin{vmatrix} 111 & 122 \\ 211 & 222 \end{vmatrix} + \begin{vmatrix} 112 & 121 \\ 212 & 221 \end{vmatrix}\right\}^2 - 4\begin{vmatrix} 111 & 121 \\ 211 & 221 \end{vmatrix} \cdot \begin{vmatrix} 112 & 122 \\ 212 & 222 \end{vmatrix},$$

$$\left\{\begin{vmatrix} 111 & 212 \\ 121 & 222 \end{vmatrix} + \begin{vmatrix} 112 & 211 \\ 122 & 221 \end{vmatrix}\right\}^2 - 4\begin{vmatrix} 111 & 211 \\ 121 & 221 \end{vmatrix} \cdot \begin{vmatrix} 112 & 212 \\ 122 & 222 \end{vmatrix},$$

$$\left\{\begin{vmatrix} 111 & 221 \\ 112 & 222 \end{vmatrix} + \begin{vmatrix} 121 & 211 \\ 122 & 212 \end{vmatrix}\right\}^2 - 4\begin{vmatrix} 111 & 211 \\ 112 & 212 \end{vmatrix} \cdot \begin{vmatrix} 121 & 221 \\ 122 & 222 \end{vmatrix},$$

which being really identical, have their common expression designated a 'complete' hyperdeterminant. The general form of H_p is not specified. Further details would here be out of place:

the one important point to be noted is the relation between hyperdeterminants and the functions of Cayley's paper of the year 1843, and this is shortly indicated by saying that the latter functions are hyperdeterminants in which $p=1$ and n is even.

In his paper of the year 1847 an altogether different generalisation was formulated, the corresponding symbol being

$$\sum \pm (1, 2, \ldots, n),$$

and one of the objects aimed at being to extend the definition of a determinant so as to include within it the Pfaffian. (See *Collected Math. Papers*, i. p. 589.)

Having thus attempted to make clear the stage which the process of generalisation had reached with Cayley prior to 1851, we are prepared to appreciate the notable advance made by him in his paper of that year. The widely embracing conception therein formulated was that of the functions called on the suggestion of Sylvester '*permutants.*' For the sake of easy exposition we shall follow him in his special usage of the words 'form,' 'blank,' 'characters,' 'symbol.' "A *form*," he says, "may be considered as composed of *blanks* which are to be filled up by inserting in them specialising *characters*, and a form the blanks of which are so filled up becomes a *symbol*." If the 'characters' (previously called by him 'nombres symboliques') be 1, 2, 3, 4, ... the 'symbol' may always be represented in the first instance, and without reference to the nature or constitution of the 'form,' by $V_{1234\ldots}$; for example $V_{1234\ldots}$ may stand for

$$P_{12}Q_3R_4\ldots, \quad \text{or} \quad P_{12}P_{34}\ldots, \quad \text{or} \quad \ldots.$$

Now, let the characters 1, 2, 3, 4, . . . in such a symbol be permuted in every possible way, and the resulting symbols have the sign + or − prefixed to them in accordance with Cramer's rule, then the aggregate of all these symbols is a '*simple permutant.*' The originating symbol being $V_{1234\ldots}$, the corresponding permutant might have been denoted by $\Sigma \pm V_{1234\ldots}$ as in his paper of the year 1847, but as a matter of fact Cayley now makes use of

$$(V_{1234\ldots\ldots}).$$

It is thus seen that, taking for shortness' sake only three characters, we have

$$(V_{123}) = V_{123} + V_{231} + V_{312} - V_{132} - V_{213} - V_{321},$$
$$(V_{123}) = (V_{231}) = -(V_{132}) = \dots .$$
$$(V_{113}) = 0.$$

These preliminaries having been grasped, "it is easy," in Cayley's own words, "to pass to the general definition of a permutant. We have only to consider the blanks as forming, not as heretofore a single set, but any number of distinct sets, and to consider the characters in each set of blanks as permutable *inter se*, and not otherwise, giving to the symbol the sign compounded of the signs corresponding to the arrangements of the characters in the different sets of blanks." Thus, if the first and second blanks form a set, and the third and fourth blanks form a set, the permutant whose originating symbol is V_{1234} is

$$V_{1234} - V_{2134} - V_{1243} + V_{2143}.$$

The idea is hereupon suggested of arranging the blanks of a compound permutant so as to show in what manner they are grouped into sets. For example, instead of doing as we have just done, viz., using V_{1234} accompanied by a verbal explanation as to its sets, we might write

$$V_{\substack{12\\34}}$$

and so obtain

$$\left(V_{\substack{12\\34}}\right) = V_{\substack{12\\34}} - V_{\substack{21\\34}} - V_{\substack{12\\43}} + V_{\substack{21\\43}}.$$

From this it is a simple step to the idea of grouping the blanks in lines and columns, that is to say, to such a symbol as

$$V_{\substack{\alpha \;\; \beta \;\; \gamma \;\; \dots\\ \alpha' \;\; \beta' \;\; \gamma' \;\; \dots\\ \dots\dots}}$$

One case of this is that in which it is viewed as a function of the symbols $V_{\alpha\beta\gamma\dots}$, $V_{\alpha'\beta'\gamma'\dots}$, etc., and a less general case that in which it is viewed as the *product*. Cayley then proceeds:—

"Upon this assumption it becomes important to distinguish the different ways in which the blanks of a set are distributed in the different lines and columns. The cases to be considered are: (A) The blanks of a single set or of single sets are situated in more than one column, (B) The blanks of each single set are situated in the same column, (C) The blanks of each single set form a separate column. The case B (which includes the case C) and the case C merit particular consideration. In fact, the case B is that of the functions which I have, in my memoir of Linear Transformations in the *Journal*, called hyperdeterminants, and the case C is the particular class of hyperdeterminants previously treated by me in the *Cambridge Philosophical Transactions*, and also particularly noticed in the memoir on Linear Transformations. The functions of the case B, I now propose to call 'Intermutants,' and those in the case C, 'Commutants.' Commutants include as a particular case 'Determinants,' which term will be used in its ordinary signification."

To arrive at the position of determinants, therefore, in the great theory of permutants, we have first to seek out the particular permutants whose originating symbol is of the form

$$\mathrm{V}\begin{matrix} \alpha & \beta & \gamma & \ldots. \\ \alpha' & \beta' & \gamma' & \ldots. \\ \ldots & \ldots & \ldots & \ldots; \end{matrix}$$

then in this restricted field to look for those in which the symbol just given is viewed as a product of symbols $\mathrm{V}_{\alpha\beta\gamma\ldots.}$, $\mathrm{V}_{\alpha'\beta'\gamma'\ldots.}$, ...; next to confine ourselves in this smaller domain to those in which the 'blanks' of each single 'set' form a separate column; and lastly to isolate those in which the number of such columns is 2.

In his paper Cayley goes on to expound the theory first of *commutants*, and then of *intermutants*. Neither exposition, however, needs attention here, because the one has already been dealt with under the year 1843, and the other is outside our subject.

SYLVESTER, J. J. (1852, Jan.).

[On the principles of the calculus of forms. *Cambridge and Dublin Math. Journ.*, vii. pp. 52–97; *Collected Math. Papers*, i. pp. 284–327.]

A postscript added by Cayley to his paper of the year 1851 makes evident that Sylvester had a share in the latest generalisation, and, as was natural, did not wish that share to be lost sight of. It made clear that the two workers had during the year been deeply engrossed in what Sylvester then called the 'calculus of forms,' that they had been in close communication with one another, and that Sylvester's discovery that the function

$$ace + 2bcd - ad^2 - b^2e - c^3$$

could be expressed as a commutant, namely,

$$\begin{pmatrix} 0 & 0 \\ 1 & 1 \\ 2 & 2 \end{pmatrix}$$

by considering $00 = a$, $01 = 10 = b$, $02 = 11 = 20 = c$, $12 = 21 = d$, $22 = e$ had led Cayley to the conception of intermutants.

The famous paper which we have now reached, and which was doubtless completed very shortly after Cayley's, contains the results—numerous and suggestive—of Sylvester's labours. The only section, however, which directly concerns the theory of determinants is the third, bearing the heading "On Commutants." It opens with a page regarding the simplest species, "the well-known common determinant," and then proceeds:—

"If, instead of two lines of umbræ, three or more be taken, the same principle of solution will continue to be applicable. Thus, if there be a matrix of any even number r of lines each of n umbræ

$$\begin{matrix} a_1 & b_1 & \ldots\ldots & l_1 \\ a_2 & b_2 & \ldots\ldots & l_2 \\ \cdot & \cdot & \cdot\;\cdot\;\cdot\;\cdot & \cdot \\ a_r & b_r & \ldots\ldots & l_r, \end{matrix}$$

the first may be supposed to remain stationary, and the remaining $r-1$ lines each be taken in $1 \cdot 2 \cdot \ldots n$ different orders: every order in each line will be accompanied by its appropriate sign + or −; and each different grouping in each line will give rise to a particular grouping of the letters read off in columns. The value of the commutant expressed by the above matrix will therefore consist of the sum of $(1 \cdot 2 \cdot \ldots n)^{r-1}$ terms, each term being the product of n quantities respectively symbolised by a group of r letters and affected with the sign + or − according as the number of negative signs in the total of the arrangements of the lines (from the columnar reading off of which each such term is derived) is even or odd.

For example, the value of

$$\begin{matrix} a & b \\ c & d \\ e & f \\ g & h \end{matrix}$$

will be found by taking the $(1\cdot 2)^3$ arrangements, as below,

$$\begin{matrix} ab & ab & ab & ab & ab & ab & ab & ab \\ cd & dc & cd & dc & cd & dc & cd & dc \\ ef & ef & fe & fe & ef & ef & fe & fe \\ gh & gh & gh & gh & hg & hg & hg & hg, \end{matrix}$$

the signs of cd, ef, gh being supposed $+$, those of dc, fe, hg will be each $-$. Consequently the sum of the terms will be expressed by

$$aceg\cdot bdfh - adeg\cdot bcfh - acfg\cdot bdeh + adfg\cdot bceh$$
$$- aceh\cdot bdfg + adeh\cdot bcfg + acfh\cdot bdeg - adfh\cdot bceg."$$

It will be observed that the example here given is the quadratic function which Cayley would have denoted by

$$\left(\begin{matrix} \dagger & & \\ \mathrm{V} & & \\ & 0\,0\,0\,0 & \\ & 1\,1\,1\,1 & \end{matrix}\right),$$

and which, on the supposition that generally $\mathrm{V}_{\alpha,\beta,\gamma,\delta} = \mathrm{V}_{\alpha+\beta+\gamma+\delta}$ and in particular that $\mathrm{V}_0 = a$, $\mathrm{V}_1 = b$, $\mathrm{V}_2 = c$, ... would represent

$$ae - bd - bd + c^2 - bd + c^2 + c^2 - db,$$

i.e.

$$ae - 4bd + 3c^2.$$

In his applications of the theory of commutants to that of 'forms,' Sylvester uses for the first time differential operators as umbræ, speaking, for example, of the commutant

$$\begin{matrix} \dfrac{\partial}{\partial x_1}, & \dfrac{\partial}{\partial y_1}, & \dfrac{\partial}{\partial z_1}, & \cdots\cdot \\ \dfrac{\partial}{\partial x_2}, & \dfrac{\partial}{\partial y_2}, & \dfrac{\partial}{\partial z_2}, & \cdots\cdot \\ \cdot & \cdot & \cdot & \cdot\,\cdot, \end{matrix}$$

this being possible from the fact that the coefficients of a 'form' u are, save for a constant factor, representable as differential-quotients of u. This we may illustrate for ourselves by the result

$$\begin{bmatrix} \frac{\partial}{\partial x} & \frac{\partial}{\partial y} \\ \frac{\partial}{\partial x} & \frac{\partial}{\partial y} \\ \frac{\partial}{\partial x} & \frac{\partial}{\partial y} \\ \frac{\partial}{\partial x} & \frac{\partial}{\partial y} \end{bmatrix} (a,\ b,\ c,\ d,\ e \between x,\ y)^4 = \left(\frac{1}{4\,!}\right)^2 (ae - 4bd + 3c^2 .$$

He also uses effectively such umbræ as

$$\alpha^n, \quad \alpha^{n-1}\beta, \quad \alpha^{n-2}\beta^2, \quad \ldots,$$

—a usage which is most suitably illustrated by taking a commutant having two lines of three quadratic umbræ each, that is to say, the determinant of the third order

$$\begin{matrix} \alpha^2 & \alpha\beta & \beta^2 \\ \alpha^2 & \alpha\beta & \beta^2. \end{matrix}$$

This, by a similar convention, is taken to represent

$$\alpha^4 \cdot \alpha^2\beta^2 \cdot \beta^4 - \alpha^4 \cdot \alpha\beta^3 \cdot \alpha\beta^3 - \alpha^3\beta \cdot \alpha^3\beta \cdot \beta^4 + \alpha^3\beta \cdot \alpha\beta^3 \cdot \alpha^2\beta^2$$
$$+ \alpha^2\beta^2 \cdot \alpha^3\beta \cdot \alpha\beta^3 - \alpha^2\beta^2 \cdot \alpha^2\beta^2 \cdot \alpha^2\beta^2,$$

and consequently if

$$(\alpha x + \beta y)^4 \equiv ax^4 + 4bx^3y + 6cx^2y^2 + 4dxy^3 + ey^4,$$

it is an expression for

$$ace + 2bcd - ad^2 - b^2e - c^3,$$

that is to say, for the above-mentioned (p. 69) *special* determinant

$$\begin{vmatrix} a & b & c \\ b & c & d \\ c & d & e \end{vmatrix}.$$

When he comes to speak of 'partial' commutants, which are identical with intermutants, he devotes a page (pp. 88–89) to the subject of his relations with Cayley. As it would appear that he was not quite satisfied with the wording of the postscript above referred to, Cayley published a modified form of it as a note, headed "Correction of the Postscript to the Paper on Permutants," and there the matter between the two friends happily rested.

BETTI, E. (1852, Feb.).

[Sulla risoluzione delle equazioni algebriche. *Annali di Sci. mat. e fis.*, iii. pp. 49–115; or *Opere mat.*]

In this important memoir dealing with the theory of substitutions, and with the application of the same towards finding the conditions of solvability of algebraic equations, the author following Sylvester* defines on page 80, in the manner afterwards so familiar, the expression 'determinant of a substitution': and on the following page there occurs the passage—

"Quindi dal noto teorema della moltiplicazione delle determinanti è facile dedurre che, se si chiama Δ la determinante del prodotto delle due sostituzioni (h) e (h') [whose determinants are D and D′], avremo

$$\Delta = \mathrm{D}\,\mathrm{D}';$$

cioè *la determinante del prodotto di due sostituzioni è eguale al prodotto delle loro determinanti.*"

BRUNO, F. FAÀ DI (1852, May).

[Démonstration d'un théorème de M. Sylvester relatif à la décomposition d'un produit de deux déterminants. *Journ.* (*de Liouville*) *de Math.* (1), xvii. pp. 190–192.]

The theorem is that which appeared in the *Philosophical Magazine* for August 1851, and which is there formulated in the umbral notation as follows:—

$$\begin{Bmatrix} a_1 & a_2 & \dots & a_n \\ b_1 & b_2 & \dots & b_n \end{Bmatrix} \times \begin{Bmatrix} \alpha_1 & \alpha_2 & \dots & \alpha_n \\ \beta_1 & \beta_2 & \dots & \beta_n \end{Bmatrix}$$

$$= \sum\left[\pm \begin{Bmatrix} a_1 & a_2 & \dots\dots\dots\dots & a_n \\ b_1 & b_2 \dots b_p & \beta_{\theta_{p+1}}\, \beta_{\theta_{p+2}} \dots & \beta_{\theta_n} \end{Bmatrix} \begin{Bmatrix} \alpha_1 & \alpha_2 & \dots\dots\dots\dots & \alpha_n \\ \beta_{\theta_1} & \beta_{\theta_2} \dots \beta_{\theta_p} & b_{p+1}\, b_{p+2} \dots & b_n \end{Bmatrix}\right]$$

where $\theta_1, \theta_2, \dots, \theta_n$ are 'disjunctively' equal to $1, 2, \dots n$. Faà di Bruno prefers to write it in the form

* See footnote to p. 52 of his paper just described.

$$\sum(\pm a_1^{\phi_1} a_2^{\phi_2} \ldots a_p^{\phi_p} \ldots a_n^{\phi_n}) \cdot \sum(\pm b_1^{\psi_1} b_2^{\psi_2} \ldots b_p^{\psi_p} \ldots b_n^{\psi_n})$$

$$= \sum\left[\sum\left(\pm a_1^{\phi_1} a_2^{\phi_2} \ldots a_p^{\phi_p} b_{\theta_{p+1}}^{\phi_{p+1}} b_{\theta_{p+2}}^{\phi_{p+2}} \ldots b_{\theta_n}^{\phi_n}\right) \cdot \sum\left(\pm b_{\theta_1}^{\psi_1} b_{\theta_2}^{\psi_2} \ldots b_{\theta_p}^{\psi_p} a_{p+1}^{\psi_{p+1}} a_{p+2}^{\psi_{p+2}} \ldots a_n^{\psi_n}\right)\right]$$

using two other sets of letters like $\theta_1, \theta_2, \ldots, \theta_n$. This change in notation being allowed for, the new proof is in general character exactly the same as the old; it is, however, more concise and more clearly set forth. It starts with the fact that any term arising from the expansion of the typical product on the right-hand side may be written

$$a_1^{\phi_1} a_2^{\phi_2} \ldots a_p^{\phi_p} a_{p+1}^{\psi_{p+1}} \ldots a_n^{\psi_n} \cdot b_{\theta_1}^{\psi_1} b_{\theta_2}^{\psi_2} \ldots b_{\theta_p}^{\psi_p} b_{\theta_{p+1}}^{\phi_{p+1}} b_{\theta_{p+2}}^{\phi_{p+2}} \ldots b_{\theta_n}^{\phi_n}.$$

Then observing the 'indices supérieures' attached to the b's in this, we are asked to consider two possible cases. In the first place, we have to note that if no one of the ψ's be identical with any one of the ϕ's, the term is a term of the expansion of the product on the left-hand side, and that the number of such terms in the expansion of each product on the right-hand side being

$$(1.2.3 \ldots n) \cdot (1.2.3 \ldots p) \cdot (1.2.3 \ldots \overline{n-p})$$

and the number of products

$$\frac{n(n-1)(n-2) \ldots . (n-p+1)}{1.2.3 \ldots p}$$

the total number of such terms is

$$(1.2.3 \ldots n)^2,$$

which is exactly the total number on the left-hand side. In the second place, if one of the ψ's be identical with one of the ϕ's, say

$$\psi_i = \phi_{p+h},$$

it is pointed out that there must exist another term in which, in place of having

$$\ldots\ldots b_{\theta_i}^{\psi_i} \ldots\ldots b_{\theta_{p+h}}^{\psi_{p+h}} \ldots\ldots$$

we shall have

$$\ldots\ldots b_{\theta_{p+h}}^{\psi_i} \ldots\ldots b_{\theta_i}^{\psi_{p+h}} \ldots\ldots$$

and that these two terms having necessarily different signs, must cancel each other.

SALMON, G. (1852).

[A Treatise on the Higher Plane Curves: . . . By the Rev. George Salmon, M.A. . . . xii+316 pp. Dublin, 1852.]

For the convenience of his readers Salmon appends a fifteen-page note on the subject of *Elimination*, and, as was natural, the note opens with a sketch (pp. 285–292) of the theory of determinants. Short and simple as this is, it contains one paragraph (§ 11) worthy of note, namely, in regard to the multiplication-theorem.

The determinant

$$\begin{vmatrix} A_1a_1+B_1b_1+C_1c_1 & A_2a_1+B_2b_1+C_2c_1 & A_3a_1+B_3b_1+C_3c_1 \\ A_1a_2+B_1b_2+C_1c_2 & A_2a_2+B_2b_2+C_2c_2 & A_3a_2+B_3b_2+C_3c_2 \\ A_1a_3+B_1b_3+C_1c_3 & A_2a_3+B_2b_3+C_2c_3 & A_3a_3+B_3b_3+C_3c_3 \end{vmatrix},$$

he says, is evidently the result of eliminating x, y, z from the equations

$$\left.\begin{array}{l} a_1S_1+b_1S_2+c_1S_3 = 0 \\ a_2S_1+b_2S_2+c_2S_3 = 0 \\ a_3S_1+b_3S_2+c_3S_3 = 0 \end{array}\right\}$$

when

$$\left.\begin{array}{l} S_1 = A_1x+A_2y+A_3z \\ S_2 = B_1x+B_2y+B_3z \\ S_3 = C_1x+C_2y+C_3z \end{array}\right\}.$$

But this elimination may be effected at once by eliminating S_1, S_2, S_3: consequently $|a_1\ b_2\ c_3|$ must be a factor of the resultant. In the second place, since a set of values of x, y, z can be found to satisfy simultaneously the given equations if a set can be found to satisfy simultaneously the equations $S_1=0$, $S_2=0$, $S_3=0$: and since the condition that the latter shall be possible is $|A_1\ B_2\ C_3| = 0$, it follows that $|A_1\ B_2\ C_3|$ must also be a factor of the result. The remaining factor being manifestly 1, the desired end, in Salmon's opinion, is attained. We only remark in passing that a little careful scrutiny of the reasoning would have suggested the need for additional support.

Salmon also proposes a fresh enunciation of the same theorem, namely, *If any set of linear equations*

$$\left.\begin{array}{l} a_1x+b_1y+c_1z+\ldots\ldots = 0 \\ a_2x+b_2y+c_2z+\ldots\ldots = 0 \\ \ldots\ldots\ldots\ldots\ldots\ldots \end{array}\right\}$$

be transformed by any linear substitution

$$\left.\begin{array}{l} x = A_1\xi+B_1\eta+C_1\zeta+\ldots\ldots \\ y = A_2\xi+B_2\eta+C_2\zeta+\ldots\ldots \\ \ldots\ldots\ldots\ldots\ldots\ldots \end{array}\right\}$$

then the determinant of the new set will be equal to the determinant of the original set multiplied by the determinant of transformation. This new wording will be recognised as a sign of the advent of the "algebra of linear transformation."

SYLVESTER, J. J. (1852, Oct.).

[On Staudt's theorems concerning the contents of polygons and polyhedrons, with a note on a new and resembling class of theorems. *Philos. Magazine* (4), iv. pp. 335–345; *Collected Math. Papers*, i. pp. 382–391.]

After a page of introduction, written in a light semi-historical, semi-critical style, Sylvester prepares the way for considering his main subject by giving as a basis two algebraical lemmas. The first he formulates as follows :—

"If the determinants represented by two square matrices are to be multiplied together, any number of columns may be cut off from the one matrix, and a corresponding number of columns from the other. Each of the lines in either one of the matrices so reduced in width as aforesaid being then multiplied by each line of the other, and the results of the multiplication arranged as a square matrix and bordered with the two respective sets of columns cut off, arranged symmetrically (the one set parallel to the new columns, the other set parallel to the new lines), the complete determinant represented by the new matrix so bordered (abstraction made of the algebraical sign) will be the product of the two original determinants."

In illustration he gives three forms for the product of

$$\begin{vmatrix} a & b \\ c & d \end{vmatrix} \quad \text{and} \quad \begin{vmatrix} \alpha & \beta \\ \gamma & \delta \end{vmatrix}$$

viz.—

$$\begin{vmatrix} a\alpha+b\beta & a\gamma+b\delta \\ c\alpha+d\beta & c\gamma+d\delta \end{vmatrix}, \quad \begin{vmatrix} a\alpha & a\gamma & b \\ c\alpha & c\gamma & d \\ \beta & \delta & . \end{vmatrix}, \quad \begin{vmatrix} 2 & 2 & a & b \\ 2 & 2 & c & d \\ \alpha & \beta & . & . \\ \gamma & \delta & . & . \end{vmatrix}.$$

In regard to the 2's which occur in the last form his remark is:—

"Any quantities might be substituted instead of 2 . . . . , as such terms do not influence the result: this figure is probably, however, the proper quantity arising from the application of the rule, because . . . . the value of the determinant represented by a matrix of *no* places is not zero but unity."

In the case where the two given determinants are of the third order, say

$$\begin{vmatrix} a & b & c \\ a' & b' & c' \\ a'' & b'' & c'' \end{vmatrix} \quad \text{and} \quad \begin{vmatrix} \alpha & \beta & \gamma \\ \alpha' & \beta' & \gamma' \\ \alpha'' & \beta'' & \gamma'' \end{vmatrix},$$

he gives only the second and third of the four $(n+1)$ possible forms, namely—

$$-\begin{vmatrix} a\alpha\ +b\beta & a\alpha'\ +b\beta' & a\alpha''\ +b\beta'' & c \\ a'\alpha\ +b'\beta & a'\alpha'\ +b'\beta' & a'\alpha''\ +b'\beta'' & c' \\ a''\alpha+b''\beta & a''\alpha'+b''\beta' & a''\alpha''+b''\beta'' & c'' \\ \gamma & \gamma' & \gamma'' & . \end{vmatrix}$$

and

$$\begin{vmatrix} a\alpha & a\alpha' & a\alpha'' & b & c \\ a'\alpha & a'\alpha' & a'\alpha'' & b' & c' \\ a''\alpha & a''\alpha' & a''\alpha'' & b'' & c'' \\ \beta & \beta' & \beta'' & . & . \\ \gamma & \gamma' & \gamma'' & . & . \end{vmatrix},$$

pointing out by way of demonstration that the former of these is arrived at by transforming the given determinants into

$$\begin{vmatrix} a & b & c & . \\ a' & b' & c' & . \\ a'' & b'' & c'' & . \\ . & . & . & 1 \end{vmatrix} \quad \text{and} \quad -\begin{vmatrix} \alpha & \beta & . & \gamma \\ \alpha' & \beta' & . & \gamma' \\ \alpha'' & \beta'' & . & \gamma'' \\ . & . & 1 & . \end{vmatrix}$$

and applying the ordinary rule of multiplication, and similarly that the latter is got by multiplying

$$\begin{vmatrix} a & b & c & . & . \\ a' & b' & c' & . & . \\ a'' & b'' & c'' & . & . \\ . & . & . & 1 & . \\ . & . & . & . & 1 \end{vmatrix} \text{ by } \begin{vmatrix} a & . & . & \beta & \gamma \\ a' & . & . & \beta' & \gamma' \\ a'' & . & . & \beta'' & \gamma'' \\ . & 1 & . & . & . \\ . & . & 1 & . & . \end{vmatrix}.$$

He thereupon leaves the subject with the remark :—

"This rule is interesting as exhibiting .... a complete scale whereby we may descend from the ordinary mode of representing the product of two determinants to the form .... where the two original determinants are made to occupy opposite quadrants of a square whose places in one of the remaining quadrants are left vacant, and shows us that under one aspect at least this latter form may be regarded as a matrix *bordered* by the two given matrices."

The second lemma is the identity—

$$\begin{vmatrix} a_{11} & a_{12} & \dots & a_{1n} & 1 \\ a_{21} & a_{22} & \dots & a_{2n} & 1 \\ . & . & . & . & . \\ a_{n1} & a_{n2} & \dots & a_{nn} & 1 \\ 1 & 1 & \dots & 1 & . \end{vmatrix} = \begin{vmatrix} A_{11} & A_{12} & \dots & A_{1n} & 1 \\ A_{21} & A_{22} & \dots & A_{2n} & 1 \\ . & . & . & . & . \\ A_{n1} & A_{n2} & \dots & A_{nn} & 1 \\ 1 & 1 & \dots & 1 & . \end{vmatrix}$$

where $A_{rs} = a_{rs} + h_r + k_s$, and the h's and k's are perfectly arbitrary quantities, the transformation being of course effected by adding multiples of the last column to the other columns, and thereafter multiples of the last row to the other rows.

The geometrical applications which follow, it may be interesting to note, are connected with the subject of Cayley's paper of 1841,—his well-known first paper on determinants.

CAUCHY, A. L. (1853, January).

[Sur les clefs algébriques. *Comptes rendus ... Acad. des Sci.* (Paris), xxxvi. pp. 70–75, 129–136; or *Œuvres complètes* (1), xi. pp. 439–445, xii. pp. 12–20.]

[Sur les avantages que présente, dans un grand nombre de questions, l'emploi des clefs algébriques. *Comptes rendus ... Acad. des Sci.* (Paris), xxxvi. pp. 161–169; or *Œuvres complètes* (1), xii. pp. 21–30.]

These papers add nothing of algebraic importance to the contents of Cauchy's memoir of the year 1847: in fact, they may be looked on as short and simply-worded abstracts of parts of that memoir. It is worthy of note, however, that even where problems of elimination are being dealt with "sommes alternées" are not now explicitly referred to.

SAINT VENANT, B. DE (1853, March).

[De l'interprétation géométrique des *clefs algébriques* et des *déterminants. Comptes rendus . . . Acad. des Sci.* (Paris), xxxvi. pp. 582–585.]

De Saint Venant's suggestion is that Cauchy's "algebraic keys" α, β, γ, . . . may be viewed as *directed magnitudes*, and this leads up to the so-called geometric interpretation of determinants. "Un déterminant du $n$ième ordre," he says, "me paraît être le produit géométrique de n sommes algébriques de n lignes ayant, chacune à chacune, les mêmes directions dans les diverses sommes: en sorte que l'on a pour celui du troisième ordre, par exemple,

$xy'z'' - xy''z' + \ldots =$ le produit $(\bar{x}+\bar{y}+\bar{z})(x'+\bar{y}'+\bar{z}')(\bar{x}''+\bar{y}''+\bar{z}'')$

où x, x', x'' ont un même direction (c'est-à-dire sont parallèles), y, y', y'' une autre direction qui est la même pour toutes trois, et z, z', z'' aussi une même troisième direction."

HESSE, O. (1853, April).

[Ueber Determinanten und ihre Anwendung in der Geometrie, ... *Crelle's Journal*, xlix. pp. 243–264; or *Werke*, pp. 319–343.]

The product of two determinants A and B being C, Hesse's professed object is to show "wie die partiellen Differentialquotienten der Determinante C nach ihren Elementen c genommen durch die partiellen Differentialquotienten der Factoren A und B nach ihren Elementen genommen sich ausdrücken lassen." We are prepared, therefore, to find his ground already pretty well covered by Joachimsthal's paper of November 1849. The latter established the result

$$\frac{\partial \mathrm{C}}{\partial c_{\kappa\lambda}} = \frac{\partial \mathrm{A}}{\partial a_{0\kappa}}\cdot\frac{\partial \mathrm{B}}{\partial b_{0\lambda}} + \frac{\partial \mathrm{A}}{\partial a_{1\kappa}}\cdot\frac{\partial \mathrm{B}}{\partial b_{1\lambda}} + \ldots + \frac{\partial \mathrm{A}}{\partial a_{n\kappa}}\cdot\frac{\partial \mathrm{B}}{\partial b_{n\lambda}},$$

and said that others could be found: Hesse established one of these others, namely,

$$\frac{\partial^2 C}{\partial c_{\kappa\lambda}\partial c_{\mu\nu}} = \frac{1}{1\cdot 2}\sum \frac{\partial^2 A}{\partial a_{p\kappa}\partial a_{q\mu}}\cdot\frac{\partial^2 B}{\partial b_{p\lambda}\partial b_{q\nu}},$$

and said that the next would be

$$\frac{\partial^3 C}{\partial c_{\kappa\lambda}\partial c_{\mu\nu}\partial c_{\rho\sigma}} = \frac{1}{1\cdot 2\cdot 3}\sum \frac{\partial^3 A}{\partial a_{p\kappa}\partial a_{q\mu}\partial a_{r\rho}}\cdot\frac{\partial^3 B}{\partial b_{p\lambda}\partial b_{q\nu}\partial b_{r\sigma}},$$

where $p, q, \ldots$ have the values $0, 1, 2, \ldots, n$.

We can only remark that the second and third results are not so simple as they ought to have been: for Hesse does not point out that (1) when p and q are identical the term vanishes; (2) putting $p, q = \alpha, \beta$ gives the same term as putting $p, q = \beta, \alpha$; and (3) therefore the second result should be

$$\frac{\partial^2 C}{\partial c_{\kappa\lambda}\partial c_{\mu\nu}} = \sum \frac{\partial^2 A}{\partial a_{p\kappa}\partial a_{q\mu}}\cdot\frac{\partial^2 B}{\partial b_{p\lambda}\partial b_{q\nu}},$$

where p has any of the values $0, 1, 2, \ldots, n-1$, and q any of the values $1, 2, \ldots, n$, subject to the condition that $p < q$. It would then agree with the extended multiplication-theorem of Binet and Cauchy, and especially with the latter's form of it.

CHIO, F. (1853, June).

[Mémoire sur les fonctions connues sous le nom de résultantes ou de déterminants. 32 pp. Turin.]

The title here is not sufficiently descriptive, almost the whole of the thirty-two pages being occupied with the consideration of determinants *whose elements are binomial.* Beginning with the "tableau"

$$\begin{matrix} a_0+m_0 & a_1+m_1 & \ldots & a_{i-1}+m_{i-1} \\ b_0+n_0 & b_1+n_1 & \ldots & b_{i-1}+n_{i-1} \\ \cdot & \cdot & \cdot & \cdot \\ l_0+t_0 & l_1+t_1 & \ldots & l_{i-1}+t_{i-1} \end{matrix}$$

Chio seeks, of course, to express its determinant as a sum of determinants with monomial elements, and thereafter applies his result to particular cases.

The first matter of real interest is reached on p. 11, where the following theorem is given: '*Soient* s *la résultante de l'ordre* i *formée avec les termes du tableau*

$$\begin{array}{cccc} a_0 & a_1 & \dots & a_{i-1} \\ b_0 & b_1 & \dots & b_{i-1} \\ \cdot & \cdot & \cdot\;\cdot\;\cdot & \cdot \\ l_0 & l_1 & \dots & l_{i-1}, \end{array}$$

et s″ *résultante de l'ordre* i−1 *formée avec les termes compris dans le tableau*

$$\begin{array}{cccc} S(\pm a_0b_1) & S(\pm a_0b_2) & \dots & S(\pm a_0b_{i-1}) \\ S(\pm a_0c_1) & S(\pm a_0c_2) & \dots & S(\pm a_0c_{i-1}) \\ \cdot\;\cdot\;\cdot & \cdot\;\cdot\;\cdot & \cdot\;\cdot\;\cdot & \cdot\;\cdot\;\cdot \\ S(\pm a_0l_1) & S(\pm a_0l_2) & \dots & S(\pm a_0l_{i-1}). \end{array}$$

La résultante s″ *sera égale à* s, *au facteur près* a_0^{i-2}, *en sorte qu'on aura* s″ $= a_0^{i-2}$ s."

This is one form of the theorem afterwards well known as effecting the transformation of any determinant into one of the next lower order. It may be viewed as a case of Hermite's result of the year 1849.

On p. 17 particular cases cease to be considered, and the multiplication of an array of i rows and $2i$ columns by a similar array is taken up, with a result in accordance with that arrived at by Binet and Cauchy in 1812. From this result, by specialisation, the ordinary multiplication-theorem is then deduced, and with it (Chio's "théorème ix.") the first part of the memoir closes.

The second part, which begins on p. 23, concerns the solving of a set of $2n$ equations of a type which will be sufficiently specified by giving the set where $n=3$, namely,

$$\left.\begin{array}{l} x\;\;+y\;\;+z\;\;= d_1 \\ x\xi\;+y\eta\;+z\zeta\;= d_2 \\ x\xi^2+y\eta^2+z\zeta^2 = d_3 \\ \cdot\;\cdot\;\cdot\;\cdot\;\cdot\;\cdot\;\cdot\;\cdot \\ x\xi^5+y\eta^5+z\zeta^5 = d_6 \end{array}\right\}.$$

The connection of this with what precedes consists in the fact, arrived at by Sylvester in his solution of the problem of the canonisation of the quintic, that ξ, η, ζ are then the roots of the equation in ω

$$\begin{vmatrix} d_2-\omega d_1 & d_3-\omega d_2 & d_4-\omega d_3 \\ d_3-\omega d_2 & d_4-\omega d_3 & d_5-\omega d_4 \\ d_4-\omega d_3 & d_5-\omega d_4 & d_6-\omega d_5 \end{vmatrix} = 0.$$

SPOTTISWOODE, W. (1853).

[Elementary theorems relating to determinants. Second edition, rewritten and much enlarged by the author. *Crelle's Journal*, li. pp. 209–271, 328–381.]

A more correct description of Spottiswoode's second edition would be *rearranged, partly rewritten, and much enlarged,* the majority of the titles of the old sections or chapters occurring again but in a different order, the majority of the sections being enlarged, and two or three new sections being inserted. Although the total increase of matter is from 71 pages to 117, there is comparatively little to be noted concerning general determinants.

In § 2, which bears the title "Addition and Subtraction of Determinants," the following appears (p. 232) for the first time:—THEOREM ix. *The sum of two determinants in which* i *rows (on a certain level) are respectively equal, is equal to the determinant whose* ith *minors on the aforesaid level are identical with the corresponding* ith *minors of each of the two given determinants, and whose* (n−i)th *complementary minors are respectively the sum of the complementary minors of the given determinants.* No instance is given where the two determinants have more than one row different.

In § 4, which deals with the multiplication of determinants, much space (pp. 238–248) is given to Sylvester's theorem of 1852 (October). Spottiswoode's own mode of treating the subject is to begin apparently with the two factors and arrive at the product, whereas in reality the opposite is the case. For example, his proof that

$$\begin{vmatrix} a & b & c \\ a' & b' & c' \\ a'' & b'' & c'' \end{vmatrix} \cdot \begin{vmatrix} a & \beta & \gamma \\ a' & \beta' & \gamma' \\ a'' & \beta'' & \gamma'' \end{vmatrix} = \begin{vmatrix} aa & aa' & aa'' & b & c \\ a'a & a'a' & a'a'' & b' & c' \\ a''a & a''a' & a''a'' & b'' & c'' \\ \beta & \beta' & \beta'' & . & . \\ \gamma & \gamma' & \gamma'' & . & . \end{vmatrix}$$

essentially consists in expanding the right-hand determinant in terms of minors formed from the first three rows and minors formed from the last two rows. His other fresh proof is dependent on the connection between determinants and simultaneous linear equations. Taking the two sets of equations

$$\left.\begin{array}{l} \alpha x+\alpha' y+\alpha'' z = u_1 \\ \beta x+\beta' y+\beta'' z = u_2 \\ \gamma x+\gamma' y+\gamma'' z = u_3 \end{array}\right\} \qquad \left.\begin{array}{l} a u_1 + b u_2 + c u_3 = v_1 \\ a' u_1 + b' u_2 + c' u_3 = v_2 \\ a'' u_1 + b'' u_2 + c'' u_3 = v_3 \end{array}\right\}$$

and substituting for u_1, u_2, u_3, in the second set and solving, there is obtained for x an expression whose denominator is known to be

$$\begin{vmatrix} a & b & c \\ a' & b' & c' \\ a'' & b'' & c'' \end{vmatrix} \cdot \begin{vmatrix} \alpha & \alpha' & \alpha'' \\ \beta & \beta' & \beta'' \\ \gamma & \gamma' & \gamma'' \end{vmatrix}.$$

In the second place, by substituting for u_1 only there is obtained

$$\left.\begin{array}{l} a\alpha x + a\alpha' y + a\alpha'' z + b u_2 + c u_3 = v_1 \\ a'\alpha x + a'\alpha' y + a'\alpha'' z + b' u_2 + c' u_3 = v_2 \\ a''\alpha x + a''\alpha' y + a''\alpha'' z + b'' u_2 + c'' u_3 = v_3 \\ \beta x + \beta' y + \beta'' z - u_2 = 0 \\ \gamma x + \gamma' y + \gamma'' z - u_3 = 0 \end{array}\right\},$$

whence comes for x an expression whose denominator is

$$\begin{vmatrix} a\alpha & a\alpha' & a\alpha'' & b & c \\ a'\alpha & a'\alpha' & a'\alpha'' & b' & c' \\ a''\alpha & a''\alpha' & a''\alpha'' & b'' & c'' \\ \beta & \beta' & \beta'' & -1 & \cdot \\ \gamma & \gamma' & \gamma'' & \cdot & -1 \end{vmatrix}.$$

A comparison of the two denominators is supposed to establish the desired result; but, although the dropping of the two negative units in the five-line determinant is quite justifiable, no allusion is made to it. It may be added that Sylvester's umbral notation is used throughout in dealing with the subjects just referred to,

$$\begin{Bmatrix} 1 & 2 & \ldots & n \\ 1 & 2 & \ldots & n \end{Bmatrix} \quad \text{or} \quad \begin{vmatrix} (11) & (12) & \ldots & (1n) \\ (21) & (22) & \ldots & (2n) \\ \cdot & \cdot & \cdot & \cdot \\ (n1) & (n2) & \ldots & (nn) \end{vmatrix}$$

being used for one of the two determinants, and

$$\begin{Bmatrix} 1' & 2' & \dots & n' \\ 1' & 2' & \dots & n' \end{Bmatrix} \quad \text{or} \quad \begin{vmatrix} (11)' & (12)' & \dots & (1n)' \\ (21)' & (22)' & \dots & (2n)' \\ \cdot & \cdot & \cdot & \cdot \\ (n1)' & (n2)' & \dots & (nn)' \end{vmatrix}$$

for the other. The reading is thus rendered tiresome, and inaccurate printing exaggerates the trouble.

In § 8 (pp. 335–337) Sylvester's proposition of the year 1850 regarding the vanishing of the minors of a non-quadrate matrix is attempted to be proved. The matrix being, for example,

$$\begin{matrix} 11 & 12 & 13 & 14 & 15 & 16 & 17 \\ 21 & 22 & 23 & 24 & 25 & 26 & 27 \\ 31 & 32 & 33 & 34 & 35 & 36 & 37 \\ 41 & 42 & 43 & 44 & 45 & 46 & 47 \end{matrix} \quad \text{or} \quad \begin{Bmatrix} 1\,2\,3\,4\,5\,6\,7 \\ 1\,2\,3\,4 \end{Bmatrix}$$

and $\{rstu\}$ being the determinant whose columns are identical with the rth, s^{th}, t^{th}, u^{th} columns of the matrix, it is required to show that if the minors $\{1234\}$, $\{1235\}$, $\{1236\}$, $\{1237\}$ vanish, the thirty-one other four-line minors of the matrix must vanish also. In support of this Spottiswoode says truly enough that if

$$A, \ B, \ C, \ D$$

stand for

$$-\begin{Bmatrix}123\\234\end{Bmatrix}, \ \begin{Bmatrix}123\\134\end{Bmatrix}, \ -\begin{Bmatrix}123\\124\end{Bmatrix}, \ \begin{Bmatrix}123\\123\end{Bmatrix},$$

we know that

$$\begin{aligned} \{1234\} &= 14\cdot A + 24\cdot B + 34\cdot C + 44\cdot D, \\ \{1235\} &= 15\cdot A + 25\cdot B + 35\cdot C + 45\cdot D, \\ \{1236\} &= 16\cdot A + 26\cdot B + 36\cdot C + 46\cdot D, \\ \{1237\} &= 17\cdot A + 27\cdot B + 37\cdot C + 47\cdot D; \end{aligned}$$

but then to this he merely adds, "and, if these vanish, it is obvious that by direct elimination all the others may be at once deduced." He notes in addition that in the case of the matrix

$$\begin{Bmatrix} 1\,2\,3\,4 \dots n \\ 1\,2 \end{Bmatrix} \quad \text{or} \quad \begin{matrix} 11 & 12 & 13 & 14 & \dots & 1n \\ 21 & 22 & 23 & 24 & \dots & 2n \end{matrix}$$

the vanishing of

$$\begin{Bmatrix}12\\12\end{Bmatrix}, \ \begin{Bmatrix}13\\12\end{Bmatrix}, \ \begin{Bmatrix}14\\12\end{Bmatrix}, \ \dots, \ \begin{Bmatrix}1n\\12\end{Bmatrix}$$

if brought about by the vanishing of 11 and 21 will not ensure the vanishing of the other 2-line minors of the matrix; but he does not see that the vanishing of {1234}, {1235}, {1236}, {1237} in the previous example, if brought about by the vanishing of A, B, C, D, will be equally ineffectual.

GRASSMANN [H.] (1854, February, April).

[Sur les différents genres de multiplication. *Crelle's Journ.*, xlix. pp. 123–141.]

[Extrait d'un mémoire de M. Grassmann. *Comptes rendus . . . Acad. des Sci.* (Paris), xxxviii. pp. 743–744.]

Grassmann, having become aware of Cauchy's three communications to the French Academy in January of 1853, claims that the principles there established and the results deduced are absolutely the same as those published by himself in 1844. He says (p. 127), "Les clefs algébriques de M. Cauchy ne sont au fond que les unités relatives; et ses facteurs symboliques conviennent, du moins dans un certain rapport, aux quantités extensives telles que je les ai définies. La différence ne consiste qu'en ce que M. Cauchy regarde les clefs algébriques seulement comme un moyen pour résoudre divers problèmes de l'analyse et de la mécanique et qui, les problèmes étant résolus, disparaissent, tandis que d'après les principes établis par moi, on est en état, à chaque pas du procédé, d'attribuer une signification indépendante aux unités relatives et aux quantités qui en sont composées, qu'elle que soit d'ailleurs la marche que l'on suive."

MAJO, L. DE (1854, March).

[Metodi e formole generali per l' eliminazione nelle equazioni di primo grado. *Memorie ... Accad. delle Sci.* (Napoli), i. pp. 101–116.]

This is a carefully written but curiously belated exposition, the author apparently being quite out of touch with the writers of his own time, and possibly not familiar with any of the older writers save Cramer, Bezout, and Hindenburg. In the first six pages he defines "il polinomio $P_m(a_1 b_2 c_3 \ldots s_m)$" after the fashion of Bezout (1764), and gives one or two very elementary properties of it. The remaining ten pages are occupied with simultaneous linear equations, and are notable as containing (§§ 15–19) a clear exposition of Bezout's peculiar rule-of-thumb process of 1779. Herein lies the value of the paper, Majo being not only the first since Hindenburg to recall attention to a neglected process of real practical value, but also the first to give (§ 16) a reason for its validity.

CAYLEY, A. (1854, May).

[Remarques sur la notation des fonctions algébriques. *Crelle's Journal*, l. pp. 282–285; or *Collected Math. Papers*, ii. pp. 185–188.]

The notation referred to is that of *matrices*, and is exemplified by

$$\begin{array}{|ccc|} \multicolumn{3}{(c)}{\alpha_1 \quad \alpha_2 \quad \alpha_3} \\ \beta_1 & \beta_2 & \beta_3 \\ \gamma_1 & \gamma_2 & \gamma_3 \end{array},$$

a matrix being defined as a system of quantities arranged in the form of a square, but otherwise quite independent. With its help the set of equations

$$\left.\begin{aligned} \xi &= a_1x+a_2y+a_3z \\ \eta &= b_1x+b_2y+b_3z \\ \zeta &= c_1x+c_2y+c_3z \end{aligned}\right\}$$

may, he says, be written in the form

$$\xi,\ \eta,\ \zeta = \begin{array}{|ccc|} \multicolumn{3}{(c)}{a_1 \quad a_2 \quad a_3} \\ b_1 & b_2 & b_3 \\ c_1 & c_2 & c_3 \end{array} \!\!\between\!\! x,\ y,\ z),$$

and consequently the solution of the set in the form

$$x,\ y,\ z = \begin{array}{|ccc|} \multicolumn{3}{(c)}{\frac{A_1}{\Delta} \quad \frac{B_1}{\Delta} \quad \frac{C_1}{\Delta}} \\ \frac{A_2}{\Delta} & \frac{B_2}{\Delta} & \frac{C_2}{\Delta} \\ \frac{A_3}{\Delta} & \frac{B_3}{\Delta} & \frac{C_3}{\Delta} \end{array} \!\!\between\!\! \xi,\ \eta,\ \zeta).$$

The latter matrix he calls the *inverse* of the former, and is naturally led to propose that it be denoted by

$$\left(\begin{array}{ccc} a_1 & a_2 & a_3 \\ b_1 & b_2 & b_3 \\ c_1 & c_2 & c_3 \end{array}\right)^{-1}.$$

Next, supposing that along with the original set there exists the set

$$x,\ y,\ z = \left(\begin{array}{ccc} \alpha_1 & \alpha_2 & \alpha_3 \\ \beta_1 & \beta_2 & \beta_3 \\ \gamma_1 & \gamma_2 & \gamma_3 \end{array} \between \mathrm{X},\ \mathrm{Y},\ \mathrm{Z}\right),$$

so that by substitution ξ, η, ζ are expressible in terms of X, Y, Z, Cayley is led by comparison of the old and the new notations to the conception of the *product* of two matrices, and to the use of

$$\left(\begin{array}{ccc} a_1 & a_2 & a_3 \\ b_1 & b_2 & b_3 \\ c_1 & c_2 & c_3 \end{array} \between \begin{array}{ccc} \alpha_1 & \alpha_2 & \alpha_3 \\ \beta_1 & \beta_2 & \beta_3 \\ \gamma_1 & \gamma_2 & \gamma_3 \end{array}\right)$$

for

$$\left(\begin{array}{ccc} (a_1a_2a_3 \between \alpha_1\beta_1\gamma_1) & (a_1a_2a_3 \between \alpha_2\beta_2\gamma_2) & (a_1a_2a_3 \between \alpha_3\beta_3\gamma_3) \\ (b_1b_2b_3 \between \alpha_1\beta_1\gamma_1) & (b_1b_2b_3 \between \alpha_2\beta_2\gamma_2) & (b_1b_2b_3 \between \alpha_3\beta_3\gamma_3) \\ (c_1c_2c_3 \between \alpha_1\beta_1\gamma_1) & (c_1c_2c_3 \between \alpha_2\beta_2\gamma_2) & (c_1c_2c_3 \between \alpha_3\beta_3\gamma_3) \end{array}\right).$$

Lastly, he explains his related notations for lineo-linear functions and quantics.* These we need only exemplify by saying that

$$\left(\begin{array}{ccc} a_1 & a_2 & a_3 \\ b_1 & b_2 & b_3 \\ c_1 & c_2 & c_3 \end{array} \between \xi,\ \eta,\ \zeta \between x,\ y,\ z\right), \qquad \left(\begin{array}{ccc} a & h & g \\ h & b & f \\ g & f & c \end{array} \between x,\ y,\ z\right)^2,$$

are made to stand for

$$\begin{array}{l} (a_1\xi + a_2\eta + a_3\zeta)x \\ +(b_1\xi + b_2\eta + b_3\zeta)y \\ +(c_1\xi + c_2\eta + c_3\zeta)z \end{array} \quad \text{and} \quad ax^2+by^2+cz^2+2fyz+2gzx+2hxy$$

respectively, and that the latter is also denoted by

$$(a,\ b,\ c,\ f,\ g,\ h \between x,\ y,\ z)^2,$$

* Cayley's first memoir on *quantics* was presented to the Royal Society of London on 20th April, and this paper on the notation of *matrices* is the first of five which appeared together in *Crelle's Journal* with the date 24th May affixed by the author.

and the binary cubics

$$ax^3 + 3bx^2y + 3cxy^2 + dy^3, \quad ax^3 + bx^2y + cxy^2 + dy^3$$

by $$(a, b, c, d \between x, y)^3, \quad (a, b, c, d \between x, y)^3$$

respectively.

We may suggest for consideration in passing the following order of ideas, as leading up to Cayley's contracted mode of writing a set of linear equations. First, a *row* of separate quantities, *e.g.* $(a, b, c, \ldots)$; second, the statement of the identity of two rows, *e.g.* $(a, b, c, \ldots) = (x, y, z, \ldots)$, or simply $a, b, c, \ldots = x, y, z, \ldots$; third, the so-called product of two rows, *e.g.* $(a, b, c, \ldots \between x, y, z, \ldots)$; fourth, a *square* of separate quantities, *i.e.* a matrix; fifth, the result of multiplying a matrix and a row being a row. It is unfortunate that, from the point of view of notation merely, this does not at once suggest, in the sixth place, the result of multiplying two matrices, where, as Cayley is careful to point out, the multiplication is row-by-column and not row-by-row.

BRIOSCHI, F. (1854).

[LA TEORICA DEI DETERMINANTI, E LE SUE PRINCIPALI APPLICAZIONI; del Dr. Francesco Brioschi; viii+116 pp.; Pavia. Translation into French, by Combescure; ix+216 pp.; Paris, 1856. Translation into German, by Schellbach; vii+102 pp.; Berlin, 1856.]

This, the second separately-published text-book on determinants, is mainly on the same lines as the first, but is marked by greater attention to verbal and logical accuracy. It consists of an historical preface and eleven short chapters or sections, seven of the latter being devoted to determinants in general, and the remaining four to special forms.

Sylvester's umbral notation is given in the form

$$| a_1 \ a_2 \ \ldots \ a_n |$$

but is not afterwards employed. The same author's term "minor" (*minore*) is adopted, this being represented in the French translation by "mineur," and in the German by "Unterdeterminante." "Complete" is used as opposed to "minor," and "principal minor" for what nowadays we call "coaxial."

Conspicuously frequent use is made of differentiation in the specification of minors; and it is well to note that, though the work in this way becomes cumbrous, there is a certain effectiveness attained by the usage. Thus, Δ standing for $\Sigma(\pm a_{11}a_{22}\ldots a_{nn})$, Brioschi, like Jacobi, obtains

$$\left.\begin{array}{l}\dfrac{\partial\Delta}{\partial a_{r1}} = 0 + a_{s2}\dfrac{\partial^2\Delta}{\partial a_{r1}\,\partial a_{s2}} + \ldots\ldots + a_{sn}\dfrac{\partial^2\Delta}{\partial a_{r1}\,\partial a_{sn}}\\ \dfrac{\partial\Delta}{\partial a_{r2}} = a_{s1}\dfrac{\partial^2\Delta}{\partial a_{r2}\,\partial a_{s1}} + 0 + \ldots\ldots + a_{sn}\dfrac{\partial^2\Delta}{\partial a_{r2}\,\partial a_{sn}}\\ \ldots\ldots\ldots\ldots\ldots\ldots\ldots\ldots\\ \dfrac{\partial\Delta}{\partial a_{rn}} = a_{s1}\dfrac{\partial^2\Delta}{\partial a_{rn}\,\partial a_{s1}} + a_{s2}\dfrac{\partial^2\Delta}{\partial a_{rn}\,\partial a_{s2}} + \ldots\ldots + 0\end{array}\right\},$$

and then, by using the multipliers $a_{r1}, a_{r2}, \ldots, a_{rn}$ and adding, finds

$$\Delta = \begin{vmatrix} a_{r1} & a_{r2} \\ a_{s1} & a_{s2} \end{vmatrix} \frac{\partial^2\Delta}{\partial a_{r1}\partial a_{s2}} + \begin{vmatrix} a_{r1} & a_{r3} \\ a_{s1} & a_{s3} \end{vmatrix} \frac{\partial^2\Delta}{\partial a_{r1}\partial a_{s3}} + \ldots + \begin{vmatrix} a_{r,n-1} & a_{rn} \\ a_{s,n-1} & a_{sn} \end{vmatrix} \frac{\partial^2\Delta}{\partial a_{r1}\partial a_{sn}},$$

which is Laplace's expansion-theorem for the case where the minors of one set are of the second order. The remaining cases, he says, can be established in the same way.

Again, having proved the multiplication-theorem (row-by-row)

$$PQ = R,$$

where

$$P = \Sigma(\pm a_{11}a_{22}\ldots a_{nn}), \quad Q = \Sigma(\pm b_{11}b_{22}\ldots b_{nn}),$$
$$R = \Sigma(\pm c_{11}c_{22}\ldots c_{nn}),$$

he obtains by differentiation with respect to elements of P,

$$\frac{\partial P}{\partial a_{rs}}Q = \frac{\partial R}{\partial c_{r1}}b_{1s} + \frac{\partial R}{\partial c_{r2}}b_{2s} + \ldots + \frac{\partial R}{\partial c_{rn}}b_{ns}, \qquad (1)$$

$$\frac{\partial^2 P}{\partial a_{rs}\partial a_{\rho\sigma}}Q = \sum_x\sum_y \begin{vmatrix} b_{x\sigma} & b_{xs} \\ b_{y\sigma} & b_{ys} \end{vmatrix} \frac{\partial^2 R}{\partial c_{ry}\partial c_{\rho x}}, \qquad (2)$$

and by twice differentiating with respect to an element of P and an element of Q,

$$\frac{\partial P}{\partial a_{rs}}\frac{\partial Q}{\partial b_{\rho\sigma}}-\frac{\partial P}{\partial a_{r\sigma}}\frac{\partial Q}{\partial b_{\rho s}}=\sum_x\sum_y\begin{vmatrix}a_{x\sigma} & a_{xs}\\ b_{y\sigma} & b_{ys}\end{vmatrix}\frac{\partial^2 R}{\partial c_{ry}\partial c_{\rho x}}. \quad (3)$$

Taking $n=4$ and $r, s, \rho, \sigma=1, 2, 3, 4$, we can best illustrate these by writing them thus:—

$$-|a_{21}\,a_{33}\,a_{44}|\cdot|b_{11}\,b_{22}\,b_{33}\,b_{44}|=\begin{vmatrix}b_{12} & b_{22} & b_{32} & b_{42}\\ c_{21} & c_{22} & c_{23} & c_{24}\\ c_{31} & c_{32} & c_{33} & c_{34}\\ c_{41} & c_{42} & c_{43} & c_{44}\end{vmatrix}$$

$$=b_{12}|c_{22}\,c_{33}\,c_{44}|-\ldots., \quad (1')$$

$$|a_{21}\,a_{43}|\cdot|b_{11}\,b_{22}\,b_{33}\,b_{44}|=\begin{vmatrix}b_{12} & b_{22} & b_{32} & b_{42}\\ c_{21} & c_{22} & c_{23} & c_{24}\\ b_{14} & b_{24} & b_{34} & b_{44}\\ c_{41} & c_{42} & c_{43} & c_{44}\end{vmatrix}$$

$$=-|b_{12}\,b_{24}|\cdot|c_{23}\,c_{44}|+\ldots., \quad (2')$$

$$\left\|\begin{matrix}|a_{21}\,a_{33}\,a_{44}| & |a_{21}\,a_{32}\,a_{43}|\\ |b_{11}\,b_{23}\,b_{44}| & |b_{11}\,b_{22}\,b_{43}|\end{matrix}\right\|=\begin{vmatrix}b_{12} & b_{22} & \cdot & b_{42}\\ c_{21} & c_{22} & a_{24} & c_{24}\\ c_{31} & c_{32} & a_{34} & c_{34}\\ c_{41} & c_{42} & a_{44} & c_{44}\end{vmatrix}-\begin{vmatrix}b_{14} & b_{24} & \cdot & b_{44}\\ c_{21} & c_{22} & a_{22} & c_{24}\\ c_{31} & c_{32} & a_{32} & c_{34}\\ c_{41} & c_{42} & a_{42} & c_{44}\end{vmatrix}$$

$$=-|a_{24}\,b_{12}|\cdot|c_{32}\,c_{44}|+\ldots. \quad (3')$$

The right-hand member of (1) is equivalent to a direction to substitute for the rth row of R the sth column of Q; similarly, the right-hand member of (2) is a direction, though not so evident, to substitute for the rth and ρ^{th} rows of R the sth and σ^{th} columns of Q; and it is clear that (1) and (2) are but the first two identities of many. On the other hand, (3) is quite diverse in character, being got by the combination of two results analogous to (2). This is best brought out by noting that in the examples the right-hand members of (1′) and (2′) are got by multiplying

$$\begin{vmatrix}\cdot & 1 & \cdot & \cdot\\ a_{21} & a_{22} & a_{23} & a_{24}\\ a_{31} & a_{32} & a_{33} & a_{34}\\ a_{41} & a_{42} & a_{43} & a_{44}\end{vmatrix}\quad\text{and}\quad\begin{vmatrix}\cdot & 1 & \cdot & \cdot\\ a_{21} & a_{22} & a_{23} & a_{24}\\ \cdot & \cdot & \cdot & 1\\ a_{41} & a_{42} & a_{43} & a_{44}\end{vmatrix}$$

respectively by $|b_{11}\,b_{22}\,b_{33}\,b_{44}|$; and that similarly the two four-line determinants on the right of (3) are got by multiplying

$$\begin{vmatrix} \cdot & 1 & \cdot & \cdot \\ a_{21} & a_{22} & a_{23} & a_{24} \\ a_{31} & a_{32} & a_{33} & a_{34} \\ a_{41} & a_{42} & a_{43} & a_{44} \end{vmatrix} \text{ by } \begin{vmatrix} b_{11} & b_{12} & b_{13} & b_{14} \\ b_{21} & b_{22} & b_{23} & b_{24} \\ \cdot & \cdot & \cdot & 1 \\ b_{41} & b_{42} & b_{43} & b_{44} \end{vmatrix}$$

and

$$\begin{vmatrix} \cdot & \cdot & \cdot & 1 \\ a_{21} & a_{22} & a_{23} & a_{24} \\ a_{31} & a_{32} & a_{33} & a_{34} \\ a_{41} & a_{42} & a_{43} & a_{44} \end{vmatrix} \text{ by } \begin{vmatrix} b_{11} & b_{12} & b_{13} & b_{14} \\ b_{21} & b_{22} & b_{23} & b_{24} \\ \cdot & 1 & \cdot & \cdot \\ b_{41} & b_{42} & b_{43} & b_{44} \end{vmatrix}.$$

It should be carefully noted also that, while in (2) the number of terms in the development is $\frac{1}{2}n(n-1)$, in (3) the number is $(n-1)^2$.

Lastly, putting

$$\left.\begin{array}{l} b_{11}\dfrac{\partial P}{\partial a_{r1}} + b_{12}\dfrac{\partial P}{\partial a_{r2}} + \ldots . + b_{1n}\dfrac{\partial P}{\partial a_{rn}} = H_{r1} \\ b_{21}\dfrac{\partial P}{\partial a_{r1}} + b_{22}\dfrac{\partial P}{\partial a_{r2}} + \ldots . + b_{2n}\dfrac{\partial P}{\partial a_{rn}} = H_{r2} \\ \cdot\quad\cdot\quad\cdot\quad\cdot\quad\cdot\quad\cdot\quad\cdot\quad\cdot\quad\cdot\quad\cdot\quad\cdot\quad\cdot\quad\cdot\quad\cdot\quad\cdot \\ b_{n1}\dfrac{\partial P}{\partial a_{r1}} + b_{n2}\dfrac{\partial P}{\partial a_{r2}} + \ldots . + b_{nn}\dfrac{\partial P}{\partial a_{rn}} = H_{rn} \end{array}\right\}$$

so that H_{rs} stands for what P becomes when its rth row is replaced by the sth row of Q, and using the multipliers $\partial Q/\partial b_{11}$, $\partial Q/\partial b_{21}, \ldots ., \partial Q/\partial b_{n1}$ prior to addition, Brioschi obtains

$$Q\frac{\partial P}{\partial a_{r1}} = H_{r1}\frac{\partial Q}{\partial b_{11}} + H_{r2}\frac{\partial Q}{\partial b_{21}} + \ldots . + H_{rn}\frac{\partial Q}{\partial b_{n1}},$$

and similarly

$$Q\frac{\partial P}{\partial a_{r2}} = H_{r1}\frac{\partial Q}{\partial b_{12}} + H_{r2}\frac{\partial Q}{\partial b_{22}} + \ldots . + H_{rn}\frac{\partial Q}{\partial b_{n2}},$$

$$\cdot\quad\cdot\quad\cdot\quad\cdot\quad\cdot\quad\cdot\quad\cdot\quad\cdot\quad\cdot\quad\cdot\quad\cdot\quad\cdot\quad\cdot\quad\cdot\quad\cdot$$

$$Q\frac{\partial P}{\partial a_{rn}} = H_{r1}\frac{\partial Q}{\partial b_{1n}} + H_{r2}\frac{\partial Q}{\partial b_{2n}} + \ldots . + H_{rn}\frac{\partial Q}{\partial b_{nn}}.$$

With this derived set of equations the multipliers $a_{r1}, a_{r2}, \ldots, a_{rn}$ are then used, and addition performed, the result being Sylvester's theorem of 1839, namely,

$$QP = H_{r1}K_{r1} + H_{r2}K_{r2} + \ldots + H_{rn}K_{rn},$$

where K_{rs} stands for what Q becomes when its sth row is replaced by the rth row of P.*

In his treatment of the minors of the adjugate determinant Brioschi (pp. 36–39) closely follows Spottiswoode; that is to say, from a set of linear equations he derives one result, then from the adjugate set another result, and finally draws a deduction from a comparison of the two. His thus obtained extension of Spottiswoode's theorem is open to the same criticism as Spottiswoode's extension of Jacobi's.

The section (§ 7) on "determinanti di determinanti" is founded on Cauchy, and contains known extensions of two or three theorems above given in the notation of differentiation.

CANTOR [M. B.] (1855, March).

[Théorème sur les déterminants Cramériens. *Nouv. Annales de Math.* (1), xiv. pp. 113–114.]

The theorem in question may be formulated thus—*If the permutations of* 1, 2, 3, . . . n *be arranged in order of magnitude as if they were integral numbers, the sign of the* kth *permutation is independent of* n. Reference is appropriately made to Reiss' paper of 1825, but the theorem is virtually contained in Hinderburg's rule of the year 1784.

Another author who dealt with the 'rule of signs' in this year was Mainardi; his paper is referred to along with a kindred one by Zehfuss of the year 1858.

---

* Brioschi does not note the independent importance of his second set of equations, which may be condensed into

$$Q\frac{\partial P}{\partial a_{rs}} = H_{r1}\frac{\partial Q}{\partial b_{1s}} + H_{r2}\frac{\partial Q}{\partial b_{2s}} + \ldots + H_{rn}\frac{\partial Q}{\partial b_{ns}},$$

and which, when $r, s = 1, 1$ and $n = 3$, is

$$\begin{vmatrix} \alpha_1 & \alpha_2 & \alpha_3 \\ \beta_1 & \beta_2 & \beta_3 \\ \gamma_1 & \gamma_2 & \gamma_3 \end{vmatrix} \cdot \begin{vmatrix} b_2 & b_3 \\ c_2 & c_3 \end{vmatrix} = \begin{vmatrix} \alpha_1 & \alpha_2 & \alpha_3 \\ b_1 & b_2 & b_3 \\ c_1 & c_2 & c_3 \end{vmatrix} \cdot \begin{vmatrix} \beta_2 & \beta_3 \\ \gamma_2 & \gamma_3 \end{vmatrix} - \begin{vmatrix} \beta_1 & \beta_2 & \beta_3 \\ b_1 & b_2 & b_3 \\ c_1 & c_2 & c_3 \end{vmatrix} \cdot \begin{vmatrix} \alpha_2 & \alpha_3 \\ \gamma_2 & \gamma_3 \end{vmatrix} + \begin{vmatrix} \gamma_1 & \gamma_2 & \gamma_3 \\ b_1 & b_2 & b_3 \\ c_1 & c_2 & c_3 \end{vmatrix} \cdot \begin{vmatrix} \alpha_2 & \alpha_3 \\ \beta_2 & \beta_3 \end{vmatrix}.$$

This, however, may be viewed also as a case of Sylvester's theorem, namely, where the first row of P is 1, 0, 0.

CAUCHY, A. L. (1856, Feb.).

[Sur une formule très-simple et très-général .... *Comptes rendus* .... *Acad. des Sci.* (Paris), xlii. pp. 366–374; or *Œuvres complètes* (1), xii. pp. 302–311.]

The main theorem concerns the set of equations

$$a_{r1}x_1 + a_{r2}x_2 + \dots + a_{rm}x_m = u_r \Big\}_{r=1}^{r=n}$$

where the number m of unknowns is greater than the number n of equations, and the other theorem deals with the case of this where the u's all vanish. The weapons employed in the discussion are the 'produit symbolique' of the u's and 'clefs algébriques,' or 'clefs anastrophiques' as they are now called. The paper may be compared with Sylvester's of the year 1839 on 'the derivation of coexistence.'

HERMITE, C. (1856).

[Sur la théorie des polynomes homogènes du second degré. Note vi. (pp. 154–190) of *Programme .... d'Arithmétique, d'Algèbre et de Géométrie Analytique,* .... par MM. Gerono et Roguet: 4e éd. 215 pp. Paris.]

This note of thirty-seven pages, which is said to be 'd'après M. Hermite,' consists of five sections, the first of which (pp. 154–157) deals expressly and the others incidentally with determinants. The latter sections concern the invariance of the discriminant (called '*the* invariant'), orthogonal transformation, etc., and are simply but suggestively written.

HEGER, I. (1856, July).

[Ueber die Auflösung eines Systemes von mehreren unbestimmten Gleichungen des ersten Grades in ganzen Zahlen. *Denkschr. d. k. Akad. d. Wiss.* (Wien): *math.-naturw. Cl.*, xiv. (2), pp. 1–122.]

Although in this lengthy paper the vanishing of the determinants of a 2-by-n array is repeatedly under consideration (*e.g.* § 24, p. 87), nothing new on the subject presents itself.

SCHLÖMILCH, O. (1856).

[Brioschi's Theorie der Determinante und ihre hauptsächlichsten Anwendungen. *Zeitschrift f. Math. u. Phys.*, I. *Literaturzeitung*, pp. 80–87.]

After a faithful account of Schellbach's translation of Brioschi's text-book, Schlömilch inveighs against the adoption of "die miserable englische Terminologie," instancing *Unterdeterminante*, *Determinante mit reciproken Elementen*, and *Hessian*, for the last of which he proposes to substitute "Inflexionsdeterminante."

RUBINI, R. (1857, May).

[Applicazione della teorica dei determinanti. *Annali de Sci. mat. e fis.*, viii. pp. 179–200.]

This resembles Chio's paper of 1853, having the same fundamental theorem, but different illustrative examples. In the mere enunciation of the theorem Rubini is the more successful. Taking the n-line determinant whose element in the place r,s is $a_{rs}+b_{rs}$, and denoting by A the determinant of the a's, and by A_r a determinant obtainable from A on substituting for r columns of a's the corresponding r columns of b's, he writes the expansion in the form

$$A+\Sigma A_1+\Sigma A_2+ \ldots . +\Sigma A_{n-1}+A_n .$$

BELLAVITIS, G. (1857, June).

[Sposizione elementare della teorica dei determinante. *Memorie . . . Istituto Veneto*, . . . vii. pp. 67–144.]

Notwithstanding its place of publication, this writing of Bellavitis' is exactly what its title implies; and as a text-book it could scarcely have failed to be useful, so simple and clear is it in style. It consists of two chapters, one on determinants in general (pp. 3–30), and one on special forms (pp. 30–72): a note of six pages on permutations appears as an appendix.

To Bellavitis we owe the modification of Laplace's notation which is now in common use. The passage introducing it is:

"Quando gli elementi sieno indicati in modo che chiaramente apparisca la loro formazione, noi porremo tra le due | | i soli elementi della *diagonale* (intendendo sempre per *diagonale* quella da sinistra verso destra descendendo). Cosi

$$|a_1 b_2 c_3 \ldots\ldots| \text{ equivalerà a } \begin{vmatrix} a_1 & b_1 & c_1 & \ldots\ldots \\ a_2 & b_2 & c_2 & \ldots\ldots \\ a_3 & b_3 & c_3 & \ldots\ldots \\ \cdot & \cdot & \cdot & \cdot\ \cdot\ \cdot \end{vmatrix}$$

$$|a_q^{(h)} a_s^{(r)}| \text{ equivalerà a } \begin{vmatrix} a_q^{(h)} & a_q^{(r)} \\ a_s^{(h)} & a_s^{(r)} \end{vmatrix}, \text{ ec.''}$$

Throughout the exposition this notation is employed. "Riga" he uses either for a "fila orizzontale" or a "fila verticale," and "colonna" for a "fila perpendicolare a quella che s'intese per riga."

Two well-known developments he specifies thus:—

$$|a_1 b_2 c_3 \ldots| = \left(a_1 \frac{\partial}{\partial a_1} + b_1 \frac{\partial}{\partial b_1} + c_1 \frac{\partial}{\partial c_1} + \ldots\right) |a_1 b_2 c_3 \ldots|,$$

$$\begin{aligned} |a_1 b_2 c_3 \ldots h_n| = \Big(a_1 \frac{\partial}{\partial a_1} &+ a_2 b_1 \frac{\partial^2}{\partial a_2 \partial b_1} + \ldots\ldots + a_2 h_1 \frac{\partial^2}{\partial a_2 \partial h_1} \\ &+ a_3 b_1 \frac{\partial^2}{\partial a_3 \partial b_1} + \ldots\ldots + a_3 h_1 \frac{\partial^2}{\partial a_3 \partial h_1} \\ &\cdot\ \cdot\ \cdot\ \cdot\ \cdot\ \cdot\ \cdot\ \cdot\ \cdot\ \cdot\ \cdot\ \cdot\ \cdot \\ &+ a_n b_1 \frac{\partial^2}{\partial a_n \partial b_1} + \ldots\ldots + a_n h_1 \frac{\partial^2}{\partial a_n \partial h_1}\Big) |a_1 b_2 c_3 \ldots h_n|. \end{aligned}$$

In reference to determinants with binomial elements (§ 13) he says: "Compiendo questo sviluppo si ottiene la formula

$$\begin{aligned} |a_1 + \alpha_1 \ b_2 + \beta_2 \ c_3 + \gamma_3| = \ & |a_1 \ b_2 \ c_3| + |a_1 \ b_2 \ \gamma_3| + |a_1 \ \beta_2 \ c_3| + |a_1 \ \beta_2 \ \gamma_3| \\ & + |\alpha_1 \ b_2 \ c_3| + |\alpha_1 \ b_2 \ \gamma_3| + |\alpha_1 \ \beta_2 \ c_3| + |\alpha_1 \ \beta_2 \ \gamma_3| \end{aligned}$$

che è facile da tenersi a memoria per la sua perfetta analogia collo sviluppo del prodotto di tre binomii."

After giving a sufficient condition for the vanishing of a determinant, he enunciates (§ 15) the converse, namely, *When a determinant vanishes, one of the rows is equal to a sum of multiples of the other rows*, basing its validity on the fact that the multipliers referred to can actually be found by solving a set of simultaneous linear equations.

The multiplication-theorem for determinants Δ_1, Δ_2 of the third order he seeks to establish (§ 31) by partitioning the product-determinant into twenty-seven determinants, and showing that the sum of the six which do not vanish is $\Delta_1 \Delta_2$.

Chio's theorem of 1853 is introduced (§ 38) by noting that the resultant of

$$a_r x + b_r y + c_r = 0 \qquad (r=1,2,3)$$

may be viewed as the resultant of

$$\left.\begin{aligned} |a_1 b_2| y + |a_1 c_2| &= 0 \\ |a_1 b_3| y + |a_1 c_3| &= 0 \end{aligned}\right\},$$

and that therefore

$$\left\| \begin{matrix} |a_1 b_2| & |a_1 c_2| \\ |a_1 b_3| & |a_1 c_3| \end{matrix} \right\| \text{ must be a multiple of } |a_1 b_2 c_3|.$$

That it is so he proves by diminishing the 2nd and 3rd columns of $| a_1 b_2 c_3 |$ by b_1/a_1 times the 1st column and c_1/a_1 times the 1st column respectively. Further, he points out (§§ 39, 40) a practical application, namely, in evaluating a determinant whose elements are given in figures.

The adjugate determinant (unfortunately renamed *associato*) is dealt with (§§ 55–58) in connection with the solution of a set of simultaneous linear equations, the special cases being considered where the determinant of the set is 1 and 0. In the former special case he notes the theorem, *The adjugate of the product of two unit determinants is identical in all its elements with the product of the adjugates of the said determinants*; and in the latter the theorem all but reached by Jacobi in 1835 and 1841, *In a zero determinant the cofactors of the elements of a row are proportional to the cofactors of the elements of any other row.*

Cauchy's "clefs algébriques" (*chiavi algebriche*) are expounded at some length (§§ 81–88).

In the last three paragraphs he draws attention to the existence of expressions which may be viewed as "determinanti simbolici," his first kind being those in which symbols of differentiation take the place of elements; *e.g.* the negative of the expression

$$P\left(\frac{\partial Q}{\partial z}-\frac{\partial R}{\partial y}\right)+Q\left(\frac{\partial R}{\partial x}-\frac{\partial P}{\partial z}\right)+R\left(\frac{\partial P}{\partial y}-\frac{\partial Q}{\partial x}\right),$$

whose vanishing is the condition for the derivability of the equation

$$P\partial x + Q\partial y + R\partial z = 0$$

from a single primitive, is denoted by

$$\left| P \ \frac{\partial}{\partial y} \ R \right|$$

—a notation which is even less satisfactory than that for which it is a contraction, namely,

$$\begin{vmatrix} P & \frac{\partial}{\partial x} & P \\ Q & \frac{\partial}{\partial y} & Q \\ R & \frac{\partial}{\partial z} & R \end{vmatrix}.$$

The other kind of expressions originated with Binet, who in 1812 gave the identities

$$\begin{aligned} \Sigma ab' &= \Sigma a \Sigma b - \Sigma ab, \\ \Sigma ab'c'' &= \Sigma a \Sigma b \Sigma c + 2\Sigma abc - \Sigma a \Sigma bc - \Sigma b \Sigma ca - \Sigma c \Sigma ab; \\ & \cdots\cdots\cdots\cdots \end{aligned}$$

but in this case, though the close resemblance of the right-hand expressions to the developments of axisymmetric determinants is pointed out, no notation founded on the fact is suggested.

As an appendix there is a note on permutations, explaining circular substitutions, interchanges (*alternazioni*), inversions of order (*rovesciamenti d'ordine*), and their relations to one another. Cauchy's sign-rule depending on the number of circular substitutions is replaced by a simpler rule, which requires the counting of only the *even* circular substitutions. Thus the permutation 3265417 being got from the standard permutation 1234567 by means of the circular substitutions

$$(316), \quad (2), \quad (54), \quad (7),$$

and only one of these being even, the sign of 3265417 is $(-)^1$. Bellavitis' enunciation is: "*Il numero delle alternazioni, con cui una disposizione può mutarsi in un' altra è pari o dispari insieme col numero di tutte le sostituzioni binomie, quadrinomie, sestinomie, ecc. che occorrono per passare da una disposizione all' altra.*"

BALTZER, R. (1857).

[THEORIE UND ANWENDUNG DER DETERMINANTEN, mit Beziehung auf die Originalquellen, dargestellt von Dr. Richard Baltzer . . .; vi+129 pp.; Leipzig. French translation by J. Houel, xii+235 pp.; Paris, 1861.]

The good qualities spoken of above as belonging to Brioschi's text-book are still more conspicuous in the German text-book of three years later, and the historical footnotes in Baltzer's give it a special additional value. The *theory* is dealt with in eight little chapters or sections, and the so-called *applications* in ten; several of the latter, however, might quite well have been classed with the former, as they are merely concerned with determinants of special form.

The first section corresponds closely in subject with Bellavitis' appendix: and in connection therewith may be noted Baltzer's remark (§ 2, 3) that *any term got from the diagonal term by substituting* $\mathrm{k}_1, \mathrm{k}_2, \ldots, \mathrm{k}_n$ *for the second suffixes* $1, 2, \ldots, \mathrm{n}$ *may also be got by substituting* $1, 2, \ldots, \mathrm{n}$ *for* $\mathrm{k}_1, \mathrm{k}_2, \ldots, \mathrm{k}_n$ *in the set of first suffixes.*

Brioschi's mode of proving Sylvester's theorem of 1839 is improved upon (§ 3, 11) by taking Q one order lower than P, and using the multipliers $\partial Q/\partial b_{11}, \partial Q/\partial b_{21}, \ldots, \partial Q/\partial b_{n-1,1}$ on the identities

$$\left.\begin{array}{l}
a_{11}\dfrac{\partial P}{\partial a_{n1}} + a_{12}\dfrac{\partial P}{\partial a_{n2}} + \ldots . + a_{1n}\dfrac{\partial P}{\partial a_{nn}} = 0 \\
a_{21}\dfrac{\partial P}{\partial a_{n1}} + a_{22}\dfrac{\partial P}{\partial a_{n2}} + \ldots . + a_{2n}\dfrac{\partial P}{\partial a_{nn}} = 0 \\
\cdot \quad \cdot \quad \cdot \quad \cdot \quad \cdot \quad \cdot \quad \cdot \quad \cdot \quad \cdot \quad \cdot \quad \cdot \quad \cdot \\
a_{n-1,1}\dfrac{\partial P}{\partial a_{n1}} + a_{n-1,2}\dfrac{\partial P}{\partial a_{n2}} + \ldots . + a_{n-1,n}\dfrac{\partial P}{\partial a_{nn}} = 0
\end{array}\right\},$$

the result of addition then being

$$\left(a_{11}\frac{\partial Q}{\partial b_{11}}+a_{21}\frac{\partial Q}{\partial b_{21}}+\ldots\ldots+a_{n-1,1}\frac{\partial Q}{\partial b_{n-1,1}}\right)\frac{\partial P}{\partial a_{n1}}$$
$$+\left(a_{12}\frac{\partial Q}{\partial b_{11}}+a_{22}\frac{\partial Q}{\partial b_{21}}+\ldots\ldots+a_{n-1,2}\frac{\partial Q}{\partial b_{n-1,1}}\right)\frac{\partial P}{\partial a_{n2}}$$
$$+\ \cdot\ \cdot\ \cdot\ \cdot\ \cdot\ \cdot\ \cdot\ \cdot\ \cdot\ \cdot\ \cdot\ \cdot\ \cdot\ \cdot\ \cdot\ \cdot\ \cdot$$
$$+\left(a_{1n}\frac{\partial Q}{\partial b_{11}}+a_{2n}\frac{\partial Q}{\partial b_{21}}+\ldots\ldots+a_{n-1,n}\frac{\partial Q}{\partial b_{n-1,1}}\right)\frac{\partial P}{\partial a_{nn}} = 0,$$

which, if we bear in mind what single determinants the expressions in brackets stand for, is seen to be Sylvester's theorem in its alternative form as the assertion of the vanishing of an aggregate of products of pairs of determinants.

Of Jacobi's theorem regarding any coaxial minor of the adjugate an obvious extension is made (§ 7, 2), namely, *Any* m^{th} *order minor of the adjugate of any determinant* Δ *is equal to the product obtained by multiplying the cofactor of the corresponding minor in* Δ *by* Δ^{m-1}. The mode of proof followed in Cayley's of 1843.

The theorem formulated by Bellavitis regarding a zero determinant is appropriately based (§ 7, 5) on the vanishing of the two-line minors of the adjugate determinant—a course suggested by what Lebesgue did in 1837.

Cayley's development of 1847 is well stated (§ 8, 6) in the form

$$D + \Sigma a_{ii}D_i + \Sigma a_{ii}a_{kk}D_{ik} + \Sigma a_{ii}a_{kk}a_{ll}D_{ikl} + \ldots + a_{11}a_{22}\ldots a_{nn},$$

where D is what the given determinant becomes when all its diagonal elements are made 0, and $D_{ik\ldots}$ is the minor of D got by deleting the ith, k^{th}, ... rows and the ith, k^{th}, ... columns; and the proof consists in showing that no term is thus neglected or repeated.

NEWMAN, F. (1857).

[On determinants, better called eliminants. *Proceedings Roy. Soc. London*, viii. pp. 426–431; or *Philos. Magazine* (4), xiv. p. 392.]

The author's object was merely to recommend the introduction of the subject into elementary text-books.

DEL GROSSO, R. (1857).

[Sulla regola secondo la quale debbono procedere i segni nello sviluppo d'un determinante in prodotti di determinanti minori. *Rendic.* . . . *Accad. Pontaniana* (Napoli), Ann. v. pp. 196–198. See also pp. 198–206.]

When a determinant is expressed in accordance with Laplace's theorem as an aggregate of products of complementary minors, Del Grosso directs that the sign of any product is to be $(-1)^\sigma$, where σ is the sum of the odd row-numbers and odd column-numbers of one of the factors. The rule is not stated with sufficient care, and the author in reaching it concludes too hastily that the simplest case is all that need be established.

JANNI, G. (1858).

[Saggio di una Teorica Elementare de' Determinanti, del Sacerdote Giuseppe Janni. . . . 40 pp. Napoli.]

Janni's professed object was to make determinants more readily accessible, previous text-books having, he says, either totally neglected demonstrations or used those of great difficulty. He speaks of the work as the first of a series, and its contents certainly look like the first five chapters of a text-book planned on a fairly large scale. The theorems, twenty-three in number, are carefully enunciated and are printed in italics; but, although the proofs receive every attention, it is very doubtful whether the object aimed at was to any extent accomplished. There is at any rate nothing sufficiently fresh in the treatment to warrant attention here.

ZEHFUSS, G. (1858).

[Ueber die Auflösung der linearen endlichen Differenzengleichungen mit variabeln Coefficienten. *Zeitschrift f. Math. u. Phys.*, iii. pp. 175–177.]

His solution suggests to Zehfuss the remark (p. 177) that every determinant can be expressed as a multiple integral. It will suffice to give the result in the case of a determinant of the 4th order. Denoting $\cos 2\pi\theta + \sqrt{-1}\sin 2\pi\theta$ by 1^θ, and putting P for

$$1^\alpha 1^\beta 1^\gamma 1^\delta (1^\delta - 1^\gamma)(1^\delta - 1^\beta)(1^\delta - 1^\alpha)(1^\gamma - 1^\beta)(1^\gamma - 1^\alpha)(1^\beta - 1^\alpha)$$

and Q for

$$\begin{aligned} &(a_1 1^{-\alpha} + b_1 1^{-\beta} + c_1 1^{-\gamma} + d_1 1^{-\delta}) \\ \times &(a_2 1^{-2\alpha} + b_2 1^{-2\beta} + c_2 1^{-2\gamma} + d_2 1^{-2\delta}) \\ \times &(a_3 1^{-3\alpha} + b_3 1^{-3\beta} + c_3 1^{-3\gamma} + d_3 1^{-3\delta}) \\ \times &(a_4 1^{-4\alpha} + b_4 1^{-4\beta} + c_4 1^{-4\gamma} + d_4 1^{-4\delta}), \end{aligned}$$

Zehfuss says that

$$\sum \pm a_1 b_2 c_3 d_4 = \overline{\iiiint}_0^1 \mathrm{PQ}\, d\alpha\, d\beta\, d\gamma\, d\delta.$$

He does not, however, note in passing that

$$\mathrm{P} = \begin{vmatrix} 1^\alpha & 1^\beta & 1^\gamma & 1^\delta \\ 1^{2\alpha} & 1^{2\beta} & 1^{2\gamma} & 1^{2\delta} \\ 1^{3\alpha} & 1^{3\beta} & 1^{3\gamma} & 1^{3\delta} \\ 1^{4\alpha} & 1^{4\beta} & 1^{4\gamma} & 1^{4\delta} \end{vmatrix}.$$

BELLAVITIS, G. (1858, June).

[Studii sulle Memorie pubblicate dal Prof. Mainardi negli Atti del' i. r. Istituto Lombardo, vol. i., 1855, p. 90. *Atti . . . Istituto Veneto . . .* An. 1857–58, pp. 623–629.]

One note (p. 627), closely following Sylvester's paper of 1840, points out that if the equations

$$\left.\begin{aligned} a + bx + cx^2 &= 0 \\ \alpha + \beta x + \gamma x^2 + \delta x^3 &= 0 \end{aligned}\right\}$$

have a common root, it is given by the equation

$$\begin{vmatrix} a + bx & c & . \\ \alpha + \beta x & \gamma & \delta \\ ax & b & c \end{vmatrix} = 0;$$

and if the equations

$$\left.\begin{aligned} a + bx + cx^2 + dx^3 &= 0 \\ \alpha + \beta x + \gamma x^2 + \delta x^3 + \epsilon x^4 &= 0 \end{aligned}\right\}$$

have two common roots, these roots are given by the equation*

$$\begin{vmatrix} a + bx + cx^2 & d & \cdot \\ \alpha + \beta x + \gamma x^2 & \delta & \epsilon \\ ax + bx^2 & c & d \end{vmatrix} = 0.$$

Another note (pp. 627–628), in view of Mainardi's 'regola' above referred to, very properly draws attention to the rule given in the appendix to Bellavitis' *Sposizione* of 1857.

ZEHFUSS, G. (1858): MAINARDI, G. (1855).

[Ueber die Zeichen der einzelnen Glieder einer Determinante. *Zeitschrift f. Math. u. Phys.*, iii. pp. 249–250.]

[Una regola per attribuire il segno proprio ad ogni parte di un determinante numerico. *Atti . . . Istituto Lombardo* (Milano), i. pp. 105–106.]

Neither of these communications is of importance. Zehfuss, using the recurrent law of formation and giving "dérangement" the very opposite of its original meaning, so that the principal term of an n-line determinant has $\frac{1}{2}n(n-1)$ derangements, seeks to show that the sign of any other term having μ derangements is $(-1)^{\frac{1}{2}n(n-1)-\mu}$.

Mainardi, employing Cauchy's "clefs algébriques," finds himself also face to face with derangements, and seriously advises that in counting them we should say, not 1, 2, 3, 4, 5, . . . , but 1, 2, 1, 2, 1, . . . , the sign being $-$ or $+$ according as we end with 1 or 2.

GALLENKAMP, W. (1858).

[Die einfachsten Eigenschaften und Anwendungen der Determinanten. 12 pp. Sch. Progr. Duisburg.]

A workmanlike twelve-page exposition.

*No reference is made by either Sylvester or Bellavitis to the two other similarly derivable quadratics

$$\begin{vmatrix} b + cx + dx^2 & a & \cdot \\ a + bx + cx^2 & \cdot & d \\ \beta + \gamma x + \delta x^2 & \alpha & \epsilon \end{vmatrix} = 0, \quad \begin{vmatrix} c + dx & a & b \\ b + cx + dx^2 & \cdot & a \\ \gamma + \delta x + \epsilon x^2 & \alpha & \beta \end{vmatrix} = 0.$$

SPERLING, I. (1858).

[Teorija opredělitelej i ėja važnějsija priloženija. Č. 1. St. Petersburg.]

This dissertation I have failed to see. In English the title is, *The Theory of Determinants and its most important applications.* The letters used here in transliterating the Russian title have German values.

ZEHFUSS, G. (1858).

[Ueber eine gewisse Determinante. *Zeitschrift f. Math. u. Phys.*, iii. pp. 298–301.]

From two sets of simultaneous linear equations

$$a_r x + b_r y + c_r z + d_r w = 0 \quad (r=1,2,3,4)$$
$$\mu_r x' + \nu_r y' = 0 \quad (r=1,2)$$

there arises by multiplication a set of eight equations in

$$xx', \; yx', \; zx', \; wx', \; xy', \; yy', \; zy', \; wy',$$

whose determinant, Δ say, must contain as factors the determinants D_4, D_2 of the original sets. This observation leads Zehfuss to consider generally such determinants as

$$\begin{vmatrix} a_1\mu_1 & a_1\nu_1 & b_1\mu_1 & b_1\nu_1 & c_1\mu_1 & c_1\nu_1 & d_1\mu_1 & d_1\nu_1 \\ a_1\mu_2 & a_1\nu_2 & b_1\mu_2 & b_1\nu_2 & c_1\mu_2 & c_1\nu_2 & d_1\mu_2 & d_1\nu_2 \\ a_2\mu_1 & a_2\nu_1 & b_2\mu_1 & b_2\nu_1 & c_2\mu_1 & c_2\nu_1 & d_2\mu_1 & d_2\nu_1 \\ a_2\mu_2 & a_2\nu_2 & b_2\mu_2 & b_2\nu_2 & c_2\mu_2 & c_2\nu_2 & d_2\mu_2 & d_2\nu_2 \\ a_3\mu_1 & a_3\nu_1 & b_3\mu_1 & b_3\nu_1 & c_3\mu_1 & c_3\nu_1 & d_3\mu_1 & d_3\nu_1 \\ a_3\mu_2 & a_3\nu_2 & b_3\mu_2 & b_3\nu_2 & c_3\mu_2 & c_3\nu_2 & d_3\mu_2 & d_3\nu_2 \\ a_4\mu_1 & a_4\nu_1 & b_4\mu_1 & b_4\nu_1 & c_4\mu_1 & c_4\nu_1 & d_4\mu_1 & d_4\nu_1 \\ a_4\mu_2 & a_4\nu_2 & b_4\mu_2 & b_4\nu_2 & c_4\mu_2 & c_4\nu_2 & d_4\mu_2 & d_4\nu_2 \end{vmatrix} \quad i.e. \quad D_{4\cdot 2}$$

where the elements of D_2 are repeated 4^2 times and each time are associated with a different element of D_4. Multiplying the first and second columns by $\partial D_4/\partial a_1$, the third and fourth by $\partial D_4/\partial b_1$, the fifth and sixth by $\partial D_4/\partial c_1$, the seventh and eighth by $\partial D_4/\partial d_1$, and performing two additions we obtain

$$\Delta_{4\cdot 2} = \begin{vmatrix} D_4\mu_1 & D_4\nu & b_1\mu_1 & \dots & d_1\nu_1 \\ D_4\mu_2 & D_4\nu_2 & b_1\mu_2 & \dots & d_1\nu_2 \\ \cdot & \cdot & b_2\mu_1 & \dots & d_2\nu_1 \\ \cdot & \cdot & b_2\mu_2 & \dots & d_2\nu_2 \\ \cdot & \cdot & b_3\mu_1 & \dots & d_3\nu_1 \\ \cdot & \cdot & b_3\mu_2 & \dots & d_3\nu_2 \\ \cdot & \cdot & b_4\mu_1 & \dots & d_4\nu_1 \\ \cdot & \cdot & b_4\mu_2 & \dots & d_4\nu_2 \end{vmatrix} \div \left(\frac{\partial D_4}{\partial a_1}\right)^2$$

$$= \frac{(D_4)^2 D_2}{\left(\frac{\partial D_4}{\partial a_1}\right)^2} \begin{vmatrix} b_2\mu_1 & \dots & d_2\nu_1 \\ b_2\mu_2 & \dots & d_2\nu_2 \\ \cdot & \cdot & \cdot \\ b_4\mu_2 & \dots & d_4\nu_2 \end{vmatrix}.$$

But since the six-line determinant here is formed from $\partial D_4/\partial a_1$ and D_2 just as $D_{4\cdot 2}$ is formed from D_4 and D_2, it follows that

$$\Delta_{4\cdot 2} = \frac{(D_4)^2 D_2}{\left(\frac{\partial D_4}{\partial a_1}\right)^2} \cdot \frac{\left(\frac{\partial D_4}{\partial a_1}\right)^2 D_2}{\left(\frac{\partial^2 D_4}{\partial a_1\, \partial b_2}\right)^2} \cdot \begin{vmatrix} c_3\mu_1 & \dots & d_3\nu_1 \\ c_3\mu_2 & \dots & d_3\nu_2 \\ c_4\mu_1 & \dots & d_4\nu_1 \\ c_4\mu_2 & \dots & d_4\nu_2 \end{vmatrix},$$

and ultimately

$$\Delta_{4\cdot 2} = \frac{(D_4)^2 D_2}{\left(\frac{\partial D_4}{\partial a_1}\right)^2} \cdot \dots\dots \frac{\left(\frac{\partial^3 D_4}{\partial a_1\, \partial b_2\, \partial c_3}\right)^2 D_2}{\left(\frac{\partial^4 D_4}{\partial a_1\, \partial b_2\, \partial c_3\, \partial d_4}\right)^2},$$

$$= (D_4)^2 D_2{}^4.$$

Had the orders of the original determinants been other than the 4th and 2nd the result which would have been reached by an exactly similar process is readily foreseen. The general result we may formulate for ourselves as follows:—*If P and Q be determinants of the* p*th and* q*th orders respectively; and if P's array of elements be written* q^2 *times, namely,* q *times in each of* q *rows, thus forming a grand array of* pq *rows and* pq *columns; and if every element of each of the sub-arrays be multiplied by one and the same element of Q, the multiplier in the case of the sub-array in the place* h, k *being the element* *which occupies the corresponding place in Q; then the determinant of the grand array is equal to* $P^q Q^p$.

SIMERKA, W. (1858).

[Bestimmte Gleichungen des ersten Grades mit n Unbekannten gelöst mittels der Permutationslehre. *Sitzungsb. . . . Acad. d. Wiss.* (Wien), xxxiii. pp. 277–281.]

The contents of this simply-written paper are quite in accord with the title. The author writes as if nothing had ever previously been done on the subject. The common denominator of the value of the x's in

$$a_{r1}x_1 + a_{r2}x_2 + \ldots + a_{rn}x_n = g_r \Big\}_{r=1}^{r=n}$$

he denotes by $$\mathfrak{P}(a_1\, a_2\, \ldots\, a_n).$$

CASORATI, F. (1858, September).

[Intorno ad alcuni punti della teoria dei minimi quadrati. *Annali di Mat.*, i. pp. 329–343.]

The title here refers only to the latter half of the paper, the other half being concerned with an auxiliary series of theorems on the product-determinant. The first of these theorems is avowedly old, being that which concerns the so-called product C of two non-quadrate arrays

$$\begin{array}{ccccc} a_{11} & a_{12} & a_{13} & \ldots & a_{1n}, \\ a_{21} & a_{22} & a_{23} & \ldots & a_{2n}, \\ \cdot & \cdot & \cdot & \cdot\;\cdot\;\cdot & \cdot \\ a_{m1} & a_{m2} & a_{m3} & \ldots & a_{mn}, \end{array} \qquad \begin{array}{ccccc} b_{11} & b_{12} & b_{13} & \ldots & b_{1n} \\ b_{21} & b_{22} & b_{23} & \ldots & b_{2n} \\ \cdot & \cdot & \cdot & \cdot\;\cdot\;\cdot & \cdot \\ b_{m1} & b_{m2} & b_{m3} & \ldots & b_{mn}, \end{array}$$

where $n > m$. The second, though not so spoken of, is only new in form, and concerns any primary minor of C. Unfortunately, Casorati does not observe that any primary minor of C is a determinant formed exactly like C after omitting a row from the first array and a row from the second, and that therefore his second theorem is unnecessary. Further, his mode of procedure leads him to an expression for a multiple of the minor, namely, for

$$(n-m+1)\frac{\partial C}{\partial c_{rs}},$$

and making an oversight similar to Hesse's of 1853, he does not divide both sides by $n-m+1$.

His third theorem,

$$C\frac{\partial C}{\partial c_{rs}} = \frac{\partial C}{\partial a_{r1}}\frac{\partial C}{\partial b_{s1}} + \frac{\partial C}{\partial a_{r2}}\frac{\partial C}{\partial b_{s2}} + \ldots . + \frac{\partial C}{\partial a_{rm}}\frac{\partial C}{\partial b_{sm}},$$

is more worthy of note. The proof of it depends essentially on substituting for C in the first factor of each term of the right-hand member its equivalent,

$$c_{r1}\frac{\partial C}{\partial c_{r1}} + c_{r2}\frac{\partial C}{\partial c_{r2}} + \ldots . + c_{rm}\frac{\partial C}{\partial c_{rm}},$$

in which, it is important to note, the differential-quotients are necessarily all independent of $a_{r1}, a_{r2}, \ldots .$ The said right-hand member can then be transformed into

$$\begin{aligned} &\frac{\partial C}{\partial b_{s1}}\left(b_{11}\frac{\partial C}{\partial c_{r1}} + b_{21}\frac{\partial C}{\partial c_{r2}} + \ldots . + b_{m1}\frac{\partial C}{\partial c_{rm}}\right) \\ + &\frac{\partial C}{\partial b_{s2}}\left(b_{12}\frac{\partial C}{\partial c_{r1}} + b_{22}\frac{\partial C}{\partial c_{r2}} + \ldots . + b_{m2}\frac{\partial C}{\partial c_{rm}}\right) \\ + &\ \cdot \quad \cdot \quad \cdot \quad \cdot \quad \cdot \quad \cdot \quad \cdot \quad \cdot \quad \cdot \quad \cdot \quad \cdot \quad \cdot \quad \cdot \\ + &\frac{\partial C}{\partial b_{sn}}\left(b_{1n}\frac{\partial C}{\partial c_{r1}} + b_{2n}\frac{\partial C}{\partial c_{r2}} + \ldots . + b_{mn}\frac{\partial C}{\partial c_{rm}}\right), \end{aligned}$$

which, if addition be performed columnwise, becomes

$$0 + 0 + \ldots + C\frac{\partial C}{\partial c_{rs}} + 0 + \ldots + 0,$$

because of the fact that the theorem

$$c_{r1}\frac{\partial C}{\partial c_{s1}} + c_{r2}\frac{\partial C}{\partial c_{s2}} + \ldots . + c_{rn}\frac{\partial C}{\partial c_{sn}} = \begin{matrix} C \\ 0 \end{matrix}\Bigg\} \quad \text{when} \quad \Bigg\{\begin{matrix} r = s \\ r \neq s \end{matrix}$$

holds in reference to the a's and b's as well as to the c's—a fact which should be noted for other purposes, and which is readily seen to be justifiable if we view C in its composite form AB and bear in mind that the operation

$$a\frac{\partial}{\partial a} + \beta\frac{\partial}{\partial b} + \gamma\frac{\partial}{\partial c} + \ldots ,$$

when performed on a homogeneous linear function of $a, b, c, \ldots$ is equivalent to a substitution.

The case where the two given arrays are identical is formulated, due care being taken with the differential-quotients because of C becoming axisymmetric.

We have only to add that the form in which this new theorem of Casorati's is stated obscures to some extent its significance. If we write the case of AB = C where $m=3$, $n=4$ in the form

$$\left\|\begin{matrix} a_1 & a_2 & a_3 & a_4 \\ b_1 & b_2 & b_3 & b_4 \\ c_1 & c_2 & c_3 & c_4 \end{matrix}\right\| \cdot \left\|\begin{matrix} l_1 & l_2 & l_3 & l_4 \\ m_1 & m_2 & m_3 & m_4 \\ n_1 & n_2 & n_3 & n_4 \end{matrix}\right\| = \left|\begin{matrix} \Sigma al & \Sigma am & \Sigma an \\ \Sigma bl & \Sigma bm & \Sigma bn \\ \Sigma cl & \Sigma cm & \Sigma cn \end{matrix}\right|,$$

then, freed from all reference to differentiation, the theorem for the case $r=2$, $s=3$ is

$$-\left|\begin{matrix} \Sigma al & \Sigma am & \Sigma an \\ \Sigma bl & \Sigma bm & \Sigma bn \\ \Sigma cl & \Sigma cm & \Sigma cn \end{matrix}\right| \cdot \left|\begin{matrix} \Sigma al & \Sigma am \\ \Sigma cl & \Sigma cm \end{matrix}\right|$$

$$= \left|\begin{matrix} \Sigma al & \Sigma am & \Sigma an \\ l_1 & m_1 & n_1 \\ \Sigma cl & \Sigma cm & \Sigma cn \end{matrix}\right| \cdot \left|\begin{matrix} \Sigma al & \Sigma am & a_1 \\ \Sigma bl & \Sigma bm & b_1 \\ \Sigma cl & \Sigma cm & c_1 \end{matrix}\right| + \left|\begin{matrix} \Sigma al & \Sigma am & \Sigma an \\ l_2 & m_2 & n_2 \\ \Sigma cl & \Sigma cm & \Sigma cn \end{matrix}\right| \cdot \left|\begin{matrix} \Sigma al & \Sigma am & a_2 \\ \Sigma bl & \Sigma bm & b_2 \\ \Sigma cl & \Sigma cm & c_2 \end{matrix}\right|$$

$$+ \left|\begin{matrix} \Sigma al & \Sigma am & \Sigma an \\ l_3 & m_3 & n_3 \\ \Sigma cl & \Sigma cm & \Sigma cn \end{matrix}\right| \cdot \left|\begin{matrix} \Sigma al & \Sigma am & a_3 \\ \Sigma bl & \Sigma bm & b_3 \\ \Sigma cl & \Sigma cm & c_3 \end{matrix}\right| + \left|\begin{matrix} \Sigma al & \Sigma am & \Sigma an \\ l_4 & m_4 & n_4 \\ \Sigma cl & \Sigma cm & \Sigma cn \end{matrix}\right| \cdot \left|\begin{matrix} \Sigma al & \Sigma am & a_4 \\ \Sigma bl & \Sigma bm & b_4 \\ \Sigma cl & \Sigma cm & c_4 \end{matrix}\right|.$$

Further, no change but substitution is necessary on passing to the case where the two original arrays are identical.

SALMON, G. (1859).

[LESSONS INTRODUCTORY TO THE MODERN HIGHER ALGEBRA. By the Rev. George Salmon, A.M. . . . xii + 147 pp. Dublin.]

The first three lessons (pp. 1–18) of this historically interesting text-book are devoted to an elementary exposition of determinants. The only fresh matter (§ 20) concerns the determinant formed from

$$\left\|\begin{matrix} a_1 & b_1 \\ a_2 & b_2 \\ a_3 & b_3 \end{matrix}\right\|, \qquad \left\|\begin{matrix} \alpha_1 & \beta_1 \\ \alpha_2 & \beta_2 \\ \alpha_3 & \beta_3 \end{matrix}\right\|$$

by row-by-row multiplication. This is shown to vanish, not by pointing out that it contains at least one zero determinant of the third order as a factor, but by partitioning it into eight determinants with monomial elements, and showing that all the eight vanish.*

Unfortunately, for *terms* of a determinant the word "elements" is used, and for *adjugate* the word "reciprocal," although the elements of the adjugate are spoken of as the "inverse constituents."

SPERLING, J. F. DE (1860, April).

[Note sur un théorème de M. Sylvester relatif à la transformation du produit de déterminants du même ordre. *Journ. (de Liouville) de Math.* . . . (2), v. pp. 121–126.]

This is a carefully formulated proof of Sylvester's theorem of 1839 and the extended theorem of 1851, the lines followed being those suggested and illustrated by Cayley in 1843. Unfortunately, however, instead of extending Cayley's method to prove directly and at once the generalisation of 1851, Sperling repeats Cayley's proof of the simpler theorem, and then uses the method of so-called mathematical induction to arrive at the generalisation.

The two determinants whose product is the subject of discus-

---

* In using the notation || || he is not more explicit than its author, Cayley. If it were explained that

$$\left\| \begin{matrix} a_1 & a_2 & a_3 \\ b_1 & b_2 & b_3 \end{matrix} \right\|$$

stands for $|a_1 b_2|, \quad |a_1 b_3|, \quad |a_2 b_3|,$

it would readily follow that the statement

$$\left\| \begin{matrix} a_1 & a_2 & a_3 \\ b_1 & b_2 & b_3 \end{matrix} \right\| = 0$$

was short for $|a_1 b_2|, \quad |a_1 b_3|, \quad |a_2 b_3| = 0, \quad 0, \quad 0;$

and that

$$\left\| \begin{matrix} a_1 & a_2 & a_3 \\ b_1 & b_2 & b_3 \end{matrix} \right\| \cdot \left\| \begin{matrix} \alpha_1 & \alpha_2 & \alpha_3 \\ \beta_1 & \beta_2 & \beta_3 \end{matrix} \right\|$$

was short for

$$\Big(|a_1 b_2|, \quad |a_1 b_3|, \quad |a_2 b_3| \,\big)\!\!\big(\, |\alpha_1 \beta_2|, \quad |\alpha_1 \beta_3|, \quad |\alpha_2 \beta_3| \Big),$$

sion being $|a_{11}\, a_{22} \dots a_{nn}|$ and $|b_{11}\, b_{22} \dots b_{nn}|$, or, say, A and B, he forms the determinant

$$\left|\begin{array}{ccccccccc}
a_{11}\; a_{12} \dots a_{1,\,n-m} & \cdot & \cdot & \cdots & \cdot & a_{1n} & b_{11}\; b_{12} \dots b_{1n} \\
a_{21}\; a_{22} \dots a_{2,\,n-m} & \cdot & \cdot & \cdots & \cdot & a_{2n} & b_{21}\; b_{22} \dots b_{2n} \\
\cdot\;\cdot\;\cdot\;\cdot\;\cdot & \cdot & \cdot & \cdot\;\cdot\;\cdot & \cdot & \cdot & \cdot\;\cdot\;\cdot\;\cdot \\
a_{n1}\; a_{n2} \dots a_{n,\,n-m} & \cdot & \cdot & \cdots & \cdot & a_{nn} & b_{n1}\; b_{n2} \dots b_{nn} \\
\cdot\;\;\cdot\;\cdots\;\;\cdot & a_{1,\,n-m+1} & a_{1,\,n-m+2} & \cdots & a_{1,\,n-1} & a_{1n} & b_{11}\; b_{12} \dots b_{1n} \\
\cdot\;\;\cdot\;\cdots\;\;\cdot & a_{2,\,n-m+1} & a_{2,\,n-m+2} & \cdots & a_{2,\,n-1} & a_{2n} & b_{21}\; b_{22} \dots b_{2n} \\
\cdot\;\cdot\;\cdot\;\cdot\;\cdot & \cdot & \cdot & \cdot\;\cdot\;\cdot & \cdot & \cdot & \cdot\;\cdot\;\cdot\;\cdot \\
\cdot\;\;\cdot\;\cdots\;\;\cdot & a_{n,\,n-m+1} & a_{n,\,n-m+2} & \cdots & a_{n,\,n-1} & a_{nn} & b_{n1}\; b_{n2} \dots b_{nn}
\end{array}\right|,$$

which, he says, is seen to vanish on trying to find Laplace's expansion of it in terms of minors formed from the last $n+1$ columns and the minors that are complementary of those. Then, noting that the like outcome is not met with when the boundary-line necessary for the application of the said expansion-theorem is horizontal and bisects the determinant, he sets about obtaining the terms of the latter development in orderly fashion. Clearly, the first factors of those terms are all alike as regards their first $n-m$ columns, but the remaining m columns may include another column of a's or may not. Making a separation of the terms in accordance with this distinction, and calling the one aggregate Σ_{m-1} and the other Σ_m, where the suffix corresponds with the number of columns of b's appearing in each first factor, and therefore also with the number of columns of a's appearing in each second factor, Sperling gives evidence that Σ_{m-1} is Sylvester's expansion for $|a_{11}\, a_{22} \dots a_{nn}| \cdot |b_{11}\, b_{22} \dots b_{nn}|$ when in the formation of it there is an interchange of $m-1$ columns, that Σ_m is the corresponding expansion due to an interchange of m columns, and that the two Σ's occur with different signs. The conclusion is thus reached that, if we have previously proved the identity $AB = \Sigma_{m-1}$, the identity $AB = \Sigma_m$ must follow.

It is important to note in passing that if Sperling had put zeros for $a_{1n}, a_{2n}, \dots, a_{nn}$ in the second half of his $2n$-line determinant, its value then would have been, when obtained in one way, $(-1)^{m-1}AB$, and in another, $(-1)^{m-1}\Sigma_m$. He would thus have made the natural extension of Cayley's simple proof.

CHAPTER III.

AXISYMMETRIC DETERMINANTS, FROM 1841 TO 1860.

UNDER the heading of axisymmetric determinants in our first volume a reminder ought to have been given that the determinant of the set of linear equations reached by Bezout in his so-called abridged method for eliminating the unknown between two equations of the same degree is axisymmetric, the first appearance of this special form being consequently thrown back to the year 1764 (see p. 317 of Bezout's *Recherches . . .*). Further, it should naturally thereafter have been recalled that the said axisymmetry had been discussed by Jacobi in 1835 (see *History*, i. p. 214, pp. 485-487) and by Cauchy in 1840 (see *History*, i. pp. 242–243).

CAYLEY, A. (1841, May).

[On a theorem in the geometry of position. *Cambridge Math. Journ.*, ii. pp. 267–271; or *Collected Math. Papers*, i. pp. 1–4.]

A general account has already been given of this interesting paper—interesting as regards the subject, and interesting as being the author's first 'prentice effort. All that remains to be noticed here is what may be called Cayley's series of vanishing axisymmetric determinants. These we may write in the short form

$$\begin{vmatrix} \cdot & (x)_{12} & (x)_{13} & 1 \\ (x)_{21} & \cdot & (x)_{23} & 1 \\ (x)_{31} & (x)_{32} & \cdot & 1 \\ 1 & 1 & 1 & \cdot \end{vmatrix},$$

$$\begin{vmatrix} . & (xy)_{12} & (xy)_{13} & (xy)_{14} & 1 \\ (xy)_{21} & . & (xy)_{23} & (xy)_{24} & 1 \\ (xy)_{31} & (xy)_{32} & . & (xy)_{34} & 1 \\ (xy)_{41} & (xy)_{42} & (xy)_{43} & . & 1 \\ 1 & 1 & 1 & 1 & . \end{vmatrix},$$

$$\begin{vmatrix} . & (xyz)_{12} & (xyz)_{13} & (xyz)_{14} & (xyz)_{15} & 1 \\ (xyz)_{21} & . & (xyz)_{23} & (xyz)_{24} & (xyz)_{25} & 1 \\ (xyz)_{31} & (xyz)_{32} & . & (xyz)_{34} & (xyz)_{35} & 1 \\ (xyz)_{41} & (xyz)_{42} & (xyz)_{43} & . & (xyz)_{45} & 1 \\ (xyz)_{51} & (xyz)_{52} & (xyz)_{53} & (xyz)_{54} & . & 1 \\ 1 & 1 & 1 & 1 & 1 & . \end{vmatrix},$$

. . . .

if we put

$$\begin{aligned} &(x)_{rs} \text{ for } (x_r - x_s)^2, \\ &(xy)_{rs} \text{ for } (x_r - x_s)^2 + (y_r - y_s)^2, \\ &(xyz)_{rs} \text{ for } (x_r - x_s)^2 + (y_r - y_s)^2 + (z_r - z_s)^2, \end{aligned}$$

. . . .

The fact that they are identically equal to zero is established by showing that each one is resolvable into two factors, of which one or both vanish; for example, that the first is equal to

$$\begin{vmatrix} x_1^2 & -2x_1 & . & 1 \\ x_2^2 & -2x_2 & . & 1 \\ x_3^2 & -2x_3 & . & 1 \\ 1 & . & . & . \end{vmatrix} \times_{rr} \begin{vmatrix} 1 & x_1 & . & x_1^2 \\ 1 & x_2 & . & x_2^2 \\ 1 & x_3 & . & x_3^2 \\ . & . & . & 1 \end{vmatrix},$$

where, for the moment, we use $\times_{rr}$ to indicate row-by-row multiplication.

It should be noted that not more than three of the series are contemplated by Cayley, as he viewed the identities mainly on their geometrical side, namely, as giving the relation between the distances of three points in a straight line, four points in a plane, and five points in three-dimensional space.*

* To one taking this point of view Sylvester's paper "On Staudt's theorems. . . ." will be of interest. See *Philos. Magazine*, iv. (1852), pp. 335–345; or *Collected Math. Papers*, i. pp. 382–391.

MOON, R. (1842).

[On elimination. *Cambridge Math. Journ.*, iii. pp. 183–184.]

What is here given is a rule-of-thumb for obtaining the development of any one of Bezout's condensed eliminants from the preceding one, the exposition being continued so far as to give the eliminant, R_{44} say, of

$$\left.\begin{aligned} ax^4 + bx^3 + cx^2 + dx + e &= 0 \\ \alpha x^4 + \beta x^3 + \gamma x^2 + \delta x + \epsilon &= 0 \end{aligned}\right\}$$

in the form

$$\{f(a\epsilon)\}^4 + \{f(a\gamma)\}\{f(c\epsilon)\}^2 + \ldots\ldots,$$

where $f(a\epsilon)$, $f(a\gamma)$, ... stand for $a\epsilon - \alpha e$, $a\gamma - \alpha c$, ...

In a note covering the same ground but free of anything empirical, Salmon drew attention ten years later to the existence of three or four incorrect terms in Moon's last expansion. (See *Higher Plane Curves*, pp. 293–294.)

HESSE, O. (1844, Jan.).

[Ueber die Elimination der Variabeln aus drei algebraischen Gleichungen vom zweiten Grade mit zwei Variabeln. *Crelle's Journal*, xxviii. pp. 68–96; or *Werke*, pp. 89–122.]

In this paper there appears the first reference to the special form of determinant which has for its elements the second differential-quotients of a function, and which consequently is axisymmetric. On account of its importance this form falls to be dealt with separately: we merely note here that Hesse himself viewed it as a special form of Jacobian, namely, the Jacobian whose originating functions are the first differential-quotients of a single function, and that he called it *the determinant of this single function*, a practice which, when the function is quadratic, is not at variance with that introduced by Gauss.

For full information the reader is referred to Chapter XIII. of the present volume.

CAYLEY, A. (1846).

[Note on the maxima and minima of functions of three variables. *Cambridge and Dub. Math. Journ.*, i. pp. 74–75; or *Collected Math. Papers*, i. pp. 228–229.]

Using Δ to stand for the determinant

$$\begin{vmatrix} a & h & g \\ h & b & f \\ g & f & c \end{vmatrix}$$

we may formulate Cayley's first theorem by saying that if $(a+b+c)\Delta$ and $A+B+C$ be positive, then $a\Delta$, $b\Delta$, $c\Delta$, A, B, C are all positive. By way of proof it is noted that the equation

$$\begin{vmatrix} A-x & H & G \\ H & B-x & F \\ G & F & C-x \end{vmatrix} = 0,$$

$$i.e. \qquad x^3 - (A+B+C)x^2 + (a+b+c)\Delta x - \Delta^2 = 0,$$

has by reason of the data all its roots positive: that therefore the roots of the equations

$$\begin{vmatrix} A-x & H \\ H & B-x \end{vmatrix} = 0, \quad \begin{vmatrix} B-x & F \\ F & C-x \end{vmatrix} = 0, \quad \begin{vmatrix} C-x & G \\ G & A-x \end{vmatrix} = 0,$$

$$i.e. \qquad x^2 - (A+B)x + c\Delta = 0, \quad x^2 - (B+C)x + a\Delta = 0,$$

$$x^2 - (C+A)x + b\Delta = 0,$$

on account of Cauchy's localisation of them in relation to the roots of the previous equation, must also be positive; and consequently that

$$A+B, \quad B+C, \quad C+A, \quad c\Delta, \quad a\Delta, \quad b\Delta$$

must be positive. This last is essentially what was to be proved; because, for example, to say that $A+B$ and $c\Delta$ are positive implies that $A+B$ and AB are positive, and therefore that A and B are positive.

Cayley also puts on record the theorem that the equation in x

$$\begin{vmatrix} A - xa & H - xh & G - xg \\ H - xh & B - xb & F - xf \\ G - xg & F - xf & C - xc \end{vmatrix} = 0$$

has all its roots real if either of the quadrics

$$\begin{array}{l} Ax^2 + By^2 + Cz^2 + 2Fyz + 2Gzx + 2Hxy, \\ ax^2 + by^2 + cz^2 + 2fyz + 2gzx + 2hxy, \end{array}$$

remain constant as to sign.

CAYLEY, A. (1846).

[Problème de géométrie analytique. *Crelle's Journal*, xxxi. pp. 227–230; or *Collected Math. Papers*, i. pp. 329–331.]

The problem in question depends on an algebraic identity which, after a little examination, is seen to be a property of axisymmetric determinants. Cayley writes the identity in the form

$$F_{pp}(U+V^2)\cdot K(U) \;-\; F_{pp}(U)\cdot K(U+V^2) \;=\; \{F_{po}(U)\}^2$$

where

$$U = Ax^2 + By^2 + Cz^2 + 2Fyz + 2Gzx + 2Hxy + 2Lxw + 2Myw + 2Nzw + Pw^2,$$
$$V = ax + \beta y + \gamma z + \delta w,$$

$$K(U) = \begin{vmatrix} A & H & G & L \\ H & B & F & M \\ G & F & C & N \\ L & M & N & P \end{vmatrix}, \qquad P_{po}(U) = \begin{vmatrix} \cdot & \xi & \eta & \zeta & \omega \\ \alpha & A & H & G & L \\ \beta & H & B & F & M \\ \gamma & G & F & C & N \\ \delta & L & M & N & P \end{vmatrix},$$

and $P_{pp}(U)$ is what is obtained from $P_{po}(U)$ on changing $\alpha, \beta, \gamma, \delta$ into ξ, η, ζ, ω respectively: but freed of all fresh notation it is nothing more nor less than

$$\begin{vmatrix} \cdot & \xi & \eta & \zeta & \omega \\ \xi & A+\alpha^2 & H+\alpha\beta & G+\alpha\gamma & L+\alpha\delta \\ \eta & H+\beta\alpha & B+\beta^2 & F+\beta\gamma & M+\beta\delta \\ \zeta & G+\gamma\alpha & F+\gamma\beta & C+\gamma^2 & N+\gamma\delta \\ \omega & L+\delta\alpha & M+\delta\beta & N+\delta\gamma & P+\delta^2 \end{vmatrix} \cdot \begin{vmatrix} A & H & G & L \\ H & B & F & M \\ G & F & C & N \\ L & M & N & P \end{vmatrix}$$

$$-\begin{vmatrix} \cdot & \xi & \eta & \zeta & \omega \\ \xi & A & H & G & L \\ \eta & H & B & F & M \\ \zeta & G & F & C & N \\ \omega & L & M & N & P \end{vmatrix} \cdot \begin{vmatrix} A+\alpha^2 & H+\alpha\beta & G+\alpha\gamma & L+\alpha\delta \\ H+\beta\alpha & B+\beta^2 & F+\beta\gamma & M+\beta\delta \\ G+\gamma\alpha & F+\gamma\beta & C+\gamma^2 & N+\gamma\delta \\ L+\delta\alpha & M+\delta\beta & N+\delta\gamma & P+\delta^2 \end{vmatrix}$$

$$= \begin{vmatrix} \cdot & \xi & \eta & \zeta & \omega \\ \alpha & A & H & G & L \\ \beta & H & B & F & M \\ \gamma & G & F & C & N \\ \delta & L & M & N & P \end{vmatrix}^2 .$$

Nothing is said about the mode of proving it.*

"H (1)" (1846, Nov.).

[Mathematical notes, ii. *Camb. and Dublin Math. Journ.*, i. p. 286.]

A correspondent signing himself as above puts on record without proof two identities which, when the notation of determinants is used, may be written in the form

$$\begin{vmatrix} aa'-bb'-cc' & ab'+ba' & ca'+ac' \\ ab'+ba' & bb'-cc'-aa' & bc'+cb' \\ ca'+ac' & bc'+cb' & ca'-aa'-bb' \end{vmatrix} = (a^2+b^2+c^2)\cdot(aa'+bb'+cc') \cdot(a'^2+b'^2+c'^2)$$

* If we note that the first determinant can be written in the form

$$-\begin{vmatrix} -1 & \cdot & \alpha & \beta & \gamma & \delta \\ \cdot & \cdot & \xi & \eta & \zeta & \omega \\ \alpha & \xi & A & H & G & L \\ \beta & \eta & H & B & F & M \\ \gamma & \zeta & G & F & C & N \\ \delta & \omega & L & M & N & P \end{vmatrix},$$

and the fourth in the form

$$-\begin{vmatrix} -1 & \alpha & \beta & \gamma & \delta \\ \alpha & A & H & G & L \\ \beta & H & B & F & M \\ \gamma & G & F & C & N \\ \delta & L & M & N & P \end{vmatrix}$$

light dawns at once, for the last three determinants of the identity are then seen to be principal minors of the first, and the identity itself to be a case of Jacobi's theorem regarding a minor of the adjugate.

and

$$\begin{vmatrix} 2aa' & ab'+ba' & ca'+ac' \\ ab'+ba' & 2bb' & bc'+cb' \\ ca'+ac' & bc'+cb' & 2cc' \end{vmatrix} = 0.$$

It is seen that both determinants are axisymmetric, that the second is expressible as the product of two vanishing determinants, and that the first is formable from the second by subtracting $aa' + bb' + cc'$ from each diagonal element,—a fact which, taken along with the vanishing of the second, shows that $aa' + bb' + cc'$ is a factor of the first.

CAYLEY, A. (1847).

[Note sur les hyperdéterminants. *Crelle's Journal*, xxxiv. pp. 148–152; or *Collected Math. Papers*, i. pp. 352–355.]

The second paragraph of this note concerns the expression

$$6abcd + 3b^2c^2 - a^2d^2 - 4ac^3 - 4b^3d, \quad \text{or } \nabla \text{ say},$$

soon afterwards (1851) to be called the "discriminant" of the binary cubic

$$ax^3 + 3bx^2y + 3cxy^2 + y^3;$$

and Cayley's proposition is that the determinant whose elements are the second differential-quotients of ∇ with respect to a, b, c, d, namely, the axisymmetric determinant

$$\begin{vmatrix} -2d^2 & 6cd & 6bd-12c^2 & 6bc-4ad \\ 6cd & 6c^2-24bd & 6ad+12bc & 6ac-12b^2 \\ 6bd-12c^2 & 6ad+12bc & 6b^2-24ac & 6ab \\ 6bc-4ad & 6ac-12b^2 & 6ab & -2a^2 \end{vmatrix},$$

is a numerical multiple of ∇^2. As a matter of fact he says the multiplier is 3; but this is because, instead of writing the determinant as here, he removes from it the factors 2, 6, 6, 2. A verificatory proof, unsatisfactory to himself, is given, the determinant being as a preliminary again altered into a multiple (by 6^4) of

$$\begin{vmatrix} a^2 & ab & 2b^2-ac & 3bc-2ad \\ ab & \frac{4}{3}ac-\frac{1}{3}b^2 & \frac{2}{3}bc+\frac{1}{3}ad & 2c^2-bd \\ 2b^2-ac & \frac{2}{3}bc+\frac{1}{3}ad & \frac{4}{3}bd-\frac{1}{3}c^2 & cd \\ 3bc-2ad & 2c^2-bd & cd & d^2 \end{vmatrix}$$

or

$$\begin{vmatrix} a^2 & ab & ac-3r & ad+9q \\ ba & b^2+2r & bc-q & bd-3p \\ ca-3r & cb-q & c^2+2p & cd \\ da+9q & db-3p & dc & d^2 \end{vmatrix},$$

where

$$p = \tfrac{2}{3}(bd-c^2), \quad q = \tfrac{1}{3}(bc-ad), \quad r = \tfrac{2}{3}(ac-b^2),$$

the last change being probably due to the fact that it was known that

$$\nabla = 9(pr-q^2)$$

and that verification would thus be easier.

CAYLEY, A. (1848).

[On geometrical reciprocity. *Cambridge and Dub. Math. Journ.*, iii. pp. 173–179; or *Collected Math. Papers*, i. pp. 377–382.]

Incidentally Cayley gives the identity

$$-\begin{vmatrix} \cdot & \xi & \eta & \zeta \\ \xi & 2a & a'+b & a''+c \\ \eta & a'+b & 2b' & b''+c' \\ \zeta & a''+c & b''+c' & 2c'' \end{vmatrix}$$

$$= 4\begin{vmatrix} a & a' & a'' \\ b & b' & b'' \\ c & c' & c'' \end{vmatrix} \cdot \left\{ax^2+b'y^2+c''z^2+(b''+c')yz+(c+a'')zx+(a'+b)xy\right\} \\ -\left[x(ab''-a''b+a'c-ac')+y(b'c-bc'+b''a'-b'a'') \\ +z(c''a'-c'a''+cb''-c''b)\right]^2,$$

where

$$\xi=ax+a'y+a''z, \quad \eta=bx+b'y+b''z, \quad \zeta=cx+c'y+c''z.$$

No proof is adduced, and it is not noted that, when the determinant $|ab'c''|$ is axisymmetric, the expression in rectangular brackets vanishes, and the identity becomes in later notation

$$-\begin{vmatrix} \cdot & \xi & \eta & \zeta \\ \xi & a & h & g \\ \eta & h & b & f \\ \zeta & g & f & c \end{vmatrix} = \begin{vmatrix} a & h & g \\ h & b & f \\ g & f & c \end{vmatrix} \cdot \begin{array}{ccc|c} x & y & z & \\ \hline a & h & g & x \\ h & b & f & y \\ g & f & c & z \end{array},$$

where
$$\xi,\ \eta,\ \zeta = (x,\ y,\ z \between \begin{matrix} a & h & g \\ h & b & f \\ g & f & c \end{matrix}) .$$

This particular result, we may note, is easily verified by performing on the four-line determinant the operation

$$\text{row}_1 - x\,\text{row}_2 - y\,\text{row}_3 - z\,\text{row}_4 .$$

SYLVESTER, J. J. (1850, Aug.).

[On the intersections, contacts and other correlations of two conics expressed by indeterminate co-ordinates. *Cambridge and Dublin Math. Journ.*, v. pp. 262–282; or *Collected Math. Papers*, i. pp. 119–137.]

In this paper an important property of axisymmetric determinants is incidentally brought to notice; but, unfortunately, the author's intended statement is almost hidden through want of care. He says (p. 270) that the determinant

$$\begin{vmatrix} A & C' & B' & l \\ C' & B & A' & m \\ B' & A' & C & n \\ l & m & n & 0 \end{vmatrix},$$

where

$$\begin{aligned} A &= \quad bc - a'^2, & B &= \quad ca - b'^2, & C &= \quad ab - c'^2, \\ A' &= -b'c' + aa', & B' &= -c'a' + bb', & C' &= -a'b' + cc', \end{aligned}$$

is "the product of the determinant

$$\begin{vmatrix} a & c' & b' \\ c' & b & a' \\ b' & a' & c \end{vmatrix}$$

by the quantity

$$al^2 + bm^2 + cn^2 - 2a'mn - 2b'ln - 2c'lm."$$

Now the said four-line determinant is not resolvable unless A′, B′, C′ be changed in sign, and even then the second factor is not as printed, but is

$$-(al^2 + bm^2 + cn^2 + 2a'mn + 2b'nl + 2c'lm).$$

The theorem, in fact, may be viewed as giving an expression for the product of a ternary quadric by its discriminant; and at a later date might have been written*

$$\begin{array}{ccc|c} x & y & z & \\ \hline a & h & g & x \\ h & b & f & y \\ g & f & c & z \end{array} \cdot \begin{vmatrix} a & h & g \\ h & b & f \\ g & f & c \end{vmatrix} = - \begin{vmatrix} \cdot & x & y & z \\ x & \mathrm{A} & \mathrm{H} & \mathrm{G} \\ y & \mathrm{H} & \mathrm{B} & \mathrm{F} \\ z & \mathrm{G} & \mathrm{F} & \mathrm{C} \end{vmatrix}.$$

No proof is given by Sylvester; but in a footnote we are told that it depends on "a theorem given by M. Cauchy, and which is included as a particular case in a theorem of my own, relating to compound determinants." What theorem of Cauchy's is thus referred to it is not easy to say. One would think that the most natural proceeding would be to show that the coefficients of x^2, y^2, . . . on the one side are identical with those on the other; and, in this case, the names to be mentioned would be Lagrange and Jacobi.

In a postscript Sylvester enunciates a theorem connected with the linear transformation of an n-ary quadric; and as this concerns the "determinant" of the quadric, or what a year later he named the "discriminant," it necessarily involves a property of axisymmetric determinants. His wording (p. 281) is:—"Let U be a quadratic function of any number of letters $x_1, x_2, \ldots, x_n$, and let any number r of linear equations of the general form

$$a_{1r}x_1 + a_{2r}x_2 + \ldots + a_{nr}x_n = 0$$

be instituted between them; and by means of these equations let U be expressed as a function of any $n-r$ of the given letters, say of $x_{r+1}, x_{r+2}, \ldots, x_n$, and let U so expressed be called M. Let

$$a_{1r}r_1 + a_{2r}x_2 + \ldots + a_{nr}x_n$$

* A comparison of this with a result just obtained from Cayley (1848) gives the curious identity

$$\begin{vmatrix} \cdot & x & y & z \\ x & \mathrm{A} & \mathrm{H} & \mathrm{G} \\ y & \mathrm{H} & \mathrm{B} & \mathrm{F} \\ z & \mathrm{G} & \mathrm{F} & \mathrm{C} \end{vmatrix} = \begin{vmatrix} \cdot & \xi & \eta & \zeta \\ \xi & a & h & g \\ \eta & h & b & f \\ \zeta & g & f & c \end{vmatrix}$$

where $\xi, \eta, \zeta = (x, y, z \between \begin{array}{ccc} a & h & g \\ h & b & f \\ g & f & c \end{array})$ and, as usual, $\mathrm{A} = bc - f^2$, ...

be called L_r. Then the determinant of M in respect to the $n-r$ letters above given is equal to the determinant of

$$U + L_1 x_{n+1} + L_2 x_{n+2} + \ldots + L_r x_{n+r}$$

considered as a function of the $n+r$ letters

$$x_1,\ x_2,\ \ldots,\ x_{n+r}$$

divided by the square of the determinant

$$\left.\begin{matrix} a_{11} & a_{21} & \ldots & a_{r1} \\ a_{12} & a_{22} & \ldots & a_{r2} \\ \cdot & \cdot & \cdot & \cdot \\ a_{1r} & a_{2r} & \ldots & a_{rr} \end{matrix}\right\}."$$

As regards this we have to remark (1) that again no proof is offered, and (2) that the discriminant of

$$U + L_1 x_{n+1} + \ldots + L_r x_{n+r}$$

is easily got by "bordering" the discriminant of U. Taking the case where U is

$$ax^2+by^2+cz^2+dw^2+2fyz+2gzx+2hxy+2pxw+2qyw+2rzw$$

with the discriminant

$$\begin{vmatrix} a & h & g & p \\ h & b & f & q \\ g & f & c & r \\ p & q & r & d \end{vmatrix},$$

and where the linear equations serving to eliminate x, y from U are

$$\mu_1 x + \mu_2 y + \mu_3 z + \mu_4 w = 0$$
$$\nu_1 x + \nu_2 y + \nu_3 z + \nu_4 w = 0,$$

we have, according to Sylvester, the discriminant of the altered U equal to

$$\begin{vmatrix} a & h & g & p & \mu_1 & \nu_1 \\ h & b & f & q & \mu_2 & \nu_2 \\ g & f & c & r & \mu_3 & \nu_3 \\ p & q & r & d & \mu_4 & \nu_4 \\ \mu_1 & \mu_2 & \mu_3 & \mu_4 & \cdot & \cdot \\ \nu_1 & \nu_2 & \nu_3 & \nu_4 & \cdot & \cdot \end{vmatrix} \div |\mu_1\nu_2|^2.$$

SYLVESTER, J. J. (1852, July).

[A demonstration of the theorem that every homogeneous quadratic polynomial is reducible by real orthogonal substitutions to the form of a sum of positive and negative squares. *Philos. Magazine* (4), iv. pp. 138–142; or *Collected Math. Papers*, i. pp. 378–381.]

What is really proved here is the important proposition in the theory of orthogonants regarding the reality of the roots of Lagrange's determinantal equation, or, as it was then called, the equation of the secular inequalities. Our present interest in the demonstration, however, lies merely in the fact that it is based on two properties of axisymmetric determinants which it is desirable to isolate and to have more carefully formulated than it was Sylvester's wont to do. They are—

(1) If $|(11)\ (22) \ldots . (nn)|$ be axisymmetric, and the result of multiplying it by itself be $|[11]\ [22] \ldots . [nn]|$; and if $f(x)$ be the determinant got from the former by adding x to each element of the principal diagonal, and $F(x)$ the determinant got similarly from the latter; then

$$f(x) \cdot f(-x) = F(-x^2).$$

(2) If $F(x)$ be expanded and arranged according to descending powers of x, so that

$$F(x) = x^n + C_1 x^{n-1} + C_2 x^{n-2} + \ldots + C_n,$$

then C_r is the sum of the squares of all the r-line minors of the original determinant, it being understood that the one-line minors are the elements and the n-line minor the determinant itself.

Both are taken for granted,—a liberty which is not so defensible in the second case as in the first; for C_r is at the outset obtained merely as the sum of the r-line coaxial minors of $|[11]\ [22] \ldots . [nn]|$, and use must thus be latently made of the not quite self-evident theorem that *if* Δ *be an axisymmetric determinant, the sum of the* r*-line coaxial minors of* Δ^2 *is the sum of the squares of all the* r*-line minors of* Δ. As an illustration of the whole, let us take the case where the given determinant and its square are

$$\begin{vmatrix} a & h & g \\ h & b & f \\ g & f & c \end{vmatrix} \quad \text{and} \quad \begin{vmatrix} L & R & Q \\ R & M & P \\ Q & P & N \end{vmatrix}$$

and where therefore

$$\left.\begin{matrix} L & R, & Q \\ & M & P \\ & & N \end{matrix}\right\} \equiv \left\{\begin{matrix} a^2+h^2+g^2 & ah+hb+gf & ag+hf+gc \\ & h^2+b^2+f^2 & hg+bf+fc \\ & & g^2+f^2+c^2. \end{matrix}\right.$$

We have then

$$\begin{vmatrix} a+x & h & g \\ h & b+x & f \\ g & f & c+x \end{vmatrix} \cdot \begin{vmatrix} a-x & h & g \\ h & b-x & f \\ g & f & c+x \end{vmatrix} = \begin{vmatrix} L-x^2 & R & Q \\ R & M-x^2 & P \\ Q & P & N-x^2 \end{vmatrix},$$

$$= -x^6 + x^4(L+M+N) - x^2\left(\begin{vmatrix} L & R \\ R & M \end{vmatrix} + \begin{vmatrix} L & Q \\ Q & N \end{vmatrix} + \begin{vmatrix} M & P \\ P & N \end{vmatrix}\right) + \begin{vmatrix} L & R & Q \\ R & M & P \\ Q & P & N \end{vmatrix},$$

$$= -x^6 + x^4\left\{\begin{matrix} a^2+h^2+g^2 \\ +h^2+b^2+f^2 \\ +g^2+f^2+c^2 \end{matrix}\right\} - x^2\left\{\begin{matrix} \begin{vmatrix} a & h \\ h & b \end{vmatrix}^2 + \begin{vmatrix} a & g \\ h & f \end{vmatrix}^2 + \begin{vmatrix} h & g \\ b & f \end{vmatrix}^2 \\ + \begin{vmatrix} a & g \\ h & f \end{vmatrix}^2 + \begin{vmatrix} a & g \\ g & c \end{vmatrix}^2 + \begin{vmatrix} h & g \\ f & c \end{vmatrix}^2 \\ + \begin{vmatrix} h & g \\ b & f \end{vmatrix}^2 + \begin{vmatrix} h & g \\ f & c \end{vmatrix}^2 + \begin{vmatrix} b & f \\ f & c \end{vmatrix}^2 \end{matrix}\right\} + \begin{vmatrix} a & h & g \\ h & b & f \\ g & f & c \end{vmatrix}^2.$$

The fact that the coefficients here are negative and positive alternately is what Sylvester utilises for his main purpose, application being made of Descartes' rule of signs.

SYLVESTER, J. J. (1852, Oct.).

[On Staudt's theorems concerning the contents of polygons and polyhedrons, with a note on a new and resembling class of theorems. *Philos. Magazine* (4), iv. pp. 335–345; or *Collected Math. Papers*, i. pp. 382–391.]

As an illustration of his mode of expressing the product of two determinants of the nth order as a determinant of the $(n+1)^{\text{th}}$ order, Sylvester gives the identity

$$\begin{vmatrix} x_1 & y_1 & z_1 & 1 \\ x_2 & y_2 & z_2 & 1 \\ x_3 & y_3 & z_3 & 1 \\ x_4 & y_4 & z_4 & 1 \end{vmatrix} \cdot \begin{vmatrix} \xi_1 & \eta_1 & \zeta_1 & 1 \\ \xi_2 & \eta_2 & \zeta_2 & 1 \\ \xi_3 & \eta_3 & \zeta_3 & 1 \\ \xi_4 & \eta_4 & \zeta_4 & 1 \end{vmatrix}$$

$$= \begin{vmatrix} \Sigma x_1\xi_1 & \Sigma x_1\xi_2 & \Sigma x_1\xi_3 & \Sigma x_1\xi_4 & 1 \\ \Sigma x_2\xi_1 & \Sigma x_2\xi_2 & \Sigma x_2\xi_3 & \Sigma x_2\xi_4 & 1 \\ \Sigma x_3\xi_1 & \Sigma x_3\xi_2 & \Sigma x_3\xi_3 & \Sigma x_3\xi_4 & 1 \\ \Sigma x_4\xi_1 & \Sigma x_4\xi_2 & \Sigma x_4\xi_3 & \Sigma x_4\xi_4 & 1 \\ 1 & 1 & 1 & 1 & . \end{vmatrix}$$

where $\Sigma x_r\xi_s$ is put for $x_r\xi_s + y_r\eta_s + z_r\zeta_s$. Then performing on the last determinant the operations which we may denote by

$$\begin{array}{lll} \text{row}_1 - \tfrac{1}{2}\Sigma x_1^2 \cdot \text{row}_5, & \text{row}_2 - \tfrac{1}{2}\Sigma x_2^2 \cdot \text{row}_5, & \ldots . \\ \text{col}_1 - \tfrac{1}{2}\Sigma \xi_1^2 \cdot \text{col}_5, & \text{col}_2 - \tfrac{1}{2}\Sigma \xi_2^2 \cdot \text{col}_5, & \ldots . \end{array}$$

he obtains

$$-\tfrac{1}{8} \begin{vmatrix} \Sigma(x_1-\xi_1)^2 & \Sigma(x_1-\xi_2)^2 & \Sigma(x_1-\xi_3)^2 & \Sigma(x_1-\xi_4)^2 & 1 \\ \Sigma(x_2-\xi_1)^2 & \Sigma(x_2-\xi_2)^2 & \Sigma(x_2-\xi_3)^2 & \Sigma(x_2-\xi_4)^2 & 1 \\ \Sigma(x_3-\xi_1)^2 & \Sigma(x_3-\xi_2)^2 & \Sigma(x_3-\xi_3)^2 & \Sigma(x_3-\xi_4)^2 & 1 \\ \Sigma(x_4-\xi_1)^2 & \Sigma(x_4-\xi_2)^2 & \Sigma(x_4-\xi_3)^2 & \Sigma(x_4-\xi_4)^2 & 1 \\ 1 & 1 & 1 & 1 & . \end{vmatrix};$$

so that, if x_r, y_r, z_r and ξ_r, η_r, ζ_r be rectangular co-ordinates of points in space, the result reached gives an expression for thirty-six times the product of the volumes of two tetrahedrons in terms of the distances of the angular points of the one from the angular points of the other. By proceeding to the case where the two tetrahedrons are coincident, and thence to the case where the four remaining points are situated in the same plane, we reach Cayley's relation connecting the mutual distances of four such points.

It is thus seen that whereas Cayley's vanishing axisymmetric determinant was originally got as a multiple of a peculiarly obtained square of the determinant

$$\begin{vmatrix} \Sigma x_1^2 & x_1 & y_1 & 0 & 1 \\ \Sigma x_2^2 & x_2 & y_2 & 0 & 1 \\ \Sigma x_3^2 & x_3 & y_3 & 0 & 1 \\ \Sigma x_4^2 & x_4 & y_4 & 0 & 1 \\ 1 & 0 & 0 & 0 & 0 \end{vmatrix}.$$

Sylvester arrives at it by squaring

$$\begin{vmatrix} x_1 & y_1 & 0 & 1 \\ x_2 & y_2 & 0 & 1 \\ x_3 & y_3 & 0 & 1 \\ x_4 & y_4 & 0 & 1 \end{vmatrix}$$

in a special fashion, and then performing certain transformations on the result.

SYLVESTER, J. J. (1852, Nov.).

[Sur une propriété nouvelle de l'équation qui sert à déterminer les inégalités séculaires des planètes. *Nouv. Annales de Math.*, xi. pp. 434–440; or *Collected Math. Papers*, i. pp. 364–366.]

This paper of composite authorship probably originated in a letter from Sylvester giving his theorem and demonstration, with a remark or two additional. To these, which were made §§ 7, 7′, 8, the editor prefixed an introduction (§§ 1–6) on determinants and determinant-multiplication.*

The theorem is an extension of one which is the basis of his paper in the *Philosophical Magazine* of the same year, and may be shortly enunciated as follows: *If* $|(11)\,(22)\ldots(nn)|$ *be axisymmetric and have* $|[11]\,[22]\ldots[nn]|$ *for its* p^{th} *power, then the roots of the equation*

$$\begin{vmatrix} [11]-x & [12] & \ldots & [1n] \\ [21] & [22]-x & \ldots & [2n] \\ \cdot & \cdot & \cdot & \cdot \\ [n1] & [n2] & \ldots & [nn]-x \end{vmatrix} = 0$$

are the pth *powers of the roots of the equation*

$$\begin{vmatrix} (11)-x & (12) & \ldots & (1n) \\ (21) & (22)-x & \ldots & (2n) \\ \cdot & \cdot & \cdot & \cdot \\ (n1) & (n2) & \ldots & (nn)-x \end{vmatrix} = 0.$$

* In the *Coll. Math. Papers* §§ 1–6 are omitted, and §§ 7, 7′, 8 are numbered §§ 6, 7, 8. The theorem of the original § 6 is incorrect.

The "demonstration" leaves a good deal to be desired. In effect it amounts to saying that if $\zeta_1, \zeta_2, \ldots, \zeta_p$ be the pth roots of unity, and D_m the determinant got from $|(11)\,(22)\ldots(nn)|$ by subtracting $\zeta_m x$ from each of the diagonal elements, then

$$D_1 D_2 D_3 \ldots D_p \equiv \begin{vmatrix} [11]-x^p & [12] & \ldots & [1n] \\ [21] & [22]-x^p & \ldots & [2n] \\ \cdots & \cdots & \cdots & \cdots \\ [n1] & [n2] & \ldots & [nn]-x^p \end{vmatrix}.$$

Now it is well known that the multiplication of $D_1, D_2, \ldots, D_p$ enables us to arrive at the equation whose roots are the pth powers in question, but this and Sylvester's statement are not by any means identical. The separate points to be established are (1) that the element in the rth row and s^{th} column of the determinant which is the product of $D_1, D_2, \ldots, D_p$ consists of $[rs]$ and a tail of other terms, (2) that this tail vanishes in the case of every non-diagonal element, (3) that in the case of the diagonal elements it reduces to $-x^p$; and Sylvester's only justificatory statements are that the product of the p determinants is independent of the order in which they are taken, and that all the terms containing a ζ in any other power than the pth will vanish.

Another true proposition made on insufficient foundation is that *the pth power of an axisymmetric determinant is itself axisymmetric.* The foundation here is the incorrect proposition of § 6.

CAYLEY, A. (1852, Dec.).

[On the rationalisation of certain algebraical questions. *Cambridge and Dub. Math. Journ.*, viii. pp. 97–101; or *Collected Math. Papers*, ii. pp. 40–44.]

The equations first considered are of the type

$$a^{\frac{1}{2}}+b^{\frac{1}{2}}+c^{\frac{1}{2}}+\ldots = 0,$$

and the fresh departure consists in viewing such an equation as the outcome of the set of equations

$$x+y+z+\ldots = 0,\quad x^2=a,\quad y^2=b,\quad z^2=c,\quad \ldots.$$

Taking the case where the number of variables in the set is three, Cayley operates on the equation $x+y+z=0$ with the multipliers 1, yz, zx, xy, thus obtaining with the help of the other equations a set from which the variables x, y, z, xyz may be eliminated, with the result*

$$\begin{vmatrix} . & 1 & 1 & 1 \\ 1 & . & c & b \\ 1 & c & . & a \\ . & b & a & . \end{vmatrix} = 0.$$

Also and, so to say, conversely he operates with the multipliers x, y, z, xyz, and eliminates 1, yz, zx, xy, with the result

$$\begin{vmatrix} . & a & b & c \\ a & . & 1 & 1 \\ b & 1 & . & 1 \\ c & 1 & 1 & . \end{vmatrix} = 0.$$

Similarly, when there are four variables, the multipliers are

$$1,\ yz,\ zx,\ xy,\ xw,\ yw,\ zw,\ xyzw,$$

and the eliminands

$$x,\ y,\ z,\ w,\ yzw,\ zwx,\ wxy,\ xyz,$$

or *vice versa*; but in this case the two resultants are not essentially distinct, the one being derivable from the other by mere transference of lines. Cayley then adds, "And in general for any *even* number of quadratic radicals the two forms are not essentially distinct,† but may be derived from each other by interchanging lines and columns, while for an *odd* number of quadratic radicals the two forms cannot be so derived from each other, but are essentially distinct."

The equations next dealt with are of the type

$$a^{\frac{1}{3}}+b^{\frac{1}{3}}+c^{\frac{1}{3}}+\ldots\ldots = 0,$$

* Already thus formed by Cayley in his first paper of all (1841).

† Observe that, although Cayley considers the two determinants of the previous case to be essentially distinct, the second is derivable from the first by multiplying the columns by abc, a, b, c respectively, and then dividing the rows by 1, bc, ca, ab respectively.

Sylvester having suggested the extension of the process. Taking the case of three variables, that is to say, when the set of equations is

$$x+y+z=0, \quad x^3=a, \quad y^3=b, \quad z^3=c,$$

he first uses the multipliers

$$1,\ xyz,\ x^2y^2z^2,\ x^2z,\ y^2x,\ z^2y,\ x^2y,\ y^2z,\ z^2x,$$

the eliminands then being

$$x,\ y,\ z,\ y^2z^2,\ x^2yz,\ y^2zx,\ z^2xy,\ z^2x^2,\ x^2y^2;$$

next he uses the said eliminands as multipliers, the new eliminands being

$$x^2,\ y^2,\ z^2,\ yz,\ zx,\ xy,\ xy^2z^2,\ yz^2x^2,\ zx^2y^2;$$

and finally using the new eliminands as multipliers, he eliminates

$$1,\ xyz,\ x^2y^2z^2,\ x^2z,\ y^2x,\ z^2y,\ x^2y,\ y^2z,\ z^2x,$$

that is to say, the first set of multipliers. Only in the case of the second elimination is the determinant axisymmetric, namely,

$$\begin{vmatrix} . & c & b & . & . & . & 1 & . & . \\ c & . & a & . & . & . & . & 1 & . \\ b & a & . & . & . & . & . & . & 1 \\ . & . & . & a & . & . & . & 1 & 1 \\ . & . & . & . & b & . & 1 & . & 1 \\ . & . & . & . & . & c & 1 & 1 & . \\ 1 & . & . & . & 1 & 1 & . & . & . \\ . & 1 & . & 1 & . & 1 & . & . & . \\ . & . & 1 & 1 & 1 & . & . & . & . \end{vmatrix}$$

which must thus be equal to $(a+b+c)^3-27abc$. In the two other cases the determinants are not essentially distinct,* the rows of the one being columns of the other; and this is said to be true of two of the three forms whatever be the number of variables.

---

* Cayley fails to notice, however, that each of these is readily transformable into the second. Thus, taking his first form, we have only to multiply the 4th, 8th, 9th columns of it by a, and divide the 2nd, 4th, 7th rows by a, when we obtain a determinant which by mere permutation of the rows $\begin{pmatrix} 1\,2\,3\,4\,5\,6\,7\,8\,9 \\ 7\,4\,1\,9\,3\,6\,8\,5\,2 \end{pmatrix}$ and subsequently of the columns $\begin{pmatrix} 1\,2\,3\,4\,5\,6\,7\,8\,9 \\ 1\,6\,5\,4\,9\,8\,7\,3\,2 \end{pmatrix}$ becomes identical with the axisymmetric form.

SYLVESTER, J. J. (1853, March).

[On the relation between the volume of a tetrahedron and the product of the sixteen algebraical values of its superficies. *Cambridge and Dub. Math. Journ.*, viii. pp. 171–178; or *Nouv. Annales de Math.*, xiii. pp. 203–209; or *Collected Math. Papers*, i. pp. 404–410.]

Denoting the vertices of a tetrahedron by a, b, c, d, its volume in terms of the edges by V, and the areas of its faces by $\frac{1}{4}\sqrt{\mathrm{F}}$, $\frac{1}{4}\sqrt{\mathrm{G}}$, $\frac{1}{4}\sqrt{\mathrm{H}}$, $\frac{1}{4}\sqrt{\mathrm{K}}$, we know that

$$\begin{aligned}144\mathrm{V}^2 = &\ (bc)^2(da)^2\{(ca)^2+(db)^2+(ab)^2+(cd)^2-(bc)^2-(da)^2\}\\ &+(ca)^2(db)^2\{(ab)^2+(cd)^2+(bc)^2+(da)^2-(ca)^2-(db)^2\}\\ &+(ab)^2(cd)^2\{(bc)^2+(da)^2+(ca)^2+(db)^2-(ab)^2-(cd)^2\}\\ &-(bc)^2(ca)^2(ab)^2-(bc)^2(db)^2(cd)^2-(ca)^2(cd)^2(da)^2-(ab)^2(da)^2(db)^2\\ = &\ \mathrm{W}\ \text{say,}\end{aligned}$$

and

$$\begin{aligned}\mathrm{F} &= -(bc)^4-(cd)^4-(db)^4+2(cd)^2(db)^2+2(db)^2(bc)^2+2(bc)^2(cd)^2\\ \mathrm{G} &= -(ac)^4-(cd)^4-(da)^4+2(cd)^2(da)^2+2(da)^2(ac)^2+2(ac)^2(cd)^2\\ \mathrm{H} &= -(ab)^4-(bd)^4-(da)^4+2(bd)^2(da)^2+2(da)^2(ab)^2+2(ab)^2(bd)^2\\ \mathrm{K} &= -(ab)^4-(bc)^4-(ca)^4+2(bc)^2(ca)^2+2(ca)^2(ab)^2+2(ab)^2(bc)^2.\end{aligned}$$

With this notation Sylvester points out that the condition for the vanishing of the surface of the tetrahedron is

$$\sqrt{\mathrm{F}}+\sqrt{\mathrm{G}}+\sqrt{\mathrm{H}}+\sqrt{\mathrm{K}} = 0,$$

and that this when freed of root-signs is

$$\sum\mathrm{F}^4 - 4\sum\mathrm{F}^3\mathrm{G} + 6\sum\mathrm{F}^2\mathrm{G}^2 + 4\sum\mathrm{F}^2\mathrm{GH} - 40\mathrm{FGHK} = 0,$$

or say

$$\mathrm{N} = 0,$$

where N consequently is of the eighth degree in the squared edges. His reasoning then is that as the vanishing of the surface and the vanishing of the volume are necessarily coincident, it follows that W, having no rational factors, must itself be a factor of N; and that, W being of the third degree in the squared edges, the quotient N/W must be of the fifth degree.

Relying on this he proceeds to determine the quotient by considering the cases where (1) $ab=0=cd$, (2) $ab=0=ac$, (3) $ab=ac=ad=bc=bd=cd=1$, his result being

$$\sum(ab)^2(bc)^2(ca)^2\left[\begin{array}{c}(da)^4+(db)^4+(dc)^4\\ -\{(da)^2+(db)^2+(dc)^2\}\{(ab)^2+(bc)^2+(ca)^2\}\\ +(ab)^2(bc)^2+(bc)^2(ca)^2+(ca)^2(ab)^2\end{array}\right]$$

$$+2\sum(ab)^2(bc)^2(cd)^2(da)^2(ac)^2,$$

where there are four expressions under the first Σ and six under the second; or

$$\sum(ab)^2(bc)^2(ca)^2\left[\begin{array}{l}(da)^4+(db)^4+(dc)^4\\ -\{(da)^2+(db)^2+(dc)^2\}\{(ab)^2+(bc)^2+(ca)^2\}\\ +(da)^2(db)^2+(db)^2(dc)^2+(dc)^2(da)^2\\ +(ab)^2(bc)^2+(bc)^2(ca)^2+(ca)^2(ab)^2\end{array}\right]$$

where there are four expressions under the Σ, one corresponding to each face.

Although there is no explicit mention here of determinants, it being unnecessary, it has now to be noted that Sylvester had the determinant-form of W before him throughout: he even says that he had tried to express the quotient as a determinant, but had been unsuccessful. Without further restriction, then, as to form, his proposition is *If* F, G, H, K *be the complementary minors of the elements in the places* 11, 22, 33, 44 *of the determinant*

$$\left|\begin{array}{ccccc} . & (ab)^2 & (ac)^2 & (ad)^2 & 1\\ (ab)^2 & . & (bc)^2 & (bd)^2 & 1\\ (ac)^2 & (bc)^2 & . & (cd)^2 & 1\\ (ad)^2 & (bd)^2 & (cd)^2 & . & 1\\ 1 & 1 & 1 & 1 & .\end{array}\right|, \quad \text{or } 2\text{W say,}$$

then the result of rationalising

$$\sqrt{\text{F}}+\sqrt{\text{G}}+\sqrt{\text{H}}+\sqrt{\text{K}}$$

is divisible by W. This is not all, however; for Sylvester having noted the analogous case connected with the relation between the perimeter and area of a triangle, namely, the fact that

$$\begin{vmatrix} . & (ab)^2 & (ac)^2 & 1 \\ (ab)^2 & . & (bc)^2 & 1 \\ (ac)^2 & (bc)^2 & . & 1 \\ 1 & 1 & 1 & . \end{vmatrix} \quad i.e. \quad \sum(ab)^4 - 2\sum(ab)^2(bc)^2$$

is a divisor of the result of rationalising

$$\sqrt{2(bc)^2} + \sqrt{2(ca)^2} + \sqrt{2(ab)^2}$$

where the radicands are the complementary minors of the elements in the places 11, 22, 33 of the determinant, boldly extends the proposition (without proof) to any triangular number of arbitrary quantities, taking occasion also to point out that when we leave geometry (ab), (ac), ... may be written for $(ab)^2$, $(ac)^2$, ...

HESSE, O. (1853, April).

[Ueber Determinanten und ihre Anwendung in der Geometrie, insbesondere auf Curven vierter Ordnung. *Crelle's Journal*, xlix. pp. 243–264; or *Werke*, pp. 319–343.]

The main subject of the first half of Hesse's paper (pp. 243–253) is a property of axisymmetric determinants required for the establishment of the geometrical results contained in the second half. In the first three pages he considers the relations between the minors of two general determinants A, B, and the minors of their product C; or, as he unfortunately feels himself compelled to put it, "wie die partiellen Differential-quotienten der Determinante C nach ihren Elementen c genommen durch die partiellen Differential-quotienten der Factoren A und B nach ihren Elementen genommen sich ausdrücken lassen." What follows thereafter may be described as the establishment of the simple identity

$$\begin{vmatrix} u_{11} & u_{12} & \alpha_1 \\ u_{21} & u_{22} & \alpha_2 \\ \alpha_1 & \alpha_2 & 0 \end{vmatrix} \begin{vmatrix} u_{11} & u_{12} & \gamma_1 \\ u_{21} & u_{22} & \gamma_2 \\ \gamma_1 & \gamma_2 & 0 \end{vmatrix} - \begin{vmatrix} u_{11} & u_{12} & \alpha_1 \\ u_{21} & u_{22} & \alpha_2 \\ \gamma_1 & \gamma_2 & 0 \end{vmatrix}^2 = \begin{vmatrix} u_{11} & u_{12} \\ u_{21} & u_{22} \end{vmatrix} \begin{vmatrix} \alpha_1 & \alpha_2 \\ \gamma_1 & \gamma_2 \end{vmatrix}^2,$$

where $u_{21}=u_{12}$, by multiplying together

$$\begin{vmatrix} a_1 & -a_2 \\ \gamma_1 & -\gamma_2 \end{vmatrix}, \quad \begin{vmatrix} u_{11} & u_{12} \\ u_{21} & u_{22} \end{vmatrix}, \quad \begin{vmatrix} -a_2 & a_1 \\ -\gamma_2 & \gamma_1 \end{vmatrix}$$

in row-by-row fashion; and then the generalisation of this identity in two different directions.

The first generalisation consists in the proposition that the two-line axisymmetric determinant

$$\begin{vmatrix} u_{11} & u_{12} & \dots & u_{1n} & a_1 \\ u_{21} & u_{22} & \dots & u_{2n} & a_2 \\ \cdot & \cdot & \cdot & \cdot & \cdot \\ u_{n1} & u_{n2} & \dots & u_{nn} & a_n \\ a_1 & a_2 & \dots & a_n & 0 \end{vmatrix} \begin{vmatrix} u_{11} & u_{12} & \dots & u_{1n} & \gamma_1 \\ u_{21} & u_{22} & \dots & u_{2n} & \gamma_2 \\ \cdot & \cdot & \cdot & \cdot & \cdot \\ u_{n1} & u_{n2} & \dots & u_{nn} & \gamma_n \\ \gamma_1 & \gamma_2 & \dots & \gamma_n & 0 \end{vmatrix} - \begin{vmatrix} u_{11} & u_{12} & \dots & u_{1n} & a_1 \\ u_{21} & u_{22} & \dots & u_{2n} & a_2 \\ \cdot & \cdot & \cdot & \cdot & \cdot \\ u_{n1} & u_{n2} & \dots & u_{nn} & a_n \\ \gamma_1 & \gamma_2 & \dots & \gamma_n & 0 \end{vmatrix}^2$$

where $u_{\kappa\lambda} = u_{\lambda\kappa}$, contains

$$\begin{vmatrix} u_{11} & u_{12} & \dots & u_{1n} \\ u_{21} & u_{22} & \dots & u_{2n} \\ \cdot & \cdot & \cdot & \cdot \\ u_{n1} & u_{n2} & \dots & u_{nn} \end{vmatrix}$$

as a factor, and that the cofactor is an integral homogeneous function of the a's and likewise of the γ's. The case where n is equal to 3 is treated as follows. The determinant

$$\begin{vmatrix} u_{11} & u_{12} & u_{13} & a_1 \\ u_{21} & u_{22} & u_{23} & a_2 \\ u_{31} & u_{32} & u_{33} & a_3 \\ \gamma_1 & \gamma_2 & \gamma_3 & \beta \end{vmatrix},$$

having $u_{\kappa\lambda}=u_{\lambda\kappa}$, is introduced and denoted by B, with the result that the two-line determinant in question is representable by

$$\begin{vmatrix} \frac{\partial B}{\partial \gamma_1}a_1+\frac{\partial B}{\partial \gamma_2}a_2+\frac{\partial B}{\partial \gamma_3}a_3 & \frac{\partial B}{\partial a_1}a_1+\frac{\partial B}{\partial a_2}a_2+\frac{\partial B}{\partial a_3}a_3 \\ \frac{\partial B}{\partial \gamma_1}\gamma_1+\frac{\partial B}{\partial \gamma_2}\gamma_2+\frac{\partial B}{\partial \gamma_3}\gamma_3 & \frac{\partial B}{\partial a_1}\gamma_1+\frac{\partial B}{\partial a_2}\gamma_2+\frac{\partial B}{\partial a_3}\gamma_3 \end{vmatrix}$$

and its predicated factor by $\partial B/\partial \beta$. It is then pointed out that the former is the differential-quotient of the product of the two determinants

$$\begin{vmatrix} \frac{\partial B}{\partial \gamma_1} & \frac{\partial B}{\partial \gamma_2} & \frac{\partial B}{\partial \gamma_3} \\ \frac{\partial B}{\partial \alpha_1} & \frac{\partial B}{\partial \alpha_2} & \frac{\partial B}{\partial \alpha_3} \\ m_1 & m_2 & m_3 \end{vmatrix}, \quad \begin{vmatrix} \alpha_1 & \alpha_2 & \alpha_3 \\ \gamma_1 & \gamma_2 & \gamma_3 \\ n_1 & n_2 & n_3 \end{vmatrix},$$

called M and N, taken with respect to $m_1n_1 + m_2n_2 + m_3n_3$; and that consequently it is equal to

$$\frac{\partial M}{\partial m_1}\frac{\partial N}{\partial n_1} + \frac{\partial M}{\partial m_2}\frac{\partial N}{\partial n_2} + \frac{\partial M}{\partial m_3}\frac{\partial N}{\partial n_3}. \qquad (\varpi)$$

Since, however, we have

$$\frac{\partial B}{\partial \gamma_1}u_{11} + \frac{\partial B}{\partial \gamma_2}u_{12} + \frac{\partial B}{\partial \gamma_3}u_{13} + \frac{\partial B}{\partial \beta}\alpha_1 = 0,$$

and other similar identities, it follows that

$$M \cdot \frac{\partial B}{\partial \beta} \quad \textit{i.e.} \quad \begin{vmatrix} \frac{\partial B}{\partial \gamma_1} & \frac{\partial B}{\partial \gamma_2} & \frac{\partial B}{\partial \gamma_3} \\ \frac{\partial B}{\partial \alpha_1} & \frac{\partial B}{\partial \alpha_2} & \frac{\partial B}{\partial \alpha_3} \\ m_1 & m_2 & m_3 \end{vmatrix} \cdot \begin{vmatrix} u_{11} & u_{12} & u_{13} \\ u_{21} & u_{22} & u_{23} \\ u_{31} & u_{32} & u_{33} \end{vmatrix}$$

$$= \begin{vmatrix} -\frac{\partial B}{\partial \beta}\alpha_1 & -\frac{\partial B}{\partial \beta}\gamma_1 & u_{11}m_1 + u_{12}m_2 + u_{13}m_3 \\ -\frac{\partial B}{\partial \beta}\alpha_2 & -\frac{\partial B}{\partial \beta}\gamma_2 & u_{21}m_1 + u_{22}m_2 + u_{23}m_3 \\ -\frac{\partial B}{\partial \beta}\alpha_3 & -\frac{\partial B}{\partial \beta}\gamma_3 & u_{31}m_1 + u_{32}m_2 + u_{33}m_3 \end{vmatrix},$$

and therefore

$$M = \frac{\partial B}{\partial \beta}\begin{vmatrix} \alpha_1 & \gamma_1 & u_{11}m_1 + u_{12}m_2 + u_{13}m_3 \\ \alpha_2 & \gamma_2 & u_{21}m_1 + u_{22}m_2 + u_{23}m_3 \\ \alpha_3 & \gamma_3 & u_{31}m_1 + u_{32}m_2 + u_{33}m_3 \end{vmatrix} = \frac{\partial B}{\partial \beta} \cdot P \text{ say.}$$

The expression (ϖ) then becomes

$$\frac{\partial B}{\partial \beta}\left\{ \frac{\partial P}{\partial m_1}\frac{\partial N}{\partial n_1} + \frac{\partial P}{\partial m_2}\frac{\partial N}{\partial n_2} + \frac{\partial P}{\partial m_3}\frac{\partial N}{\partial n_3} \right\},$$

and all that remains is the evaluation of the bracketed factor after the equivalents of P and N have been substituted therein. The final result is

$$\begin{vmatrix} u_{11} & u_{12} & u_{13} \\ u_{21} & u_{22} & u_{23} \\ u_{31} & u_{32} & u_{33} \end{vmatrix} \cdot \left\{ \begin{aligned} u_{11}(\alpha_2\gamma_3-\alpha_3\gamma_2)^2+u_{22}(\alpha_3\gamma_1-\alpha_1\gamma_3)^2+u_{33}(\alpha_1\gamma_2-\alpha_2\gamma_1)^2 \\ +(u_{23}+u_{32})(\alpha_3\gamma_1-\alpha_1\gamma_3)(\alpha_1\gamma_2-\alpha_2\gamma_1) \\ +(u_{31}+u_{13})(\alpha_1\gamma_2-\alpha_2\gamma_1)(\alpha_2\gamma_3-\alpha_3\gamma_2) \\ +(u_{12}+u_{21})(\alpha_2\gamma_3-\alpha_3\gamma_2)(\alpha_3\gamma_1-\alpha_1\gamma_3) \end{aligned} \right\}.$$

The case where $n=4$ is similarly dealt with; but as it is necessarily more complicated, it is not carried quite so far, the cofactor of $|u_{11}\ u_{22}\ u_{33}\ u_{44}|$ being merely stated to be of the desired form and easily calculable. "Sie hat aber zu viele Glieder, um sie berechnet hinzuschreiben." The general proposition, as above given, is then formally enunciated.

The other generalisation made of the case where $n=2$ is to the effect that *the product of an axisymmetric determinant by the square of any other determinant is expressible as an axisymmetric determinant.* In connection with this the interesting point is the notation used for the elements of the product-determinants. Since the differential-quotient of

$$\tfrac{1}{2}\{u_{11}x_1{}^2+u_{22}x_2{}^2+\ .\ .\ .\ +u_{nn}x_n{}^2+2u_{12}x_1x_2+\ .\ .\ .\ \},$$

$$\text{or}\quad F(x_1,\, x_2,\, .\ .\ .,\, x_n)\quad \text{say},$$

with respect to x_p is

$$u_{1p}x_1+u_{2p}x_2+\ .\ .\ .\ +u_{pp}x_p+\ .\ .\ .\ +u_{np}x_n$$

the result of annexing q as a second suffix to the x's in this may be suitably denoted by

$$F'(x_{pq});$$

so that in accordance with this the product of

$$\begin{vmatrix} u_{11} & u_{12} & \dots & u_{1n} \\ u_{21} & u_{22} & \dots & u_{2n} \\ \dots & \dots & \dots & \dots \\ u_{n1} & u_{n2} & \dots & u_{nn} \end{vmatrix}_{u_{\kappa\lambda}=u_{\lambda\kappa}} \quad \text{and} \quad \begin{vmatrix} x_{11} & x_{21} & \dots & x_{n1} \\ x_{12} & x_{22} & \dots & x_{n2} \\ \dots & \dots & \dots & \dots \\ x_{1n} & x_{2n} & \dots & x_{nn} \end{vmatrix}$$

will be

$$\begin{vmatrix} F'(x_{11}) & F'(x_{21}) & \dots & F'(x_{n1}) \\ F'(x_{12}) & F'(x_{22}) & \dots & F'(x_{n2}) \\ \dots & \dots & \dots & \dots \\ F'(x_{1n}) & F'(x_{2n}) & \dots & F'(x_{nn}) \end{vmatrix}$$

" Multiplicirt man diese Determinante nochmals mit der vorhergehenden und setzt:

$$F_{pq} = x_{1p}F'(x_{1q}) + x_{2p}F'(x_{2q}) + \ . \ . \ . \ + x_{np}F'(x_{nq}),$$

so erhält man

$$\begin{vmatrix} u_{11} & u_{12} & \dots & u_{1n} \\ u_{21} & u_{22} & \dots & u_{2n} \\ \cdot & \cdot & \cdot & \cdot \\ u_{n1} & u_{n2} & \dots & u_{nn} \end{vmatrix} \cdot \begin{vmatrix} x_{11} & x_{21} & \dots & x_{n1} \\ x_{12} & x_{22} & \dots & x_{n2} \\ \cdot & \cdot & \cdot & \cdot \\ x_{1n} & x_{2n} & \dots & x_{nn} \end{vmatrix}^2 = \begin{vmatrix} F_{11} & F_{12} & \dots & F_{1n} \\ F_{21} & F_{22} & \dots & F_{2n} \\ \cdot & \cdot & \cdot & \cdot \\ F_{n1} & F_{n2} & \dots & F_{nn} \end{vmatrix}.$$

Da aber $F_{pq} = F_{qp}$ ist, so ist die letzte Determinante wieder *symmetrisch.*"

BRIOSCHI, F. (1853, July).

[Sur une propriété d'un produit de facteurs linéaires. *Cambridge and Dub. Math. Journ.*, ix. pp. 137–144; or *Opere mat.*, v. pp. 121–128.]

This paper of Brioschi's is avowedly an outcome of Cayley's " On the rationalisation of certain algebraical equations," published two months earlier. Starting with the equation $x+y+z=0$, and using on it the multipliers 1, yz, zx, xy in succession, he obtains

$$\left.\begin{array}{rl} x + y + z & = 0 \\ xyz + \qquad + z^2.y + y^2.z & = 0 \\ xyz + z^2.x \qquad + x^2.z & = 0 \\ xyz + y^2.x + x^2.y \qquad & = 0 \end{array}\right\}$$

whence on elimination of xyz, x, y, z there results

$$0 = \begin{vmatrix} . & 1 & 1 & 1 \\ 1 & . & z^2 & y^2 \\ 1 & z^2 & . & x^2 \\ 1 & y^2 & x^2 & . \end{vmatrix} = \Delta \text{ say};$$

and as this equation has its origin in the equation $x+y+z=0$, he concludes that the determinant Δ must have $x+y+z$ for a factor. Then, since any one of the three other equations

$$x+y-z=0, \quad x-y+z=0, \quad -x+y+z=0$$

gives rise to the same result, it is easily seen how he reaches the identity

$$-\begin{vmatrix} \cdot & 1 & 1 & 1 \\ 1 & \cdot & z^2 & y^2 \\ 1 & z^2 & \cdot & x^2 \\ 1 & y^2 & x^2 & \cdot \end{vmatrix} = (x+y+z)(x+y-z)(x-y+z)(-x+y+z).$$

An alternative form of Δ is introduced by the words " Observons qu'on a évidemment*

$$\Delta = \frac{1}{x^2y^2z^2}\begin{vmatrix} \cdot & x & y & z \\ x & \cdot & z^2xy & y^2xz \\ y & z^2xy & \cdot & x^2yz \\ z & y^2xz & x^2yz & \cdot \end{vmatrix} = \begin{vmatrix} \cdot & x & y & z \\ x & \cdot & z & y \\ y & z & \cdot & x \\ z & y & x & \cdot \end{vmatrix}\text{"}.$$

Similarly from the equation $x+y+z+w=0$ or any one of its seven relatives by using the multipliers

$$1,\ zw,\ yw,\ xw,\ zy,\ zx,\ yx,\ xyzw$$

and eliminating

$$xyz,\ xyw,\ xzw,\ yzw,\ x,\ y,\ z,\ w$$

there is obtained the result

$$\begin{vmatrix} \cdot & \cdot & \cdot & \cdot & 1 & 1 & 1 & 1 \\ \cdot & \cdot & 1 & 1 & \cdot & \cdot & w^2 & z^2 \\ \cdot & 1 & \cdot & 1 & \cdot & w^2 & \cdot & y^2 \\ \cdot & 1 & 1 & \cdot & w^2 & \cdot & \cdot & x^2 \\ 1 & \cdot & \cdot & 1 & \cdot & z^2 & y^2 & \cdot \\ 1 & \cdot & 1 & \cdot & z^2 & \cdot & x^2 & \cdot \\ 1 & 1 & \cdot & \cdot & y^2 & x^2 & \cdot & \cdot \\ w^2 & z^2 & y^2 & x^2 & \cdot & \cdot & \cdot & \cdot \end{vmatrix}$$

$$= (x+y+z+w)\cdot(-x+y+z+w)(x-y+z+w)(x+y-z+w)(x+y+z-w)$$
$$\cdot(-x-y+z+w)(-x+y-z+w)(-x+y+z-w),$$

* It would seem preferable to multiply the columns of Δ in order by xyz, x, y, z, the quantities just eliminated; and then divide the rows by the multipliers 1, yz, zx, xy used in obtaining the set of equations. The advantage of this method would be still greater in the next case.

an alternative form being stated to be the axisymmetric determinant

$$\begin{vmatrix} \cdot & \cdot & \cdot & \cdot & x & y & z & w \\ \cdot & \cdot & x & y & \cdot & \cdot & w & z \\ \cdot & x & \cdot & z & \cdot & w & \cdot & y \\ \cdot & y & z & \cdot & w & \cdot & \cdot & x \\ x & \cdot & \cdot & w & \cdot & z & y & \cdot \\ y & \cdot & w & \cdot & z & \cdot & x & \cdot \\ z & w & \cdot & \cdot & y & x & \cdot & \cdot \\ w & z & y & x & \cdot & \cdot & \cdot & \cdot \end{vmatrix}.$$

The general theorem is then formulated as follows: "Si en général on considère n éléments $x_1, x_2, \ldots, x_n$, en posant

$$X_n = x_1 + x_2 + \ldots + x_n,$$

et en désignant par

$$|X_n(1, 2, \ldots, m)|$$

le produit des facteurs linéaires qu'on deduit de X_n en changeant les signes à m des éléments $x_1, x_2, \ldots, x_n$; en aura pour n impair

$$X_n \cdot |X_n(1)| \cdot |X_n(1,2)| \ldots\ldots |X_n(1, 2, 3, \ldots, \tfrac{1}{2}(n-1)| = -\Delta,$$

et pour n pair

$$X_n \cdot |X_n(1)| \cdot |X_n(1,2)| \ldots\ldots |X_n(1,2,\ldots, \tfrac{1}{2}(n-2)| \cdot |X_n(\bar{1},2,3,\ldots,\tfrac{1}{2}n)| = \Delta$$

où le symbole

$$X_n(\dot{1}, 2, 3, \ldots, \tfrac{1}{2}n)$$

dénote que dans ce produit l'élément x_1 entre toujours parmi les éléments auxquels on a changé de signe. Le déterminant Δ résultera de la multiplication successive de l'équation $X_n = 0$ par l'unité, et par chacune des combinaisons deux à deux, quatre à quatre, ..., $(n-1)$ à $(n-1)$ si n est impair: n à n si n est pair, des éléments $x_1, x_2, \ldots, x_n$."

In addition, it is shown under this head that the number of equations in the set which originates the determinant is 2^{n-1}, and, a little unnecessarily, that the number of linear factors in the product is the same. It is also noted that if $x_1, x_2, \ldots, x_n$ be quadratic radicals, the product of the linear factors is rational.

Following Cayley's paper still further, Brioschi similarly makes clear that one of the nine-line determinants there obtained, namely

$$\left|\begin{array}{ccccccccc} \cdot & \cdot & \cdot & \cdot & \cdot & \cdot & 1 & 1 & 1 \\ \cdot & \cdot & 1 & \cdot & \cdot & 1 & \cdot & z^3 & \cdot \\ \cdot & \cdot & 1 & \cdot & 1 & \cdot & z^3 & \cdot & \cdot \\ \cdot & 1 & \cdot & \cdot & \cdot & 1 & \cdot & \cdot & y^3 \\ \cdot & 1 & \cdot & 1 & \cdot & \cdot & y^3 & \cdot & \cdot \\ 1 & \cdot & \cdot & \cdot & 1 & \cdot & \cdot & \cdot & x^3 \\ 1 & \cdot & \cdot & 1 & \cdot & \cdot & \cdot & x^3 & \cdot \\ \cdot & \cdot & \cdot & z^3 & y^3 & x^3 & \cdot & \cdot & \cdot \\ 1 & 1 & 1 & \cdot & \cdot & \cdot & \cdot & \cdot & \cdot \end{array}\right|,$$

may be viewed as originating in any one of the nine equations

$$\begin{array}{lll} x + y + z = 0, & x + y + \alpha z = 0, & x + y + \beta z = 0, \\ x + \alpha y + \beta z = 0, & x + \alpha y + z = 0, & x + \beta y + z = 0, \\ x + \beta y + \alpha z = 0, & \alpha x + y + z = 0, & \beta x + y + z = 0, \end{array}$$

where α, β are the imaginary cube roots of unity, and could thus be shown to be equal to the product of the nine left-hand members of those equations.

SPOTTISWOODE (1851, 1853).

[Elementary theorems relating to determinants. Second edition, rewritten and much enlarged by the author. *Crelle's Journal*, li. pp. 209–271, 328–381.]

The information given by Spottiswoode regarding axisymmetric determinants appears under a variety of headings. What little the first edition contained (pp. 33–34) as a part of § vi. on "Inverse Systems" is placed in the second edition under "Compound Determinants" (pp. 368–372). Sylvester's mode of reaching Cayley's determinants connected with the mutual distances of points is given under "Multiplication" (pp. 250–53); and the chapter or "section" (§ iv.) on "Homogeneous Functions," which of course has to deal with quadrics, goes so far as to assign the name *determinant of a quadratic form* to any determinant possessed of axisymmetry.* (See pp. 329, 336.)

That the mere number of available coefficients in a quantic may suggest a form of determinant which has no connection with the properties of the said quantic is enforced when in the same chapter (pp. 331–333) the ternary quantic is reached, and it is pointed out that the number of coefficients, $\frac{1}{2}(n+1)(n+2)$, is the same as in the case of the $(n+1)$-ary quadric. Thus the axisymmetric determinant

$$\begin{vmatrix} a & h & g & f' \\ h & b & f & g' \\ g & f & c & h' \\ f' & g' & h' & k \end{vmatrix}$$

which is formed from the coefficients of the ternary cubic

$$ax^3+by^3+cz^3+3(fy^2z+gz^2x+hx^2y+f'yz^2+g'zx^2+h'xy^2)+6kxyz$$

is "foreign to the nature of that function" although intimately associated with the properties of the quaternary quadric

$$ax^2 + by^2 + cz^2 + kw^2 + 2(fyz+gzx+hxy+f'xw+g'yw+h'zw).$$

The most interesting matter, however, is found in the last section of all, § 11, the contents of which are miscellaneous. There, on pages 376–380, the determinants

$$\begin{vmatrix} 1 & . & a_1 & a_2 \\ . & 1 & b_1 & b_2 \\ a_1 & b_1 & 1 & . \\ a_2 & b_2 & . & 1 \end{vmatrix}, \qquad \begin{vmatrix} 1 & . & . & a_1 & a_2 & a_3 \\ . & 1 & . & b_1 & b_2 & b_3 \\ . & . & 1 & c_1 & c_2 & c_3 \\ a_1 & b_1 & c_1 & 1 & . & . \\ a_2 & b_2 & c_2 & . & 1 & . \\ a_3 & b_3 & c_3 & . & . & 1 \end{vmatrix},$$

are considered, but, to one's regret, only with reference to the case where $|a_1b_2c_3|$ is an orthogonant. The first of the two determinants is given in the form

$$1 - a_1^2 - b_1^2 - a_2^2 - b_2^2 + (a_1b_2 - a_2b_1)^2;$$

* With this view in one's mind, the occurrence of particular axisymmetric determinants might suggest the construction of fresh quadratic forms. The form known as a Bezoutiant is actually defined in this way, the axisymmetric determinant then being Bezout's condensed eliminant of two equations of like degree. (See above, p. 109, and below, p. 138.)

and similar non-determinant forms are given for the whole of the thirty-six primary minors and for the first fifteen of the secondary minors of the second determinant. Thus, the primary minors which are the cofactors of the elements in the places (1, 1), (1, 3), (1, 5) are

$$1-b_1^2-b_2^2-b_3^2-c_1^2-c_2^2-c_3^2+(b_2c_3-b_3c_2)^2+(b_3c_1-b_1c_3)^2+(b_1c_2-b_2c_1)^2,$$
$$(a_1c_1+a_2c_2+a_3c_3)(1-b_1^2-b_2^2-b_3^2)+(b_1c_1+b_2c_2+b_3c_3)(a_1b_1+a_2b_2+a_3b_3),$$
$$|a_1b_2c_3|\cdot|b_1c_3|-a_1(1-b_1^2-b_2^2-b_3^2-c_1^2-c_2^2-c_3^2)-b_1(a_1b_1+a_2b_2+a_3b_3)$$
$$-c_1(a_1c_1+a_2c_2+a_3c_3),$$

all the others being similar in form to one or other of these three; and in like manner the secondary minors are exemplified by

$$1-c_1^2-c_2^2-c_3^2,$$
$$-(b_1c_1+b_2c_2+b_3c_3),$$
$$-b_1(1-c_1^2-c_2^2-c_3^2)+c_1(b_1c_1+b_2c_2+b_3c_3),$$
$$a_1(b_1c_1+b_2c_2+b_3c_3)-b_1(c_1a_1+c_2a_2+c_3a_3),$$
$$-c_1|a_1b_2c_3|+|a_2b_3|.$$

CAYLEY, A. (1853*).

[Note sur la méthode d'élimination de Bezout. *Crelle's Journ.*, liii. pp. 366–367; or *Collected Math. Papers*, iv. pp. 38–39.]

The determinant, to which Bezout's so-called 'abridged method' leads, we have already seen dealt with by Jacobi (1835) and Cauchy (1840).† Cayley's noteworthy contribution to it is contained in a single sentence, namely, "Pour éliminer les variables x, y entre deux équations du $n^{\text{ième}}$ degré

$$\left.\begin{array}{l}(a, \ldots \between x, y)^n = 0\\ (a', \ldots \between x, y)^n = 0\end{array}\right\}$$

on n'a qu'à former l'équation identique

$$\frac{(a, \ldots \between x, y)^n\cdot(a', \ldots \between \lambda, \mu)^n-(a', \ldots \between x, y)^n\cdot(a, \ldots \between \lambda, \mu)^n}{\mu x-\lambda y}$$

* The author's date is April 1855; but the rule was published as Cayley's by Sylvester in 1853. See § 62 of the memoir 'On a theory of the syzygetic relations . . .'; and note also that Sylvester there first introduced the closely related idea of the *Bezoutiant*. (See footnote on p. 137 above.)

† See *History*, i. pp. 214, 243, 485–7.

$$= \left(\begin{vmatrix} a_{00} & a_{01} & \cdots & a_{0,\,n-1} \\ a_{10} & a_{11} & \cdots & a_{1,\,n-1} \\ \cdot & \cdot & \cdot & \cdot \\ a_{n-1,\,0} & a_{n-1,\,1} & \cdots & a_{n-1,\,n-1} \end{vmatrix}\!\!\!\left(\!\!\right. x,\, y)^{n-1}(\lambda,\, \mu)^{n-1};$$

le résultat de l'élimination sera

$$\begin{vmatrix} a_{00} & a_{01} & \cdots & a_{0,\,n-1} \\ a_{10} & a_{11} & \cdots & a_{1,\,n-1} \\ \cdot & \cdot & \cdot & \cdot \\ a_{n-1,\,0} & a_{n-1,\,1} & \cdots & a_{n-1,\,n-1} \end{vmatrix} = 0."$$

Instead of paraphrasing this by means of a more familiar notation, as the editor of *Crelle* found it necessary to do, it will be better to apply it to a simple case, namely, where the given equations are

$$\left.\begin{aligned} a_3x^3 + a_2x^2 + a_1x + a_0 &= 0 \\ b_3x^3 + b_2x^2 + b_1x + b_0 &= 0 \end{aligned}\right\},$$

being careful to proceed in such a way as also to justify the 'rule.' Beginning with

$$\begin{vmatrix} a_0+a_1x+a_2x^2+a_3x^3 & a_0+a_1y+a_2y^2+a_3y^3 \\ b_0+b_1x+b_2x^2+b_3x^3 & b_0+b_1y+b_2y^2+b_3y^3 \end{vmatrix} \div (y-x),$$

which vanishes for all values of y, we change it into

$$\left.\begin{matrix} |a_0b_0| + |a_0b_1|y + |a_0b_2|y^2 + |a_0b_3|y^3 \\ + |a_1b_0|x + |a_1b_1|xy + \cdot\ \cdot\ \cdot\ \cdot\ \cdot\ \cdot\ \cdot \\ + \cdot\ \cdot\ \cdot\ \cdot\ \cdot\ \cdot\ \cdot\ \cdot\ \cdot\ \cdot\ \cdot\ \cdot \end{matrix}\right\} \div (y-x),$$

then into

$$\begin{aligned} |a_0b_1| + |a_0b_2|(y+x) &+ |a_0b_3|(y^2+yx+x^2) \\ + |a_1b_2|xy \quad &+ |a_1b_3|xy(y+x) \\ &+ |a_2b_3|x^2y^2, \end{aligned}$$

and finally into

$$\begin{array}{ccc|l} 1 & x & x^2 & \\ \hline |a_0b_1| & |a_0b_2| & |a_0b_3| & 1 \\ |a_0b_2| & |a_0b_3| + |a_1b_2| & |a_1b_3| & y \\ |a_0b_3| & |a_1b_3| & |a_2b_3| & y^2, \end{array}$$

where the elements of the square array are those of Bezout's condensed eliminant, as we have already found in a footnote when dealing with Jacobi's paper of 1835.

BRIOSCHI, F. (1854, March).

[LA TEORICA DEI DETERMINANTI, E LE SUE PRINCIPALI APPLICAZIONI. viii+116 pp. Pavia.]

Unlike Spottiswoode, Brioschi in methodical manner defines "un determinante simmetrico," and gives four known properties expressed in clear language, all within the space of one page (p. 70).

BRIOSCHI, F. (1854, Dec.; 1855, Apr.).

[Sur quelques questions de la géométrie de position. *Crelle's Journal*, l. pp. 233–238; or *Opere mat.*, v. pp. 259–265.]

[Relations de distance entre des points. *Nouv. Annales de Math.*, xiv. pp. 172–173; or *Opere mat.*, v. pp. 123–124.]

The first title here recalls that of Cayley's maiden effort; and, as a matter of fact, the paper of 1854 had its origin in the paper of 1841. Cayley, it will be remembered, obtained the relation connecting the mutual distances of five points in space by multiplying the two determinants

$$\begin{vmatrix} \Sigma x_1^2 & -2x_1 & -2y_1 & -2z_1 & -2w_1 & 1 \\ \Sigma x_2^2 & -2x_2 & -2y_2 & -2z_2 & -2w_2 & 1 \\ \cdot & \cdot & \cdot & \cdot & \cdot & \cdot \\ \Sigma x_5^2 & -2x_5 & -2y_5 & -2z_5 & -2w_5 & 1 \\ 1 & \cdot & \cdot & \cdot & \cdot & \cdot \end{vmatrix}, \quad \begin{vmatrix} 1 & x_1 & y_1 & z_1 & w_1 & \Sigma x_1^2 \\ 1 & x_2 & y_2 & z_2 & w_2 & \Sigma x_2^2 \\ \cdot & \cdot & \cdot & \cdot & \cdot & \cdot \\ 1 & x_5 & y_5 & z_5 & w_5 & \Sigma x_5^2 \\ \cdot & \cdot & \cdot & \cdot & \cdot & 1 \end{vmatrix},$$

the first of which is -16 times the second, and then putting the w's equal to zero. Brioschi now follows on the same lines, but with an interesting difference. Having shown that the determinant

$$\begin{vmatrix} x_1^2+y_1^2+z_1^2 & (x_6-x_1)^2+(y_6-y_1)^2+(z_6-z_1)^2 & 1 & -2x_1 & -2y_1 & -2z_1 \\ x_2^2+y_2^2+z_2^2 & (x_6-x_2)^2+(y_6-y_2)^2+(z_6-z_2)^2 & 1 & -2x_2 & -2y_2 & -2z_2 \\ \cdot & \cdot & \cdot & \cdot & \cdot & \cdot \\ x_5^2+y_5^2+z_5^2 & (x_6-x_5)^2+(y_6-y_5)^2+(z_6-z_5)^2 & 1 & -2x_5 & -2y_5 & -2z_5 \\ 1 & 1 & \cdot & \cdot & \cdot & \cdot \end{vmatrix}$$

vanishes identically, the simple fact being that the second

column is a sum of multiples of the other columns, he multiplies it by $\frac{1}{8}$ of itself, namely by

$$\begin{vmatrix} 1 & \Sigma(x_6-x_1)^2 & \Sigma x_1^2 & x_1 & y_1 & z_1 \\ 1 & \Sigma(x_6-x_2)^2 & \Sigma x_2^2 & x_2 & y_2 & z_2 \\ \cdot & \cdot & \cdot & \cdot & \cdot & \cdot \\ 1 & \Sigma(x_6-x_5)^2 & \Sigma x_5^2 & x_5 & y_5 & z_5 \\ \cdot & 1 & 1 & \cdot & \cdot & \cdot \end{vmatrix}$$

and, putting $d_{61}, \ldots$ for $\Sigma(x_6-x_1)^2, \ldots,$ obtains the relation

$$\begin{vmatrix} d_{61}{}^2 & d_{61}d_{62}+d_{12} & \ldots & d_{61}d_{65}+d_{15} & d_{61}+1 \\ d_{61}d_{62}+d_{12} & d_{62}{}^2 & \ldots & d_{62}d_{65}+d_{25} & d_{62}+1 \\ \cdot & \cdot & \cdot & \cdot & \cdot \\ d_{61}d_{65}+d_{15} & d_{62}d_{65}+d_{25} & \ldots & d_{65}{}^2 & d_{65}+1 \\ d_{61}+1 & d_{62}+1 & \ldots & d_{65}+1 & 1 \end{vmatrix},$$

which degenerates into Cayley's result when we put

$$x_6,\ y_6,\ z_6 = x_1,\ y_1,\ z_1,$$

and make certain easy transformations.

In a similar manner the relation between the distances of five points on an ellipsoid is found, and the relation "entre les plus courtes distances respectives et les inclinaisons mutuelles de sept lignes quelconques."

The second paper contains nothing new.

FERRERS, N. M. (1855, Dec.).

[Two elementary theorems in determinants. *Quarterly Journ. of Math.*, i. p. 364; or *Nouv. Annales de Math.*, xvi. pp. 402–403, xvii. pp. 190–191.]

The first theorem referred to is

$$\begin{vmatrix} 1 & 1 & 1 & \ldots & 1 \\ 1 & 1+a_1 & 1 & \ldots & 1 \\ 1 & 1 & 1+a_2 & \ldots & 1 \\ \cdot & \cdot & \cdot & \cdot & \cdot \\ 1 & 1 & 1 & \ldots & 1+a_n \end{vmatrix} = a_1a_2\ldots a_n,$$

the proof being dependent on the fact that, if any one of the a's be put equal to 0, the determinant vanishes. The second is

$$\begin{vmatrix} 1+a_1 & 1 & 1 & \ldots & 1 \\ 1 & 1+a_2 & 1 & \ldots & 1 \\ 1 & 1 & 1+a_3 & \ldots & 1 \\ \cdot & \cdot & \cdot & \cdot & \cdot \\ 1 & 1 & 1 & \ldots & 1+a_n \end{vmatrix}$$
$$= a_1a_2\ldots a_n\left(1+\frac{1}{a_1}+\frac{1}{a_2}+\ldots+\frac{1}{a_n}\right),$$

which is made to rest mainly on the fact that if any one of the a's be put equal to 0 the determinant takes the form of the preceding determinant. Note is taken that Sylvester's theorem on p. 55 of the same volume is a special case of this second result.

BRUNO, F. FAA DI (1855, Dec.).

[Addizione alla nota inserita nel fascicolo di ottobre ultimo. *Annali di Sci. mat. e fis.*, vi. pp. 476–479; or § vi. of his Théorie générale de l'élimination, x+224 pp., Paris, 1859.]

The note referred to in the title professed to be "Sulle funzioni simmetriche delle radici di un' equazione," and contained, besides other things, the final expansions of the resultants of two quadrics, two cubics, and two quartics. The "addizione," on the other hand, draws attention to the axisymmetric determinants which represent those resultants, the author being apparently unaware that Jacobi had already done this in 1835 and Cauchy in 1840. His rule of formation is the same as Sylvester's of 1853 (June).*

In his "Théorie Générale de l'Élimination" the matter is gone into in greater detail, the rule of formation occupying a full page (pp. 55–56). He there also devotes a section (§ ix. pp. 66, 67) to an account of Jacobi's relations between the elements of Bezout's condensed eliminant.

* See our chapter on Persymmetric Determinants.

BRIOSCHI, F. (1856, Jan.).

[Sur une nouvelle propriété du resultant de deux équations algébriques. *Crelle's Journ.*, liii. pp. 372–376; or *Opere mat.*, v. pp. 277–282.]

Bezout's condensed eliminant of the equations

$$\left.\begin{aligned} a_0x^n + a_1x^{n-1} + \ldots + a_n &= 0 \\ b_0x^n + b_1x^{n-1} + \ldots + b_n &= 0 \end{aligned}\right\}$$

being denoted by $|\beta_{11}\ \beta_{22}\ \ldots\ \beta_{nn}|$, or B say, the new property referred to is

$$\sum_{r=1}^{r=n} \beta_{mr} \frac{\partial \mathrm{B}}{\partial a_r} = -(n-m+1)b_{m-1}\mathrm{B}.$$

The proof given is of no interest in connection with the theory of determinants.

CAYLEY, A. (1856, March).

[Note upon a result of elimination. *Philos. Magazine* (4), xi. pp. 378–379; or *Collected Math. Papers*, iii. pp. 214–215.]

If the quadric

$$ax^2+by^2+cz^2+2fyz+2gzx+2hxy$$

have a linear factor, $\xi x+\eta y+\zeta z$ say, it must vanish identically where we make the substitution

$$x,\ y,\ z = \beta\zeta-\gamma\eta,\quad \gamma\xi-\alpha\zeta,\quad \alpha\eta-\beta\xi,$$

where α, β, γ are any quantities whatever; and consequently the co-factors of $\alpha^2, \beta^2, \ldots$ in the result must vanish,—that is to say, we must have

$$\left.\begin{array}{llllll} & c\eta^2 + b\zeta^2 & - 2f\eta\zeta & & & = 0 \\ c\xi^2 & + a\zeta^2 & & - 2g\zeta\xi & & = 0 \\ b\xi^2 + a\eta^2 & & & & - 2h\xi\eta & = 0 \\ -2f\xi^2 & & - 2a\eta\zeta & + 2h\zeta\xi & + 2g\xi\eta & = 0 \\ & - 2g\eta^2 & + 2h\eta\zeta & - 2b\zeta\xi & + 2f\xi\eta & = 0 \\ & - 2h\zeta^2 & + 2g\eta\zeta & + 2f\zeta\xi & - 2c\xi\eta & = 0 \end{array}\right\}$$

and therefore

$$\begin{vmatrix} \cdot & c & b & -2f & \cdot & \cdot \\ c & \cdot & a & \cdot & -2g & \cdot \\ b & a & \cdot & \cdot & \cdot & -2h \\ -2f & \cdot & \cdot & -2a & 2h & 2g \\ \cdot & -2g & \cdot & 2h & -2b & 2f \\ \cdot & \cdot & -2h & 2g & 2f & -2c \end{vmatrix} = 0,$$

or, say, $8\Delta = 0$.

But on account of the breaking up of the quadric into linear factors the discriminant must likewise vanish. It is thus suggested that the discriminant is a factor of Δ; and by actual trial it is found that

$$\Delta = -2 \begin{vmatrix} a & h & g \\ h & b & f \\ g & f & c \end{vmatrix}^2.$$

It has only in addition to be recalled that Δ originally made its appearance in Sylvester's second paper on dialytic elimination.

BRUNO, F. FAÀ DI (1857, April).

[Sopra il volume della piramide triangolare. *Annali di Sci. mat. e fis.*, viii. pp. 77–78.]

With an eye on Sylvester's paper of October 1852, Faà di Bruno first expresses the volume (V) of a tetrahedron in the form

$$\tfrac{1}{6}\begin{vmatrix} x_1-x_2 & y_1-y_2 & z_1-z_2 \\ x_1-x_3 & y_1-y_3 & z_1-z_3 \\ x_1-x_4 & y_1-y_4 & z_1-z_4 \end{vmatrix},$$

and then by squaring obtains the result

$$288V^2 = \begin{vmatrix} 2d_{12}^2 & d_{12}^2+d_{13}^2-d_{23}^2 & d_{12}^2+d_{14}^2-d_{24}^2 \\ d_{12}^2+d_{13}^2-d_{23}^2 & 2d_{13}^2 & d_{13}^2+d_{14}^2-d_{34}^2 \\ d_{12}^2+d_{14}^2-d_{24}^2 & d_{13}^2+d_{14}^2-d_{34}^2 & 2d_{14}^2 \end{vmatrix}$$

as a consequence of the relation

$$2\sum(x_1-x_2)(x_1-x_3) = \sum(x_1-x_2)^2+\sum(x_1-x_3)^2-\sum(x_2-x_3)^2.$$

Another form of the relation between the mutual distances of four points in a plane is thus brought to light.

RUBINI, R. (1857).

[Applicazione della teorica dei determinanti. *Annali di Sci. mat. e fis.*, viii. pp. 179–200.]

Rubini starts with the theorem which expresses a determinant with binomial elements as a sum of determinants with monomial elements, and then considers a long series of special cases. Among these Ferrers' theorems of the year 1855 occupy the first place (§§ 2, 3, pp. 181–184).

BELLAVITIS, G. (1857, June).

[Sposizione elementare della teorica dei determinanti. *Memorie .... Istituto Veneto* .... vii. pp. 67–144.]

Besides the paragraphs (§§ 42, 43, 44) specifically devoted in the second half of the exposition to axisymmetric determinants, there are two others (§§ 9, 35) connected with the same subject in the first half. One of the latter (§ 35) draws attention to the fact that any coaxial minor of the axisymmetric determinant which is the square of a determinant is expressible as a sum of squares. What is new in the former is the definite reference to determinants which are doubly axisymmetric ("doppiamente simmetrici"), the examples given being *

$$\begin{vmatrix} a & b \\ b & a \end{vmatrix}, \quad \begin{vmatrix} a & b & c \\ b & d & b \\ c & b & a \end{vmatrix}, \quad \begin{vmatrix} a & b & c & d \\ b & a & d & c \\ c & d & a & b \\ d & c & b & a \end{vmatrix},$$

* The third, by reason of its central two-line minor, which might have been $\begin{vmatrix} \alpha & \delta \\ \delta & \alpha \end{vmatrix}$, is more specialised than a doubly axisymmetric determinant.

the last of which is noted as being equal to

$$-(a+b+c-d)(a+b-c+d)(a-b+c+d)(-a+b+c+d),$$

whereas in reality it is equal to

$$(a+b+c+d)(a+b-c-d)(a-b+c-d)(a-b-c+d).$$

BALTZER, R. (1857).

[THEORIE UND ANWENDUNG DER DETERMINANTEN, mit . . . . . vi+129 pp. Leipzig, 1857.]

In five different sections (§ 3, 8, 9; § 5, 2; § 6, 2, 5; § 7, 5; § 18, 12) Baltzer gives attention to determinants whose elements a_{ik}, a_{ki} are identical. In § 3 he notes that conjugate minors are equal, and proves Jacobi's theorem regarding the differential-quotient of a determinant with respect to a non-diagonal element by using the fact that if u be a function of x and y, and y be a function of x, then

$$\frac{du}{dx} = \frac{\partial u}{\partial x} + \frac{\partial u}{\partial y}\cdot\frac{dy}{dx},$$

the whole matter being

$$\frac{\partial\Delta}{\partial a_{1k}} = \mathrm{A}_{ik} + \mathrm{A}_{ki}\frac{\partial a_{ki}}{\partial a_{ik}} = \mathrm{A}_{ik} + \mathrm{A}_{ki} = 2\mathrm{A}_{ik},$$

where it will be observed ∂'s only are employed. In § 7, 5 is given the result which we have already seen in Lebesgue's paper of 1837, namely, that for a vanishing axisymmetric determinant

$$\mathrm{A}_{ik} = \sqrt{\mathrm{A}_{ii}\mathrm{A}_{kk}};$$

and there is thence deduced

$$\mathrm{A}_{i1} : \mathrm{A}_{i2} : \mathrm{A}_{i3} : \ldots\ldots = \sqrt{\mathrm{A}_{11}} : \sqrt{\mathrm{A}_{22}} : \sqrt{\mathrm{A}_{33}} : \ldots\ldots$$

Lastly, in § 18, 12 he applies this to Cayley's vanishing determinants of the year 1841; for example, to the determinant

$$\begin{vmatrix} \cdot & 1 & 1 & 1 & 1 \\ 1 & \cdot & d_{12} & d_{13} & d_{14} \\ 1 & d_{12} & \cdot & d_{23} & d_{24} \\ 1 & d_{13} & d_{23} & \cdot & d_{34} \\ 1 & d_{14} & d_{24} & d_{34} & \cdot \end{vmatrix}.$$

This being equal to zero, if we write $[r, s]$ for the cofactor of the element in the place (r, s), we have of course

$$[12] + [13] + [14] + [15] = 0,$$

and consequently also

$$\sqrt{[22]} + \sqrt{[33]} + \sqrt{[44]} + \sqrt{[55]} = 0,$$

—a result hitherto only obtained independently from geometry.*

BORCHARDT, C. W. (1859, May).

[Ueber eine der Interpolation entsprechende Darstellung der Eliminations-Resultante. *Crelle's Journ.*, lvii. pp. 111–121; or *Monatsb. d. Akad. d. Wiss.* (Berlin), pp. 376–388; or *Gesammelte Werke*, pp. 133–144; also abstract in *Annali di Mat.* . . . , ii. pp. 262–264.]

Following a suggestion of Rosenhain's, Borchardt seeks to obtain an expression for the resultant of two equations of the nth degree $\phi(x) = 0$, $\psi(x) = 0$ in terms of the values† which $\phi(x)$ and $\psi(x)$ assume when x receives any $n+1$ values $a_0, a_1, \ldots a_n$.

---

* Baltzer gives (p. 20) Cayley's determinant form for

$$-(\sqrt{a}+\sqrt{b}+\sqrt{c})\cdot(-\sqrt{a}+\sqrt{b}+\sqrt{c})\cdot(\sqrt{a}-\sqrt{b}+\sqrt{c})\cdot(\sqrt{a}+\sqrt{b}-\sqrt{c}),$$

placing in front of it what looks like a generalisation, namely

$$\begin{vmatrix} \cdot & a_1 & b_1 & c_1 \\ a_1 & \cdot & c_2 & b_2 \\ b_1 & c_2 & \cdot & a_2 \\ c_1 & b_2 & a_2 & \cdot \end{vmatrix},$$

but is not really such. We can easily show that if a_1, b_1, c_1 be multiplied and a_2, b_2, c_2 be divided by x, y, z respectively, the determinant is unaltered; consequently it

$$= \begin{vmatrix} \cdot & a_1a_2 & b_1b_2 & c_1c_2 \\ a_1a_2 & \cdot & 1 & 1 \\ b_1b_2 & 1 & \cdot & 1 \\ c_1c_2 & 1 & 1 & \cdot \end{vmatrix} = \begin{vmatrix} \cdot & 1 & 1 & 1 \\ 1 & \cdot & c_1c_2 & b_1b_2 \\ 1 & c_1c_2 & \cdot & a_1a_2 \\ 1 & b_1b_2 & a_1a_2 & \cdot \end{vmatrix} = \begin{vmatrix} \cdot & \sqrt{a_1a_2} & \sqrt{b_1b_2} & \sqrt{c_1c_2} \\ \sqrt{a_1a_2} & \cdot & \sqrt{c_1c_2} & \sqrt{b_1b_2} \\ \sqrt{b_1b_2} & \sqrt{c_1c_2} & \cdot & \sqrt{a_1a_2} \\ \sqrt{c_1c_2} & \sqrt{b_1b_2} & \sqrt{a_1a_2} & \cdot \end{vmatrix} = \ldots$$

† He does not mean in terms of these alone, but in terms of these and $a_0, a_1, \ldots a_n$.

The basis of his procedure is Cayley's result that

$$\begin{vmatrix} \phi(x) & \phi(y) \\ \psi(x) & \psi(y) \end{vmatrix} \div (y-x) = \begin{array}{c|l} 1 \quad x \quad \ldots \quad x^{n-1} & \\ \hline & 1 \\ & y \\ \mathrm{D} & \vdots \\ & y^{n-1} \end{array}$$

where D is the array of Bezout's condensed eliminant $|\mathrm{D}|$. Denoting the left-hand member of this by $\mathrm{F}(x, y)$ or $\mathrm{F}(y, x)$, we see that

$$\begin{vmatrix} 1 & x_1 & x_1^2 & \ldots & x_1^{n-1} \\ 1 & x_2 & x_2^2 & \ldots & x_2^{n-1} \\ \cdot & \cdot & \cdot & \cdot & \cdot \\ 1 & x_n & x_n^2 & \ldots & x_n^{n-1} \end{vmatrix} \cdot |\mathrm{D}| \cdot \begin{vmatrix} 1 & 1 & \ldots & 1 \\ y_1 & y_2 & \ldots & y^n \\ \cdot & \cdot & \cdot & \cdot \\ y_1^{n-1} & y_2^{n-1} & \ldots & y_n^{n-1} \end{vmatrix} = \begin{vmatrix} \mathrm{F}(x_1, y_1) & \mathrm{F}(x_1, y_2) & \ldots & \mathrm{F}(x_1, y_n) \\ \mathrm{F}(x_2, y_1) & \mathrm{F}(x_2, y_2) & \ldots & \mathrm{F}(x_2, y_n) \\ \cdot & \cdot & \cdot & \cdot \\ \mathrm{F}(x_n, y_1) & \mathrm{F}(x_n, y_2) & \ldots & \mathrm{F}(x_n, y_n) \end{vmatrix}$$

and $\therefore$ $|\mathrm{D}| = |\mathrm{F}(a_1, a_1) \ \mathrm{F}(a_2, a_2) \ \ldots \ \mathrm{F}(a_n, a_n)| \div |a_1^0 \ a_2^1 \ \ldots \ a_n^{n-1}|^2$

which, were it not for the illusory elements $\mathrm{F}(a_r, a_r)$ in the diagonal, might be viewed as constituting a solution of the problem.

To obtain unobjectionable expressions for these, the interpolational forms of $\phi(x)$, $\psi(x)$ are necessarily taken, namely

$$\frac{\phi(a_0)\cdot f(x)}{f'(a_0)\cdot(x-a_0)} + \ldots + \frac{\phi(a_n)\cdot f(x)}{f'(a_n)\cdot(x-a_n)} \quad \text{or} \quad \sum_{i=0}^{i=n} \frac{f(x)/(x-a_i)}{f'(a_i)} \phi(a_i),$$

$$\frac{\psi(a_0)\cdot f(x)}{f'(a_0)\cdot(x-a_0)} + \ldots + \frac{\psi(a_n)\cdot f(x)}{f'(a_n)\cdot(x-a_n)} \quad \text{or} \quad \sum_{i=0}^{i=n} \frac{f(x)/(x-a_i)}{f'(a_i)} \psi(a_i),$$

where $f(x)$ stands for $(x-a_0)(x-a_1)\ldots(x-a_n)$. The resulting expression for $\mathrm{F}(x, y)$,

$$\frac{\sum \frac{f(x)/(x-a_i)}{f'(a_i)}\phi(a_i)\cdot \sum \frac{f(y)/(y-a_i)}{f'(a_i)}\psi(a_i) - \sum \frac{f(y)/(y-a_i)}{f'(a_i)}\phi(a_i)\cdot \sum \frac{f(x)/(x-a_i)}{f'(a_i)}\psi(a_i)}{y-x}$$

is then developed with a view to the actual performance of the division by $y-x$. Each of the two multiplications in the numerator gives $(n+1)^2$ terms; but, as the product of any term of $\phi(x)$ by the corresponding term of $\psi(y)$ is cancelled by the product of the corresponding terms of $\phi(y)$ and $\psi(x)$, the number of terms then remaining is

$$2(n+1)^2 - 2(n+1), \quad i.e. \quad 2n(n+1).$$

Next, taking the product of the $(r+1)^{\text{th}}$ term of $\phi(x)$ by the $(s+1)^{\text{th}}$ of $\psi(y)$ and subtracting the product of the $(r+1)^{\text{th}}$ of $\phi(y)$ by the $(s+1)^{\text{th}}$ of $\psi(x)$, we have

$$\frac{f(x)/(x-a_r)}{f'(a_r)}\phi(a_r)\cdot\frac{f(y)/(y-a_s)}{f'(a_s)}\psi(a_s) - \frac{f(y)/(y-a_r)}{f'(a_r)}\phi(a_r)\cdot\frac{f(x)/(x-a_s)}{f'(a_s)}\psi(a_s),$$

$$i.e. \quad \frac{f(x)\cdot f(y)\cdot\phi(a_r)\cdot\psi(a_s)}{f'(a_r)\cdot f'(a_s)}\left\{\frac{1}{(x-a_r)(y-a_s)} - \frac{1}{(y-a_r)(x-a_s)}\right\},$$

$$i.e. \quad \frac{f(x)\cdot f(y)\cdot\phi(a_r)\cdot\psi(a_s)}{f'(a_r)\cdot f'(a_s)}\cdot\frac{(y-x)(a_r-a_s)}{(x-a_r)(x-a_s)(y-a_r)(y-a_s)},$$

in which $y-x$ is a visible factor. From this, by the mere interchange of r and s, we secure the combination of another pair of terms; and by adding the two results we see that $\mathrm{F}(x, y)$ or $\mathrm{F}(y, x)$ can be expressed as a sum of $\frac{1}{2}n(n+1)$ terms of the form

$$\frac{f(x)\cdot f(y)}{f'(a_r)\cdot f'(a_s)}\left\{\phi(a_r)\cdot\psi(a_s)-\phi(a_s)\cdot\psi(a_r)\right\}\frac{a_r-a_s}{(x-a_r)(x-a_s)(y-a_r)(y-a_s)},$$

the individual terms of the sum being got by giving r, s the values

$$\begin{array}{llll} 0, 1; & 0, 2; & \ldots; & 0, n \\ & 1, 2; & \ldots; & 1, n \\ & \cdot\quad\cdot & \cdot\quad\cdot & \cdot\quad\cdot \\ & & & n-1, n. \end{array}$$

We thus arrive at

$$\mathrm{F}(x, x) = \sum_{r,\, s = 0,1;\, \ldots;\, n-1,n} \frac{-\{f(x)\}^2\cdot\mathrm{F}(a_r, a_s)\cdot(a_r-a_s)^2}{f'(a_r)\cdot f'(a_s)\cdot(x-a_r)^2(x-a_s)^2},$$

and so see that if x be put equal to one of the a's, say a_i, all the terms under Σ which have r and s both different from i must vanish because of the presence of the first factor of the numerator. We have only therefore to consider the terms got by putting

$$r, s = i, 0;\ i, 1;\ \ldots;\ i, i-1;\ i, i+1;\ \ldots;\ i, n,$$

and thus have

$$F(a_i, a_i) = \sum_{r,\,s=i,0;\;i,1;\;\ldots;\;i,i-1}^{r,\,s=i,\,i+1;\;\ldots;\;i,\,n} \frac{-\{f(a_i)\}^2 \cdot F(a_r, a_s)\cdot(a_r-a_s)^2}{f'(a_r)\cdot f'(a_s)\cdot(a_i-a_s)^2(a_i-a_s)^2}$$

$$= -\sum_{s=0,1,\ldots,i-1}^{s=i+1,\ldots,n} \frac{f'(a_i)}{f'(a_s)} F(a_i, a_s).$$

What was desired has thus been fully attained; it happens, however, that because of the peculiar constitution of this expression for $F(a_i, a_i)$, it being such that

$$\frac{F(a_i, a_i)}{f'(a_i)\cdot f'(a_i)} = -\sum_{s=0,1,\ldots,i-1}^{s=i+1,\ldots,n} \frac{F(a_i, a_s)}{f'(a_i)\cdot f'(a_s)},$$

we can proceed a little further and throw $|D|$ into a more elegant form. Thus, dividing the rows of

$$|F(a_1, a_1)\cdot F(a_2, a_2)\cdot \ldots F(a_n, a_n)|$$

by $f'(a_1)$, $f'(a_2)$, ..., $f'(a_n)$ respectively, and thereafter the columns by the same, we have, on putting

$$(rs) \quad \text{for} \quad F(a_r, a_s)/f'(a_r)\cdot f'(a_s),$$

$$|D| = \frac{|(11)\,(22)\ldots(nn)|\cdot\{f'(a_1)\cdot f'(a_2)\ldots f'(a_n)\}^2}{|\overset{0}{a_1}\overset{1}{a_2}\ldots \overset{n-1}{a_n}|^2}$$

$$= |(11)\,(22)\ldots(nn)|\cdot|\overset{0}{a_0}\overset{1}{a_1}\ldots\overset{n}{a_n}|^2,$$

where

$$(ii) = -\Big\{(i0)+(i1)+\ldots+(i,i-1)+(i,i+1)+\ldots+(in)\Big\},$$

that is to say, where each diagonal element (ii) with its sign changed is equal to the sum of all the other elements of its row together with the additional element $(i0)$.

Borchardt's treatment of this peculiar axisymmetric determinant is dealt with elsewhere. (See Chapter XVI. under Unisignants.) Meanwhile a remark incidentally made (p. 114) should be noted, namely, that *Any axisymmetric determinant having the sum of every row equal to zero has all its primary minors equal.*

CAYLEY, A. (1859, June).

[Note on the value of certain determinants, the terms of which are the squared distances of points in a plane or in space. *Quart. Journ. of Math.*, iii. pp. 275–277; or *Collected Math. Papers*, iv. pp. 460–462.]

The determinants referred to are those occurring in his first paper of the year 1841; but the expansions of them which are given do not assume that $\overline{12}=\overline{21}$, etc. Sylvester's related paper of March 1853 is also referred to, the 2W of which is put in the form

$$\begin{vmatrix} . & c & b & f & 1 \\ c & . & a & g & 1 \\ b & a & . & h & 1 \\ f & g & h & . & 1 \\ 1 & 1 & 1 & 1 & . \end{vmatrix},$$

with the result that Q takes the form

$$\begin{aligned} & a^2b^2c^2\{f^4+g^4+h^4+g^2h^2+h^2f^2+f^2g^2+b^2c^2+c^2a^2+a^2b^2-(f^2+g^2+h^2)(a^2+b^2+c^2)\} \\ +\ & a^2g^2h^2\{f^4+b^4+c^4+b^2c^2+c^2f^2+f^2b^2+g^2h^2+h^2a^2+a^2g^2-(f^2+b^2+c^2)(a^2+g^2+h^2)\} \\ +\ & b^2h^2f^2\{g^4+c^4+a^4+c^2a^2+a^2g^2+g^2c^2+h^2f^2+f^2b^2+b^2h^2-(g^2+c^2+a^2)(b^2+h^2+f^2)\} \\ +\ & c^2f^2g^2\{h^4+a^4+b^4+a^2b^2+b^2h^2+h^2a^2+f^2g^2+g^2c^2+c^2f^2-(h^2+a^2+b^2)(c^2+f^2+g^2)\}. \end{aligned}$$

SALMON, G. (1859, July).

[On the relation which connects the mutual distances of five points in space. *Quart. Journ. of Math.*, iii. pp. 282–288.]

Salmon starts from the elimination of a, b, c from the set of trigonometrical equations

$$\left.\begin{aligned} a &= b\cos C + c\cos B \\ b &= c\cos A + a\cos C \\ c &= a\cos B + b\cos A \end{aligned}\right\}$$

and proceeding to similar eliminants of higher order reaches Cayley's results of 1841 and others related to them. The interest of the paper is mainly geometrical. Use is made of the proposition that *if the quadric*

$$ax^2 + by^2 + cz^2 + dw^2$$
$$+ 2lyz + 2mzx + 2nxy + 2pxw + 2qyw + 2rzw$$

be increased by $(\alpha x+\beta y+\gamma z+\delta w)^2$, *the discriminant*

$$\begin{vmatrix} a & n & m & p \\ n & b & l & q \\ m & l & c & r \\ p & q & r & d \end{vmatrix}$$

is increased by *

$$-\begin{vmatrix} \cdot & \alpha & \beta & \gamma & \delta \\ \alpha & a & n & m & p \\ \beta & n & b & l & q \\ \gamma & m & l & c & r \\ \delta & p & q & r & d \end{vmatrix}.$$

SALMON, G. (1859).

[LESSONS INTRODUCTORY TO THE MODERN HIGHER ALGEBRA. .... xii+147 pp. Dublin.]

The evectant (or first evectant) of an invariant I of a quantic $(a, b, c, \ldots \between x, y, z, \ldots)^n$ being

$$\left(\frac{\partial \mathrm{I}}{\partial a}, \frac{\partial \mathrm{I}}{\partial b}, \frac{\partial \mathrm{I}}{\partial c}, \ldots \between \xi, \eta, \zeta, \ldots\right)^n,$$

* The new discriminant being

$$\begin{vmatrix} a+\alpha^2 & n+\alpha\beta & m+\alpha\gamma & p+\alpha\delta \\ n+\alpha\beta & b+\beta^2 & l+\beta\gamma & q+\beta\delta \\ m+\alpha\gamma & l+\beta\gamma & c+\gamma^2 & r+\gamma\delta \\ p+\alpha\delta & q+\beta\delta & r+\gamma\delta & d+\delta^2 \end{vmatrix}$$

is easily seen to be equal to

$$-\begin{vmatrix} -1 & \alpha & \beta & \gamma & \delta \\ \alpha & a & n & m & p \\ \beta & n & b & l & q \\ \gamma & m & l & c & r \\ \delta & p & q & r & d \end{vmatrix},$$

and therefore to be separable into the two expressions referred to.

it is readily seen that, in the special case where $n=2$ and where therefore I is the discriminant of the quantic, the evectant is expressible as an axisymmetric determinant, namely, that obtained by 'bordering' the discriminant by the contragredient variables; for example, taking the ternary quadric

$$\begin{array}{ccc|c} x & y & z & \\ \hline a & h & g & x \\ h & b & f & y \\ g & f & c & z \end{array}$$

we have for the evectant of its discriminant

$$-\begin{vmatrix} \cdot & \xi & \eta & \zeta \\ \xi & a & h & g \\ \eta & h & b & f \\ \zeta & g & f & c \end{vmatrix}.$$

In recalling this fact at the beginning of his Fifteenth Lesson and a further fact regarding the form of the evectant of a discriminant which vanishes, Salmon digresses for a moment to put on record (§ 155) an implicated property of axisymmetric determinants, namely, that *if the cofactor of the element in the place* (1, 1) *of an axisymmetric determinant vanishes, the determinant is expressible as 'a perfect square,'* his meaning being, *the square of a linear function of the elements of the first row.*

It is not uninteresting to observe here that this unpretentious but important manual of Salmon's and the very useful *Théorie Générale de l'Élimination* of Faà di Bruno above and elsewhere referred to appeared almost simultaneously in the year 1859, and that the one was dedicated to Cayley and Sylvester and the other to Cayley.

CHAPTER IV.

ALTERNANTS, FROM 1832 TO 1860.

NOTE ought to be taken of the fact that the first three papers here to be dealt with, namely, the papers by Murphy (1832), Binet (1837), and Haedenkamp (1841) belong to the previous period. Nothing, however, appearing in the former chapter on the subject requires modification on that account.

Further, as exaggerated statements regarding Vandermonde's contribution to the subject have been widely accepted, it seems desirable to point out the exact foundation on which such statements rest. In a paper* read in November 1770 Vandermonde says (p. 369), "Or $a^2b+b^2c+c^2a-a^2c-b^2a-c^2b$, qui égale $(a-b)(a-c)(b-c)$ a pour carré $a^4b^2+\ldots.$" This is the whole matter.†

MURPHY, R. (1832, Nov.).

[On elimination between an indefinite number of unknown quantities. *Transac. Cambridge Philos. Soc.*, v. pp. 65–76.]

Murphy's third example in illustration of his method is the set of equations

$$\left.\begin{array}{l} 1+\ x_1+\ \ x_2+\ldots\ldots+\ \ x_n=0 \\ 1+2x_1+2^2x_2+\ldots\ldots+2^nx_n=0 \\ 1+3x_1+3^2x_2+\ldots\ldots+3^nx_n=0 \\ \cdot\quad\cdot\quad\cdot\quad\cdot\quad\cdot\quad\cdot\quad\cdot\quad\cdot\quad\cdot\quad\cdot\quad\cdot\quad\cdot \\ 1+nx_1+n^2x_2+\ldots\ldots+n^nx_n=0 \end{array}\right\}$$

* VANDERMONDE, N. Mémoire sur la résolution des équations. *Mém. de l'Acad. des Sci.* (Paris), Année 1771, pp. 365–416.

† See *Nouv. Annales de Math.*, xix. p. 181 footnote.

which he neatly and easily solves, giving the value of x_m, and thus in effect evaluating

$$(-)^m \begin{vmatrix} 1 & 1 & 1 & \dots & 1 & 1 & \dots & 1 \\ 1 & 2 & 2^2 & \dots & 2^{m-1} & 2^{m+1} & \dots & 2^n \\ 1 & 3 & 3^2 & \dots & 3^{m-1} & 3^{m+1} & \dots & 3^n \\ \cdot & \cdot & \cdot & \cdot & \cdot & \cdot & \cdot & \cdot \\ 1 & n & n^2 & \dots & n^{m-1} & n^{m+1} & \dots & n^n \end{vmatrix} \div \begin{vmatrix} 1 & 1 & \dots & 1 \\ 2 & 2^2 & \dots & 2^n \\ 3 & 3^2 & \dots & 3^n \\ \cdot & \cdot & \cdot & \cdot \\ n & n^2 & \dots & n^n \end{vmatrix}.$$

His connection with our subject is thus seen to be similar to Prony's.

It should be carefully noted, however, in passing, that Prony's set of equations is not the same as Murphy's, the determinant of the one being conjugate to that of the other.* When the use of determinants is debarred or avoided, this difference is far from unimportant,—a fact which might readily be surmised from the present instance, since Murphy's mode of procedure, though strikingly effective upon his own set, is quite inapplicable to Prony's. It should also be observed that the solution of Murphy's set is not essentially different from the solution of the familiar interpolation-problem *to determine* $a_1, a_2, \dots, a_n$, *so that* $a_1+a_2x+a_3x^2+\dots+a_nx^{n-1}$ *or* y *may have the values* $y_1, y_2, \dots, y_n$ *when* x *has the values* $x_1, x_2, \dots, x_n$ *respectively*,—a problem which had been solved in one way by Newton (1687), in another way by Lagrange (1795), and in a third way to a certain extent by Cauchy (1812).†

* The two sets of equations are

$$a_r^1x_1 + a_r^2x_2 + \dots + a_r^nx_n = u_r\Big\}_{r=1}^{r=n} \qquad \text{(I)}$$

and

$$a_1^rx_1 + a_2^rx_2 + \dots + a_n^rx_n = u_r\Big\}_{r=1}^{r=n} \qquad \text{(J)}$$

The former is substantially the set of the interpolation-problem which goes back to Newton, and which may therefore for distinction's sake be associated with his name: the latter being first found solved by Lagrange (Recherches sur les suites récurrentes . . . *Mém. de l'acad. de Berlin*, 1775, pp. 183–272; 1792, pp. 247–257: or *Œuvres complètes*, iv. pp. 149–251; v. pp. 625–641) may be called Lagrange's set, provided we remember that he also gave a solution of the other. The first to deal with both of them in more or less general form by means of determinants was Cauchy (1812)—see also his *Resumés Analytiques*, p. 19, . . . 4° Turin, 1834—but in saying so a mental reservation must be made in view of Cramer's mode (1750) of continuing Newton's work.

† NEWTON, *Principia*, lib. iii. lemma v.: also *Arithmetica Universalis*, probl. lxi. LAGRANGE, *Journ. de l'éc. polyt.*, ii. cah. 8, 9, pp. 276, 277; or *Œuvres complètes*, vii. pp. 285, 286. CAUCHY, *Journ. de l'éc. polyt.*, x. cah. 17, pp. 73–74; or *Œuvres complètes*, 2e sér. i. pp. 133–134.

BINET, [J. P. M.] (1837).

[Observations sur des théorèmes de Géométrie, énoncées page 160 de ce volume et page 222 du volume précédent. *Journ.* (*de Liouville*) *de Math.*, ii. pp. 248-252; or, in abstract, *Nouv. Annales de Math.*, v. pp. 164, 165.]

The main object of this short paper of Binet's was to draw attention to the fact that a theorem regarding homofocal surfaces which Lamé had just published was originally given by Binet in 1811. He thus has occasion to say that the form under which he had considered the equation of homofocal surfaces was

$$\frac{a^2}{K-A} + \frac{b^2}{K-B} + \frac{c^2}{K-C} = 1,$$

where a, b, c are the co-ordinates of any point on the surface, A, B, C are positive constants such that $A>B>C$, and K is a quantity which may be of any magnitude greater than C. And as Lamé had obtained expressions for the co-ordinates in terms of three values given to K, Binet intimates that many years before he had not only done the same but had extended the solution to the case of n equations. It is this purely algebraical problem which is of interest to us, and fortunately Binet gives it in full.

Taking the set of equations in the form

$$\begin{aligned} \frac{a}{K-A} + \frac{b}{K-B} + \frac{c}{K-C} + \dots\dots &= 1, \\ \frac{a}{K_1-A} + \frac{b}{K_1-B} + \frac{c}{K_1-C} + \dots\dots &= 1, \\ \frac{a}{K_2-A} + \frac{b}{K_2-B} + \frac{c}{K_2-C} + \dots\dots &= 1, \\ \dots\dots\dots\dots\dots\dots \end{aligned}$$

he introduces, for temporary purposes, two functions, $F(x)$, $f(x)$, the former being

$$(x-A)(x-B)(x-C)\dots\dots$$

and therefore of the nth degree, and the latter being any integral function of a degree less than n. He then recalls the fact that

$f(x) \div F(x)$ can be partitioned into n fractions having $x-A$, $x-B$, $x-C$, . . . for denominators, the result as given by Euler being

$$\frac{f(x)}{F(x)} = \frac{f(A)}{(x-A)f'(A)} + \frac{f(B)}{(x-B)f'(B)} + \frac{f(C)}{(x-C)f'(C)} + \ldots.$$

Substituting successively K, K_1, K_2, . . . for x in this, a set of equations is obtained from which it is seen that the solution of the set

$$\frac{a}{K-A} + \frac{b}{K-B} + \frac{c}{K-C} + \ldots. = \frac{f(K)}{F(K)},$$

$$\frac{a}{K_1-A} + \frac{b}{K_1-B} + \frac{c}{K_1-C} + \ldots. = \frac{f(K_1)}{F(K_1)},$$

$$\frac{a}{K_2-A} + \frac{b}{K_2-B} + \frac{c}{K_2-C} + \ldots. = \frac{f(K_2)}{F(K_2)},$$

. . . .

is

$$a = \frac{f(A)}{F'(A)} = \frac{f(A)}{(A-B)(A-C)(A-D)\ldots.},$$

$$b = \frac{f(B)}{F'(B)} = \frac{f(B)}{(B-A)(B-C)(B-D)\ldots.},$$

. . . .

Now the set of equations here solved is more general than that with which we started, the latter being the particular case of the former where

$$\frac{f(K)}{F(K)} = \frac{f(K_1)}{F(K_1)} = \ldots. = 1.$$

To effect this specialisation it is only necessary to make the arbitrary function $f(x)$ equal to

$$(x-A)(x-B)(x-C)\ldots. - (x-K)(x-K_1)(x-K_2)\ldots.$$

or equal to

$$F(x) - \mathrm{f}(x) \text{ say};$$

(where, be it observed, the condition as to the degree of $f(x)$ is fulfilled); for then, since $\mathrm{f}(x)$ vanishes when $x = K, K_1, K_2, \ldots.$ we have

$$f(K) = F(K), \quad f(K_1) = F(K_1), \quad f(K_2) = F(K_2), \quad \ldots.$$

The corresponding change in the values of the unknowns is easily made: for example, in the case of a, we have only to substitute F(A)−f(A),—or, what is the same thing, −f(A),—for f(A) in the numerator, the result being

$$a = -\frac{(A-K)(A-K_1)(A-K_2)\ldots\ldots}{(A-B)(A-C)\ldots\ldots}$$

We have thus as Binet's theorem:—

The solution of the set of equations

$$\left.\begin{array}{l} \dfrac{x_1}{b_1-\beta_1}+\dfrac{x_2}{b_1-\beta_2}+\dfrac{x_3}{b_1-\beta_3}+\ldots\ldots+\dfrac{x_n}{b_1-\beta_n}=1 \\ \dfrac{x_1}{b_2-\beta_1}+\dfrac{x_2}{b_2-\beta_2}+\dfrac{x_3}{b_2-\beta_3}+\ldots\ldots+\dfrac{x_n}{b_2-\beta_n}=1 \\ \cdot\quad\cdot\quad\cdot\quad\cdot\quad\cdot\quad\cdot\quad\cdot\quad\cdot\quad\cdot\quad\cdot\quad\cdot\quad\cdot \\ \dfrac{x_1}{b_n-\beta_1}+\dfrac{x_2}{b_n-\beta_2}+\dfrac{x_3}{b_n-\beta_3}+\ldots\ldots+\dfrac{x_n}{b_n-\beta_n}=1 \end{array}\right\}$$

is

$$x_1 = -\frac{(\beta_1-b_1)(\beta_1-b_2)\ldots\ldots(\beta_1-b_n)}{(\beta_1-\beta_2)\ldots\ldots(\beta_1-\beta_n)},$$

$$x_2 = -\frac{(\beta_2-b_1)(\beta_2-b_2)\ldots\ldots(\beta_2-b_n)}{(\beta_2-\beta_1)\ldots\ldots(\beta_2-\beta_n)},$$

$$\cdot\quad\cdot\quad\cdot\quad\cdot\quad\cdot\quad\cdot\quad\cdot\quad\cdot\quad\cdot\quad\cdot\quad\cdot\quad\cdot$$

the binomial factors of the numerator in the case of x_r *being got by subtracting from* β_r *all the* b's *in succession, and the similar factors of the denominator by subtracting from* β_r *all the other* β's.

Remembering that Binet had originally been an expert in working with determinants, it is not a little curious to note that he did not compare with these expressions for $x_1, x_2, x_3, \ldots\ldots$ the expressions in terms of determinants, viz.—

$$x_1 = \begin{vmatrix} 1 & (b_1-\beta_2)^{-1} & \ldots & (b_1-\beta_n)^{-1} \\ 1 & (b_2-\beta_2)^{-1} & \ldots & (b_2-\beta_n)^{-1} \\ \cdot & \cdot & \cdot & \cdot \\ 1 & (b_n-\beta_2)^{-1} & \ldots & (b_n-\beta_n)^{-1} \end{vmatrix} \div \begin{vmatrix} (b_1-\beta_1)^{-1} & (b_1-\beta_2)^{-1} & \ldots & (b_1-\beta_n)^{-1} \\ (b_2-\beta_1)^{-1} & (b_2-\beta_2)^{-1} & \ldots & (b_2-\beta_n)^{-1} \\ \cdot & \cdot & \cdot & \cdot \\ (b_n-\beta_1)^{-1} & (b_n-\beta_2)^{-1} & \ldots & (b_n-\beta_n)^{-1} \end{vmatrix},$$

$$x_2 = \cdot\quad\cdot\quad\cdot\quad\cdot\quad\cdot\quad\cdot\quad\cdot\quad\cdot\quad\cdot\quad\cdot\quad\cdot\quad\cdot$$

Had he done so he would undoubtedly have reached a result which was not brought to light until four years later by Cauchy.

HAEDENKAMP, H. (1841).

[Ueber Transformation vielfacher Integrale. *Crelle's Journ.*, xxii. pp. 184–192.*]

The transformation referred to in the title has its origin in a special equation of the nth degree in y, viz.—

$$\frac{x_1}{a_1 - y} + \frac{x_2}{a_2 - y} + \dots + \frac{x_n}{a_n - y} = 1;$$

and, as Haedenkamp gives the values of $x_1, x_2, \dots, x_n$ in terms of the n roots $y_1, y_2, \dots, y_n$ of this equation he may of course be viewed as having solved the set of linear equations—

$$\left.\begin{array}{l} \dfrac{x_1}{a_1 - y_1} + \dfrac{x_2}{a_2 - y_1} + \dots + \dfrac{x_n}{a_n - y_1} = 1, \\ \dfrac{x_1}{a_1 - y_2} + \dfrac{x_2}{a_2 - y_2} + \dots + \dfrac{x_n}{a_n - y_2} = 1, \\ \dots\dots\dots\dots\dots\dots \end{array}\right\}$$

which Binet had explicitly dealt with four years before.

BORCHARDT, C. W. (1845, Jan.).

[Neue Eigenschaft der Gleichung, mit deren Hülfe man die seculären Störungen der Planeten bestimmt. *Crelle's Journ.*, xxx. pp. 38–45; or, in extended form, *Journ. (de Liouville) de Math.*, xii. pp. 50–67; or *Gesammelte Werke*, pp. 3–13.]

For the present this paper is only noteworthy as containing the square of the difference-product in the form of a determinant of the particular type soon after to be named "persymmetric" by Sylvester. M being used to stand for

$$\left\{\sum \pm g_1^0 g_2^1 g_3^2 \dots g_n^{n-1}\right\}^2$$

* See also *Crelle's Journ.*, xxv. pp. 178–183 (1842), and *Grunert's Archiv d. Math. u. Phys.*, xxiii. pp. 235, 236 (1854).

Borchardt says "Es wird also M die Determinante aus dem System

$$\begin{matrix} s_0 & s_1 & s_2 & \dots & s_{n-1} \\ s_1 & s_2 & s_3 & \dots & s_n \\ s_2 & s_3 & s_4 & \dots & s_{n+1} \\ \cdot & \cdot & \cdot & \cdot & \cdot \\ s_{n-1} & s_n & s_{n+1} & \dots & s_{2n-2} \end{matrix}$$

wo

$$s_m = g_1^m + g_2^m + \dots + g_n^m."$$

ROSENHAIN, G. (1845, Sept.).

[Neue Darstellung der Resultante der Elimination von z aus zwei algebraische Gleichungen $f(z)=0$ und $\phi(z)=0$, ... *Crelle's Journ.*, xxx. pp. 157–165.]

Although the subject of alternating functions is incidentally dealt with in Rosenhain's paper (p. 161), nothing of importance occurs. The identity *

$$\zeta^{\frac{1}{2}}(a_1, a_2, \dots, a_n) = \zeta^{\frac{1}{2}}(a_1, a_2, \dots, a_m) \cdot \zeta^{\frac{1}{2}}(a_{m+1}, a_{m+2}, \dots, a_n) . \prod_{s=1,2,\dots,m}^{r=m+1,\dots,n}(a_r - a_s)$$

appears in the form

$$\Pi(a_1, a_2, \dots, a_n) = \Pi(a_1, a_2, \dots, a_m) \cdot \Pi(a_{m+1}, a_{m+2}, \dots, a_n) \cdot M_{1,2,\dots,m},$$

where the mode of denoting the rectangular array of differences cannot be commended.

---

* A still better form for the right-hand member is

$$\zeta^{\frac{1}{2}}(a_1, a_2, \dots, a_m) \cdot \prod_{s=1,2,\dots,m}^{r=m+1,\dots,n}(a_r - a_s) \cdot \zeta^{\frac{1}{2}}(a_{m+1}, a_{m+2}, \dots, a_n)$$

where the suffixes are seen to run twice from 1 to n. Another identity, just as worthy of note, is

$$\zeta^{\frac{1}{2}}(a_1, a_2, \dots, a_n) = \zeta^{\frac{1}{2}}(a_1, a_2, \dots, a_m) \cdot \prod_{s=1,2,\dots,m-1}^{r=m+1,\dots,n}(a_r - a_s) \cdot \zeta^{\frac{1}{2}}(a_m, a_{m+1}, \dots, a_n).$$

The one is exemplified by the partition

$$\begin{array}{cc|ccc} a_2 - a_1 & a_3 - a_1 & a_4 - a_1 & a_5 - a_1 & a_6 - a_1 \\ & a_3 - a_2 & a_4 - a_2 & a_5 - a_2 & a_6 - a_2 \\ & & a_4 - a_3 & a_5 - a_3 & a_6 - a_3 \\ \hline & & & a_5 - a_4 & a_6 - a_4 \\ & & & & a_6 - a_5, \end{array}$$

the other when instead of this the right-to-left dotted line is made to separate the third row of differences from the second. The former is that to which we have drawn attention when dealing with Jacobi's memoir of 1841.

What the title refers to is a new variant of a form of the resultant given by Euler in 1748, namely, the variant

$$\zeta^{\frac{1}{2}}(\beta_1, \beta_2, \ldots, \beta_n, a_1, a_2, \ldots, a_n) \div \zeta^{\frac{1}{2}}(\beta_1, \beta_2, \ldots, \beta_n) \cdot \zeta^{\frac{1}{2}}(a_1, a_2, \ldots, a_n),$$

where the a's are the roots of $f(z)=0$ and the β's the roots of $\phi(z)=0$.

STURM, R. (1845); TERQUEM, O. (1846).

[Cours d'analyse de l'École Polytechnique. 4to, lithogr., Paris.*]

[Sur la résolution d'une certaine classe d'équations à plusieurs inconnues du premier degré. *Nouv. Annales de Math.*, v. pp. 67–68, 162–165.]

Employing the method of "undetermined multipliers" Sturm here supplies the want left by Prony, namely the solution of

$$a^r x_1 + a_2^r x_2 + \ldots + a_n^r x_n = b^r \Big\}_{r=0}^{r=n-1}.$$

The said method may be generally described as making the solution of a set of n equations dependent on the solution of a set of $n-1$ equations, the latter set being related to the former in having its determinant conjugate to a primary minor of the determinant of the other set. Thus the given set being

$$\left.\begin{array}{l} a_1x_1 + a_2x_2 + a_3x_3 + a_4x_4 = a_5 \\ b_1x_1 + b_2x_2 + b_3x_3 + b_4x_4 = b_5 \\ c_1x_1 + c_2x_2 + c_3x_3 + c_4x_4 = c_5 \\ d_1x_1 + d_2x_2 + d_3x_3 + d_4x_4 = d_5 \end{array}\right\}$$

we conclude therefrom that the equation

$$(a_1+\lambda b_1+\mu c_1+\nu d_1)x_1 + (a_2+\lambda b_2+\mu c_2+\nu d_2)x_2 + \ldots\ldots = a_5 + \lambda b_5 + \mu c_5 + \nu d_5$$

holds for all values of λ, μ, ν; and in order to obtain the value of x_1, we have to solve the set

$$\left.\begin{array}{l} a_2 + b_2\lambda + c_2\mu + d_2\nu = 0 \\ a_3 + b_3\lambda + c_3\mu + d_3\nu = 0 \\ a_4 + b_4\lambda + c_4\mu + d_4\nu = 0 \end{array}\right\},$$

* Not the posthumous book with this title edited by Prouhet and published in 1857.

where the determinant of the coefficients of the unknowns is the conjugate of the complementary minor of a_1 in $|a_1b_2c_3d_4|$. With this fact in view, and along with it the nature of the relation of Murphy's set to Prony's, it will be readily seen that both sets appear in Sturm's procedure.

Terquem follows Sturm, and extends his method to the set of n equations

$$\left.\begin{array}{l} x_1 + 0.x_2 \;\; + x_3 \;\; + \ldots + x_n \;\; = b_0 \\ a_1x_1 + 1.x_2 \;\; + a_3x_3 + \ldots + a_nx_n = b_1 \\ a_1^2x_1 + 2\,a_1x_2 + a_3^2x_3 + \ldots + a_n^2x_n = b_2 \\ a_1^3x_1 + 3\,a_1x_2 + a_3^3x_3 + \ldots + a_n^3x_n = b_3 \\ \cdot\;\cdot\;\cdot\;\cdot\;\cdot\;\cdot\;\cdot\;\cdot\;\cdot\;\cdot\;\cdot\;\cdot\;\cdot\;\cdot\;\cdot\;\cdot \end{array}\right\}$$

where the coefficients of x_2 are the differential-quotients of the corresponding coefficients of x_1. The possibility of this solution rests on selecting x_2 as the first unknown to be determined, and on the set being thus reducible to one of the previous type.

CAYLEY, A. (1846, Aug.).

[Note sur les fonctions de M. Sturm. *Journ. (de Liouville) de Math.*, xi. pp. 297–299; *Collected Math. Papers*, i. pp. 306–308.]

The functions referred to, which are really Sylvester's substitutes * for Sturm's functions, are introduced in the form—

$$\begin{array}{l} f(x) = (x-a_1)(x-a_2)(x-a_3)\ \ldots\ (x-a_n) \\ f_1(x) = \Sigma\,(x-a_2)(x-a_3)(x-a_4)\ \ldots . \\ f_2(x) = \Sigma\,(a_1-a_2)^2.\,(x-a_3)(x-a_4)\ \ldots . \\ f_3(x) = \Sigma\,(a_1-a_2)^2(a_2-a_3)^2(a_3-a_1)^2.\,(x-a_4)\ \ldots . \\ \cdot\;\cdot \\ f_m(x) = f(x).\sum\dfrac{\mathrm{P}^2}{(x-a_1)(x-a_2)\ \ldots\ (x-a_m)}, \end{array}$$

---

* SYLVESTER. On rational derivation from equations of existence, ... *Philos. Mag.*, xv. (1839), pp. 428–435: *Collected Math. Papers*, i. pp. 40–46.

STURM. Démonstration d'un théorème d'algèbre de M. Sylvester. *Journ. (de Liouville) de Math.*, vii. (1842), pp. 356–368.

where P stands for the difference-product of $a_1, a_2, \ldots, a_m$, or for what Sylvester afterwards denoted by $\zeta^{\frac{1}{2}}(a_1, a_2, \ldots, a_m)$: and the problem is professedly to express $f_m(x)$ "par les coefficients de $f(x)$," but in reality to express it as a series arranged according to descending powers of x.

This is accomplished by partitioning $\mathrm{P}/(x-a_1)(x-a_2)\ldots(x-a_m)$ into an aggregate of fractions having $x-a_1, x-a_2, \ldots$ for denominators,* namely,

$$(-)^{m-1}\frac{\zeta^{\frac{1}{2}}(a_1, a_2, \ldots, a_m)}{(x-a_1)(x-a_2)\ldots(x-a_m)} = \frac{\zeta^{\frac{1}{2}}(a_2, a_3, \ldots, a_m)}{x-a_1} - \frac{\zeta^{\frac{1}{2}}(a_1, a_3, \ldots, a_m)}{x-a_2} + \ldots\ldots,$$

so that the coefficient of x^{-r} is seen to be

$$a_1^{r-1}\zeta^{\frac{1}{2}}(a_2, a_3, \ldots, a_m) - a_2^{r-1}\zeta^{\frac{1}{2}}(a_1, a_3, \ldots, a_m) + \ldots\ldots$$

and therefore to be

$$(-)^{m-1}\begin{vmatrix} 1 & a_1 & a_1^2 & \ldots & a_1^{m-2} & a_1^{r-1} \\ 1 & a_2 & a_2^2 & \ldots & a_2^{m-2} & a_2^{r-1} \\ \cdot & \cdot & \cdot & \cdot & \cdot & \cdot \\ 1 & a_m & a_m^2 & \ldots & a_m^{m-2} & a_m^{r-1} \end{vmatrix},$$

where $a_1, a_2, \ldots, a_m$ are the first m a's chosen from $a_1, a_2, \ldots, a_n$.

Multiplying both sides by P and performing the requisite summation we find that the coefficient of x^{-r} in $f_m(x) \div f(x)$ is

$$\begin{vmatrix} s_0 & s_1 & \ldots & s_{m-2} & s_{r-1} \\ s_1 & s_2 & \ldots & s_{m-1} & s_r \\ \cdot & \cdot & \cdot & \cdot & \cdot \\ s_{m-1} & s_m & \ldots & s_{2m-3} & s_{r+m-2} \end{vmatrix}, \quad \text{or } \mathrm{V}_{r-1} \text{ say,}$$

where s_q is the sum of the qth powers of *all* the a's; in other words, that

$$f_m(x) \div f(x) = x^{-m}\cdot\mathrm{V}_{m-1} + x^{-m-1}\cdot\mathrm{V}_m + \ldots\ldots$$

---

*It may be noted in this connection that

$$\zeta^{\frac{1}{2}}(a_1, a_2, \ldots, a_n) \div \phi'(a_k) = (-)^{n-k}\zeta^{\frac{1}{2}}(a_1, a_2, \ldots, a_{k-1}, a_{k+1}, \ldots, a_n)$$

if $\phi(x) = (x-a_1)(x-a_2)\ldots(x-a_n)$.

It only remains now to multiply by $f(x)$ in the form

$$x^n - p_1 x^{n-1} + p_2 x^{n-2} - \ldots\ldots$$

obtaining

$$\begin{aligned} f_m(x) = x^{n-m}\cdot \mathrm{V}_{m-1} &+ x^{n-m-1}(\mathrm{V}_m - p_1\mathrm{V}_{m-1}) \\ &+ x^{n-m-2}(\mathrm{V}_{m+1} - p_1\mathrm{V}_m + p_2\mathrm{V}_{m-1}) \\ &+ \ldots\ldots\ldots\ldots\ldots \end{aligned}$$

and then to condense the coefficients,—an easy operation, since all the V's are identical save in their last columns: for example

$$\mathrm{V}_{m+1} - p_1\mathrm{V}_m + p_2\mathrm{V}_{m-1} = \begin{vmatrix} s_0 & s_1 & \ldots & s_{m-2} & s_{m+1} - p_1 s_m + p_2 s_{m-1} \\ s_1 & s_2 & \ldots & s_{m-1} & s_{m+2} - p_1 s_{m+1} + p_2 s_m \\ \cdot & \cdot & \cdot & \cdot & \cdot \\ s_{m-1} & s_m & \ldots & s_{2m-3} & s_{2m} - p_1 s_{2m-1} + p_2 s_{2m-2} \end{vmatrix}.$$

CHELINI, D. (1846); LIOUVILLE, J. (1846).

[Determinazione geometrica in coordinate ellittiche .... *Raccolta sci. di Palomba,* ii. pp. 109–113, 126–131; see also v. pp. 227–263, 333–374.]

[Sur une classe d'équations du premier degré. *Journ.* (*de Liouville*) *de Math.*, xi. pp. 466–467; or *Nouv. Annales de Math.*, vi. pp. 129–131; or *Archiv d. Math. u. Phys.*, xxii. pp. 226–228.]

The set of equations referred to is that dealt with by Binet in 1837. Chelini and Liouville arrived at a new solution, much simpler than Binet's, and related to that used by Murphy in 1832 in solving other sets of linear equations.

GRUNERT, J. A. (1847).

[Vollständige independente Auflösung der n Gleichungen der ersten Grades ... *Archiv d. Math. u. Phys.*, x. pp. 284–302.]

The equations are

$$\mathrm{A}_1 + \mathrm{A}_2 a_r + \mathrm{A}_3 a_r^2 + \ldots + \mathrm{A}_n a_r^{n-1} = a_r \Big\}_{r=1}^{r=n},$$

that is to say, are of the type to which Murphy's belong, and with which a problem in interpolation is connected; and the solution, rather tardily reached (p. 301), is

$$
\begin{aligned}
(-1)^m \mathrm{A}_{n-m} = & \frac{\overset{m}{\mathrm{K}}(a_2, a_3, a_4, \ldots, a_n)}{(a_1-a_2)(a_1-a_3)(a_1-a_4)\ldots(a_1-a_n)} a_1 \\
& + \frac{\overset{m}{\mathrm{K}}(a_1, a_3, a_4, \ldots, a_n)}{(a_2-a_1)(a_2-a_3)(a_2-a_4)\ldots(a_2-a_n)} a_2 \\
& + \frac{\overset{m}{\mathrm{K}}(a_1, a_2, a_4, \ldots, a_n)}{(a_3-a_1)(a_3-a_2)(a_3-a_4)\ldots(a_3-a_n)} a_3 \\
& + \ldots\ldots\ldots\ldots \\
& + \frac{\overset{m}{\mathrm{K}}(a_1, a_2, a_3, \ldots, a_{n-1})}{(a_n-a_1)(a_n-a_2)(a_n-a_3)\ldots,(a_n-a_{n-1})} a_n ,
\end{aligned}
$$

where by $\overset{m}{\mathrm{K}}$ is denoted "die *m*te Klasse der Kombinationen ohne Wiederholungen."

ROSENHAIN, G. (1849).

[Auszug mehrerer Schreiben .... über die hyperelliptischen Transcendenten. No. IV. *Crelle's Journ.*, xl. pp. 347–360.]

In the course of an investigation regarding the relation between two Abelian integrals Rosenhain is brought up against the determinant

$$\sum \pm \frac{1}{t_1-a_1} \cdot \frac{1}{t_2-a_2} \cdot \ldots \cdot \frac{1}{t_{n-1}-a_{n-1}}$$

already dealt with by Cauchy in 1841, and afterwards known as "Cauchy's double alternant." The multiple integrals in question have to suffer transformation of the variables, and as a preliminary it is ascertained that the Jacobian [not yet so called]

$$\sum \pm \frac{\partial x_1}{\partial t_1} \cdot \frac{\partial x_2}{\partial t_2} \cdot \ldots \cdot \frac{\partial x_{n-1}}{\partial t_{n-1}} = \mathrm{C} \cdot \sum \pm \frac{1}{t_1-a_1} \cdot \frac{1}{t_2-a_2} \cdot \ldots \cdot \frac{1}{t_{n-1}-a_{n-1}}$$

and

$$\sum \pm \frac{\partial t_1}{\partial x_1} \cdot \frac{\partial t_2}{\partial x_2} \cdot \ldots \cdot \frac{\partial t_{n-1}}{\partial x_{n-1}} = \mathrm{D} \cdot \sum \pm \frac{1}{t_1-a_1} \cdot \frac{1}{t_2-a_2} \cdot \ldots \cdot \frac{1}{t_{n-1}-a_{n-1}}$$

where C and D are specified functions of the a's and t's. From this by multiplication it follows that

$$\left\{\sum \pm \frac{1}{t_1 - a_1} \cdot \frac{1}{t_2 - a_2} \cdot \cdots \frac{1}{t_{n-1} - a_{n-1}}\right\}^2 = \frac{1}{\text{CD}},$$

and thence ultimately that

$$\sum \pm \frac{1}{t_1 - a_1} \cdot \frac{1}{t_2 - a_2} \cdot \cdots \frac{1}{t_{n-1} - a_{n-1}}$$
$$= \frac{(-1)^{\frac{1}{2}n(n-1)}\Pi(a_1 \cdot a_2, \ldots, a_{n-1}) \cdot \Pi(t_1, t_2, \ldots, t_{n-1})}{\phi(a_1) \cdot \phi(a_2) \ldots \phi(a_{n-1})}$$

where $\phi(\xi) = (\xi - t_1)(\xi - t_2) \ldots (\xi - t_{n-1})$.

There is then added a simple verificatory proof which consists in noting (1) the double alternating character of the function *

$$\phi(a_1) \cdot \phi(a_2) \ldots \phi(a_{n-1}) \cdot \sum \pm \frac{1}{t_1 - a_1} \cdot \frac{1}{t_2 - a_2} \cdots \frac{1}{t_{n-1} - a_{n-1}};$$

(2) its degree in any one of the a's or t's; (3) the sign of any one of its terms. The exact words are—

"Der Beweis der obigen Formel ergiebt sich durch die Betrachtung, dass

$$\phi(a_1) \cdot \phi(a_2) \ldots \phi(a_{n-1}) \cdot \sum \pm \frac{1}{t_1 - a_1} \cdot \frac{1}{t_2 - a_2} \cdot \cdots \frac{1}{t_{n-1} - a_{n-1}}$$

eine ganze rationale alternirende Function sowohl in Bezug auf die $n-1$ Grössen t_p als auch in Bezug auf die $n-1$ Grössen a_m ist. Es übersteigt aber in dieser Function weder eine der Grössen t_p noch eine

* The product $\phi(a_1) . \phi(a_2) \ldots \phi(a_{n-1})$ is arrangeable as a square array of binomial factors, being in fact, save as to sign, the product of all the denominators in the double alternant, and is thus seen to be symmetrical with respect both to the a's and to the t's. If therefore we multiply each row of the alternant by the product of the denominators of the row, or each column by the product of the denominators of the column, we multiply the alternant by

$$(-1)^{(n-1)(n-2)}\phi(a_1) \cdot \phi(a_2) \ldots \phi(a_{n-1}).$$

The two determinants thus resulting have elements which are the product of $n-2$ binomial factors, and are equal to

$$(-1)^{\frac{1}{2}(n-1)(n-2)}\Pi(a_1, a_2, \ldots, a_{n-1}) \cdot \Pi(t_1, t_2, \ldots, t_{n-1}).$$

If, on the other hand, we multiply each element $1/(a_r - t_s)$ of the alternant by $a_r^{n-1} - t_s^{n-1}$ we obtain the product of the two Π's as reached by the ordinary multiplication-theorem of determinants.

der Grössen a_m den $(n-2)$ten Grad: daher ist sie nicht bloss durch das Product

$$\Pi(a_1, a_2, \ldots, a_{n-1}) \cdot \Pi(t_1, t_2, \ldots, t_{n-1})$$

theilbar, sondern, abgesehen vom Zeichen, diesem Producte selbst gleich, da ihre einzelnen Terme keine andern Zahlencoefficienten haben, als ± 1. Da nun die Determinante positiv sein soll, so musste rechts vom Gleichheitszeichen noch der Factor $(-1)^{\frac{1}{2}n(n-1)}$ hinzugefügt werden."

CAYLEY, A. (1853).

[Note on the transformation of a trigonometrical expression. *Cambridge and Dublin Math. Journ.*, ix. pp. 61, 62; or *Collected Math. Papers*, ii. pp. 45, 46.]

In order to show that the vanishing of the alternating function

$$\begin{vmatrix} 1 & x & (a+x)\sqrt{c+x} \\ 1 & y & (a+y)\sqrt{c+y} \\ 1 & z & (a+z)\sqrt{c+z} \end{vmatrix}$$

implies the vanishing of

$$\tan^{-1}\sqrt{\frac{a-c}{c+x}} + \tan^{-1}\sqrt{\frac{a-c}{c+y}} + \tan^{-1}\sqrt{\frac{a-c}{c+z}}$$

the determinant is proved to contain the factor

$$\sqrt{\frac{a-c}{c+x}} + \sqrt{\frac{a-c}{c+y}} + \sqrt{\frac{a-c}{c+z}} - \sqrt{\frac{a-c}{c+x}}\sqrt{\frac{a-c}{c+y}}\sqrt{\frac{a-c}{c+z}}$$

with the cofactor

$$-\frac{(c+x)^{\frac{3}{2}}(c+y)^{\frac{3}{2}}(c+z)^{\frac{3}{2}}}{(a-c)^2}, \quad \begin{vmatrix} 1 & \sqrt{\frac{a-c}{c+x}} & \frac{a-c}{c+x} \\ 1 & \sqrt{\frac{a-c}{c+y}} & \frac{a-c}{c+y} \\ 1 & \sqrt{\frac{a-c}{c+z}} & \frac{a-c}{c+z} \end{vmatrix}.$$

This is done by writing ξ, η, ζ for $\sqrt{\frac{a-c}{c+x}}, \sqrt{\frac{a-c}{c+y}}, \sqrt{\frac{a-c}{c+z}}$

respectively,* and so changing the given determinant into

$$\begin{vmatrix} 1 & (a-c)(1+\xi^{-2}) & (a-c)^{\frac{3}{2}}(\xi^{-1}+\xi^{-3}) \\ 1 & (a-c)(1+\eta^{-2}) & (a-c)^{\frac{3}{2}}(\eta^{-1}+\eta^{-3}) \\ 1 & (a-c)(1+\zeta^{-2}) & (a-c)^{\frac{3}{2}}(\zeta^{-1}+\zeta^{-3}) \end{vmatrix},$$

thence into

$$(a-c)^{\frac{5}{2}}\xi^{-3}\eta^{-3}\zeta^{-3}\begin{vmatrix} \xi^3 & \xi^3+\xi & \xi^2+1 \\ \eta^3 & \eta^3+\eta & \eta^2+1 \\ \zeta^3 & \zeta^3+\zeta & \zeta^2+1 \end{vmatrix},$$

and finally into

$$-(a-c)^{\frac{5}{2}}\xi^{-3}\eta^{-3}\zeta^{-3}\begin{vmatrix} 1 & \xi & \xi^2 \\ 1 & \eta & \eta^2 \\ 1 & \zeta & \zeta^2 \end{vmatrix}(\xi+\eta+\zeta-\xi\eta\zeta).$$

BRIOSCHI, F. (1854).

[La Teorica dei Determinanti, e le sue principali applicazioni. viii+116 pp. Pavia.]

Brioschi devotes the 9th section of his text-book (pp. 73–84) to "determinanti delle radici delle equazioni algebriche," viewing the difference-product and its allies as arising when the roots of the equation

$$x^n + A_{n-1}x^{n-1} + A_{n-2}x^{n-2} + \ldots . + A_1x + A_0 = 0$$

are substituted for x, and the values of A_{n-1}, A_{n-2}, . . . , are to

---

*It is simpler still to express the given determinant in terms of alternants having $\sqrt{c+x}$, $\sqrt{c+y}$, $\sqrt{c+z}$ for variables. Thus the given determinant

$$= \begin{vmatrix} 1 & x & (c+x+a-c)\sqrt{c+x} \\ 1 & y & (c+y+a-c)\sqrt{c+y} \\ 1 & z & (c+z+a-c)\sqrt{c+z} \end{vmatrix},$$

$$= \begin{vmatrix} 1 & c+x & (c+x)\sqrt{c+x} \\ 1 & c+y & (c+y)\sqrt{c+y} \\ 1 & c+z & (c+z)\sqrt{c+z} \end{vmatrix} + (a-c)\begin{vmatrix} 1 & c+x & \sqrt{c+x} \\ 1 & c+y & \sqrt{c+y} \\ 1 & c+z & \sqrt{c+z} \end{vmatrix},$$

$$= \begin{vmatrix} 1 & \mu^2 & \mu^3 \\ 1 & \nu^2 & \nu^3 \\ 1 & \zeta^2 & \zeta^3 \end{vmatrix} - (a-c)\begin{vmatrix} 1 & \mu & \mu^2 \\ 1 & \nu & \nu^2 \\ 1 & \zeta & \zeta^2 \end{vmatrix} \text{ say,}$$

$$= \begin{vmatrix} 1 & \mu & \mu^2 \\ 1 & \nu & \nu^2 \\ 1 & \zeta & \zeta^2 \end{vmatrix} \cdot \{\mu\nu+\nu\zeta+\zeta\mu-(a-c)\}.$$

be determined from the n equations thus resulting. His proof, obtained in this way, that the common denominator of the A's is resolvable into binomial factors is not of consequence. It is more important to note that, as an alternative, he proceeds "facendo uso di sole proprieta dei determinanti," obtaining in the first place

$$\begin{vmatrix} 1 & 1 & \dots & 1 \\ a_1 & a_2 & \dots & a_n \\ a_1^2 & a_2^2 & \dots & a_n^2 \\ \cdot & \cdot & \cdot & \cdot \\ a_1^{n-1} & a_2^{n-1} & \dots & a_n^{n-1} \end{vmatrix} = (-1)^{n-1} \begin{vmatrix} a_1 - a_2 & a_2 - a_3 & \dots & a_{n-1} - a_n \\ a_1^2 - a_2^2 & a_2^2 - a_3^2 & \dots & a_{n-1}^2 - a_n^2 \\ \cdot & \cdot & \cdot & \cdot \\ a_1^{n-1} - a_2^{n-1} & a_2^{n-1} - a_3^{n-1} & \dots & a_{n-1}^{n-1} - a_n^{n-1} \end{vmatrix},$$

from which he removes the factors $a_1 - a_2$, $a_2 - a_3$, ...; then repeating the first set of operations he removes the factors $a_1 - a_3$, $a_2 - a_4$, . . . , and so on.

After this an application is made to the solution of a set of linear equations which differs from Prony's set by having $z^0, z^1, z^2, \dots$ in place of $z_0, z_1, z_2, \dots$, and where therefore, as Cauchy in 1812 had pointed out, the numerators of the unknowns, as well as the common denominator, are resolvable into binomial factors. Borchardt's and Cayley's persymmetric determinants in $s_0, s_1, s_2, \dots$, got by multiplication, are also given. The remaining pages (77–84) contain illustrations.

JOACHIMSTHAL, F. (1854, May).

[Bemerkungen über den Sturm'schen Satz. *Crelle's Journ.*, xlviii. pp. 386–416.]

In the course of his investigations Joachimsthal evaluates (§ 5) the determinant

$$\begin{vmatrix} s_0 & s_1 & s_2 & 1 \\ s_1 & s_2 & s_3 & x \\ s_2 & s_3 & s_4 & x^2 \\ s_3 & s_4 & s_5 & x^3 \end{vmatrix}$$

where $s_q = x_1^q + x_2^q + x_3^q$. Using the fact that by reason of the trinomial elements the determinant is partitionable into twenty-seven determinants with monomial elements, he shows next that all of the twenty-seven except six vanish; that the six contain the common factor

$$(x-x_1)(x-x_2)(x-x_3) \cdot (x_3-x_1)(x_3-x_2)(x_2-x_1);$$

that the aggregate of the cofactors is

$$x_3^2x_2 - x_3^2x_1 + x_2^2x_1 - x_2^2x_3 + x_1^2x_3 - x_1^2x_2$$

or

$$(x_3-x_1)(x_3-x_2)(x_2-x_1);$$

and that therefore finally the given determinant is equal to

$$\left\{(x_3-x_1)(x_3-x_2)(x_2-x_1)\right\}^2 \cdot (x-x_1)(x-x_2)(x-x_3).$$

This is followed by the assertion that if s_q were made to stand for $s_1^q+s_2^q+\ldots+s_n^q$ the determinant could be partitioned into n^3 determinants, of which $n(n-1)(n-2)$ would be non-evanescent; and that these could be grouped into sets of six and condensed, the ultimate result being

$$\begin{vmatrix} s_0 & s_1 & s_2 & 1 \\ s_1 & s_2 & s_3 & x \\ s_2 & s_3 & s_4 & x^2 \\ s_3 & s_4 & s_5 & x^3 \end{vmatrix} = \sum\left\{(x_3-x_2)(x_3-x_1)(x_2-x_1)\right\}^2(x-x_1)(x-x_2)(x-x_3).$$

A large generalisation is then made, the exact words being "Genau eben so beweist man folgenden allgemeinen Satz: Bezeichnet man die Potenzsumme $x_1^i+x_2^i+\ldots+s_n^i$ durch s_i; ferner das Quadrat des Productes, welches aus den $\frac{1}{2}i(i-1)$ Differenzen der i Grössen $x_1, x_2, \ldots, x_i$ gebildet ist, durch $\delta(x_1, x_2, \ldots, x_i)$, so ist

$$\begin{vmatrix} s_0 & s_1 & s_2 & \ldots\ldots & s_{a-1} & 1 \\ s_1 & s_2 & s_3 & \ldots\ldots & s_a & x \\ s_2 & s_3 & s_4 & \ldots\ldots & s_{a+1} & x^2 \\ \ldots & \ldots & \ldots & \ldots\ldots & \ldots & \ldots \\ s_a & s_{a+1} & s_{a+2} & \ldots\ldots & s_{2a-1} & x^a \end{vmatrix} = \sum\delta(x_1, x_2, \ldots, x_a)\cdot(x-x_1)\ldots(x-x_a)$$

wo die rechte Seite eine Summe von $\dfrac{n(n-1)\ldots n-a+1)}{1\,.\,2\ldots a}$ ähnlich gebildeten Glieder enthält."

The known result (Cayley's) obtained from this by equating coefficients of x^a is pointed out: also the extension

$$\begin{vmatrix} S_0 & S_1 & S_2 & \ldots\ldots & S_{a-1} & 1 \\ S_1 & S_2 & S_3 & \ldots\ldots & S_a & x \\ S_2 & S_3 & S_4 & \ldots\ldots & S_{a+1} & x_2 \\ \ldots & \ldots & \ldots & \ldots & \ldots & \ldots \\ S_a & S_{a+1} & S_{a+2} & \ldots\ldots & S_{2a-1} & x^a \end{vmatrix} = \sum \epsilon_1\epsilon_2\ldots\epsilon_a\,.\,\delta(x_1, x_2, \ldots, x_a) \\ (x-x_1)(x-x_2)\ldots\ldots(x-x_a)$$

where $S_i = \epsilon_1 x_1^i + \epsilon_2 x_2^i + \ldots + \epsilon_n x_n^i$. It may be noted that the reason for discussing such determinants is that the series of them obtained by giving a the values n, $n-1$, $n-2$, ..., 2, 1, 0 is put forward (p. 400) as a substitute for Sturm's series of functions.*

Towards the end of the paper (§ 17, p. 414) the determinant

$$\begin{vmatrix} (a_1+b_1)^{-1} & (a_1+b_2)^{-1} & \ldots\ldots & (a_1+b_n)^{-1} \\ (a_2+b_1)^{-1} & (a_2+b_2)^{-1} & \ldots\ldots & (a_2+b_n)^{-1} \\ \ldots & \ldots & \ldots & \ldots \\ (a_n+b_1)^{-1} & (a_n+b_2)^{-1} & \ldots\ldots & (a_n+b_n)^{-1} \end{vmatrix}, \quad \text{or } \Delta_n \text{ say,}$$

is evaluated. The process consists at the outset in subtracting the first column from each column after the first, removing the factor

$$\frac{(b_1-b_2)(b_1-b_3)\ldots(b_1-b_n)}{(a_1+b_1)(a_2+b_1)\ldots(a_n+b_1)},$$

and writing the cofactor in the form

$$\begin{vmatrix} 1 & (a_1+b_2)^{-1} & \ldots\ldots & (a_1+b_n)^{-1} \\ 1 & (a_2+b_2)^{-1} & \ldots\ldots & (a_2+b_n)^{-1} \\ \ldots & \ldots & \ldots & \ldots \\ 1 & (a_n+b_2)^{-1} & \ldots\ldots & (a_n+b_n)^{-1} \end{vmatrix}.$$

The latter determinant is then transformed by subtracting the

* In this connection papers by Cayley (1846) and Borchardt (1845) are referred to by Joachimsthal, but no mention is made of Sylvester's (1839).

first row from each row after the first, when it is found that the factor

$$\frac{(a_1-a_2)(a_1-a_3)\ldots(a_1-a_n)}{(a_1+b_2)(a_1+b_3)\ldots(a_1+b_n)}$$

can be removed, and that the cofactor is a determinant similar to the original but of the $(n-1)^{\text{th}}$ order, namely, the determinant which is the cofactor of the element in the place (1, 1) of the original. The final result thus obtained agrees with Cauchy's save in having no sign-factor, the latter being only necessary when the b's are all made negative.

BRIOSCHI, F. (1854, Oct.).

[Intorno ad alcune formole per la risoluzione delle equazioni algebriche. *Annali di Sci. mat. e fis.*, v. pp. 416–421; also reprinted as half (§ 2) of the last note in the French translation of Brioschi's text-book; or *Opere mat.*, i. pp. 157–161.]

All that occurs in this paper in connection with our subject is the statement

$$\left(\frac{\partial\Delta}{\partial x^{n-1}}\right)^2 = \begin{vmatrix} s_0 & s_1 & \ldots\ldots & s_{n-2} & 1 \\ s_1 & s_2 & \ldots\ldots & s_{n-1} & x \\ s_2 & s_3 & \ldots\ldots & s_n & x^2 \\ \cdot & \cdot & \cdot\;\cdot\;\cdot\;\cdot & \cdot & \cdot \\ s_{n-2} & s_{n-1} & \ldots\ldots & s_{2n-4} & x^{n-2} \\ 1 & x & \ldots\ldots & x^{n-2} & 1 \end{vmatrix}$$

where Δ is the determinant-form of the difference-product of $x_1, x_2, \ldots, x_n$. No explanation of the statement is given, nor the mode of arriving at it. All is made clear, however, if we note first that by x on the right hand is meant any x chosen at will from the set $x_1, x_2, \ldots, x_n$: second that the differential-quotient on the left is intended to stand for the cofactor of the $(n-1)^{\text{th}}$ power of that particular x in Δ, and therefore merely denotes the difference-product of a certain $n-1$ of the x's. What the statement thus gives us is an alternative form for the square of the difference-product of $n-1$ quantities.

If we use column-by-column multiplication, and put s_r for $\alpha^r+\beta^r+\gamma^r+\delta^r$, we clearly have

$$\begin{vmatrix} 4 & s_1 & s_2 & 1 \\ s_1 & s_2 & s_3 & \beta \\ s_2 & s_3 & s_4 & \beta^2 \\ 1 & \beta & \beta^2 & 1 \end{vmatrix} = \begin{vmatrix} 1 & \alpha & \alpha^2 & . \\ 1 & \beta & \beta^2 & 1 \\ 1 & \gamma & \gamma^2 & . \\ 1 & \delta & \delta^2 & . \end{vmatrix}^2,$$

$$= \begin{vmatrix} 1 & \alpha & \alpha^2 \\ 1 & \gamma & \gamma^2 \\ 1 & \delta & \delta^2 \end{vmatrix}^2,$$

and so the result is established. It will be observed that the chosen letter β which occurs most conspicuously in the new form thus obtained for $\zeta(\alpha,\gamma,\delta)$ is one which the expression is quite independent of. Further, by performing on this new form the operations

$$\text{col}_1-\text{col}_4,\quad \text{col}_2-\beta\,\text{col}_4,\quad \text{col}_3-\beta^2\,\text{col}_4,$$

we return to the more natural form

$$\zeta(\alpha,\gamma,\delta) = \begin{vmatrix} 3 & \alpha+\gamma+\delta & \alpha^2+\gamma^2+\delta^2 \\ \alpha+\gamma+\delta & \alpha^2+\gamma^2+\delta^2 & \alpha^3+\gamma^3+\delta^3 \\ \alpha^2+\gamma^2+\delta^2 & \alpha^3+\gamma^3+\delta^3 & \alpha^4+\gamma^4+\delta^4 \end{vmatrix}.$$

BORCHARDT, C. W. (1855, March).

[Bestimmung der symmetrischen Verbindungen vermittelst ihrer erzeugenden Funktion. *Monatsb. . . . . Akad. d. Wiss.* (Berlin), 1855, pp. 165–171; *Crelle's Journ.*, liii. pp. 193–198; *Gesammelte Werke*, pp. 97–105.]

The generating function in question is

$$\sum \frac{1}{t-a}\cdot\frac{1}{t_1-a_1}\cdots\frac{1}{t_n-a_n},$$

or T say, the sign of summation being meant to indicate that of the two series of elements the one is to remain unaltered and the other is to be permuted in every possible way. The development of this function according to descending powers of $t, t_1, t_2, \ldots, t_n$ leads to those simplest types of integral symmetric functions of

$a, a_1, a_2, \ldots, a_n$ which originate by permutation from a single product of integral powers of the said variables. The determination of such functions is thus reduced to the problem of transforming T so as to have no longer occurring therein the single elements $a, a_1, a_2, \ldots, a_n$, but instead those combinatory sums of them which are the coefficients of the powers of z in the development of $(z-a)(z-a_1)(z-a_2)\ldots(z-a_n)$ or $f(z)$ say.

Without further preparatory statement the announcement is made that the solution is readily reached when the relation of T to the determinants

$$\Sigma \pm \frac{1}{t-a}\cdot\frac{1}{t_1-a_1}\cdots\frac{1}{t_n-a_n} \quad \text{or} \quad \Delta,$$

$$\Sigma \pm \frac{1}{(t-a)^2}\cdot\frac{1}{(t_1-a_1)^2}\cdots\frac{1}{(t_n-a_n)^2} \quad \text{or} \quad \mathrm{D},$$

is known, namely, the relation

$$\mathrm{D} = \mathrm{T}\cdot\Delta.$$

In proof of this relation it is pointed out that

$$\{f(t)\cdot f(t_1)\cdot f(t_2)\ldots f(t_n)\}^2\cdot \mathrm{D}$$

being an integral alternating function both with respect to the elements $t, t_1, t_2, \ldots, t_n$ and with respect to the elements $a, a_1, a_2, \ldots, a_n$ is exactly divisible by the two difference-products

$$\Pi(t, t_1, t_2, \ldots, t_n), \quad \Pi(a, a_1, a_2, \ldots, a_n),$$

and that although we cannot with equal promptness tell the remaining factor, we are able to determine it from knowing a sufficient number of its special values, namely, those values got by putting each t equal to one of the a's. Since the number of ways in which the $n+1$ a's can be taken when repetitions are allowed is $(n+1)^{n+1}$, this gives us $(n+1)^{n+1}$ values, of which, however, only two are different, namely, the value $(-1)^{\frac{1}{2}n(n+1)}\cdot f'(a)\cdot f'(a_1)\cdot f'(a_2)\ldots f'(a_n)$ obtained in the $n\,!$ cases where all the a's used are different, and the value 0 obtained in every other case. The determination, we are told, can be made by using an extension of Lagrange's interpolation-formula, the outcome of the work being

$$D = T \cdot (-1)^{\frac{1}{2}n(n+1)} \frac{\Pi(t, t_1, t_2, \ldots, t_n) \cdot \Pi(a, a_1, a_2, \ldots, a_n)}{f(t) \cdot f(t_1) \cdot f(t_2) \ldots\ldots f(t_n)}$$

which, of course, gives us

$$D = T \cdot \Delta.$$

This relation having been established, Borchardt then proceeds in a line or two to use it for the main purpose of his paper. As the determinant D, he says, arises out of the determinant Δ by performance of successive differentiation with respect to all the variables $t, t_1, t_2, \ldots t_n$, there is obtained at once an alternative expression for T, namely,

$$T = (-1)^{n+1} \frac{f(t) \cdot f(t_1) \ldots f(t_n)}{\Pi(t, t_1, \ldots, t_n)} \cdot \frac{\partial}{\partial t} \frac{\partial}{\partial t_1} \cdots \frac{\partial}{\partial t_n} \left(\frac{\Pi(t, t_1, \ldots, t_n)}{f(t) \cdot f(t) \ldots f(t_n)} \right)$$

or say rather

$$T = (-1)^{n+1} \frac{\dfrac{\partial}{\partial t} \cdot \dfrac{\partial}{\partial t_1} \cdots \dfrac{\partial}{\partial t_n} \dfrac{\Pi(t, t_1, \ldots, t_n)}{f(t) \cdot f(t_1) \ldots f(t_n)}}{\dfrac{\Pi(t, t_1, \ldots, t_n)}{f(t) \cdot f(t_1) \ldots f(t_n)}};$$

and so the transformation aimed at is accomplished.

PROUHET, E. (1856, March).

[Note sur quelques identités. *Nouv. Annales de Math.*, xv. pp. 86–91.]

In order to generalise certain algebraical identities published by O. Werner in Grunert's *Archiv*, xxii. p. 353, Prouhet first establishes the theorem in alternants foreshadowed by Prony and Cauchy, and readily derivable from Schweins' first multiplication-theorem. His mode of treatment may be concisely stated as follows:—

To say that a, b, c, d, e, f are the roots of

$$x^6 - p_1 x^5 + p_2 x^4 - p_3 x^3 + p_4 x^2 - p_5 x + p_6 = 0$$

implies that

$$p_1 = \Sigma a, \quad p_2 = \Sigma ab, \quad p_3 = \Sigma abc, \quad \ldots;$$

and as it also means that

$$\left.\begin{array}{l} a^6 - p_1a^5 + p_2a^4 - \ldots + p_6 = 0 \\ b^6 - p_1b^5 + p_2b^4 - \ldots + p_6 = 0 \\ \cdot \quad \cdot \quad \cdot \quad \cdot \quad \cdot \quad \cdot \quad \cdot \quad \cdot \quad \cdot \\ f^6 - p_1f^5 + p_2f^4 - \ldots + p_6 = 0 \end{array}\right\}$$

from which we have

$$p_1 = \begin{vmatrix} a_6 & a_4 & a_3 & a_2 & a_1 & 1 \\ b_6 & b_4 & b_3 & b_2 & b_1 & 1 \\ \cdot & \cdot & \cdot & \cdot & \cdot & \cdot \\ f_6 & f_4 & f_3 & f_2 & f_1 & 1 \end{vmatrix} \div \begin{vmatrix} a_5 & a_4 & a_3 & a_2 & a_1 & 1 \\ b_5 & b_4 & b_3 & b_2 & b_1 & 1 \\ \cdot & \cdot & \cdot & \cdot & \cdot & \cdot \\ f_5 & f_4 & f_3 & f_2 & f_1 & 1 \end{vmatrix},$$

$$p_2 = \begin{vmatrix} a_5 & a_6 & a_3 & a_2 & a_1 & 1 \\ b_5 & b_6 & b_3 & b_2 & b_1 & 1 \\ \cdot & \cdot & \cdot & \cdot & \cdot & \cdot \\ f_5 & f_6 & f_3 & f_2 & f_1 & 1 \end{vmatrix} \div \begin{vmatrix} a_5 & a_4 & a_3 & a_2 & a_1 & 1 \\ b_5 & b_4 & b_3 & b_2 & b_1 & 1 \\ \cdot & \cdot & \cdot & \cdot & \cdot & \cdot \\ f_5 & f_4 & f_3 & f_2 & f_1 & 1 \end{vmatrix}.$$

. . . .

it follows that in later notation

$$\begin{array}{l} |\, a^0b^1c^2d^3e^4f^6 \,| \div |\, a^0b^1c^2d^3e^4f^5 \,| = \Sigma a, \\ |\, a^0b^1c^2d^3e^5f^6 \,| \div |\, a^0b^1c^2d^3e^4f^5 \,| = \Sigma ab, \\ |\, a^0b^1c^2d^4e^5f^6 \,| \div |\, a^0b^1c^2d^3e^4f^5 \,| = \Sigma abc, \\ \cdot \quad \cdot \quad \cdot \quad \cdot \quad \cdot \quad \cdot \quad \cdot \quad \cdot \quad \cdot \end{array}$$

with similar results when the alternants are of any other order.

JOACHIMSTHAL, F. (1856, Sept.).

[De aequationibus quarti et sexti gradus quae in theoria linearum et superficierum secundi gradus occurrunt. *Crelle's Journ.*, liii. pp. 149–172.]

The problem of finding the normals drawn from an external point (ξ, η, ζ) to the surface

$$\frac{x^2}{a} + \frac{y^2}{b} + \frac{z^2}{c} = 1$$

being dependent on the solution of the sixth-degree equation

$$\frac{a\xi^2}{(a+u)^2} + \frac{b\eta^2}{(b+u)^2} + \frac{c\zeta^2}{(c+u)^2} = 1$$

the relation between four of the six roots is evidently

$$\begin{vmatrix} \frac{1}{(a+u_1)^2} & \frac{1}{(b+u_1)^2} & \frac{1}{(c+u_1)^2} & 1 \\ \frac{1}{(a+u_2)^2} & \frac{1}{(b+u_2)^2} & \frac{1}{(c+u_2)^2} & 1 \\ \frac{1}{(a+u_3)^2} & \frac{1}{(b+u_3)^2} & \frac{1}{(c+u_3)^2} & 1 \\ \frac{1}{(a+u_4)^2} & \frac{1}{(b+u_4)^2} & \frac{1}{(c+u_4)^2} & 1 \end{vmatrix} = 0.$$

Joachimsthal knowing this, and having obtained by an entirely different process the result

$$\sum \frac{1}{(a+u_1)(b+u_2)(c+u_3)} + \sum \frac{1}{(a+u_1)(b+u_2)(c+u_4)}$$
$$+ \sum \frac{1}{(a+u_1)(b+u_3)(c+u_4)} + \sum \frac{1}{(a+u_2)(b+u_3)(c+u_4)} = 0,$$

where the sign of summation refers to permutation of the u's, is naturally led to inquire into the connection between the two results, and to extend the inquiry to the higher cases of the same kind, including, therefore, the evaluation of the determinant

$$\begin{vmatrix} \frac{1}{(a_1+u_1)^2} & \frac{1}{(a_2+u_1)^2} & \cdots & \frac{1}{(a_n+u_1)^2} & 1 \\ \frac{1}{(a_1+u_2)^2} & \frac{1}{(a_2+u_2)^2} & \cdots & \frac{1}{(a_n+u_2)^2} & 1 \\ \cdot & \cdot & \cdot & \cdot & \cdot \\ \frac{1}{(a_1+u_{n+1})^2} & \frac{1}{(a_2+u_{n+1})^2} & \cdots & \frac{1}{(a_n+u_{n+1})^2} & 1 \end{vmatrix} \quad \text{or} \quad J.$$

The investigation of the determinant occupies the fifth and sixth sections (pp. 164–169) of his paper, the third and fourth being devoted to obtaining the other form of the resultant

$$\sum \frac{1}{(a_1+u_1)(a_2+u_2)\ldots(a_n+u_n)} + \sum \frac{1}{(a_1+u_1)\ldots(a_{n-1}+u_{n-1})(a_n+u_{n+1})}$$
$$+ \cdot \cdot \cdot \cdot \cdot \cdot \cdot \cdot \cdot \cdot \cdot \cdot + \sum \frac{1}{(a_1+u_2)(a_2+u_3)\ldots(a_n+u_{n+1})},$$

or, say,

$$[1, 2, \ldots, n+1].$$

On multiplying each row of J by the product of all the denominators occurring in the row there is obtained a determinant V whose r^{th} row consists of elements which are expressible as polynomials arranged according to descending powers of u_r, the index of the highest power of u_r being $2n-2$ in all the places except the last where it is $2n$. V, which is equal to

$$J \cdot A_1^2 A_2^2 \ldots A_n^2$$

if we put

$$A_s = (a_s+u_1)(a_s+u_2)\ldots(a_s+u_{n+1}),$$

can thus be partitioned into $(2n-1)^n(2n+1)$ determinants, each expressible in the form

$$\text{a} \cdot \begin{vmatrix} u_1^{a_1} & u_1^{a_2} & \ldots & u_1^{a_n} & u_1^{a_{n+1}} \\ u_2^{a_1} & u_2^{a_2} & \ldots & u_2^{a_n} & u_2^{a_{n+1}} \\ \cdot & \cdot & \cdot & \cdot & \cdot \\ u_{n+1}^{a_1} & u_{n+1}^{a_2} & \ldots & u_{n+1}^{a_n} & u_{n+1}^{a_{n+1}} \end{vmatrix}$$

where a is an integral function of the a's. Further, V in this way is seen to be not of higher order with respect to the u's than the determinant

$$\begin{vmatrix} u_1^{n-1} & u_1^{n} & u_1^{n+1} & \ldots & u_1^{2n-3} & u_1^{2n-2} & u_1^{2n} \\ u_2^{n-1} & u_2^{n} & u_2^{n+1} & \ldots & u_2^{2n-3} & u_2^{2n-2} & u_2^{2n} \\ \cdot & \cdot & \cdot & \cdot & \cdot & \cdot & \cdot \\ u_{n+1}^{n-1} & u_{n+1}^{n} & u_{n+1}^{n+1} & \ldots & u_{n+1}^{2n-3} & u_{n+1}^{2n-2} & u_{n+1}^{2n} \end{vmatrix},$$

that is to say, its order-number cannot exceed $\frac{1}{2}n(3n+1)$; and as it is exactly divisible by the difference-product of the u's, which is of the order $\frac{1}{2}n(n+1)$, it follows that

$$V = \Delta(u_1, u_2, \ldots, u_{n+1}) \cdot V_1$$

where V_1 is a function whose order-number is not greater than n^2. Noting now that the other form of the resultant, namely $[1, 2, \ldots, n+1]$, can by addition be transformed into

$$\frac{U}{A_1 A_2 \ldots A_n}$$

where U cannot contain any of the differences of the u's, and in its order-number cannot exceed $n(n+1)-n$, i.e. n^2, Joachimsthal concludes that V_1 and U can only differ by a factor dependent on the a's. He thus has the two results

$$J \cdot A_1^2 A_2^2 \ldots A_n^2 = \Delta(u_1, u_2, \ldots, u_{n+1}) \cdot V_1$$

and $$V_1 = \zeta \cdot U = \zeta \cdot A_1 A_2 \ldots A_n [1, 2, \ldots, n+1]$$

where ζ is a rational function of the a's: and by combining the two there is deduced

$$J = \zeta \cdot [1, 2, \ldots, n+1] \cdot \frac{\Delta(u_1, u_2, \ldots, u_{n+1})}{A_1 A_2 \ldots A_n}.$$

At this stage, we are told, the investigation rested for five years until the publication, in 1855, of Borchardt's paper in the Berlin Monatsbericht. Taking a hint from this, Joachimsthal, in order to determine ζ, multiplied both sides of this result by the product of all the denominators occurring in the diagonal of J, and then put $u_1 = -a_1$, $u_2 = -a_2$, ..., $u_n = -a_n$. The left-hand side was thus changed into

$$\begin{vmatrix} 1 & 0 & \ldots & 0 & 0 \\ 0 & 1 & \ldots & 0 & 0 \\ \cdot & \cdot & \cdot & \cdot & \cdot \\ 0 & 0 & \ldots & 1 & 0 \\ \dfrac{1}{(a_1+u_{n+1})^2} & \dfrac{1}{(a_2+u_{n+1})^2} & \ldots & \dfrac{1}{(a_n+u_{n+1})^2} & 1 \end{vmatrix}$$

or 1; the second factor of the right-hand side, being equal to

$$(u_1 - u_{n+1})(u_2 - u_{n+1}) \ldots (u_n - u_{n+1}) \cdot \Delta(u_1, u_2, \ldots, u_n),$$

was changed into

$$(-1)^n (a_1 + u_{n+1})(a_2 + u_{n+1}) \ldots (a_n + u_{n+1}) \cdot (-1)^{\frac{1}{2}n(n-1)} \Delta(a_1, a_2, \ldots, a_n);$$

and the third factor $[1, 2, \ldots, n+1]/A_1 A_2 \ldots A_n$ into a fraction with the numerator 1 and with the denominator

$$\begin{array}{l} (a_1 - a_2)(a_1 - a_3) \ldots (a_1 - a_n) \quad (a_1 + u_{n+1}) \\ \cdot\, (a_2 - a_1)(a_2 - a_3) \ldots (a_2 - a_n) \quad (a_2 + u_{n+1}) \\ \cdot\, (a_3 - a_1)(a_3 - a_2) \ldots (a_3 - a_n) \quad (a_3 + u_{n+1}) \\ \cdot \quad \cdot \quad \cdot \quad \cdot \quad \cdot \quad \cdot \quad \cdot \quad \cdot \\ \cdot\, (a_n - a_1)(a_n - a_2) \ldots (a_n - a_{n-1})(a_n + u_{n+1}) \end{array}$$

or

$$(-1)^{\frac{1}{2}n(n-1)}\Delta(a_1, a_2, \ldots, a_n)^2 \cdot (a_1+u_{n+1})(a_2+u_{n+1}) \ldots (a_n+u_{n+1}).$$

The result of the whole change was therefore

$$1 = \zeta \cdot \frac{(-1)^n}{\Delta(a_1, a_2, \ldots, a_n)},$$

whence it followed that

$$\zeta = (-1)^n \Delta(a_1, a_2, \ldots, a_n);$$

and so the longed-for result was reached

$$J = (-1)^n \frac{\Delta(a_1, a_2, \ldots, a_n) \cdot \Delta(u_1, u_2, \ldots, u_{n+1})}{A_1 A_2 \ldots A_n}[1, 2, \ldots, n+1].$$

Thereupon additional results came with ease. First we are told that in a similar manner the determinant got from J by changing the second power in the denominator of every element into the first power* is found equal to

$$(-1)^n \frac{\Delta(a_1, a_2, \ldots, a_n) \cdot \Delta(u_1, u_2, \ldots, u_{n+1})}{A_1 A_2 \ldots A_n}.$$

Then "E combinatione aequationum prodit

$$\frac{\det. \left\{ \frac{1}{(a_1+u)^2}, \frac{1}{(a_2+u)^2}, \ldots, \frac{1}{(a_n+u)^2}, 1 \right\}}{\det. \left\{ \frac{1}{a_1+u}, \frac{1}{a_2+u}, \ldots, \frac{1}{a_n+u}, 1 \right\}}_{u=u_1, =u_2, =\ldots, =u_{n+1}} = [1, 2, \ldots, n+1].$$

Faciendo u_{n+1} = quantitati infinite magnae, aequatio in relationem a cl. Borchardt inventam transit, scilicet in

$$\frac{\det. \left\{ \frac{1}{(a_1+u)^2}, \frac{1}{(a_2+u)^2}, \ldots, \frac{1}{(a_n+u)^2} \right\}}{\det. \left\{ \frac{1}{a_1+u}, \frac{1}{a_2+u}, \ldots, \frac{1}{a_n+u} \right\}}_{u=u_1, =u_2, =\ldots, =u_n} = \sum \frac{1}{(a_1+u_1)(a_2+u_2)\ldots(a_n+u_n)},$$

* Previous suggestions of such a determinant appear in Binet's paper of 1837 and Joachimsthal's of 1854.

BELLAVITIS, G. (1857, June).

[Sposizione elementare della teorica dei determinanti. *Memorie .... Istituto Veneto* ..., vii. pp. 67–144.]

Bellavitis reaches the subject of the difference-product first in § 7 and then in § 47 of his exposition, and his proof of the results dealt with in the preceding year by Prouhet is his own and interesting. Denoting the equation whose roots are $a_1, a_2, \ldots, a_n$ by

$$x^n - p_1 x^{n-1} + p_2 x^{n-2} - \ldots\ldots = 0$$

and the difference-product of the roots by Π, he multiplies both sides of the identity

$$(x-a_1)(x-a_2)\ldots(x-a_n) = x^n - p_1 x^{n-1} + p_2 x^{n-2} - \ldots.$$

by Π; and as the result on the left-hand side is evidently* the difference-product of $a_1, a_2, \ldots, a_n, x$, he obtains

$$| a_1^0 a_2^1 \ldots a_n^{n-1} x^n | = (x^n - p_1 x^{n-1} + \ldots.)\Pi.$$

It only remains then to equate like powers of x and there results

$$| a_1^0 a_2^1 a_3^2 \ldots. a_{n-1}^{n-2} a_n^n | = p_1 \Pi,$$
$$| a_1^0 a_2^1 \ldots a_{n-2}^{n-3} a_{n-1}^{n-1} a_n^n | = p_2 \Pi,$$
$$\ldots\ldots\ldots\ldots$$
$$| a_1^1 a_2^2 a_3^3 \ldots. a_{n-1}^{n-1} a_n^n | = p_n \Pi.$$

He points out also that as an alternative to this we may begin with $| a_1^0 a_2^1 a_3^2 \ldots a_n^{n-1} x^n |$, express it as a determinant of the next lower order, remove the factors $(x-a_1), (x-a_2), \ldots, (x-a_n)$, change the product of these into $x^n - px^{n-1} + \ldots.$, and then equate coefficients of like powers of x as before.

Multiplying again by Π he has of course

$$| a_1^0 a_2^1 a_3^2 \ldots a_n^{n-1} x^n | \cdot \Pi = (x^n - p_1 x^{n-1} + \ldots)\Pi^2,$$

---

* See footnote to page 163 above.

and by changing Π on the left into

$$\begin{vmatrix} 1 & 1 & 1 & \ldots\ldots & 1 & 0 \\ a_1 & a_2 & a_3 & \ldots\ldots & a_n & 0 \\ a_1^2 & a_2^2 & a_3^2 & \ldots\ldots & a_n^2 & 0 \\ \cdot & \cdot & \cdot & \cdot\cdot\cdot\cdot & \cdot & \cdot \\ a_1^{n-1} & a_2^{n-1} & a_3^{n-1} & \ldots\ldots & a_n^{n-1} & 0 \\ 0 & 0 & 0 & \ldots\ldots & 0 & 1 \end{vmatrix}$$

and twice using the multiplication-theorem there is obtained

$$\begin{vmatrix} s_0 & s_1 & \ldots\ldots & s_{n-1} & 1 \\ s_1 & s_2 & \ldots\ldots & s_n & x \\ \cdot & \cdot & \cdot\cdot\cdot\cdot & \cdot & \cdot \\ s_{n-1} & s_n & \ldots\ldots & s_{2n-2} & x^{n-1} \\ s_n & s_{n+1} & \ldots\ldots & s_{2n-1} & x^n \end{vmatrix} = (x^n - p_1x^{n-1} + \ldots\ldots) \begin{vmatrix} s_0 & s_1 & \ldots\ldots & s_{n-1} \\ s_1 & s_2 & \ldots\ldots & s_n \\ \cdot & \cdot & \cdot\cdot\cdot\cdot & \cdot \\ s_{n-1} & s_n & \ldots\ldots & s_{2n-2} \end{vmatrix},$$

a result already reached by Joachimsthal, and which by the equatement of like powers of x gives "i coefficienti p espressi da rapporti di determinanti di n[esimo] grado."

BETTI, E. (1857, June).

[Sur les fonctions symétriques des racines des équations. *Crelle's Journ.*, liv. pp. 98–100; or *Opere mat.*]

Betti recalls Borchardt's result of the year 1855, namely, that the symmetric function

$$\sum x_1^{a_1} x_2^{a_2} \ldots x_n^{a_n}$$

where $x_1, x_2, \ldots, x_n$ are the roots of the equation

$$\begin{aligned} 0 &= x^n - p_1x^{n-1} + p_2x^{n-2} - \ldots, \\ &= f(x) \text{ say,} \end{aligned}$$

is the coefficient of $t_1^{-(a_1+1)} t_2^{-(a_2+1)} \ldots t_n^{-(a_n+1)}$ in the development of

$$(-1)^n \frac{f(t_1)\cdot f(t_2) \ldots f(t_n)}{\Pi(t_1, t_2, \ldots t_n)} \cdot \frac{\partial}{\partial t_1} \frac{\partial}{\partial t_2} \cdots \frac{\partial}{\partial t_n} \left(\frac{\Pi(t_1, t_2, \ldots t_n)}{f(t_1)\cdot f(t_2) \ldots f(t_n)} \right)$$

according to descending powers of the t's. He then gives an observation of his own, namely, that the said symmetric function is likewise the coefficient of $t_1^{-(a_1+1)} t_2^{-(a_2+1)} \ldots t_n^{-(a_n+1)}$ in the similar development of

$$\frac{f'(t_1) \cdot f'(t_2) \ldots f'(t_n) \cdot \Pi^2(t_1, t_2, \ldots, t_n)}{f(t_1) \cdot f(t_2) \ldots f(t_n) \cdot \Pi^2(x_1, x_2, \ldots, x_n)};$$

and by comparison of the two results draws the conclusion that if Borchardt's generating function be denoted by

$$\theta(t_1, t_2, \ldots, t_n),$$

and his own after removal of $\Pi^2(x_1, x_2, \ldots, x_n)$ from the denominator be denoted by

$$\phi(t_1, t_2, \ldots, t_n)$$

the squared difference-product of the x's is equal to

$$\frac{\left\{\phi(t_1, t_2, \ldots, t_n)\right\}_{t_1^{-(a_1+1)} t_2^{-(a_2+1)} \ldots t_n^{-(a_n+1)}}}{\left\{\theta(t_1, t_2, \ldots, t_n)\right\}_{t_1^{(a_1+1)} t_2^{-(a_2+1)} \ldots t_n^{-(a\ +1)}}},$$

where the notation used is sufficiently explained by saying that in accordance with it the coefficient of x^r in the expansion of $F(x)$ is denoted by

$$\left\{F(x)\right\}_{x^r}.$$

BALTZER, R. (1857).

[THEORIE UND ANWENDUNG DER DETERMINANTEN, ... vi+129 pp., Leipzig.]

The section (§ 12) dealing with the "Product aller Differenzen von gegebenen Grössen" belongs to the second part of Baltzer's text-book, that is to say, the part concerning "applications." It occupies eleven pages, those devoted strictly to alternants being the first three (pp. 50–53).

At the outset he establishes the determinant form for the difference-product $P(a_1, a_2, \ldots, a_n)$: then he gives two determinant-forms for $P(\alpha_1, \alpha_2, \ldots, \alpha_n).P(\beta_1, \beta_2, \ldots, \beta_n)$: passes thence to the persymmetric determinants in $s_0, s_1, s_2, \ldots$: and finally gives Cauchy's evaluation of the double alternant $|(\alpha_1-\beta_1)^{-1}(\alpha_2-\beta_2)^{-1}\ldots(\alpha_n-\beta_n)^{-1}|$. The applications, which come next, concern the solution of Lagrange's set of linear equations, Sylvester's transformation of a binary quantic of odd degree into canonical form, and the discussion of the equality of two roots of the equation $a_nx^n+a_{n-1}x^{n-1}+\ldots+a_0=0$, or say $f(x)=0$, viewed in connection with what, following Salmon, he calls the "determinant" of the equation, although Sylvester's use of the word "discriminant" is explained a page or two later.

Under this last head an interesting transformation falls to be noted. Calling the roots of the said equation $\alpha_1, \alpha_2, \ldots, \alpha_n$, and taking the determinant which is the square of their difference-product, namely,

$$\begin{vmatrix} s_0 & s_1 & \cdots & s_{n-1} \\ s_1 & s_2 & \cdots & s_n \\ \cdot & \cdot & \cdot & \cdot \\ s_{n-1} & s_n & \cdots & s_{2n-2} \end{vmatrix}, \quad \text{or Z say,}$$

he substitutes for it a determinant of the $(2n-2)^{\text{th}}$ order

$$\begin{vmatrix} 1 & 0 & 0 & \ldots. & 0 & 0 & 0 & \ldots. & 0 \\ 0 & 1 & 0 & \ldots. & 0 & 0 & 0 & \ldots. & 0 \\ 0 & 0 & 1 & \ldots. & 0 & 0 & 0 & \ldots. & 0 \\ \cdot & \cdot & \cdot & \cdot & \cdot & \cdot & \cdot & \cdot & \cdot \\ 0 & 0 & 0 & \ldots. & s_0 & s_1 & s_2 & \ldots. & s_{n-1} \\ 0 & 0 & 0 & \ldots. & s_1 & s_2 & s_3 & \ldots. & s_n \\ \cdot & \cdot & \cdot & \cdot & \cdot & \cdot & \cdot & \cdot & \cdot \\ 0 & s_0 & s_1 & \ldots. & s_{n-3} & s_{n-2} & s_{n-1} & \ldots. & s_{2n-4} \\ s_0 & s_1 & s_2 & \ldots. & s_{n-2} & s_{n-1} & s_n & \ldots. & s_{2n-3} \\ s_1 & s_2 & s_3 & \ldots. & s_{n-1} & s_n & s_{n+1} & \ldots. & s_{2n-2} \end{vmatrix}$$

where the first $n-2$ rows do not contain an s, and the rows following contain all the s's in descending order from right to left, beginning with s_{n-1} in the last place of the $(n-1)^{\text{th}}$ row, with s_n

in the last place of the nth row, and so on. He then multiplies every row by a_n, and performs the operations which we may indicate by

$$\text{col}_2 + \frac{a_{n-1}}{a_n}\text{col}_1,$$

$$\text{col}_3 + \frac{a_{n-1}}{a_n}\text{col}_2 + \frac{a_{n-2}}{a_n}\text{col}_1,$$

$$\text{col}_4 + \frac{a_{n-1}}{a_n}\text{col}_3 + \frac{a_{n-2}}{a_n}\text{col}_2 + \frac{a_{n-3}}{a_n}\text{col}_1,$$

$$\cdots\cdots\cdots\cdots$$

thus obtaining

$$Z \cdot a_n^{2n-2} = \begin{vmatrix} a_n & a_{n-1} & a_{n-2} & \cdots \\ 0 & a_n & a_{n-1} & \cdots \\ 0 & 0 & a_n & \cdots \\ \cdots & \cdots & \cdots & \cdots \\ 0 & a_n s_0 & a_n s_1 + a_{n-1}s_0 & \cdots \\ a_n s_0 & a_n s_1 + a_{n-1}s_0 & a_n s_2 + a_{n-1}s_1 + a_{n-2}s_0 & \cdots \\ a_n s_1 & a_n s_2 + a_{n-1}s_1 & a_n s_3 + a_{n-1}s_2 + a_{n-2}s_1 & \cdots \end{vmatrix},$$

and by using Newton's relations

$$\begin{aligned} na_n &= a_n s_0, \\ (n-1)a_{n-1} &= a_n s_1 + a_{n-1}s_0, \\ (n-2)a_{n-2} &= a_n s_2 + a_{n-1}s_1 + a_{n-2}s_0, \\ (n-3)a_{n-3} &= a_n s_3 + a_{n-1}s_2 + a_{n-2}s_1 + a_{n-3}s_0, \\ &\cdots\cdots\cdots\cdots \end{aligned}$$

the elements of the last n rows of the right-hand determinant, we are told, can be so changed that in each there will occur only one of the a's and that in the first power. Thereupon the conclusion is formally announced that the determinant with which we started can be expressed as a rational integral function of the $(2n-2)^{\text{th}}$ degree in the quantities

$$\frac{a_0}{a_n}, \frac{a_1}{a_n}, \ldots, \frac{a_{n-1}}{a_n},$$

and that the said function becomes homogeneous on multiplication by a_n^{2n-2}. The actual result is not given, but in the second edition (1864) it is stated to be

$$-\mathrm{Z} \cdot a_n^{2n-2} = \begin{vmatrix} a_n & a_{n-1} & a_{n-2} & \cdots \\ 0 & a_n & a_{n-1} & \cdots \\ 0 & 0 & a_n & \cdots \\ \cdot & \cdot & \cdot & \cdot \\ 0 & na_n & (n-1)a_{n-1} & \cdots \\ na_n & (n-1)a_{n-1} & (n-2)a_{n-2} & \cdots \\ a_{n-1} & 2a_{n-2} & 3a_{n-3} & \cdots \end{vmatrix},$$

"eine Determinante $(2n-2)$ten Grades, bei welcher die $m-2$ ersten und die $m-1$ folgenden Zeilen in Bezug auf die nicht verschwindenden Elemente übereinstimmen."

Part of the object which Baltzer had here in view was to establish the relation between two forms of the discriminant of the given equation; namely, that obtained by squaring the determinant-form of the difference-product and that obtained as the eliminant of the equations

$$f'(x) = 0, \quad nf(x)-xf'(x) = 0,$$

or the equations

$$\left.\begin{array}{l} \dfrac{\partial}{\partial x}(a_n x^n + a_{n-1}x^n y + \ldots + a_0 y^n) = 0 \\ \dfrac{\partial}{\partial y}(a_n x^n + a_{n-1}x^n y + \ldots + a_0 y^n) = 0 \end{array}\right\}.$$

Now a glance at the final determinant suffices to show that it is not the eliminant sought, there being in it three types of rows, whereas the two equations giving rise to the said eliminant being both of the $(n-1)^{\text{th}}$ degree, the coefficients of the one must occur in as many rows of the eliminant as the coefficients of the other. Further, since the coefficients of the equation $f'(x) = 0$ are seen to occur in their full number of rows, and those of the other equation in the last row only, it is therefore the first $n-2$ rows that need to be changed. The set of operations requisite to effect this is

$$n \cdot \text{row}_1 - \text{row}_{2n-3},$$
$$n \cdot \text{row}_2 - \text{row}_{2n-4},$$
$$\cdot \quad \cdot \quad \cdot \quad \cdot \quad \cdot \quad \cdot \quad \cdot \quad \cdot$$
$$n \cdot \text{row}_{n-2} - \text{row}_n.$$

BRIOSCHI, F. (1857, Oct.).

[Sullo svillippo di un determinante. *Annali di Mat.*, i. pp. 9–11; or *Opere mat.*, i. pp. 273–275.]

Brioschi enunciates without proof the proposition that the even-ordered determinant

$$\left|\begin{array}{ccccccc} \frac{1}{x_1-a_1} & \frac{1}{(x_1-a_1)^2} & \frac{1}{x_1-a_2} & \frac{1}{(x_1-a_2)^2} & \cdots & \frac{1}{x_1-a_n} & \frac{1}{(x_1-a_n)^2} \\ \frac{1}{x_2-a_1} & \frac{1}{(x_2-a_1)^2} & \frac{1}{x_2-a_2} & \frac{1}{(x_2-a_2)^2} & \cdots & \frac{1}{x_2-a_n} & \frac{1}{(x_2-a_n)^2} \\ \cdot & \cdot & \cdot & \cdot & \cdot & \cdot & \cdot \\ \frac{1}{x_{2n}-a_1} & \frac{1}{(x_{2n}-a_1)^2} & \frac{1}{x_{2n}-a_2} & \frac{1}{(x_{2n}-a_2)^2} & \cdots & \frac{1}{x_{2n}-a_n} & \frac{1}{(x_{2n}-a_n)^2} \end{array}\right|,$$

which is seen to be a function of $2n$ x's and n a's, is equal to

$$(-1)^n \frac{\Pi^4(a_1, a_2, \ldots, a_n) \cdot \Pi(x_1, x_2, \ldots, x_{2n})}{\phi^2(a_1) \cdot \phi^2(a_2) \ldots \phi^2(a_n)},$$

where $\phi(x) = (x-x_1)(x-x_2) \ldots (x-x_{2n})$. He then obtains similar expressions for the principal minors, namely, (1) for the cofactor of any element in an odd-numbered column, and (2) for the cofactor of any element in an even-numbered column, his procedure being to express the minor in question in terms of determinants like the original but of the order $2n-2$ and then to make the substitutions which are thus rendered possible.

PROUHET, E. (1857, Nov.).

[Questions 410, 411. *Nouv. Annales de Math.*, (1) xvi. pp. 403, 404; xvii. pp. 187–190.]

By reason of the existence of the identity

$$2^{s-1} \cos^s a = \cos sa + s \cos(s-2)a + \tfrac{1}{2}s(s-1)\cos(s-4)a + \ldots$$

where the number of terms on the right is s and the last term has to be halved when s is odd, it is clear that the determinant

$$\begin{vmatrix} \cos n\alpha_0 & \cos(n-1)\alpha_0 & \cos(n-2)\alpha_0 & \dots & \cos 0.\alpha_0 \\ \cos n\alpha_1 & \cos(n-1)\alpha_1 & \cos(n-2)\alpha_1 & \dots & \cos 0.\alpha_1 \\ \cos n\alpha_2 & \cos(n-1)\alpha_2 & \cos(n-2)\alpha_2 & \dots & \cos 0.\alpha_2 \\ \dots & \dots & \dots & \dots & \dots \\ \cos n\alpha_n & \cos(n-1)\alpha_n & \cos(n-2)\alpha_n & \dots & \cos 0.\alpha_n \end{vmatrix}$$

may be transformed into

$$\begin{vmatrix} 2^{n-1}\cos^n\alpha_0 & 2^{n-2}\cos^{n-1}\alpha_0 & 2^{n-3}\cos^{n-2}\alpha_0 & \dots & \cos^0\alpha_0 \\ 2^{n-1}\cos^n\alpha_1 & 2^{n-2}\cos^{n-1}\alpha_1 & 2^{n-3}\cos^{n-2}\alpha_1 & \dots & \cos^0\alpha_1 \\ 2^{n-1}\cos^n\alpha_2 & 2^{n-2}\cos^{n-1}\alpha_2 & 2^{n-3}\cos^{n-2}\alpha_2 & \dots & \cos^0\alpha_2 \\ \dots & \dots & \dots & \dots & \dots \\ 2^{n-1}\cos^n\alpha_n & 2^{n-2}\cos^{n-1}\alpha_n & 2^{n-3}\cos^{n-2}\alpha_n & \dots & \cos^0\alpha_n \end{vmatrix}$$

by increasing the 1st column by multiples of the 3rd, 5th, 7th, . . . , the 2nd column by multiples of the 4th, 6th, 8th, . . . and so forth. In this way there is deduced the result

$$\Delta_1 = 2^{\frac{1}{2}n(n-1)} . \mathrm{D},$$

where Δ_1 is the first determinant, and D is the determinant got from Δ_1 by changing the multipliers of α_0, α_1, α_2, . . . into indices of powers of $\cos\alpha_0$, $\cos\alpha_1$, $\cos\alpha_2$, ...

Again by using the similar expansion for

$$\sin(s+1)\alpha \div \sin\alpha$$

on every element of the determinant

$$\begin{vmatrix} \sin(n+1)\alpha_0 & \sin n\alpha_0 & \dots & \sin\alpha_0 \\ \sin(n+1)\alpha_1 & \sin n\alpha_1 & \dots & \sin\alpha_1 \\ \dots & \dots & \dots & \dots \\ \sin(n+1)\alpha_n & \sin n\alpha_n & \dots & \sin\alpha_n \end{vmatrix}, \quad \text{or } \Delta_2 \text{ say,}$$

it is seen that the factors $\sin\alpha_0$, $\sin\alpha_1$, ... can be removed from the rows in order, and that the determinant so produced is simplifiable into a multiple of D: so that there is obtained the second result

$$\Delta_2 = 2^{\frac{1}{2}n(n-1)} . \sin\alpha_0 \sin\alpha_1 \dots \sin\alpha_n \cdot \mathrm{D}.$$

The two results are Prouhet's, who set them for proof by others.

CAYLEY, A. (1858, Feb.).

[A fifth memoir upon quantics. *Philos. Transac. R. Soc.* (London), cxlviii. pp. 429–460; or *Collected Math. Papers*, ii. pp. 527–557.]

When dealing with the subject of the equivalence of two anharmonic ratios, Cayley gives (p. 538) the result

$$\begin{vmatrix} 1 & \alpha+\alpha' & \alpha\alpha' \\ 1 & \beta+\beta' & \beta\beta' \\ 1 & \gamma+\gamma' & \gamma\gamma' \end{vmatrix} \cdot \begin{vmatrix} u^2 & -u & 1 \\ v^2 & -v & 1 \\ w^2 & -w & 1 \end{vmatrix}$$

$$= \begin{vmatrix} (u-\alpha)(u-\alpha') & (v-\alpha)(v-\alpha') & (w-\alpha)(w-\alpha') \\ (u-\beta)(u-\beta') & (v-\beta)(v-\beta') & (w-\beta)(w-\beta') \\ (u-\gamma)(u-\gamma') & (v-\gamma)(v-\gamma') & (w-\gamma)(w-\gamma') \end{vmatrix}.$$

This, if we put $\alpha', \beta', \gamma' = \alpha, \beta, \gamma$, becomes

$$\begin{vmatrix} (u-\alpha)^2 & (v-\alpha)^2 & (w-\alpha)^2 \\ (u-\beta)^2 & (v-\beta)^2 & (w-\beta)^2 \\ (u-\gamma)^2 & (v-\gamma)^2 & (w-\gamma)^2 \end{vmatrix} = 2 \begin{vmatrix} 1 & \alpha & \alpha^2 \\ 1 & \beta & \beta^2 \\ 1 & \gamma & \gamma^2 \end{vmatrix} \cdot \begin{vmatrix} 1 & u & u^2 \\ 1 & v & v^2 \\ 1 & w & w^2 \end{vmatrix},$$

or, in later notation,

$$|(u-\alpha)^2(v-\beta)^2(w-\gamma)^2| = 2\zeta^{\frac{1}{2}}(\alpha, \beta, \gamma) \cdot \zeta^{\frac{1}{2}}(u, v, w).$$

Cayley also gives (p. 539) the result

$$\begin{vmatrix} 1 & \alpha & \alpha' & \alpha\alpha' \\ 1 & \beta & \beta' & \beta\beta' \\ 1 & \gamma & \gamma' & \gamma\gamma' \\ 1 & \delta & \delta' & \delta\delta' \end{vmatrix} \cdot \begin{vmatrix} ss' & -s' & -s & 1 \\ tt' & -t' & -t & 1 \\ uu' & -u' & -u & 1 \\ vv' & -v' & -v & 1 \end{vmatrix}$$

$$= \begin{vmatrix} (s-\alpha)(s'-\alpha') & (s-\beta)(s'-\beta') & (s-\gamma)(s'-\gamma') & (s-\delta)(s'-\delta') \\ (t-\alpha)(t'-\alpha') & (t-\beta)(t'-\beta') & (t-\gamma)(t'-\gamma') & (t-\delta)(t'-\delta') \\ (u-\alpha)(u'-\alpha') & (u-\beta)(u'-\beta') & (u-\gamma)(u'-\gamma') & (u-\delta)(u'-\delta') \\ (v-\alpha)(v'-\alpha') & (v-\beta)(v'-\beta') & (v-\gamma)(v'-\gamma') & (v-\delta)(v'-\delta') \end{vmatrix},$$

which is seen in similar fashion to include the identity

$$|(s-\alpha)^2(t-\beta)^2(u-\gamma)^2(v-\delta)^2| = 0.$$

These specialisations, however, are not referred to by Cayley.

ZEHFUSS, G. (1859).

[Ueber die Determinante $Q_p = \Sigma \pm (a_0+b_0)^p(a_1+b_1)^p \ldots (a_n+b_n)^p$. *Zeitschrift f. Math. u. Phys.*, iv. pp. 233–236.]

Zehfuss having drawn attention to the known results when $p = -1$ and $p = -2$, and having asserted that there are no similar simple results when p is less than -2, confines himself to the cases where p is a positive integer. He first shows in the usual way that Q_p is divisible by the difference-product of $a_0, a_1, \ldots, a_n$ and by the difference-product of $b_0, b_1, \ldots, b_n$. Then he points out that Q_p is of the pth degree in a_0 while the difference-product of $a_0, a_1, \ldots, a_n$ is of the nth degree, and thence draws the conclusion

$$\text{if } p < n \qquad Q_p = 0,$$

giving the example deduced above from Cayley,

$$\Sigma \pm (a_0+b_0)^2(a_1+b_1)^2(a_2+b_2)^2(a_3+b_3)^2 = 0.$$

The case where $p = n$ he deals with differently, namely, as might have been suggested by Cayley's paper of the preceding year. By the multiplication-theorem he derives

$$Q_n = \begin{vmatrix} 1 & n_1a_0 & n_2a_0^2 & \ldots\ldots & a_0^n \\ 1 & n_1a_1 & n_2a_1^2 & \ldots\ldots & a_1^n \\ 1 & n_1a_2 & n_2a_2^2 & \ldots\ldots & a_2^n \\ \cdot & \cdot & \cdot & \cdot\cdot\cdot\cdot & \cdot \\ 1 & n_1a_n & n_2a_n^2 & \ldots\ldots & a_n^n \end{vmatrix} \cdot \begin{vmatrix} b_0^n & b_0^{n-1} & \ldots\ldots & 1 \\ b_1^n & b_1^{n-1} & \ldots\ldots & 1 \\ b_2^n & b_2^{n-1} & \ldots\ldots & 1 \\ \cdot & \cdot & \cdot\cdot\cdot\cdot & \cdot \\ b_n^n & b_n^{n-} & \ldots\ldots & 1 \end{vmatrix}$$

where $n_r = n(n-1)(n-2)\ldots.(n-r+1)/1\cdot2\cdot3\ldots r$; and from this there is readily obtained

$$Q_n = (-1)^{\frac{1}{2}n(n-1)}n_1n_2n_3\ldots\zeta^{\frac{1}{2}}(a_0a_1\ldots a_n)\cdot\zeta^{\frac{1}{2}}(b_0b_1\ldots b_n)$$

the illustrating example being again that which we have deduced above.

CHAPTER V.

COMPOUND DETERMINANTS, UP TO 1860.

DETERMINANTS whose elements are themselves determinants made their appearance at a very early stage in the history of the subject, the first foreshadowing of them being contained in Lagrange's "équation identique et très remarquable" of 1773, namely,

$$\xi\eta'\zeta'' + \eta\zeta'\xi'' + \zeta\xi'\eta'' - \xi\zeta'\eta'' - \eta\xi'\zeta'' - \zeta\eta'\xi'' = (xy'z'' + yz'x'' + zx'y'' - xz'y'' - yx'z'' - zy'x'')^2,$$

where

$$\xi,\quad \eta,\quad \zeta,\ \ldots = y'z'' - y''z',\quad z'x'' - z''x',\quad x'y'' - x''y',\ \ldots$$

This, viewed as a result in determinants, is a case of Cauchy's theorem of 1812 regarding the adjugate, and the adjugate of course is an instance of the special form to which we have now come. Jacobi's theorem regarding any minor of the adjugate has a like history and may be similarly classified. Passing from the case of the adjugate, where each element is a primary minor of the original determinant, Cauchy also considered the determinants, of other "systèmes dérivés," that is to say, the determinants whose elements are the secondary, ternary, . . . , minors of the original, and gave the theorem that the product of the determinants of two "complementary derived systems" is a power of the original determinant, the index of the power being

$$n(n-1)(n-2)\ldots(n-p+1)\big/1\cdot 2\cdot 3\ldots p,$$

where n is the order of the original determinant and p the order of each element of one of the "derived systems." He also in the same memoir established the theorem that *the determinant of any "derived system" of a product-determinant is equal to the product of the determinants of the corresponding "derived systems" of the two factors.*

Those are all the general results that fall to be noted prior to the middle of the nineteenth century; and, as is readily seen, they all concern what at a later date came to be called the "compounds" of $|a_{1n}|$. With one exception they are due to Cauchy.*

The fact has also to be recalled, however, that compound determinants of a *special* type were considered by Jacobi in 1841, namely, those whose elements are *functional* determinants, his main theorem being

$$\sum \pm \mathrm{J}_1^{(1)}\mathrm{J}_2^{(2)}\cdots\mathrm{J}_m^{(m)} = \left\{\sum \pm \frac{\partial f_1}{\partial x_1}\cdot\frac{\partial f_2}{\partial x_2}\cdots\frac{\partial f_n}{\partial x_n}\right\}^{m-1}\cdot\sum \pm \frac{\partial f_1}{\partial x_1}\cdot\frac{\partial f_2}{\partial x_2}\cdots\frac{\partial f_{n+m}}{\partial x_{n+m}},$$

where $f_1, f_2, \ldots, f_{n+m}$ are functions of $x_1, x_2, \ldots, x_{n+m}$, and

$$\mathrm{J}_r^{(s)} = \sum \pm \frac{\partial f_1}{\partial x_1}\cdot\frac{\partial f_2}{\partial x_2}\cdots\frac{\partial f_n}{\partial x_n}\cdot\frac{\partial f_{n+s}}{\partial x_{n+r}}.$$

This theorem and certain deductions therefrom have been already dealt with in another connection (see pp. 381–385 of *History*, i.).

SYLVESTER, J. J. (1850).

[On the intersections, contacts, and other correlations of two conics expressed by indeterminate co-ordinates. *Cambridge and Dub. Math. Journ.*, v. pp. 262–282; or *Collected Math. Papers*, i. pp. 119–137.]

In a footnote to this paper the name "*Compound* Determinants" first appears. The passage is (p. 270): ". . . a theorem given by M. Cauchy, and which is included as a particular case in a theorem of my own relating to Compound Determinants, *i.e.* Determinants of Determinants, which will take its place as an immediate consequence of my fundamental theorem given

* They are numbered xx., xxi., xli., xlii. in *History*, i.

in a memoir about to appear. The well-known rule for the Multiplication of Determinants is also a direct and simple consequence from my theorem on Compound Determinants, which indeed comprises, I believe, in one glance all the heretofore existing doctrine of determinants."

It will be of interest as we advance to try to identify the theorems of this perfervid statement, namely, (a) Sylvester's "fundamental theorem"; (b) his widely general "theorem on compound determinants" deduced therefrom, and including as a particular case a theorem of Cauchy's, and giving rise to the multiplication-theorem and many others as corollaries.

SYLVESTER, J. J. (1851, March).

[On the relation between the minor determinants of linearly equivalent quadratic functions. *Philos. Magazine* (4), i. pp. 295–305, 415; or *Collected Math. Papers*, i. pp. 241–250, 251.]

As we have already had occasion to note,* there is here given, by way of illustrating the power of the umbral notation, a theorem regarding a compound determinant, namely, the theorem which Sylvester writes in the form

$$\left\{\begin{matrix} \overline{a_1\, a_2 \ldots a_r\, a_{r+1}} & \overline{a_1\, a_2 \ldots a_r\, a_{r+2}} & \ldots\ldots & \overline{a_1\, a_2 \ldots a_r\, a_{r+s}} \\ \alpha_1\, \alpha_2 \ldots \alpha_r\, \alpha_{r+1} & \alpha_1\, \alpha_2 \ldots \alpha_r\, \alpha_{r+2} & \ldots\ldots & \alpha_1\, \alpha_2 \ldots \alpha_r\, \alpha_{r+s} \end{matrix}\right\}$$

$$= \left\{\begin{matrix} a_1\, a_2 \ldots a_r \\ \alpha_1\, \alpha_2 \ldots \alpha_r \end{matrix}\right\}^{s-1} \times \left\{\begin{matrix} a_1\, a_2 \ldots a_r\, a_{r+1}\, a_{r+2} \ldots a_{r+s} \\ \alpha_1\, \alpha_2 \ldots \alpha_r\, \alpha_{r+1}\, \alpha_{r+2} \ldots \alpha_{r+s} \end{matrix}\right\},$$

but which would now be better understood in the slightly modified form

$$\left\| \begin{vmatrix} a_1\, a_2 \ldots a_r\, a_{r+1} \\ b_1\, b_2 \ldots b_r\, b_{r+1} \end{vmatrix} \; \begin{vmatrix} a_1\, a_2 \ldots a_r\, a_{r+2} \\ b_1\, b_2 \ldots b_r\, b_{r+2} \end{vmatrix} \ldots\ldots \begin{vmatrix} a_1\, a_2 \ldots a_r\, a_{r+s} \\ b_1\, b_2 \ldots b_r\, b_{r+s} \end{vmatrix} \right\|$$

$$= \begin{vmatrix} a_1\, a_2 \ldots a_r \\ b_1\, b_2 \ldots b_r \end{vmatrix}^{s-1} \cdot \begin{vmatrix} a_1\, a_2 \ldots a_{r+s} \\ b_1\, b_2 \ldots b_{r+s} \end{vmatrix}.$$

* See above, pp. 59–60.

No proof of it is given. At a later date it would have been viewed as the "extensional" of the manifest identity

$$\begin{vmatrix} \begin{matrix} a_{r+1} & a_{r+1} & \ldots & a_{r+1} \\ b_{r+1} & b_{r+2} & & b_{r+s} \end{matrix} \\[1em] \begin{matrix} a_{r+2} & a_{r+2} & \ldots & a_{r+2} \\ b_{r+1} & b_{r+2} & & b_{r+s} \end{matrix} \\ \cdot\quad\cdot\quad\cdot\quad\cdot\quad\cdot\quad\cdot\quad\cdot \\ \begin{matrix} a_{r+s} & a_{r+s} & \ldots & a_{r+s} \\ b_{r+1} & b_{r+2} & & b_{r+s} \end{matrix} \end{vmatrix} = \begin{vmatrix} a_{r+1} & a_{r+2} & \ldots & a_{r+s} \\ b_{r+1} & b_{r+2} & & b_{r+s} \end{vmatrix}.$$

Later on in the same paper Sylvester gives for a particular purpose what he calls an "important generalisation." His words are (p. 304): "Suppose two sets of umbræ

$$\begin{matrix} a_1 & a_2 & \ldots & a_{m+n} \\ b_1 & b_2 & \ldots & b_{m+n}, \end{matrix}$$

and let r be any number less than n, and let any r-ary combination of the m numbers $1, 2, 3, \ldots, m$ be expressed by ${}^q\theta_1, {}^q\theta_2, \ldots, {}^q\theta_m$, where q goes through all the values intermediate between 1 and μ, μ being

$$\frac{m(m-1)\ldots(m-r+1)}{1\cdot 2 \quad \ldots \quad r};$$

then I say that the compound determinant

$$\begin{matrix} \overbrace{\begin{matrix} a_{1_{\theta_1}} & a_{1_{\theta_2}} & \ldots & a_{1_{\theta_m}} & a_{m+1} & a_{m+2} & \ldots & a_{m+n} \\ b_{1_{\theta_1}} & b_{1_{\theta_2}} & \ldots & b_{1_{\theta_m}} & b_{m+1} & b_{m+2} & \ldots & b_{m+n} \end{matrix}} & \overbrace{\begin{matrix} a_{2_{\theta_1}} & a_{2_{\theta_2}} & \ldots & a_{2_{\theta_m}} & a_{m+1} & a_{m+2} & \ldots & a_{m+n} \\ b_{2_{\theta_1}} & b_{2_{\theta_2}} & \ldots & b_{2_{\theta_m}} & b_{m+1} & b_{m+2} & \ldots & b_{m+n} \end{matrix}} \\ \cdot\;\cdot\;\cdot\;\cdot\;\cdot\;\cdot\;\cdot\;\cdot\;\cdot\;\cdot\;\cdot & \overbrace{\begin{matrix} a_{\mu_{\theta_1}} & a_{\mu_{\theta_2}} & \ldots & a_{\mu_{\theta_m}} & a_{m+1} & a_{m+2} & \ldots & a_{m+n} \\ b_{\mu_{\theta_1}} & b_{\mu_{\theta_2}} & \ldots & b_{\mu_{\theta_m}} & b_{m+1} & b_{m+2} & \ldots & b_{m+n} \end{matrix}} \end{matrix}$$

is equal to the following product:

$$\overbrace{\begin{matrix} a_{m+1} & a_{m+2} & \ldots & a_{m+n} \\ b_{m+1} & b_{m+2} & \ldots & b_{m+n} \end{matrix}}^{\mu'} \cdot \overbrace{\begin{matrix} a_1 & a_2 & \ldots & a_{m+n} \\ b_1 & b_2 & \ldots & b_{m+n}, \end{matrix}}^{\mu''}$$

where

$$\mu'' = \frac{(m-1)(m-2)\ldots(m-r+1)}{1\cdot 2 \quad \ldots \quad (r-1)},$$

and

$$\mu' = \frac{(m-1)(m-2)\ldots(m-r)}{1\cdot 2 \quad \ldots \quad r}.$$

In reference to this one must remark at the outset on the inappropriateness of the notation

$$^q\theta_1,\ ^q\theta_2,\ \ldots,\ ^q\theta_m$$

for the qth combination of r integers taken from $1, 2, \ldots, m$. Manifestly

$$^q\theta_1,\ ^q\theta_2,\ \ldots,\ ^q\theta_r,$$

though equally awkward, would have been less misleading. Indeed, as there is one clear misprint in the enunciation, namely, "less than n" for "less than m"; and as Sylvester is known to have been inaccurate in the correction of proofs, we might suspect θ_m to be a misprint for θ_r, were it not that θ_m occurs four times in the short passage and θ_r not once.* For $^q\theta_p$ it would have been much more convenient to write pq, which would thus have stood for "the pth integer of the qth combination"; and Sylvester's theorem might then have been written

$$\left\| \begin{vmatrix} a_{11}\, a_{21} \ldots a_{r1}\, a_{m+1}\, a_{m+2} \ldots a_{m+n} \\ b_{11}\, b_{21} \ldots b_{r1}\, b_{m+1}\, b_{m+2} \ldots b_{m+n} \end{vmatrix} \quad \begin{vmatrix} a_{12}\, a_{22} \ldots a_{r2}\, a_{m+1}\, a_{m+2} \ldots a_{m+n} \\ b_{12}\, b_{22} \ldots b_{r2}\, b_{m+1}\, b_{m+2} \ldots b_{m+n} \end{vmatrix} \right.$$

$$\left. \ldots \begin{vmatrix} a_{1\mu}\, a_{2\mu} \ldots a_{r\mu}\, a_{m+1}\, a_{m+2} \ldots a_{m+n} \\ b_{1\mu}\, b_{2\mu} \ldots b_{r\mu}\, b_{m+1}\, b_{m+2} \ldots b_{m+n} \end{vmatrix} \right\|$$

$$= \begin{vmatrix} a_{m+1}\, a_{m+2} \ldots a_{m+n} \\ b_{m+1}\, b_{m+2} \ldots b_{m+n} \end{vmatrix}^{C_{m-1,r}} \cdot \begin{vmatrix} a_1\, a_2 \ldots a_{m+n} \\ b_1\, b_2 \ldots b_{m+n} \end{vmatrix}^{C_{m-1,r-1}},$$

μ being used as before for $C_{m,r}$. To help towards clearness let us illustrate by means of the case where $m=4$, $n=3$, $r=2$. We then have $\mu=6$; each set of θ's equal to a binary combination of the first four integers, that is to say,

$$^1\theta_1{}^1\theta_2,\ ^2\theta_1{}^2\theta_2,\ ^3\theta_1{}^3\theta_2,\ \ldots,\ ^6\theta_1{}^6\theta_2$$

equal to

$$12,\ 13,\ 14,\ 23,\ 24,\ 34;$$

* θ_m was actually a misprint. Sylvester himself had to draw attention to it a year later in the *Cambridge and Dub. Math. Journ.*, viii. p. 61.

and the theorem in the form

$$\left\|\begin{array}{cccc}
\begin{vmatrix} a_1 & a_2 & a_5 & a_6 & a_7 \\ b_1 & b_2 & b_5 & b_6 & b_7 \end{vmatrix} &
\begin{vmatrix} a_1 & a_2 & a_5 & a_6 & a_7 \\ b_1 & b_3 & b_5 & b_6 & b_7 \end{vmatrix} &
\begin{vmatrix} & \\ & \end{vmatrix} \cdots\cdots &
\begin{vmatrix} a_1 & a_2 & a_5 & a_6 & a_7 \\ b_3 & b_4 & b_5 & b_6 & b_7 \end{vmatrix} \\
\begin{vmatrix} a_1 & a_3 & a_5 & a_6 & a_7 \\ b_1 & b_2 & b_5 & b_6 & b_7 \end{vmatrix} &
\begin{vmatrix} a_1 & a_3 & a_5 & a_6 & a_7 \\ b_1 & b_3 & b_5 & b_6 & b_7 \end{vmatrix} &
\begin{vmatrix} & \\ & \end{vmatrix} \cdots\cdots &
\begin{vmatrix} a_1 & a_3 & a_5 & a_6 & a_7 \\ b_3 & b_4 & b_5 & b_6 & b_7 \end{vmatrix} \\
\cdots & \cdots & \cdots & \cdots \\
\begin{vmatrix} a_3 & a_4 & a_5 & a_6 & a_7 \\ b_1 & b_2 & b_5 & b_6 & b_7 \end{vmatrix} &
\begin{vmatrix} a_3 & a_4 & a_5 & a_6 & a_7 \\ b_1 & b_3 & b_5 & b_6 & b_7 \end{vmatrix} &
\begin{vmatrix} & \\ & \end{vmatrix} \cdots\cdots &
\begin{vmatrix} a_3 & a_4 & a_5 & a_6 & a_7 \\ b_3 & b_4 & b_5 & b_6 & b_7 \end{vmatrix}
\end{array}\right\|$$

$$= \begin{vmatrix} a_5 & a_6 & a_7 \\ b_5 & b_6 & b_7 \end{vmatrix}^3 \cdot \begin{vmatrix} a_1 & a_2 & a_3 & a_4 & a_5 & a_6 & a_7 \\ b_1 & b_2 & b_3 & b_4 & b_5 & b_6 & b_7 \end{vmatrix}^3.$$

No proof is given by Sylvester: attention is merely drawn by him to the fact that when r is put equal to 1 we obtain the theorem with which his paper commences. It is rather remarkable that he should not have singled out the case where $n=0$. For then the theorem becomes

$$\left\|\begin{array}{cccc}
\begin{vmatrix} a_{11} & a_{21} & \ldots & a_{r1} \\ b_{11} & b_{21} & \ldots & b_{r1} \end{vmatrix} &
\begin{vmatrix} a_{12} & a_{22} & \ldots & a_{r2} \\ b_{12} & b_{22} & \ldots & b_{r2} \end{vmatrix} &
\begin{vmatrix} & \\ & \end{vmatrix} \cdots\cdots &
\begin{vmatrix} a_{1\mu} & a_{2\mu} & \ldots & a_{r\mu} \\ b_{1\mu} & b_{2\mu} & \ldots & b_{r\mu} \end{vmatrix}
\end{array}\right\|$$

$$= \begin{vmatrix} a_1 & a_2 & \ldots & a_m \\ b_1 & b_2 & \ldots & b_m \end{vmatrix}^{C_{m-1,r-1}},$$

where, as before, the subscript pq denotes the pth integer of the qth set of r integers taken from $1, 2, \ldots, m$, and μ stands for $C_{m,r}$; and this is the theorem well known at a later date in the form: *The* rth *compound of a determinant of the* mth *order is a power of the said determinant, the index of the power being* $C_{m-1,r-1}$. Cauchy, it will be remembered, only got the length of a similar theorem in reference to the *product* of two complementary compounds. Now, since the complementary of the rth compound is the $(m-r)^{\text{th}}$ compound, the product of the two must be that power of the original determinant whose index is

$$C_{m-1,r-1} + C_{m-1,m-r-1},$$

$$\textit{i.e.} \qquad C_{m-1,r-1} + C_{m-1,r},$$

$$\textit{i.e.} \qquad C_{m,r},$$

which agrees of course with Cauchy's result.

We thus learn that Sylvester's general result may be accurately described in later phraseology as *the "extensional" of the theorem regarding the* r*th* *compound of a determinant,* and that the discovery of both the said theorem and of its "extensional" is almost certainly due to him. At the same time it is hard to believe that this "extensional" is the all-embracing theorem referred to by him in a previous paper: for by no stretch of imagination could we see comprised in it "all the heretofore existing doctrine of determinants." His last words thereanent are: "This very general theorem is itself several degrees removed from my still unpublished Fundamental Theorem, which is a theorem for the expansion of products of determinants."

SYLVESTER, J. J. (1852, Dec.).

[On a theorem concerning the combination of determinants. *Cambridge and Dub. Math. Journ.*, viii. pp. 60–62; or *Collected Math. Papers*, i. pp. 399–401.]

The statement of the theorem referred to in the title unfortunately shows want of proper care,* with the result that it is unnecessarily lengthy. It may be recast as follows:—

If from the array

$$\begin{matrix} a_{11} & a_{12} & \dots & a_{1n} \\ a_{21} & a_{22} & \dots & a_{2n} \\ \cdot & \cdot & \cdot & \cdot \\ a_{m1} & a_{m2} & \dots & a_{mn} \end{matrix} \quad \textit{or } \mathrm{A}, \textit{ say},$$

we form every possible array of r *rows* ($\mathrm{r} \not> \mathrm{m} < \mathrm{n}$), *calling the said arrays* $\mathrm{A}_1, \mathrm{A}_2, \dots, \mathrm{A}_\mu$, *where of course* $\mu = \mathrm{C}_{\mathrm{m},\mathrm{r}}$; *and if the corresponding arrays formed from*

$$\begin{matrix} b_{11} & b_{12} & \dots & b_{1n} \\ b_{21} & b_{22} & \dots & b_{2n} \\ \cdot & \cdot & \cdot & \cdot \\ b_{m1} & b_{m2} & \dots & b_{mn} \end{matrix} \quad \textit{or } \mathrm{B}, \textit{ say},$$

* See especially line 8 from bottom of p. 61, where in every case m should be $m-1$.

be denoted by $B_1, B_2, \ldots, B_\mu$; *then*

$$\begin{vmatrix} A_1 \cdot B_1 . & A_1 \cdot B_2 & \ldots & A_1 \cdot B_\mu \\ A_2 \cdot B_1 & A_2 \cdot B_2 & \ldots & A_2 \cdot B_\mu \\ \cdot & \cdot & \cdot & \cdot \\ A_\mu \cdot B_1 & A_\mu \cdot B_2 & \ldots & A_\mu \cdot B_\mu \end{vmatrix} = (A \cdot B)^{C_{m-1, r-1}}.$$

By way of proof, Sylvester merely states that it is obtainable from his general theorem of March 1851, "by making

$$\left.\begin{matrix} a_{m+1} & a_{m+2} & \ldots & a_{m+n} \\ b_{m+1} & b_{m+2} & \ldots & b_{m+n} \end{matrix}\right\}$$

represent a determinant all whose terms (*i.e.* elements) are zeros except those which lie in one of the diagonals, these latter being all units."

His only other remark is that when $r=1$ and when $r=m$ the right-hand members are identical, and that the equating of the two left-hand members which is thus legitimised gives Cauchy's extended multiplication-theorem.

The former remark, unfortunately, is another troublesome instance of inaccuracy. The specialisation given therein is only one of two which are needed, the other being that every element of the determinant

$$\begin{vmatrix} a_1 \, a_2 \, a_3 & \ldots & a_m \\ b_1 \, b_2 \, b_3 & \ldots & b_m \end{vmatrix}$$

be made 0. To make the matter clear, let us take the case of the general theorem where $m=4$, $n=5$, $r=2$, and perform the requisite specialisations. The general theorem then is

$$\left\| \begin{matrix} \begin{vmatrix} a_1 a_2 a_5 a_6 \ldots a_9 \\ b_1 b_2 b_5 b_6 \ldots b_9 \end{vmatrix} & \begin{vmatrix} a_1 a_2 a_5 a_6 \ldots a_9 \\ b_1 b_3 b_5 b_6 \ldots b_9 \end{vmatrix} & \ldots\ldots & \begin{vmatrix} a_1 a_2 a_5 a_6 \ldots a_9 \\ b_3 b_4 b_5 b_6 \ldots b_9 \end{vmatrix} \\ \begin{vmatrix} a_1 a_3 a_5 a_6 \ldots a_9 \\ b_1 b_2 b_5 b_6 \ldots b_9 \end{vmatrix} & \begin{vmatrix} a_1 a_3 a_5 a_6 \ldots a_9 \\ b_1 b_3 b_5 b_6 \ldots b_9 \end{vmatrix} & \ldots\ldots & \begin{vmatrix} a_1 a_3 a_5 a_6 \ldots a_9 \\ b_3 b_4 b_5 b_6 \ldots b_9 \end{vmatrix} \\ \cdot & \cdot & \cdot & \cdot \\ \begin{vmatrix} a_3 a_4 a_5 a_6 \ldots a_9 \\ b_1 b_2 b_5 b_6 \ldots b_9 \end{vmatrix} & \begin{vmatrix} a_3 a_4 a_5 a_6 \ldots a_9 \\ b_1 b_3 b_5 b_6 \ldots b_9 \end{vmatrix} & \ldots\ldots & \begin{vmatrix} a_3 a_4 a_5 a_6 \ldots a_9 \\ b_3 b_4 b_5 b_6 \ldots b_9 \end{vmatrix} \end{matrix} \right\|$$

$$= \begin{vmatrix} a_5 \, a_6 \ldots a_9 \\ b_5 \, b_6 \ldots b_9 \end{vmatrix}^3 . \begin{vmatrix} a_1 \, a_2 \ldots a_9 \\ b_1 \, b_2 \ldots b_9 \end{vmatrix}^3 .$$

By changing now the matrix of the determinant

$$\begin{vmatrix} a_5\, a_6 \ldots a_9 \\ b_5\, b_6 \ldots b_9 \end{vmatrix}$$

into matrix unity, and the matrix of the determinant

$$\begin{vmatrix} a_1\, a_2\, a_3\, a_4 \\ b_1\, b_2\, b_3\, b_4 \end{vmatrix}$$

into matrix zero, the first element of the compound determinant becomes, if we write $a_r b_s$ in place of $\begin{matrix} a_r \\ b_s \end{matrix}$,

$$\begin{vmatrix} \cdot & \cdot & a_1b_5 & a_1b_6 & \ldots & a_1b_9 \\ \cdot & \cdot & a_2b_5 & a_2b_6 & \ldots & a_2b_9 \\ a_5b_1 & a_5b_2 & 1 & \cdot & \ldots & \cdot \\ a_6b_1 & a_6b_2 & \cdot & 1 & \ldots & \cdot \\ \cdot & \cdot & \cdot & \cdot & \cdot & \cdot \\ a_9b_1 & a_9b_2 & \cdot & \cdot & \ldots & 1 \end{vmatrix}$$

which from Laplace's expansion-theorem we know to be equal to

$$\sum \begin{vmatrix} a_1b_5 & a_1b_6 \\ a_2b_5 & a_2b_6 \end{vmatrix} \cdot \begin{vmatrix} a_5b_1 & a_6b_1 \\ a_5b_2 & a_6b_2 \end{vmatrix}$$

or

$$\begin{vmatrix} a_1b_5 & a_1b_6 & \ldots & a_1b_9 \\ a_2b_5 & a_2b_6 & \ldots & a_2b_9 \end{vmatrix} \cdot \begin{vmatrix} a_5b_1 & a_6b_1 & \ldots & a_9b_1 \\ a_5b_2 & a_6b_2 & \ldots & a_9b_2 \end{vmatrix}.$$

Further, it is seen that the other elements of the compound determinant take like forms, and that in fact the said determinant is

$$\begin{vmatrix} A_1 \cdot B_1 & A_1 \cdot B_2 & \ldots & A_1 \cdot B_6 \\ A_2 \cdot B_1 & A_2 \cdot B_2 & \ldots & A_2 \cdot B_6 \\ \cdot & \cdot & \cdot & \cdot \\ A_6 \cdot B_1 & A_6 \cdot B_2 & \ldots & A_6 \cdot B_6 \end{vmatrix}$$

if A and B be taken to denote the arrays

$$\begin{matrix} a_1b_5 & a_1b_6 & \ldots & a_1b_9 \\ a_2b_5 & a_2b_6 & \ldots & a_2b_9 \\ a_3b_5 & a_3b_6 & \ldots & a_3b_9 \\ a_4b_5 & a_4b_6 & \ldots & a_4b_9 \end{matrix} \qquad \begin{matrix} a_5b_1 & a_6b_1 & \ldots & a_9b_1 \\ a_5b_2 & a_6b_2 & \ldots & a_9b_2 \\ a_5b_3 & a_6b_3 & \ldots & a_9b_3 \\ a_5b_4 & a_6b_4 & \ldots & a_9b_4. \end{matrix}$$

As for the right-hand member of the general identity, the first determinant in it becomes 1, and the second becomes

$$\begin{vmatrix} . & . & . & . & a_1b_5 & a_1b_6 & \ldots & a_1b_9 \\ . & . & . & . & a_2b_5 & a_2b_6 & \ldots & a_2b_9 \\ . & . & . & . & a_3b_5 & a_3b_6 & \ldots & a_3b_9 \\ . & . & . & . & a_4b_5 & a_4b_6 & \ldots & a_4b_9 \\ a_5b_1 & a_5b_2 & a_5b_3 & a_5b_4 & 1 & . & \ldots & . \\ a_6b_1 & a_6b_2 & a_6b_3 & a_6b_4 & . & 1 & \ldots & . \\ \ldots & \ldots & \ldots & \ldots & \ldots & \ldots & \ldots & \ldots \\ a_9b_1 & a_9b_2 & a_9b_3 & a_9b_4 & . & . & \ldots & 1 \end{vmatrix},$$

which equals A·B; so that the member in question becomes

$$(\mathrm{A}\cdot\mathrm{B})^3$$

as it ought.

The important thing to note in connection with the deduction here made is the fact that Sylvester must at this date have known how to express the product of two m-by-n arrays as a determinant of the $(m+n)^{\text{th}}$ order.* (See footnote on p. 57 above.)

---

* And knowing this he *might* have indicated another mode of proving Cauchy's extended multiplication-theorem. For example:

$$\begin{vmatrix} a_1 & a_2 \\ b_1 & b_2 \end{vmatrix}\cdot\begin{vmatrix} x_1 & x_2 \\ y_1 & y_2 \end{vmatrix} + \begin{vmatrix} a_1 & a_3 \\ b_1 & b_3 \end{vmatrix}\cdot\begin{vmatrix} x_1 & x_3 \\ y_1 & y_3 \end{vmatrix} + \begin{vmatrix} a_2 & a_3 \\ b_2 & b_3 \end{vmatrix}\cdot\begin{vmatrix} x_2 & x_3 \\ y_2 & y_3 \end{vmatrix}$$

$$= \begin{vmatrix} . & . & a_1 & a_2 & a_3 \\ . & . & b_1 & b_2 & b_3 \\ x_1 & y_1 & 1 & . & . \\ x_2 & y_2 & . & 1 & . \\ x_3 & y_3 & . & . & 1 \end{vmatrix}$$

$$= \begin{vmatrix} -a_1x_1-a_2x_2-a_3x_3 & -a_1y_1-a_2y_2-a_3y_3 & . & . & . \\ -b_1x_1-b_2x_2-b_3x_3 & -b_1y_1-b_2y_2-b_3y_3 & . & . & . \\ x_1 & y_1 & 1 & . & . \\ x_2 & y_2 & . & 1 & . \\ x_3 & y_3 & . & . & 1 \end{vmatrix}$$

$$= \begin{vmatrix} a_1x_1+a_2x_2+a_3x_3 & a_1y_1+a_2y_2+a_3y_3 \\ b_1x_1+b_2x_2+b_3x_3 & b_1y_1+b_2y_2+b_3y_3 \end{vmatrix}.$$

SPOTTISWOODE, W. (1853).

[Elementary theorems relating to determinants. Second edition, rewritten and much enlarged by the author. *Crelle's Journal*, li. pp. 209–271, 328–381.]

In his first edition Spottiswoode had a section (§ vi.) headed *On Inverse Systems and Determinants of Determinants*; but in it, as we have seen, he dealt merely with the adjugate determinant and its minors. Now, this is supplanted by a section (§ x.) of considerably greater extent (pp. 350–372) with the short Sylvesterian title *On Compound Determinants.*

Although it is the original definition of a compound determinant which is given on starting, the name afterwards seems to be unconsciously limited to compound determinants whose elements are minors of a given determinant of the nth order. This limitation leads to the introduction of the word *class* in connection with compound determinants to indicate "the degree of minority of the constituents" (*i.e.* elements), a compound determinant of the ith class being one whose elements are minors of the $(n-i)^{\text{th}}$ order; thus, the adjugate determinant is a compound determinant of the nth order and 1st class. If a compound determinant of the ith class be of the highest possible order, namely, the

$$\frac{n(n-1)\ldots(n-i+1)}{1\cdot 2\quad\ldots\quad i}\text{th},$$

—that is to say, contains *all* the minors of the $(n-i)^{\text{th}}$ order—Spottiswoode calls it the "*complete* determinant of the ith class," a name equivalent therefore to the more modern "$(n-i)^{\text{th}}$ compound."

One of the notations employed is essentially the same as Sylvester's—that is to say, he uses

$$\left\{\begin{matrix} \overline{1_1\, 2_1 \ldots u_1} & \overline{1_2\, 2_2 \ldots u_2} & \ldots & \overline{1_n\, 2_n \ldots u_n} \\ 1_1\, 2_1 \ldots u_1 & 1_2\, 2_2 \ldots u_2 & & 1_n\, 2_n \ldots u_n \end{matrix}\right\}$$

for what at a later date would have been written

$$\left\|\begin{matrix} \begin{vmatrix} 1_1\, 2_1 \ldots u_1 \\ 1_1\, 2_1 \ldots u_1 \end{vmatrix} & \begin{vmatrix} 1_2\, 2_2 \ldots u_2 \\ 1_2\, 2_2 \ldots u_2 \end{vmatrix} & \ldots\ldots & \begin{vmatrix} 1_n\, 2_n \ldots u_n \\ 1_n\, 2_n \ldots u_n \end{vmatrix} \end{matrix}\right\|.$$

The other notation is his own, and is worthy of careful note. It differs from Sylvester's in making use of row-numbers and column-numbers not of the retained elements but of the elements omitted, the said numbers being enclosed in brackets for the purpose of recalling this difference. "Thus," he says (p. 352), "the complete compound determinant of the first class may be written

$$\left\{\binom{1}{1}\binom{2}{2}\cdots\cdots\binom{n}{n}\right\};$$

that of the second class

$$\left\{\begin{pmatrix}1 & 2\\1 & 2\end{pmatrix}\begin{pmatrix}1 & 3\\1 & 3\end{pmatrix}\cdots\cdots\begin{pmatrix}2 & 3\\2 & 3\end{pmatrix}\cdots\cdots\right\}$$

and generally that of the ith class

$$\left\{\begin{pmatrix}1_1 & 1_2 & \ldots & 1_i\\1_1 & 1_2 & \ldots & 1_i\end{pmatrix}\begin{pmatrix}2_1 & 2_2 & \ldots & 2_i\\2_1 & 2_2 & \ldots & 2_i\end{pmatrix}\cdots\cdots\begin{pmatrix}\mu_1 & \mu_2 & \ldots & \mu_i\\\mu_1 & \mu_2 & \ldots & \mu_i\end{pmatrix}\right\},$$

where $$\mu = \frac{n(n-1)\ldots(n-i+1)}{1\cdot 2 \quad \ldots \quad i};"$$

and where, it should have been added, r_s stands for the sth integer in the rth combination of i integers taken from $1, 2, \ldots, n$.

These preliminaries having been attended to, a discussion of the properties follows. The first five pages (pp. 353–358) and two later pages (pp. 366–368) are mainly concerned with compound determinants of the first class (that is to say, with the adjugate determinant), and they do not break fresh ground. The same, however, cannot be said with reference to the next two pages (pp. 358–360), which concern those of the second class. The result first reached is that the complete determinant of this class, namely,

$$\left\{\begin{pmatrix}1 & 2\\1 & 2\end{pmatrix}\begin{pmatrix}1 & 3\\1 & 3\end{pmatrix}\cdots\cdots\begin{pmatrix}2 & 3\\2 & 3\end{pmatrix}\cdots\cdots\right\}, \quad \text{or } \Delta_2, \text{ say,}$$

$$= \begin{Bmatrix}1 & 2 & \ldots & n\\1 & 2 & \ldots & n\end{Bmatrix}^{\nu},$$

where $\nu = \frac{1}{2}(n-1)(n-2)$. Although there is a semblance of reasoning, no real proof is given. Passing then to any first minor of Δ_2, say the first minor got by leaving out $\binom{1\ 2}{1\ 2}$ from the detailed symbol for Δ_2, he finds that

$$\left\{\binom{1\ 3}{1\ 3}\binom{1\ 4}{1\ 4}\cdots\cdots\binom{2\ 3}{2\ 3}\cdots\cdots\right\} = \begin{Bmatrix}1\ 2\\1\ 2\end{Bmatrix}\begin{Bmatrix}1\ 2\ \ldots\ldots\ n\\1\ 2\ \ldots\ldots\ n\end{Bmatrix}^{\nu-1}.$$

It is next pointed out that if we proceed to the second minors of Δ_2 it becomes necessary to distinguish two cases,—to distinguish, for example, the case where we leave out $\binom{1\ 2}{1\ 2}$, $\binom{1\ 3}{1\ 3}$ from the case where we leave out $\binom{1\ 2}{1\ 2}\binom{3\ 4}{3\ 4}$. In the latter case the row-numbers 1, 2, 3, 4 are all different, and the result is

$$\left\{\binom{5\ 6}{5\ 6}\binom{7\ 8}{7\ 8}\cdots\cdots\binom{1\ 3}{1\ 3}\binom{2\ 4}{2\ 4}\cdots\cdots\right\} = \begin{Bmatrix}1\ 2\ 3\ 4\\1\ 2\ 3\ 4\end{Bmatrix}\begin{Bmatrix}1\ 2\ \ldots\ n\\1\ 2\ \ldots\ n\end{Bmatrix}^{\nu-2};$$

in the former case the row-number 1 occurs twice, and the result is

$$\left\{\binom{1\ 4}{1\ 4}\binom{1\ 5}{1\ 5}\cdots\cdots\binom{2\ 3}{2\ 3}\cdots\cdots\right\} = \begin{Bmatrix}1\\1\end{Bmatrix}\begin{Bmatrix}1\ 2\ 3\\1\ 2\ 3\end{Bmatrix}\begin{Bmatrix}1\ 2\ \ldots\ n\\1\ 2\ \ldots\ n\end{Bmatrix}^{\nu-2}.$$

Generally, if $\binom{1\ 2}{1\ 2}$, $\binom{3\ 4}{3\ 4}$, $\binom{5\ 6}{5\ 6}$, . . . , $\binom{2i-1\ 2i}{2i-1\ 2i}$ be left out, we have

$$\left\{\binom{2i+1\ 2i+2}{2i+1\ 2i+2}\binom{2i+3\ 2i+4}{2i+3\ 2i+4}\cdots\cdots\binom{1\ 3}{1\ 3}\binom{2\ 4}{2\ 4}\cdots\cdots\right\} = \begin{Bmatrix}1\ 2\ldots 2i\\1\ 2\ldots 2i\end{Bmatrix}\begin{Bmatrix}1\ 2\ldots n\\1\ 2\ldots n\end{Bmatrix}^{\nu-i};$$

and if $\binom{1\ 2}{1\ 2}$, $\binom{1\ 3}{1\ 3}$, . . . , $\binom{1\ i}{1\ i}$ be left out, we have

$$\left\{\binom{1\ i+1}{1\ i+1}\binom{1\ i+2}{1\ i+2}\cdots\cdots\binom{2\ 3}{2\ 3}\cdots\cdots\right\} = \begin{Bmatrix}1\\1\end{Bmatrix}^{i-2}\begin{Bmatrix}1\ 2\ \ldots\ i\\1\ 2\ \ldots\ i\end{Bmatrix}\begin{Bmatrix}1\ 2\ \ldots\ n\\1\ 2\ \ldots\ n\end{Bmatrix}^{\nu-i+1}.$$

Again, a fresh variety of minor may be got by leaving out $\binom{1\ 2}{1\ 2}$, $\binom{1\ 3}{1\ 3}$, $\binom{2\ 3}{2\ 3}$, the row-numbers being then two 1's, two 2's, and two 3's, and the result is

$$\left\{\binom{1\ 4}{1\ 4}\binom{1\ 5}{1\ 5}\cdots\cdots\binom{2\ 4}{2\ 4}\cdots\cdots\right\} = \begin{Bmatrix}1\ 2\ 3\\1\ 2\ 3\end{Bmatrix}\begin{Bmatrix}1\ 2\ \ldots\ n\\1\ 2\ \ldots\ n\end{Bmatrix}^{\nu-3},$$

"and so on."

From these cases of compound determinants of the second class, the author passes to those of the ith class, but contents himself with stating only two results. The first is that the complete determinant of this class, denoted as above by

$$\left\{\begin{pmatrix}1_1 & 1_2 & \ldots & 1_i\\ 1_1 & 1_2 & \ldots & 1_i\end{pmatrix}\begin{pmatrix}2_1 & 2_2 & \ldots & 2_i\\ 2_1 & 2_2 & \ldots & 2_i\end{pmatrix}\ \ldots\ldots\ \begin{pmatrix}\mu_1 & \mu_2 & \ldots & \mu_i\\ \mu_1 & \mu_2 & \ldots & \mu_i\end{pmatrix}\right\},\ \text{or } \Delta_i,\ \text{say},$$

$$= \begin{Bmatrix}1 & 2 & \ldots & n\\ 1 & 2 & \ldots & n\end{Bmatrix}^{\nu'},$$

where $\nu' = \dfrac{(n-1)(n-2)\ldots(n-i+1)}{1\cdot 2\ldots i-1}$, a result which agrees with that obtained on putting $n=0$ in Sylvester's general theorem of March 1851; and the other is that any first minor of Δ_i, say the minor

$$\left\{\begin{pmatrix}2_1 & 2_2 & \ldots & 2_i\\ 2_1 & 2_2 & \ldots & 2_i\end{pmatrix}\begin{pmatrix}3_1 & 3_2 & \ldots & 3_i\\ 3_1 & 3_2 & \ldots & 3_i\end{pmatrix}\ \ldots\ldots\ \begin{pmatrix}\mu_1 & \mu_2 & \ldots & \mu_i\\ \mu_1 & \mu_2 & \ldots & \mu_i\end{pmatrix}\right\}$$

$$= \begin{Bmatrix}1_1 & 1_2 & \ldots & 1_i\\ 1_1 & 1_2 & \ldots & 1_i\end{Bmatrix}\begin{Bmatrix}1 & 2 & \ldots & n\\ 1 & 2 & \ldots & n\end{Bmatrix}^{\nu'-1}.$$

He appends, however, the words "and so on," and tells us that "other formulæ may be written as required."

Two theorems of Sylvester's are next given, the one being the general theorem just alluded to, and the other that contained in the paper of 16th December, 1852. In the case of the former he varies the notation, and probably by reason of the above-mentioned serious misprint of an m for an r in the original he misses Sylvester's meaning, and makes an incorrect statement. In the case of the other no risk of this kind is incurred, because he takes the unusual course of reproducing Sylvester's words letter for letter to the extent of almost two pages (pp. 361–363). The original two pages, however, being, as we have seen, not without evidence of Sylvester's carelessness, this course also was unsafe.

Lastly, he shows how the multiplication-theorem may be deduced from Sylvester's first theorem of March 1851 regarding compound determinants, by taking as an example of the latter the identity

$$\left\| \begin{array}{cc} \begin{vmatrix} 1 & . & a \\ . & 1 & a' \\ a & \beta & . \end{vmatrix} & \begin{vmatrix} 1 & . & b \\ . & 1 & b' \\ a & \beta & . \end{vmatrix} \\ \begin{vmatrix} 1 & . & a \\ . & 1 & a' \\ a' & \beta' & . \end{vmatrix} & \begin{vmatrix} 1 & . & b \\ . & 1 & b' \\ a' & \beta' & . \end{vmatrix} \end{array} \right\| = \begin{vmatrix} 1 & . \\ . & 1 \end{vmatrix}^{2-1} \cdot \begin{vmatrix} 1 & . & a & b \\ . & 1 & a' & b' \\ a & \beta & . & . \\ a' & \beta' & . & . \end{vmatrix},$$

and pointing out that this is evidently the same as saying

$$\begin{vmatrix} aa+a'\beta & ba+b'\beta \\ aa'+a'\beta' & ba'+b'\beta' \end{vmatrix} = \begin{vmatrix} a & \beta \\ a' & \beta' \end{vmatrix} \cdot \begin{vmatrix} a & b \\ a' & b' \end{vmatrix}.$$

In later phraseology, it may be said that the specialisation necessary for the purpose is

$$\left. \begin{array}{l} s = r, \\ \text{matrix of} \begin{vmatrix} a_1 \; a_2 \dots a_r \\ b_1 \; b_2 \dots b_r \end{vmatrix} = 1, \\ \text{matrix of} \begin{vmatrix} a_{r+1} \; a_{r+2} \dots a_{2r} \\ b_{r+1} \; b_{r+2} \dots b_{2r} \end{vmatrix} = 0.* \end{array} \right\}$$

BRIOSCHI, FR. (1854, March).

[LA TEORICA DEI DETERMINANTI, E LE SUE PRINCIPALI APPLICAZIONI. viii+116 pp. Pavia.]

After proving (p. 100) Jacobi's theorem, above referred to, regarding a compound determinant whose elements are functional determinants, Brioschi bids the reader make the functions f_1, f_2, ... linear, say

$$f_r = a_{r1}x_1 + a_{r2}x_2 + \dots + a_{r,n+m}x_{n+m},$$

and note that the outcome is

$$\sum \pm A^{(1)}A_2^{(2)} \dots A_m^{(m)} = \left\{\sum \pm a_{11}a_{22} \dots a_{nn}\right\}^{m-1} \cdot \sum \pm a_{11}a_{22} \dots a_{n+m,n+m},$$

* Some of the pages of Spottiswoode dealt with in the foregoing are, by reason of misprints and other neglects, not easy reading. On p. 360 there are at least *nine* misprints.

where

$$A_r^{(s)} = \sum \pm a_{11}a_{22} \ldots a_{nn}a_{n+s,n+r},$$

that is to say, is Sylvester's first result of March 1851.

The same course is followed by Bellavitis in his *Sposizione* of 1857 (see §§ 71, 72, p. 56).

BAZIN, H. (1854, April).

[Sur une question relative aux déterminants. *Journ. (de Liouville) de Math.*, xvi. pp. 145–160.]

Bazin's problem being to find an array of n rows and $n+h$ columns such that its n-line minors shall have given values, he notes at the outset that the said values cannot in general be independent of one another. There thus arises the preliminary task of finding the connecting equations, and his first conclusion is (pp. 148–149) that *if the value of one of the minors be given, and also the values of the* nh *other minors obtainable from the said minor by interchanging any one of its* n *columns with any one of the remaining columns, then the values of all the other minors are known.* This is established by actually showing how to obtain the expression for any one of the unknown minors in terms of a number of the known. As the underlying theorem is one of considerable importance in regard to compound determinants, we separate it from its surroundings and enunciate it in our own way, namely, *If the first of the* n*-line minors of an array of* n *rows and* n+h *columns* (h$\not>$n) *be denoted by* D, *and the minor got from* D *by supplanting its first* h *columns by the last* h *columns of the array be denoted by* E, *and if further there be formed a square array of new determinants obtained by supplanting separately and in order each of the first* h *columns of* D *by each of the first* h *columns of* E, *the determinant of this square array is equal to* D^{h-1} E.* Thus, using *rstu* to stand for

* In connection with this see the footnote on pp. 384, 385 of *Hist.*, i. Note also how from the second example with the help of the first we have

$$1274(1234 \cdot 5634) - 1734(1324 \cdot 5624) + 7234(2134 \cdot 5614) = (1234)^2 \cdot 5674,$$

the determinant whose columns are the rth, s^{th}, t^{th}, u^{th} columns of the array, if the array consist of four rows and six columns, we have

$$\begin{vmatrix} 5234 & 6234 \\ 1534 & 1634 \end{vmatrix} = 1234 \cdot 5634;$$

and if it consist of four rows and seven columns, we have

$$\begin{vmatrix} 5234 & 6234 & 7234 \\ 1534 & 1634 & 1734 \\ 1254 & 1264 & 1274 \end{vmatrix} = (1234)^2 \cdot (5674).$$

K being the compound determinant involved, Bazin's mode of proof consists in expressing K/D^h in the form of a determinant of the same order as D, then multiplying D by K/D^h, and with the help of a lemma finding the product to be E. This at once gives

$$K = D^{h-1}E$$

as desired. Thus, taking the first of the two examples just given, we have

$$\begin{vmatrix} a_1 & a_2 & a_3 & a_4 \\ b_1 & b_2 & b_3 & b_4 \\ c_1 & c_2 & c_3 & c_4 \\ d_1 & d_2 & d_3 & d_4 \end{vmatrix} \cdot \begin{vmatrix} 5234/D & 6234/D & . & . \\ 1534/D & 1634/D & . & . \\ 1254/D & 1264/D & 1 & . \\ 1235/D & 1236/D & . & 1 \end{vmatrix}$$

$$= \begin{vmatrix} a_5 & a_6 & a_3 & a_4 \\ b_5 & b_6 & b_3 & b_4 \\ c_5 & c_6 & c_3 & c_4 \\ d_5 & d_6 & d_3 & d_4 \end{vmatrix},$$

that is to say,

$$D \cdot \begin{vmatrix} 5234 & 6234 \\ 1534 & 1634 \end{vmatrix} \cdot D^{-2} = 5634,$$

and practically nothing more is wanted.

---

and thence

$$1274 \cdot 3564 + 1734 \cdot 2564 - 7234 \cdot 5614 = 1234 \cdot 5674,$$

or, on removing the 'extension,'

$$127 \cdot 356 - 137 \cdot 256 - 237 \cdot 156 = 123 \cdot 567.$$

A relation is thus established between two very unlike theorems.

The lemma referred to we may formulate for ourselves as follows: *If from any determinant* D *a row of new determinants be formed by supplanting in order each column of* D *by one and the same new column, the product of this row of determinants by any row of* D *is equal to the product of* D *by the corresponding element of the repeatedly substituted new column.* This is little else than the elementary fact regarding a determinant with two rows identical. Thus

$$a_1\cdot 5234 + a_2\cdot 1534 + a_3\cdot 1254 + a_4\cdot 1235$$

$$= -\begin{vmatrix} a_1 & a_2 & a_3 & a_4 & \cdot \\ a_1 & a_2 & a_3 & a_4 & a_5 \\ b_1 & b_2 & b_3 & b_4 & b_5 \\ c_1 & c_2 & c_3 & c_4 & c_5 \\ d_1 & d_2 & d_3 & d_4 & d_5 \end{vmatrix} = -\begin{vmatrix} a_1 & a_2 & a_3 & a_4 & \cdot \\ \cdot & \cdot & \cdot & \cdot & a_5 \\ b_1 & b_2 & b_3 & b_4 & b_5 \\ c_1 & c_2 & c_3 & c_4 & c_5 \\ d_1 & d_2 & d_3 & d_4 & d_5 \end{vmatrix} = a_5\cdot 1234.$$

From what has been said above it will be understood that Bazin views the theorem $K = D^{h-1}E$ in the form $E = K/D^{h-1}$, that is to say, as expressing the determinant E in terms of h^2+1 other determinants, namely, D and the h^2 elements of K.

For the present the rest of the paper is not of interest.

C(AYLEY,). A. (1854).

[Mathematical Notes. (No. 5.) *Cambridge and Dub. Math. Journ.*, ix. p. 171.]

To eliminate x, y from the equations

$$\frac{ax+by}{\alpha x+\beta y} = \frac{a'x+b'y}{\alpha' x+\beta' y} = \frac{a''x+b''y}{\alpha'' x+\beta'' y},$$

$-\lambda$ is introduced to stand for the equivalent of each fraction, μ for the negative reciprocal of λ, and ξ, η for any quantities whatever, with the result that from the four equations

$$\left.\begin{array}{l} ax\ +by\ +\lambda(\alpha x\ +\beta y)\ = 0 \\ a'x\ +b'y\ +\lambda(\alpha' x\ +\beta' y)\ = 0 \\ a''x+b''y+\lambda(\alpha'' x+\beta'' y) = 0 \\ \xi x\ +\eta y\ +\lambda(\mu\xi x+\mu\eta y) = 0 \end{array}\right\},$$

there is obtained

$$\begin{vmatrix} \xi & \eta & \mu\xi & \mu\eta \\ a & b & \alpha & \beta \\ a' & b' & \alpha' & \beta' \\ a'' & b'' & \alpha'' & \beta'' \end{vmatrix} = 0,$$

or, say $$A\xi + B\eta + C\mu\xi + D\mu\eta = 0.$$

As this holds for all values of ξ, η, we have

$$A + C\mu = 0, \quad B + D\mu = 0,$$

and ∴ $$AD - BC = 0,$$

that is,

$$\begin{Vmatrix} |ab'\alpha''| & |ab'\beta''| \\ |\alpha'\beta''a| & |\alpha'\beta''b| \end{Vmatrix} = 0.$$

CHAPTER VI.

RECURRENTS, FROM 1841 TO 1860.

LIKE Wronskians, and for the same reason, *recurrents* were dealt with in our first volume among "Miscellaneous Special Forms": their previous history is thus to be found under Wronski 1812, Scherk 1825, and Schweins 1825 in the chapter so entitled. (*History*, i. pp. 472–474, 478–481.)

The name is quite recent, having apparently been first used by E. Pascal in 1907 in a paper published in the *Rendiconti . . . Ist. Lombardo* (2), xl. pp. 293-305.

SPOTTISWOODE, W. (1853, August).

[Elementary theorems relating to determinants. Second edition, . . . *Crelle's Journ.*, li. pp. 209–271, 328–381.]

In the last chapter or section (§ xi.), which is headed "Miscellaneous Instances of Determinants," Spottiswoode gives (pp. 373–374) an expression for the nth differential-quotient of u/v in terms of the nth and lower differential-quotients of u and of v. The first four cases are

$$\left(\frac{u}{v}\right)' = \begin{vmatrix} v & v' \\ u & u' \end{vmatrix} v^{-2}, \qquad \left(\frac{u}{v}\right)'' = -\begin{vmatrix} \cdot & v & 2v' \\ v & v' & v'' \\ u & u' & u'' \end{vmatrix} v^{-3},$$

$$\left(\frac{u}{v}\right)''' = \begin{vmatrix} \cdot & \cdot & v & 3v' \\ \cdot & v & 2v' & 3v'' \\ v & v' & v'' & v''' \\ u & u' & u'' & u''' \end{vmatrix} v^{-4}, \qquad \left(\frac{u}{v}\right)^{\text{iv}} = -\begin{vmatrix} \cdot & \cdot & \cdot & v & 4v' \\ \cdot & \cdot & v & 3v' & 6v'' \\ \cdot & v & 2v' & 3v'' & 4v''' \\ v & v' & v'' & v''' & v^{\text{iv}} \\ u & u' & u'' & u''' & u^{\text{iv}} \end{vmatrix} v^{-5},$$

where the arithmetical coefficients appearing in the elements of the determinants are those of the binomial theorem.

He also notes that any binary quantic may be expressed as a determinant: thus $a_0x^n+a_1x^{n-1}y+x_2a^{n-2}y^2+\ldots\ldots$ is written by him in the form

$$\left|\begin{array}{cccccccc} a_0 & a_1 & a_2 & a_3 & \ldots & a_{n-1} & a_n \\ y & x & \cdot & \cdot & \ldots & \cdot & \cdot \\ \cdot & y & x & \cdot & \ldots & \cdot & \cdot \\ \cdot & \cdot & y & x & \ldots & \cdot & \cdot \\ \cdot & \cdot & \cdot & \cdot & \cdot\cdot\cdot & \cdot & \cdot \\ \cdot & \cdot & \cdot & \cdot & \ldots & y & x \end{array}\right| .$$

BRIOSCHI, F. (1854, 1855, February).

[Sur deux formules relatives à la théorie de la décomposition des fractions rationnelles. *Crelle's Journ.*, l. pp. 239–242; or *Opere mat.*, vol. v. pp. 267–276. See also *Nouv. Annales de Math.*, xiii. p. 352.]

In a review of the second edition of Serret's *Cours d'Algèbre Supérieure*, Terquem, the editor of the *Nouvelles Annales*, takes occasion to enunciate the theorem that if s_r denote the sum of the rth powers of the roots of the equation

$$x^n + a_1x^{n-1} + a_2x^{n-2} + \ldots\ldots + a_n = 0,$$

then

$$s_r = (-1)^r \left|\begin{array}{cccccc} a_1 & 1 & \cdot & \ldots\ldots & \cdot \\ 2a_2 & a_1 & 1 & \ldots\ldots & \cdot \\ 3a_3 & a_2 & a_1 & \ldots\ldots & \cdot \\ \cdot & \cdot & \cdot & \cdot\cdot\cdot\cdot & \cdot \\ ra_r & a_{r-1} & a_{r-2} & \ldots\ldots & a_1 \end{array}\right| ,$$

attributing it to Brioschi, and indicating that it had been arrived at by the solution of a "suite indéfinie d'équations périodiques du premier degré." * The origin of the determinant is thus exactly

* By this, of course, is meant the set of identities known as "Newton's formulæ,"—

$$s_1 + a_1 = 0$$
$$s_2 + a_1s_1 + 2a_2 = 0$$
$$\cdot\ \cdot\ \cdot\ \cdot\ \cdot\ \cdot\ \cdot\ \cdot$$

(See NEWTON, *Arith. Univ.*, Tom. ii., cap. iii., § 8.)

similar to that of the first determinant of like kind, namely, that occurring in the statement of Wronski's "loi générale des séries."

A few months later we find on p. 240 of vol. l. of *Crelle's Journal* Brioschi himself enunciating and proving with some trouble the theorem that if

$$\phi(x) = c_0x^n + c_1x^{n-1} + \ldots + c_n,$$

and

$$\begin{aligned} f(x) &= a_0x^n + a_1x^{n-1} + \ldots + a_n, \\ &= a_0(x-x_1)(x-x_2)\ldots(x-x_n), \end{aligned}$$

then

$$\frac{x_1{}^r\phi(x_1)}{f'(x_1)} + \frac{x_2{}^r\phi(x_2)}{f'(x_2)} + \ldots\ldots + \frac{x_n\phi(x_n)}{f'(x_n)}$$

$$= (-1)^{r+1}\frac{1}{a_0^{r+2}}\begin{vmatrix} c_0 & a_0 & . & \ldots & . \\ c_1 & a_1 & a_0 & \ldots & . \\ . & . & . & . & . \\ c_r & a_r & a_{r-1} & \ldots & a_0 \\ c_{r+1} & a_{r+1} & a_r & \ldots & a_1 \end{vmatrix}.$$

The subject, however, is not pursued further.

FAURE, [H.] (1855, March).

[Théorème sur la somme des puissances semblables des racines. *Nouv. Annales de Math.* (1), xiv. pp. 94–97; or pp. 172–175 of Combescure's translation of Brioschi's *Teorica dei Determinanti.*]

Having seen Brioschi's result regarding s_r, Faure takes up the subject and succeeds in throwing on it fresh light. His fundamental proposition is not connected with the roots of equations at all, being to the effect that if

$$\phi(x) = c_0x^m + c_1x^{m-1} + \ldots\ldots + c_m,$$

$$f(x) = a_0x^n + a_1x^{n-1} + \ldots\ldots + a_n$$

and

$$\phi(x) \div f(x) = A_0x^{m-n} + A_1x^{m-n-1} + A_2x^{m-n-2} + \ldots\ldots,$$

then

$$A_r = (-1)^r \frac{1}{a_0^{r+1}} \begin{vmatrix} c_0 & a_0 & \cdot & \cdots & \cdot \\ c_1 & a_1 & a_0 & \cdots & \cdot \\ c_2 & a_2 & a_1 & \cdots & \cdot \\ \cdot & \cdot & \cdot & \cdots & \cdot \\ c_{r-1} & a_{r-1} & a_{r-2} & \cdots & a_0 \\ c_r & a_r & a_{r-1} & \cdots & a_1 \end{vmatrix}.$$

Having stated this he recalls the theorem * that if

$$f(x) = a_0(x-x_1)(x-x_2)\ldots(x-x_n)$$

we have

$$\psi(x_1)+\psi(x_2)+\ldots+\psi(x_n) = \text{coeff. of } x^{-1} \text{ in } \frac{f'(x)\psi(x)}{f(x)},$$

and therefore as a special case

$$x_1^r + x_2^r + \ldots + x_n^r = \text{coeff. of } x^{-r-1} \text{ in } \frac{f'(x)}{f(x)}.$$

It is thus seen that to obtain a determinant expression for s_r we have only to make $\phi(x)$ identical with $f'(x)$,—in other words, put $m=n-1$, $c_0=na_0$, $c_1=(n-1)a_1$, $c_2=(n-2)a_2$, . . . ,—and find the coefficient of x^{-r-1}. Doing this we obtain

$$(-1)^r \frac{1}{a_0^{r+1}} \begin{vmatrix} na_0 & a_0 & \cdot & \cdots & \cdot \\ (n-1)a_1 & a_1 & a_0 & \cdots & \cdot \\ (n-2)a_2 & a_2 & a_1 & \cdots & \cdot \\ \cdot & \cdot & \cdot & \cdots & \cdot \\ (n-r)a_r & a_r & a_{r-1} & \cdots & a_1 \end{vmatrix}$$

which is readily reduced to

$$(-1)^r \frac{1}{a_0^{r}} \begin{vmatrix} a_1 & a_0 & \cdot & \cdots & \cdot \\ 2a_2 & a_1 & a_0 & \cdots & \cdot \\ \cdot & \cdot & \cdot & \cdots & \cdot \\ ra & a_{r-1} & a_{r-2} & \cdots & a_1 \end{vmatrix},$$

and so agrees with the result obtained from Newton's relations between the a's and s's.

* Said to be first given by Cauchy in his *Exercices de Math.* for 1826.

Here Faure leaves the subject, but he might equally easily have established Brioschi's more general result. Instead of specialising by putting $\psi(x)=x^r$ he might have made $\psi(x)=x^r\phi(x)f'(x)$ and so have got

$$\frac{x_1{}^r\phi(x_1)}{f'(x_1)}+\frac{x_2{}^r\phi(x_2)}{f'(x_2)}+\ldots\ldots+\frac{x_n{}^r\phi(x_n)}{f'(x_n)} = \text{coeff. of } x^{-r-1} \text{ in } \frac{\phi(x)}{f(x)}.$$

Bearing in mind that $\phi(x)$ as used by Brioschi was of the nth degree, we have from Faure's fundamental theorem the said coefficient

$$= \mathrm{A}_{r+1} = (-1)^{r+1}\frac{1}{a_0^{r+1}}\begin{vmatrix} c_0 & a_0 & \cdot & \cdots & \cdot \\ c_1 & a_1 & a_0 & \cdots & \cdot \\ \cdot & \cdot & \cdot & \cdot & \cdot \\ c_r & a_r & a_{r-1} & \cdots & a_0 \\ c_{r+1} & a_{r+1} & a_r & \cdots & a_1 \end{vmatrix}$$

as it ought to be.

BRUNO, F. FAÀ DI (1855, December).

[Note sur une nouvelle formule du calcul différentiel. *Quart. Journ. of Math.*, i. pp. 359–360; or, with a different title, *Annali di Sci. mat. e fis.*, vi. pp. 479–480.]

The formula referred to is

$$\frac{\partial^{n+1}}{\partial x^{n+1}}\phi\{\psi(x)\} = \begin{vmatrix} \psi'\phi & n\psi''\phi & \frac{1}{2}n(n-1)\psi'''\phi & \cdots & \psi^{(n+1)}\phi \\ -1 & \psi'\phi & (n-1)\psi''\phi & \cdots & \psi^{(n)}\phi \\ \cdot & -1 & \psi'\phi & \cdots & \psi^{(n-1)}\phi \\ \cdot & \cdot & -1 & \cdots & \psi^{(n-2)}\psi \\ \cdot & \cdot & \cdot & \cdot & \cdot \\ \cdot & \cdot & \cdot & \cdots & \psi'\phi \end{vmatrix}$$

where the coefficients in the rth row are those of the expansion of $(a+b)^{n-r+1}$, and where after development of the determinant ϕ^r is to be taken as meaning the rth differential-quotient of ϕ with respect to ψ.*

* An opportunity was here lost by Bruno of noting that a recurrent with the elements in its zero-bordered diagonal all negative has all its terms positive.

BRUNO, F. FAÀ DI (1856, February).

[Sulle funzioni isobariche. *Annali di Sci. mat. e fis.*, vii. pp. 76–89.]

On p. 81 Bruno enunciates, as having been recently ("ultimamente") discovered by him, a theorem which is essentially Faure's, the A_r of Faure being given by Bruno as the coefficient of x^r in the expansion of

$$\frac{c_0+c_1x+c_2x^2+\ldots\ldots}{a_0+a_1x+a_2x^3+\ldots\ldots}.$$

The reason for A_r being the same in both is evident on putting $m=n=0$ in Faure's.

ALLEGRET, A. (1857).

[Solutions de quelques problèmes curieux d'arithmétique. *Nouv. Annales de Math.*, xvi. pp. 136–139.]

In the course of Allegret's work, the determinant

$$\begin{vmatrix} 1 & . & . & . & 1 \\ 1 & a_1 & . & . & . \\ 1 & 1 & a_2 & . & . \\ 1 & . & 1 & a_3 & . \\ 1 & . & . & 1 & a_4 \end{vmatrix}$$

appears, which he says

$$= a_1a_2a_3a_4 + \begin{vmatrix} 1 & a_1 & . & . \\ 1 & 1 & a_2 & . \\ 1 & . & 1 & a_3 \\ 1 & . & . & 1 \end{vmatrix} = a_1a_2a_3a_4 - \begin{vmatrix} 1 & . & . & 1 \\ 1 & a_1 & . & . \\ 1 & 1 & a_2 & . \\ 1 & . & 1 & a_3 \end{vmatrix}$$

and, the four-line determinant now reached being similar in form to the original, he concludes that the final expansion of the latter must be

$$a_1a_2a_3a_4 - a_1a_2a_3 + a_1a_2 - a_1 + 1.$$

By passing the first row over the others to occupy the last place the determinant is recognisable as a special case of recurrent, and it is seen that the expansion in terms of the elements of the last row and their cofactors leads at once to Allegret's result.

BRIOSCHI, F. (1857).

[Solution de la question 350 (Wronski). *Nouv. Annales de Math.*, xvi. pp. 248–249.]

The problem having been set to find what Wronski called the "Aleph" functions * of the roots $x_1, x_2, \ldots, x_n$ of the equation

$$a_0x^n + a_1x^{n-1} + a_2x^{n-2} + \ldots + a_n = 0$$

in terms of the coefficients, Brioschi begins by saying that the rth of the said functions, being the complete homogeneous function of degree r, is the coefficient of z^r in the product

$$(1+x_1z+x_1^2z^2+\ldots)(1+x_2z+x_2^2z^2+\ldots)\ldots(1+x_nz+x_n^2z^2+\ldots)$$

i.e.

$$\frac{1}{(1-x_1z)(1-x_2z)\ldots(1-x_nz)}, \quad \text{or} \quad \frac{1}{\phi(z)} \text{ say.}$$

He thus has

$$\frac{1}{\phi(z)} = 1+\aleph_1z+\aleph_2z^2+\ldots\ldots,$$

and therefore by differentiation

$$-\frac{\phi'(z)}{\phi(z)} = \frac{\aleph_1+2\aleph_2z+3\aleph_3z^2+\ldots}{1+\aleph_1z+\aleph_2z^2+\ldots}.$$

But having also by a well-known theorem

$$\begin{aligned} -\frac{\phi'(z)}{\phi(z)} &= \frac{x_1}{1-x_1z}+\frac{x_2}{1-x_2z}+\ldots+\frac{x_n}{1-x_nz}, \\ &= \left.\begin{array}{l} \;\;x_1+x_1^2z+x_1^3z^2+\ldots\ldots \\ +x_2+x_2^2z+x_2^3z^2+\ldots\ldots \\ +\;\ldots\ldots\ldots\ldots\ldots \\ +x_n+x_n^2z+x_n^3z^2+\ldots\ldots \end{array}\right\} \\ &= s_1+s_2z+s_3z^2+\ldots\ldots, \end{aligned}$$

---

* WRONSKI, H. *Introduction à la Philosophie des Mathématiques* . . . . . . (pp. 65, . . .) vi+270 pp., Paris, 1811.

he deduces

$$s_1+s_2z+s_3z^2+\ldots = \frac{\aleph_1+2\aleph_2z+3\aleph_3z^2+\ldots\ldots}{1+\aleph_1z+\aleph_2z^2+\ldots\ldots},$$

whence by equating like coefficients of z there results

$$\begin{aligned}\aleph_1 &= s_1,\\ 2\aleph_2 &= s_2+\aleph_1s_1,\\ 3\aleph_3 &= s_3+\aleph_1s_2+\aleph_2s_1,\\ &\cdots\cdots\cdots\\ r\aleph_r &= s_r+\aleph_1s_{r-1}+\ldots\ldots+\aleph_{r-1}s_1.\end{aligned}$$

Multiplying now by $a_{r-1}, a_{r-2}, \ldots, a_0$ and adding, he has, on using Newton's relations between the a's and s's,

$$a_{r-1}\aleph_1+2a_{r-2}\aleph_2+\ldots\ldots+ra_0\aleph_r = -\{ra_r+(r-1)a_{r-1}\aleph_1+\ldots\ldots+a_1\aleph_{r-1}\},$$

whence comes Wronski's relation

$$a_r+a_{r-1}\aleph_1+\ldots\ldots+a_0\aleph_r = 0;$$

and, on solution of the set of relations obtained from this by putting $r=1, 2, \ldots, r$,

$$\aleph_r = \frac{(-1)^r}{a_0^{\ r}}\begin{vmatrix} a_1 & a_0 & \cdot & \cdots\cdot & \cdot\\ a_2 & a_1 & a_0 & \cdots\cdot & \cdot\\ \cdot & \cdot & \cdot & \cdots & \cdot\\ a_{r-1} & a_{r-2} & a_{r-3} & \cdots\cdot & a_0\\ a_r & a_{r-1} & a_{r-2} & \cdots\cdot & a_1\end{vmatrix}.$$

CATALAN, E. (1857).

[Note sur la question 350 (Wronski). *Nouv. Annales de Math.*, xvi. pp. 416–417.]

Catalan, under the anagrammatic signature of "M. Ange le Taunéac," points out a simplification. Having got as far as

$$\frac{1}{(1-x_1z)(1-x_2z)\ldots(1-x_nz)} = 1+\aleph_1z+\aleph_2z^2+\ldots$$

he merely draws attention to the fact that the denominator on the left being

$$= (a_0+a_1z+a_2z^2+ \ldots . +a_nz^n)\frac{1}{a_0},$$

there results

$$a_0 = (1+\aleph_1z+\aleph_2z^2+ \ldots)(a_0+a_1z+a_2z^2+ \ldots +a_nz^n),$$

whence Wronski's set of relations follows at once.

This, of course, is much preferable to Brioschi's procedure. It has to be noted, however, that by taking a roundabout way Brioschi came across the equations connecting the $\aleph$'s and the s's.*

---

* After all, it is this set of equations and the two other similar sets that are worth knowing, namely,

Newton's,	$a_0s_r+a_1s_{r-1}+a_2s_{r-2}+ \ldots +a_{r-1}s_1+ra_r = 0.$
Wronski's,	$a_r+\aleph_1a_{r-1}+\aleph_2a_{r-2}+ \ldots +\aleph_ra_0 = 0.$
Brioschi's,	$s_r+\aleph_1s_{r-1}+\aleph_2s_{r-2}+ \ldots \aleph_{r-1}s_1 = r\aleph_r.$

CHAPTER VII.

WRONSKIANS, FROM 1838 TO 1860.

THE previous history of Wronskians being not at all lengthy, was included in the chapter on "Miscellaneous Special Forms" (*History*, i. chap. xvi.), and is to be found there under Wronski 1812, Wronski 1815, Wronski 1816–17, and Schweins 1825 (pp. 472–478, 482–485).

The name dates only from 1882, being first suggested on p. 224 of my Text-book on Determinants.*

LIOUVILLE, J. (1838).

[Note sur la théorie de la variation des constantes arbitraires. *Journ.* (*de Liouville*) *de Math.*, iii. pp. 342–347.]

The Wronskian which incidentally appears here is of a special kind, namely, that in which the originating functions are in the so-called relation of being first differential-quotients of one and the same function, for example, in later notation,

$$\begin{vmatrix} \dfrac{\partial x}{\partial a} & \dfrac{\partial x}{\partial b} & \dfrac{\partial x}{\partial c} \\ \dfrac{\partial^2 x}{\partial a \partial t} & \dfrac{\partial^2 x}{\partial b \partial t} & \dfrac{\partial^2 x}{\partial c \partial t} \\ \dfrac{\partial^3 x}{\partial a \partial t^2} & \dfrac{\partial^3 x}{\partial b \partial t^2} & \dfrac{\partial^3 x}{\partial c \partial t^2} \end{vmatrix}.$$

It is worthy of note also that the expression for the differential-

* MUIR, TH. *A Treatise on the Theory of Determinants*, . . . viii + 240 pp., London.

quotient of this with respect to t is obtained in the form which accords with the case of Schweins' theorem of 1825 (*Hist.*, i. p. 484).

$$Zd\Big|\Big|(Zd)^a A_1 \cdot (Zd)^{a+1} A_2 \;\ldots\ldots\; (Zd)^{a+n-1} A_n\Big)$$
$$= \Big|\Big|(Zd)^a A_1 \cdot (Zd)^{a+1} A_2 \;\ldots\ldots\; (Zd)^{a+n-2} A_{n-1} \cdot (Zd)^{a+n} A_n\Big)$$

where $Z=1$ and $a=0$.

MALMSTEN, C. J. (1849).

[Moyens pour trouver l'expression de la n-ième intégrale particulière de l'équation $y^{(n)}+Py^{(n-1)}+\ldots+Sy^{(1)}+Ty=0$ à l'aide des $n-1$ valeurs $y_1, y_2, \ldots, y_{n-1}$ qui satisfont à celle équation. *Crelle's Journ.*, xxxix. pp. 91–98; or abstract in *Cambridge and Dub. Math. Journ.*, iv. pp. 286–288.]

The result obtained, after an introductory note on Determinants, is

$$y_n = z_1 y_1 + z_2 y_2 + \ldots + z_{n-1} y_{n-1},$$

where

$$z_r = (-1)^{n-1}\int \frac{dR}{dy_r^{(n-2)}} \cdot e^{-\int P dx} dx, \quad \text{and} \quad R = 1\Big/\sum \pm y_1 y_2' y_3'' \ldots y_{n-1}^{(n-2)}.$$

Only one special property of the Wronskian is used, namely, that regarding its differential-quotient.

MAINARDI, G. (1849, December).

[Sulla integrazione dell' equazioni differenziali. *Annali di Sci. mat. e fis.*, i. pp. 50–89.]

Mainardi in speaking unapprovingly (p. 70) of Libri's expressions for the coefficients of a linear differential equation in terms of its particular integrals gives his own instead, his statement being that

$$\sum_{r=1}^{r=m}(-1)^{m-r} y^{(r)} S\big(p_1' p_2'' p_3''' \ldots p_{r-1}^{(r-1)} p_r^{(r+1)} p_{r+1}^{(r+2)} \ldots p_{m-1}^{(m)}\big) = 0$$

is the linear differential equation of the mth order whose particular integrals are $p_1, p_2, p_3, \ldots, p_m$. With the explanation that the upper indices represent differentiations and that S( ) stands for the same as Cauchy's S(±) a glance suffices to raise doubts as to the accuracy of the statement. For one thing p_m does not occur in the equation, and there are other objections equally serious; but in any case it is clear that Mainardi had in his mind the correct result, his process probably being in the case of

$$y'''+Ay''+By'+Cy = 0$$

to solve the equations

$$\left.\begin{array}{l} p_1''' + Ap_1'' + Bp_1' + Cp_1 = 0 \\ p_2''' + Ap_2'' + Bp_2' + Cp_2 = 0 \\ p_3''' + Ap_3'' + Bp_3' + Cp_3 = 0 \end{array}\right\}$$

for A, B, C and then substitute in the original. This of course might be accomplished by eliminating A, B, C from the four equations, with the result

$$|\, y''' \, p_1'' \, p_2' \, p_3 \,| = 0,$$

just as when α, β, γ are the roots of $x^3+Ax^2+Bx+C=0$ we have

$$|\, x^3 \, \alpha^2 \, \beta^1 \, \gamma^0 \,| = 0.$$

In the analogy, therefore, between linear differential equations and ordinary algebraical equations the Wronskian is the analogue of the alternant.*

It may be added that the identities given on the same page by Mainardi as examples of his discoveries in the general theory of determinants are included in an old identity of Desnanot's numbered xlv. in *History*, i. p. 145.

PUISEUX, V. (1851).

[Sur la ligne dont les deux courbures ont entre elles un rapport constant. *Journ. (de Liouville) de Math.*, vi. pp. 208–211.]

At the close of his paper Puiseux remarks that his proof would have been shortened by using the theorem, "*Les lettres*

* We may appropriately note that in the same volume of the *Annali* are two short papers by P. Tardy on Malmsten's theorem just referred to (pp. 136–139, 337–341).

t, u, v, . . . , w *désignant* n *variables, si le déterminant du système de* n *quantités*

$$\begin{array}{cccc} t & u & \ldots\ldots & w \\ dt & du & \ldots\ldots & dw \\ d^2t & d^2u & \ldots\ldots & d^2w \\ \cdot & \cdot & \cdot & \cdot \\ d^{n-1}t & d^{n-1}u & \ldots\ldots & d^{n-1}w \end{array}$$

est égal à zéro, on a necessairement l'équation

$$at+bu+cv+\ldots+gw = 0$$

où a, b, c, . . . , g *sont des constantes.*" The theorem is spoken of as known, but no reference is given.

TISSOT, A. (1852).

[Sur un déterminant d'intégrales définies. *Journ.* (*de Liouville*) *de Math.*, xvii. pp. 177–185.]

Tissot incidentally asserts that if $y, y_1, y_2, \ldots, y_n$ all satisfy a linear differential equation similar to that dealt with by Malmsten above, then

$$\sum(\pm yy_1'y_2'' \ldots y_n^{(n)}) = \gamma\epsilon^{-\int \mathrm{P}dx}$$

where γ is independent of x. It is further stated that Liouville proved this in vol. x. of his journal, but such is not the case, the subject being not even referred to in that volume.

PROUHET, E. (1852).

[Mémoire sur quelques formules générales d'analyse. *Journ.* (*de Liouville*) *de Math.* (2), i. pp. 321–344.]

The third and last section (§§ 40–45) of Prouhet's memoir is headed "Théorèmes sur quelques déterminants de fonctions." The first theorem proved is that just mentioned as having been used by Malmsten. He then takes the set of $m+1$ equations,

$$\begin{array}{l} x_0 \cdot d^0(u\phi^0) + x_1 \cdot d^1(u\phi^0) + \ldots + x_m \cdot d^m(u\phi^0) = d^{m+1}(u\phi^0) \\ x_0 \cdot d^0(u\phi^1) + x_1 \cdot d^1(u\phi^1) + \ldots + x_m \cdot d^m(u\phi^1) = d^{m+1}(u\phi^1) \\ \cdot \quad \cdot \quad \cdot \quad \cdot \quad \cdot \quad \cdot \quad \cdot \quad \cdot \\ x_0 \cdot d^0(u\phi^m) + x_1 \cdot d^1(u\phi^m) + \ldots + x_m \cdot d^m(u\phi^m) = d^{m+1}(u\phi^m), \end{array}$$

and with the help of the said theorem obtains at once

$$x_m = \frac{d\Delta}{\Delta} = d(\log \Delta),$$

where Δ stands for

$$| d^0(u\phi^0) \cdot d^1(u\phi^1) \cdot \; \ldots \; \cdot d^m(u\phi^m) |.$$

Next, by using in connection with the same sets of equations the multipliers

$$\pm\phi^m, \quad \mp m\phi^{m-1}, \quad \ldots, \quad \tfrac{1}{2}m(m+1)\phi^2, \quad -m\phi^1, \quad \phi^0$$

and performing addition, the coefficients of $x_0, x_1, \ldots, x_{m-1}$ are found to vanish, with the result

$$\left\{x_m m!\, u(d\phi)^m\right\} = \tfrac{1}{2}m(m+1)\cdot m!\, u(d\phi)^{m-1}d^2\phi + (m-1)!\, du\cdot(d\phi^m),$$

this being due to the theorem in differentiation* that the expression of $m+1$ terms

$$d^r(u\phi^m) - m\phi \,.\, d^r(u\phi^{m-1}) + \tfrac{1}{2}m(m+1)\phi^2 \,.\, d^r(u\phi^{m-2}) - \ldots .$$

has the values

$$0, \quad m!\, u(d\phi)^m, \quad \tfrac{1}{2}m(m+1)\cdot m!\, u(d\phi^{m-1})d^2\phi + (m+1)!\, du\cdot(d\phi^m)$$

according as $r < m$, $= m$, or $= m+1$. An alternative value for x_m is thus found, namely,

$$\tfrac{1}{2}m(m+1)\frac{d^2\phi}{d\phi} + (m+1)\frac{du}{u} \quad \text{i.e.} \quad d\log\left[u^{m+1}(d\phi)^{\frac{1}{2}m(m+1)}\right];$$

and from the two values it follows that

$$\Delta = u^{m+1}(d\phi)^{\frac{1}{2}m(m+1)} \times \text{a constant},$$

the constant being determined to be

$$\Sigma[\pm 1^0 2^1 3^2 \ldots (m+1)^m], \text{ or } 1!\,2!\,3!\ldots m!$$

by considering the particular case where $u = \phi = e^x$. The final result thus is

$$\begin{vmatrix} u\phi^0 & d(u\phi^0) & \ldots. & d^m(u\phi^0) \\ u\phi^1 & d(u\phi^1) & \ldots. & d^m(u\phi^1) \\ \cdot & \cdot & \cdot & \cdot \\ u\phi^m & d(u\phi^m) & \ldots. & d^m(u\phi^m) \end{vmatrix} = 1!\,2!\,3!\ldots m!\, u^{m+1}(d\phi)^{\frac{1}{2}m(m+1)},$$

* Attributed in part to Lexell (1772) and to Arbogast (1800).

which on putting $u=1$ becomes Wronski's result of the year 1816, and therefore a case of Schweins' generalisation of 1825.

A[BADIE, T.] (1852).

[Sur la différentiation des fonctions de fonctions: séries de Burmann, de Lagrange, de Wronski. *Nouv. Annales de Math.*, xi. pp. 376–383; or French translation of Brioschi's *Teorica dei Determinanti*, pp. 182–193.]

By a method similar to Prouhet's, namely, by solving a set of equations in two different ways, Abadie obtains

$$\frac{\dfrac{d^{n-1}}{dh^{n-1}}\Big[\theta^{-n}\mathrm{F}'\big(n+h\big)\Big]_{h=0}}{1\cdot 2\cdot 3\ldots n} = \frac{\sum[\pm \mathrm{D}^1\phi\cdot\mathrm{D}^2\phi^2\cdot\ \ldots\ \cdot\mathrm{D}^{n-1}\phi^{n-1}\cdot\mathrm{D}^n\mathrm{F}]}{\sum[\pm \mathrm{D}^1\phi\cdot\mathrm{D}^2\phi^2\cdot\ \ldots\ \cdot\mathrm{D}^{n-1}\phi^{n-1}\cdot\mathrm{D}^n\phi^n]}$$

where θ and $\mathrm{D}^r\phi^s$ stand for

$$\frac{\phi(x+h)-\phi(x)}{h} \quad \text{and} \quad \frac{d^r}{dx^r}\Big\{\phi(x)\Big\}^s$$

respectively. Thence, by equating coefficients of $\mathrm{D}^n\mathrm{F}$, Wronski's result

$$\sum[\pm \mathrm{D}^1\phi\cdot\mathrm{D}^2\phi^2\cdot\ \ldots\ \cdot\mathrm{D}^n\phi^n] = 1!\,2!\,3!\ldots n!\,\{\phi'(x)\}^{\,n(n+1)}$$

is reached; and this in its turn is then used to simplify the result which has just originated it, another of Wronski's formulæ being thus arrived at.

BRIOSCHI, F. (1855).

[Sur une propriété d'un déterminant fonctionnel. *Quart. Journ. of Math.*, i. pp. 365–367; or *Opere mat.*, v. pp. 389–392.]

In later phraseology the property in question is that *if the Wronskian*, W *say, of* $y_1, y_2, \ldots y_n$ *be a known function of the independent variable* x, *then any one of the* y's, y_r *say,* can be *expressed in terms of the others and* W. Denoting the sth differential-quotient of y_r by $y_r^{(s)}$ we have of course

$$W \equiv \begin{vmatrix} y_1 & y_2 & \dots & y_n \\ y_1^{(1)} & y_2^{(1)} & \dots & y_n^{(1)} \\ \cdot & \cdot & \cdot & \cdot \\ y_1^{(n-1)} & y_2^{(n-1)} & \dots & y_n^{(n-1)} \end{vmatrix}$$

and using "cof" for "cofactor of" we readily see that since

$$\frac{d}{dx}(\operatorname{cof} y_s^{(n-1)}) = -\operatorname{cof} y_s^{(n-2)}$$

we have

$$\frac{d}{dx}\left(\frac{\operatorname{cof} y_s^{(n-1)}}{\operatorname{cof} y_r^{(n-1)}}\right) = \frac{-\operatorname{cof} y_r^{(n-1)} \operatorname{cof} y_s^{(n-2)} + \operatorname{cof} y_s^{(n-1)} \operatorname{cof} y_r^{(n-2)}}{(\operatorname{cof} y_r^{(n-2)})^2}$$

$$= \frac{W.\operatorname{cof}(y_s^{(n-1)}\, y_r^{(n-2)})}{(\operatorname{cof} y_r^{(n-1)})^2}.$$

Integration of both sides with respect to x then gives

$$\operatorname{cof} y_s^{(n-1)} = \operatorname{cof} y_r^{(n-1)} \int \frac{W.\operatorname{cof}(y_s^{(n-1)} y_r^{(n-2)})}{(\operatorname{cof} y_r^{(n-1)})^2}\, dx;$$

and it is seen that for the cofactor of any element in the last row of W except the rth there is an expression in which the function y_r never explicitly occurs. It only remains then to take the well-known identity

$$0 = y_1 \operatorname{cof} y_1^{(n-1)} + y_2 \operatorname{cof} y_2^{(n-1)} + \dots + y_r \operatorname{cof} y_r^{(n-1)} + \dots + y_n \operatorname{cof} y_n^{(n-1)},$$

write therein for $\operatorname{cof} y_s^{(n-1)}$ the substitute thus provided, and divide both sides by $\operatorname{cof} y_r^{(n-1)}$; for, this being done, there results

$$0 = y_1 I_1 + y_2 I_2 + \dots + y_r + \dots + y_n I_n$$

where I stands for the integral above written.

Brioschi then proceeds to establish Malmsten's theorem of 1849.

CHRISTOFFEL, E. B. (1857).

[Ueber die lineare Abhängigkeit von Functionen einer einzigen Veränderlichen. *Crelle's Journ.*, lv. pp. 281–299.]

The results directly bearing on the subject specified in the title of the paper are summed up in five propositions (pp. 293–294), the one comparable with Puiseux's of 1851 being that *If* $\Sigma\pm f(x)\cdot f_1'(x)\ldots f_n^{(n)}(x)$ *vanishes for all values of* x *from* $x=x_0$ *to* $x=x_1$, *where* $x_0<x_1$, *then the functions* $f(x), f_1(x), \ldots, f_n(x)$ *are for those values linearly dependent.* In a concluding section are brought together (pp. 297–299) the properties of the determinants which had been used in reaching the said results. The first theorem is

$$\sum \pm uu_1'u_2'' \ldots u_n^{(n)} = \sum \pm u\cdot\Delta u'\cdot\Delta^2 u'' \ldots \Delta^n u^{(n)} \qquad (1)$$

where

$$\Delta u_\nu^{(\mu)} = u_{\nu+1}^{(\mu)} - u_\nu^{(\mu)}, \qquad \Delta^2 u_\nu^{(\mu)} = \Delta u_{\nu+1}^{(\mu)} - \Delta u_\nu^{(\mu)}, \quad \ldots .$$

This, it is stated, can be proved in the same way as the theorem

$$\sum \pm f(m)\cdot f_1(m+1)\ldots f_n(m+n) = \sum \pm f(m)\cdot \Delta f_1(m) \ldots . \Delta^n f_n(m) \quad (1')$$

given earlier in the paper (p. 293), namely, by evolving the left-hand side from the right-hand side after substituting for the differences in the latter their equivalents obtainable from the identity

$$\Delta^i f_k(m) = f_k(m+i) - if_k(m+i-1) + \frac{i(i-1)}{1.2} f_k(m+i-2) - \ldots .$$

There is next derived the theorem

$$\sum \pm vu\cdot\Delta(vu')\cdot\Delta^2(vu'')\ldots\Delta^n(vu^{(n)})$$
$$= vv_1v_2\ldots v_n\cdot\sum \pm u\cdot\Delta u'\cdot\Delta^2 u'' \ldots . \Delta^n u^{(n)} \quad (2)$$

and from this again by putting $v_r = 1/u_r$ there is obtained

$$\sum \pm u\cdot\Delta u'\cdot\Delta^2 u'' \ldots \Delta^n u^{(n)} = uu_1u_2\ldots u_n \cdot \sum \pm \Delta\left(\frac{u'}{u}\right)\cdot\Delta^2\left(\frac{u''}{u}\right)\ldots\Delta^n\left(\frac{u^{(n)}}{u}\right).$$

Then, by using this last theorem on itself, and continuing in like manner, there is finally reached the result

$$\sum \pm u\cdot\Delta u'\cdot\Delta^2 u'' \ldots \Delta^n u^{(n)}$$
$$= (uu_1\ldots u_n)(u^{10}u_1^{10}\ldots u_{n-1}^{10})(u^{21}u_1^{21}\ldots u_{n-2}^{21})\ldots .(u^{n-1,n-2}u_1^{n-1,n-2})u^{n,n-1} \quad (3)$$

where

$$u^{\mu 0} = \Delta\frac{u^{(\mu)}}{u}, \quad u^{\mu,1} = \Delta\frac{u^{\mu,0}}{u^{1,0}}, \quad u^{\mu,2} = \Delta\frac{u^{\mu,1}}{u^{2,1}}, \quad \ldots .$$

In order to pass from differences to differentials Christoffel then puts

$$u_\mu^{(\nu)} = f_\nu(m\epsilon+\mu\epsilon), \quad v_\mu = \phi(m\epsilon+\mu\epsilon), \quad m\epsilon = x, \quad \epsilon = \partial x,$$

the result obtained from (2) being

$$\sum \pm \phi f \cdot \frac{\partial \phi f_1}{\partial x} \cdot \frac{\partial^2 \phi f_2}{\partial x^2} \cdots \frac{\partial^n \phi f_n}{\partial x^n} = \phi^{n+1} \sum \pm f \cdot \frac{\partial f_1}{\partial x} \cdot \frac{\partial^2 f_2}{\partial x^2} \cdots \frac{\partial^n f_n}{\partial x^n} \quad (4)$$

and from (3) being

$$\sum \pm f \cdot \frac{\partial f_1}{\partial x} \cdot \frac{\partial^2 f_2}{\partial x^2} \cdots \frac{\partial^n f_n}{\partial x^n} = f^{n+1} \cdot f_{1,0}^n \cdot f_{2,1}^{n-1} \cdots f_{n,n-1} \quad (5)$$

where

$$f_{\mu,0} = \frac{\partial}{\partial x}\left(\frac{f_\mu}{f}\right), \quad f_{\mu,1} = \frac{\partial}{\partial x}\left(\frac{f_{\mu,0}}{f_{1,0}}\right), \quad f_{\mu,2} = \frac{\partial}{\partial x}\left(\frac{f_{\mu,1}}{f_{2,1}}\right), \quad \ldots .$$

There is next given a result dealing with change of variable, namely,

$$\sum \pm f \cdot \frac{\partial f_1}{\partial x} \cdot \frac{\partial^2 f_2}{\partial x^2} \cdots \frac{\partial^n f_n}{\partial x^n} = \left(\frac{\partial t}{\partial x}\right)^{\frac{1}{2}n(n+1)} \sum \pm f \cdot \frac{\partial f_1}{\partial t} \cdot \frac{\partial^2 f_2}{\partial t^2} \cdots \frac{\partial^n f_n}{\partial t^n}$$

which is proved like (1) and (1′), the identity used for substitution purposes being now

$$\frac{\partial^r f}{\partial x^r} = \frac{\partial^r f}{\partial t^r}\left(\frac{\partial t}{\partial x}\right)^r + a_1^{(r)} \frac{\partial^{r-1} f}{\partial t^{r-1}} + a_2^{(r)} \frac{\partial^{r-2} f}{\partial t^{r-2}} + \ldots$$

where $a_1^{(r)}, a_2^{(r)}, \ldots$ do not involve differential-quotients of f.

Lastly, denoting the cofactors of $f^{(n)}, f_1^{(n)}, \ldots, f_n^{(n)}$ in

$$\sum \pm f \cdot \frac{\partial f_1}{\partial x} \cdot \frac{\partial^2 f_2}{\partial x^2} \cdots \frac{\partial^n f_n}{\partial x^n}$$

by $W_0, W_1, \ldots, W_n$ he announces the set of results

$$\begin{aligned}
\sum \pm W_0 W_1' W_2'' \ldots W_n^{(n)} &= W^n, \\
\sum \pm W_0 W_1' W_2'' \ldots W_{n-1}^{(n-1)} &= (-1)^n W^{n-1} f_n, \\
\sum \pm W_0 W_1' W_2'' \ldots W_{n-2}^{(n-2)} &= (-1)^{2n} W^{n-2} \cdot \sum \pm f_{n-1} f_n', \\
\sum \pm W_0 W_1' W_2'' \ldots W_{n-3}^{(n-3)} &= (-1)^{3n} W^{n-3} \cdot \sum \pm f_{n-2} f'_{-1} f_n'', \\
\cdots & \cdots \\
\sum \pm W_0 W_1' &= (-1)^{(n-1)n} W \cdot \sum \pm f_2 f_3' \ldots f_n^{(n-2)}, \\
W_0 &= (-1)^{nn} \sum \pm f_1 f_2' f_3'' \ldots f_n^{(n-1)}.
\end{aligned}$$

In regard to these we may note in passing that as

$$W_0, \quad W_1, \quad \ldots, \quad W_n$$

are themselves Wronskians, the name "Compound Wronskian" would not be inappropriate for the determinants on the left. Also, that in form the results bear a resemblance to those included in Jacobi's theorem regarding any minor of the adjugate of a general determinant.

HESSE, O. (1857).

[Ueber die Criterien des Maximums und Minimums der einfachen Integrale. *Crelle's Journ.*, liv. pp. 227–273; or *Werke*, pp. 413–467.]

In the course of his investigation Hesse pauses (p. 249) to enunciate the result which we have numbered (4) in dealing with Christoffel's paper. He also gives Christoffel's fifth result, having arrived at it, however, by a different method, namely, by the repeated use of the immediately preceding result. Thus, in the case of the 3rd order, the first part of his procedure would be

$$\begin{vmatrix} u & u' & u'' \\ v & v' & v'' \\ w & w' & w'' \end{vmatrix} = \begin{vmatrix} u\cdot 1 & (u\cdot 1)' & (u\cdot 1)'' \\ u\cdot\frac{v}{u} & \left(u\cdot\frac{v}{u}\right)' & \left(u\cdot\frac{v}{u}\right)'' \\ u\cdot\frac{w}{u} & \left(u\cdot\frac{w}{u}\right)' & \left(u\cdot\frac{w}{u}\right)'' \end{vmatrix} = u^3 \begin{vmatrix} 1 & (1)' & (1)'' \\ \frac{v}{u} & \left(\frac{v}{u}\right)' & \left(\frac{v}{u}\right)'' \\ \frac{w}{u} & \left(\frac{w}{u}\right)' & \left(\frac{w}{u}\right)'' \end{vmatrix}$$

$$= u^3 \begin{vmatrix} \left(\frac{v}{u}\right)' & \left(\frac{v}{u}\right)'' \\ \left(\frac{w}{u}\right)' & \left(\frac{w}{u}\right)'' \end{vmatrix};$$

and he would next treat this two-line determinant in similar fashion, the result being

$$\sum \pm uv'w'' = u^3\cdot\left\{\left(\frac{v}{u}\right)'\right\}^2\cdot\left\{\frac{\left(\frac{w}{u}\right)'}{\left(\frac{v}{u}\right)'}\right\}'.$$

MONTFERRIER, A. S. DE (1858).

[ENCYCLOPÉDIE MATHÉMATIQUE, ou exposition complète de toutes les branches des mathématiques d'après les principes . . . . de Hoëne Wronski. Première Partie: Mathématiques Pures. Tomes i.–iv., Paris.*]

Although, from the nature of this work, it cannot be expected to contain fresh results, it would be a mistake to undervalue it, as in the matter of exposition the disciple had more skill than his master. Whether, therefore, as a substitute for, or a commentary on, the original, it deserves attention. It is only in the *third* volume that the "Schin" functions appear, a short general account being given in §§ 1041, 1042 (pp. 423–428), and special instances dealt with under the headings "Les Séries" (§§ 953–960, pp. 267–276) and "La Loi Suprême" (§§ 1006–1023, pp. 358–391).

* None of the four volumes is dated, but they appeared in 1856-59: they do not complete the First Part.

CHAPTER VIII.

JACOBIANS, FROM 1841 TO 1860.

WHEN one recalls the exceptional fullness of Jacobi's memoir on this special form and the fact that it belongs really to the first year of the period now under discussion, one cannot be surprised that during the remainder of the period little was added save commentaries and suggested amendments. Even Bertrand's contribution, the longest and most interesting, is practically of this character.

JACOBI, C. G. J. (1844–1845).

[Theoria novi multiplicatoris systemati æquationum differentialium vulgarium applicandi. *Crelle's Journ.*, xxvii. pp. 199–268; xxix. pp. 218–279, 333–376; or *Math. Werke* (1846), i. pp. 47–226; or *Gesammelte Werke*, iv. pp. 317–509.]

The portion of this long memoir which is of interest to us in the present connection is the first section (pp. 201–209) of the first chapter, the heading being "Lemma fundamentale eiusque varii usus: de determinantibus functionalibus partialibus." Passing over the treatment of the first two cases of the lemma we come upon the general enunciation of it, which is—

If $A, A_1, A_2, \ldots, A_n$ *be the cofactors of* $\frac{\partial f}{\partial x}, \frac{\partial f}{\partial x_1}, \frac{\partial f}{\partial x_2}, \ldots, \frac{\partial f}{\partial x_n}$ *in the determinant* $\sum\left(\pm\frac{\partial f}{\partial x}\frac{\partial f_1}{\partial x_1}\frac{\partial f_2}{\partial x_2}\cdots\frac{\partial f_n}{\partial x_n}\right)$, *then*

$$\frac{\partial A}{\partial x}+\frac{\partial A_1}{\partial x_1}+\frac{\partial A_2}{\partial x_2}+\ldots+\frac{\partial A_n}{\partial x_n}=0.$$

Preparatory for the proof it is pointed out that since

$$\frac{\partial f}{\partial x}A + \frac{\partial f}{\partial x_1}A_1 + \ldots + \frac{\partial f}{\partial x_n}A_n = \sum\left(\pm\frac{\partial f}{\partial x}\frac{\partial f_1}{\partial x_1}\cdots\frac{\partial f_n}{\partial x_n}\right)$$

an alternative form of the lemma is

$$\frac{\partial(fA)}{\partial x} + \frac{\partial(fA_1)}{\partial x_1} + \ldots + \frac{\partial(fA_n)}{\partial x_n} = \sum\left(\pm\frac{\partial f}{\partial x}\frac{\partial f_1}{\partial x_1}\cdots\frac{\partial f_n}{\partial x_n}\right).$$

Then calling the given determinant R, and noting that A, $A_1, \ldots, A_n$ are themselves functional determinants, A_i being the determinant of $f_1, f_2, \ldots, f_n$ with respect to $x, x_1, \ldots, x_{i-1}, x_{i+1}, \ldots, x_n$, Jacobi seeks to prove the lemma true in the case of R from assuming it true in the case of A_i. To be able to formulate it in the latter case he takes each element of the first row of R along with each element of the second row, thus forming $(n+1)^2$ products whose cofactors in the determinant he denotes by

$$\begin{array}{ccccc} (00) & (01) & (02) & \ldots & (0n) \\ (10) & (11) & (12) & \ldots & (1n) \\ (20) & (21) & (22) & \ldots & (2n) \\ \cdot & \cdot & \cdot & \cdot & \cdot \\ (n0) & (n1) & (n2) & \ldots & (nn), \end{array}$$

that is to say, he puts

$$(\iota\kappa) \text{ for the cofactor of } \frac{\partial f}{\partial x_i}\frac{\partial f_1}{\partial x_k} \text{ in R,}$$

a notation which necessitates

$$(ik) = -(ki) \quad \text{and} \quad (ii) = 0.$$

It follows on this that the cofactors of

$$\frac{\partial f_1}{\partial x}, \frac{\partial f_1}{\partial x_1}, \ldots, \frac{\partial f_1}{\partial x_{i-1}}, \frac{\partial f_1}{\partial x_{i+1}}, \ldots, \frac{\partial f_1}{\partial x_n}$$

in A_i are $(i, 0), (i, 1), \ldots, (i, n)$, and thus the assumption above made takes the form

$$\frac{\partial(i, 0)}{\partial x} + \frac{\partial(i, 1)}{\partial x_1} + \ldots. + \frac{\partial(i, n)}{\partial x_n} = 0,$$

$$\text{or} \quad \frac{\partial\{(i, 0)f_1\}}{\partial x} + \frac{\partial\{(i, 1)f_1\}}{\partial x_1} + \ldots. + \frac{\partial\{(i, n)f_1\}}{\partial x_n} = A_i.$$

Since, however, $(i, k)f_1 = -(k, i)f_1$ and $(i, i)f_1 = 0$, we can apply to the latter the general proposition that *if* a_{ik} *be any quantities whatever such that* $a_{ik} = -a_{ki}$, $a_{ii} = 0$, *and* H_i *stand for*

$$\frac{\partial a_{i0}}{\partial x} + \frac{\partial a_{i1}}{\partial x_1} + \ldots + \frac{\partial a_{in}}{\partial x_n}$$

then

$$\frac{\partial H}{\partial x} + \frac{\partial H_1}{\partial x_1} + \ldots + \frac{\partial H_n}{\partial x_n} = 0,*$$

and so reach the first of our aims as desired. There only then remains to show that the lemma holds in the case of two variables, and this is unnecessary because it is then identical with the familiar proposition

$$\frac{\partial f_1}{\partial x\,\partial y} = \frac{\partial f_1}{\partial y\,\partial x}.$$

In addition to this gradational proof Jacobi gives one of a different kind. Since A_i, he says, does not involve differential-coefficients with respect to x_i, it follows that $\frac{\partial A_i}{\partial x_i}$ and $\sum \frac{\partial A_i}{\partial x_i}$ cannot involve differential-coefficients taken twice with respect to any one variable. Further, second differential-coefficients taken with respect to different variables x_i, x_k cannot occur anywhere save in†

$$\frac{\partial A_i}{\partial x_i} + \frac{\partial A_k}{\partial x_k}.$$

All we have got to show therefore is that the cofactor of $\frac{\partial^2 f_m}{\partial x_i \partial x_k}$ in $\frac{\partial A_i}{\partial x_i} + \frac{\partial A_k}{\partial x_k}$ vanishes. To do this we express A_i in terms of the elements of one column and their cofactors, say

$$A_i = a_1 \frac{\partial f_1}{\partial x_k} + a_2 \frac{\partial f_2}{\partial x_k} + \ldots\ldots + a_n \frac{\partial f_n}{\partial x_k},$$

---

* The reason for this, of course, is that $\dfrac{\partial \frac{\partial a_{ik}}{\partial x_k}}{\partial x_i} + \dfrac{\partial \frac{\partial a_{ki}}{\partial x_i}}{\partial x_k} = 0.$

† It would have been well to make clear here that every term of the final expansion of $\sum \frac{\partial A_i}{\partial x_i}$ contains one and only one second differential-coefficient.

and thus know as above that

$$A_k = -\alpha_1 \frac{\partial f_1}{\partial x_i} - \alpha_2 \frac{\partial f_2}{\partial x_i} - \ \ldots\ldots - \alpha_n \frac{\partial f_n}{\partial x_i},$$

where $\alpha_1, \alpha_2, \ldots$ involve no differential-coefficients taken with respect to x_i or with respect to x_k.

The observation made in the course of the first proof that $A, A_1, \ldots, A_n$ are themselves functional determinants leads Jacobi to the conception of "partial functional determinants" on the analogy of partial differential-quotients. The fundamental lemma then becomes viewable as the analogue of

$$\frac{\partial \frac{\partial f_1}{\partial y}}{\partial x} - \frac{\partial \frac{\partial f_1}{\partial x}}{\partial y} = 0,$$

or, in Jacobi's words, "gravissimam manifestat analogiam determinantium functionalium et quotientium differentialium partialium."

Apparently this recalls to Jacobi another analogy of the same kind, which he had omitted to draw attention to in his paper of 1830, when the first two cases of the lemma had been originally enunciated by him. The proposition involving the said analogy he now generalises thus:—*If* f, $f_1, f_2, \ldots, f_n$ *be expressible as series the terms of which involve only powers of the variables* x, $x_1, x_2, \ldots, x_n$, *the functional determinant does not involve a term in* $x^{-1} x_1^{-1} x_2^{-1} \ldots\ldots x_n^{-1}$. In support of it he has only to point out that the functional determinant is equal to

$$\frac{\partial(fA)}{\partial x} + \frac{\partial(fA_1)}{\partial x_1} + \ldots + \frac{\partial(fA_n)}{\partial x_n},$$

and that the development of the kth term of this expansion cannot contain a term in $1/x_{k-1}$.

After referring to a possible application of the lemma in connection with definite multiple integrals, Jacobi concludes § 2 by returning to the lemma itself and throwing it into a third form originally announced in 1841 (*De determ. funct.* § 9).

Viewing $x, x_1, x_2, \ldots . x_n$ as functions of $f, f_1, f_2, \ldots, f_n$ he obtains of course (*De determ. funct.* § 8)

$$\frac{\partial x}{\partial f}=\frac{\mathrm{A}}{\mathrm{R}}, \quad \frac{\partial x_1}{\partial f}=\frac{\mathrm{A}_1}{\mathrm{R}}, \quad \ldots\ldots, \quad \frac{\partial x_n}{\partial f}=\frac{\mathrm{A}_n}{\mathrm{R}},$$

so that by substitution the lemma becomes

$$\begin{aligned}
0 &= \frac{\partial\left(\mathrm{R}\frac{\partial x}{\partial f}\right)}{\partial x}+\frac{\partial\left(\mathrm{R}\frac{\partial x_1}{\partial f}\right)}{\partial x_1}+\ldots\ldots+\frac{\partial\left(\mathrm{R}\frac{\partial x_n}{\partial f}\right)}{\partial x_n}, \\
&= \left.\begin{matrix} \frac{\partial x}{\partial f}\frac{\partial \mathrm{R}}{\partial x} \\ +\mathrm{R}\frac{\partial\frac{\partial x}{\partial f}}{\partial x}\end{matrix}\right\} + \left.\begin{matrix} \frac{\partial x_1}{\partial f}\frac{\partial \mathrm{R}}{\partial x_1} \\ +\mathrm{R}\frac{\partial\frac{\partial x_1}{\partial f}}{\partial x_1}\end{matrix}\right\} + \ldots\ldots + \left.\begin{matrix} \frac{\partial x_n}{\partial f}\frac{\partial \mathrm{R}}{\partial x_n} \\ +\mathrm{R}\frac{\partial\frac{\partial x_n}{\partial f}}{\partial x_n}\end{matrix}\right\}, \\
&= \frac{\partial \mathrm{R}}{\partial f}+\mathrm{R}\left\{\frac{\partial\frac{\partial x}{\partial f}}{\partial x}+\frac{\partial\frac{\partial x_1}{\partial f}}{\partial x_1}+\ldots\ldots+\frac{\partial\frac{\partial x_n}{\partial f}}{\partial x_n}\right\},
\end{aligned}$$

or

$$0 = \frac{\partial \log \mathrm{R}}{\partial f}+\frac{\partial\frac{\partial x}{\partial f}}{\partial x}+\frac{\partial\frac{\partial x_1}{\partial f}}{\partial x_1}+\ldots\ldots+\frac{\partial\frac{\partial x_n}{\partial f}}{\partial x_n},$$

where firstly R and the x's have to be viewed as functions of the f's and all differentiated with respect to f, and secondly the differential-coefficients thus obtained have to be expressed in terms of the x's preparatory to performing the final set of differentiations.

Here the consideration of functional determinants would have come to an end in the present memoir, had it not been that a theorem on the subject which had been given incorrectly in 1841 (*De determ. funct.* § 14) was now wanted in § 3 for use. Two and a half pages (pp. 215–217) of matter are consequently intercalated in order to enunciate the theorem correctly, to prove it, and to elucidate it by a commentary. The enunciation is— *If* $\mathrm{f}_1=a_1$, $\mathrm{f}_2=a_2$, $\ldots\ldots$, $\mathrm{f}_\mathrm{n}=a_\mathrm{n}$, *where the a's are constants, the functional determinant*

$$\sum\left(\pm\frac{\partial f_1}{\partial x_1}\frac{\partial f_2}{\partial x_2}\cdots\frac{\partial f_n}{\partial x_n}\right)$$

will not be altered in value if before performing the differentiations every function f_i *be transformed in any way whatever by means of the equations* $\mathrm{f}_{\mathrm{i}+1}=a_{\mathrm{i}+1}$, $\mathrm{f}_{\mathrm{i}+2}=a_{\mathrm{i}+2}$, $\ldots\ldots$, $\mathrm{f}_\mathrm{n}=a_\mathrm{n}$. On looking back it will be seen that Jacobi had previously not excluded the use of the first $i-1$ equations in making transformation of f_i. His proof now depends on taking the a's to the same side as the f's; denoting the resulting equations by $\phi_1=0$, $\phi_2=0$, $\ldots\ldots$, $\phi_n=0$; applying his theorem (*De determ. funct.* § 10) to obtain the desired determinant in the form

$$(-1)^n\frac{\sum\left(\pm\frac{\partial\phi_1}{\partial x_1}\frac{\partial\phi_2}{\partial x_2}\cdots\frac{\partial\phi_n}{\partial x_n}\right)}{\sum\left(\pm\frac{\partial\phi_1}{\partial a_1}\frac{\partial\phi_2}{\partial a_2}\cdots\frac{\partial\phi_n}{\partial a_n}\right)};$$

and then showing that the denominator of this is $(-1)^n$. His commentary closes by assuming as allowable that

$$\phi_i = \lambda_1^{(i)}f_1 + \lambda_2^{(i)}f_2 + \ldots + \lambda_n^{(i)}f_n,$$

and having thus obtained

$$\sum\left(\pm\frac{\partial\phi_1}{\partial x_1}\frac{\partial\phi_2}{\partial x_2}\cdots\frac{\partial\phi_n}{\partial x_n}\right) = \sum(\pm\lambda_1^{(1)}\lambda_2^{(2)}\ldots\lambda_n^{(n)}).\sum\left(\pm\frac{\partial f_1}{\partial x_1}\frac{\partial f_2}{\partial x_2}\cdots\frac{\partial f_n}{\partial x_n}\right)$$

he concludes that the condition for the equality of the two functional determinants is that the determinant of the λ's shall be equal to 1.

HESSE, O. (1844).

[Ueber die Elimination der Variabeln aus drei algebraischen Gleichungen vom zweiten Grade mit zwei Variabeln. *Crelle's Journ.*, xxviii. pp. 68–96; or *Werke*, pp. 89–122.]

Having demonstrated that for the elimination of x_1, x_2, x_3 from the three quadrics f_1, f_2, f_3 it was important to discover a function of the third degree possessing certain properties, Hesse proceeds (§ 11) to show how such a function may be obtained.

From a well-known theorem regarding the differentiation of homogeneous functions he has, on putting $u_\kappa^{(\lambda)}$ for $\frac{\partial f_\kappa}{\partial x_\lambda}$,

$$\left.\begin{aligned} x_1u_1^{(1)} + x_2u_1^{(2)} + x_3u_1^{(3)} &= 2f_1 \\ x_1u_2^{(1)} + x_2u_2^{(2)} + x_3u_2^{(3)} &= 2f_2 \\ x_1u_3^{(1)} + x_2u_3^{(2)} + x_3u_3^{(3)} &= 2f_3 \end{aligned}\right\};$$

and, ϕ being Jacobi's determinant of f_1, f_2, f_3 with respect to x_1, x_2, x_3, there thus follows

$$(\alpha)\ x_1\phi = 2f_1(u_2^{(2)}u_3^{(3)} - u_2^{(3)}u_3^{(2)}) + 2f_2(u_3^{(2)}u_1^{(3)} - u_3^{(3)}u_1^{(2)}) + 2f_3(u_1^{(2)}u_2^{(3)} - u_1^{(3)}u_2^{(2)}),$$

with similar expressions for $x_2\phi$, $x_3\phi$; so that any set of values of x_1, x_2, x_3 which makes f_1, f_2, f_3 vanish will make ϕ vanish also. The formal enunciation of this result is then given, and it is pointed out that the like theorem holds when there are n homogeneous functions all of n variables and of the rth degree.

From (α) by differentiation there is next obtained

$$\begin{aligned} & x_1\frac{\partial\phi}{\partial x_1} + \phi \\ = {} & 2u_1^{(1)}(u_2^{(2)}u_3^{(3)} - u_2^{(3)}u_3^{(2)}) + 2u_2^{(1)}(u_3^{(2)}u_1^{(3)} - u_3^{(3)}u_1^{(2)}) + 2u_3^{(1)}(u_1^{(2)}u_2^{(3)} - u_1^{(3)}u_2^{(2)}) \\ & + 2f_1\frac{\partial}{\partial x_1}(u_2^{(2)}u_3^{(3)} - u_2^{(3)}u_3^{(2)}) + 2f_2\frac{\partial}{\partial x_1}(u_3^{(2)}u_1^{(3)} - u_3^{(3)}u_1^{(2)}) + 2f_3\frac{\partial}{\partial x_1}(u_1^{(2)}u_2^{(3)} - u_1^{(3)}u_2^{(2)}), \end{aligned}$$

and thence

$$\begin{aligned} & x_1\frac{\partial\phi}{\partial x_1} - \phi \\ = {} & 2f_1\frac{\partial}{\partial x_1}(u_2^{(2)}u_3^{(3)} - u_2^{(3)}u_3^{(2)}) + 2f_2\frac{\partial}{\partial x_1}(u_3^{(2)}u_1^{(3)} - u_3^{(3)}u_1^{(2)}) + 2f_3\frac{\partial}{\partial x_1}(u_1^{(2)}u_2^{(3)} - u_1^{(3)}u_2^{(2)}), \end{aligned}$$

so that any set of values x_1, x_2, x_3 which makes the ternary quadrics f_1, f_2, f_3 vanish, will make the first differential-quotients of the determinant of f_1, f_2, f_3 vanish also,—a second theorem which is asserted to hold when the number of homogeneous functions is n, the number of variables n, and each function of the rth degree in the said variables.

Combining the two results, and using later phraseology, we

may say that *When* n n-*ary* r[th]-*ics vanish, their Jacobian and each of its first differential-quotients will vanish also.*

The connection of this with the problem of elimination can be indicated in a few words. Jacobi's determinant ϕ being of the third degree in x_1, x_2, x_3, its first differential-quotients are like f_1, f_2, f_3 linear in x_1^2, x_2^2, x_3^2, x_2x_3, x_3x_1, x_1x_2; and consequently the resultant is at once obtained as a six-line determinant.

CAYLEY, A. (1847, February).

[On the differential equations which occur in dynamical problems. *Cambridge and Dub. Math. Journ.*, ii. pp. 210–219; or *Collected Math. Papers*, i. pp. 276–284.]

This is a short exposition of Jacobi's elaborate memoir of 1844 with considerable variation in the details. The portion (§ 1) which concerns us is of course that referring to the "fundamental lemma." This is established in its third form, the proof, like that originally given by Jacobi, being dependent on the theorem

$$\frac{\partial R}{\partial \frac{\partial f_i}{\partial x_k}} = R\frac{\partial x_k}{\partial f_i},$$

but differing in appearance, mainly because of the use of differentials.

BERTRAND, J. (1851, February).

[Mémoire sur le déterminant d'un système de fonctions. *Journ. (de Liouville) de Math.*, xvi. pp. 212–227; abstract in *Comptes Rendus .... Acad. des Sci.* (Paris), xxxii. pp. 134–135.]

Recalling how Jacobi had insisted on the marked analogy between a functional-determinant and a differential-coefficient, Bertrand at once intimates the adoption of a new definition of the former, which in his opinion makes the analogy still more striking, and from which the properties of the determinant are deducible like mere corollaries.

Save that Δ and δ are used where Bertrand without distinction uses d, the following is the definition:—If f_1, f_2, . . . , f_n be functions of $x_1, x_2, \ldots, x_n$, and the latter receive n distinct sets of increments

$$\begin{array}{cccc} \Delta_1 x_1 & \Delta_1 x_2 & \ldots\ldots & \Delta_1 x_n \\ \Delta_2 x_1 & \Delta_2 x_2 & \ldots\ldots & \Delta_2 x_n \\ \cdot & \cdot & \cdot\cdot\cdot\cdot & \cdot \\ \Delta_n x_1 & \Delta_n x_2 & \ldots\ldots & \Delta_n x_n \end{array}$$

with the result that the corresponding increments of the functions are

$$\begin{array}{cccc} \Delta_1 f_1 & \Delta_1 f_2 & \ldots\ldots & \Delta_1 f_n \\ \Delta_2 f_1 & \Delta_2 f_2 & \ldots\ldots & \Delta_2 f_n \\ \cdot & \cdot & \cdot\cdot\cdot\cdot & \cdot \\ \Delta_n f_1 & \Delta_n f_2 & \ldots\ldots & \Delta_n f_n , \end{array}$$

then the limiting value of the ratio of the determinant of the second array to the determinant of the first array as the elements of the latter array are indefinitely diminished is called the determinant of the n functions. Since in the circumstances mentioned

$$\Delta_k f_i = \frac{\partial f_i}{\partial x_1} \Delta_k x_i + \frac{\partial f_i}{\partial x_2} \Delta_k x_2 + \ldots\ldots + \frac{\partial f_i}{\partial x_n} \Delta_k x_n$$

for all values of k and i not greater than n, it follows from the multiplication-theorem that the aforesaid limiting value is equal to

$$\begin{vmatrix} \frac{\partial f_1}{\partial x_1} & \frac{\partial f_1}{\partial x_2} & \ldots\ldots & \frac{\partial f_1}{\partial x_n} \\ \frac{\partial f_2}{\partial x_1} & \frac{\partial f_2}{\partial x_2} & \ldots\ldots & \frac{\partial f_2}{\partial x_n} \\ \cdot & \cdot & \cdot\cdot\cdot\cdot & \cdot \\ \frac{\partial f_n}{\partial x_1} & \frac{\partial f_n}{\partial x_2} & \ldots\ldots & \frac{\partial f_n}{\partial x_n} \end{vmatrix};$$

and, this determinant being independent of the increments given to the independent variables, it is held that the definition is legitimised. It might have been added that the name assigned to the limiting value is also thereby justified.

The more important of Jacobi's results, eight or nine in number, are then re-established, precedence being given to those regarding the vanishing of the determinant. Supposing, first, that the functions are independent of one another, he asserts that $x_1, x_2, \ldots, x_n$ may be conceived as expressed in terms of $f_1, f_2, \ldots, f_n$; and, the latter being viewed as independent variables, the determinant of their increments can be considered as completely arbitrary and can thus have a value different from zero. Further, in relation to this determinant the determinant of the increments of $x_1, x_2, \ldots, x_n$ cannot be infinitely great, because the terms of both determinants have the same number of infinitesimal factors of the first order. It thus follows that their quotient—that is, the functional determinant—is not zero. Next, supposing that the functions are not all independent of one another, but that $f_{p+1}, f_{p+2}, \ldots, f_n$ are functions of $f_1, f_2, \ldots, f_p$, and that the latter alone are mutually independent, Bertrand asserts that we may suppose

$$\Delta_1 f_1 = 0, \quad \Delta_1 f_2 = 0, \quad \ldots\ldots, \quad \Delta_1 f_p = 0,$$

this in fact being possible in an infinite number of ways, because only p relations are thereby established between the increments $\Delta_1 x_1, \Delta_1 x_2, \ldots, \Delta_1 x_n$. It will then result that the increments of $f_{p+1}, f_{p+2}, \ldots, f_n$ being sums of multiples of the increments of $f_1, f_2, \ldots, f_p$ will also be zero; and thus the whole of the first row

$$\Delta_1 f_1 \quad \Delta_1 f_2 \quad \ldots\ldots \quad \Delta_1 f_n$$

will be composed of zeros, and the determinant to which it belongs will vanish. On the other hand the determinant of the increments of the x's can at the same time be made different from zero, the increments in the first row being not necessarily all zero and those in the other rows being what we please. The ratio of the two determinants therefore vanishes.

Following this are given the theorem regarding the relation of

$$\Sigma\left(\pm\frac{\partial f_1}{\partial x_1}\frac{\partial f_2}{\partial x_2}\cdots\frac{\partial f_n}{\partial x_n}\right) \quad \text{to} \quad \Sigma\left(\pm\frac{\partial x_1}{\partial f_1}\frac{\partial x_2}{\partial f_2}\cdots\frac{\partial x_n}{\partial f_n}\right):$$

the theorem for finding the functional determinant when the f's are given *mediately* as functions of the x's, that is to say, as functions of

$$\phi_1(x_1, x_2, \ldots, x_n), \quad \phi_2(x_1, x_2, \ldots, x_n), \quad \ldots\ldots, \quad \phi_p(x_1, x_2, \ldots, x_n):$$

and the corresponding theorem when the functions are only *implicitly* given, that is to say, by means of connecting equations

$$\begin{aligned} F_1(x_1, x_2, \ldots, x_n, f_1, f_2, \ldots, f_n) &= 0, \\ \cdot\quad\cdot\quad\cdot\quad\cdot\quad\cdot\quad\cdot\quad\cdot & \\ F_n(x_1, x_2, \ldots, x_n, f_1, f_2, \ldots, f_n) &= 0. \end{aligned}$$

The mode of treatment will be readily guessed from what has gone before.

The same cannot, however, be confidently affirmed in connection with the theorem which expresses the functional determinant as a single product. This is found grouped under the heading " Diverses formes que l'on peut donner à un déterminant " (fonctionnel), the said forms being obtainable on varying the systems of increments assigned to the variables. In the first example, the array of increments of the x's is taken to be

$$\begin{matrix} \Delta_1 x_1 & 0 & \ldots. & 0 \\ 0 & \Delta_2 x_2 & \ldots. & 0 \\ \cdot & \cdot & \cdot & \cdot \\ 0 & 0 & \ldots. & \Delta_n x_n \end{matrix}$$

which necessitates the other array being

$$\begin{matrix} \frac{\partial f_1}{\partial x_1}\Delta_1 x_1 & \frac{\partial f_2}{\partial x_1}\Delta_1 x_1 & \ldots. & \frac{\partial f_n}{\partial x_1}\Delta_1 x_1 \\ \frac{\partial f_1}{\partial x_2}\Delta_2 x_2 & \frac{\partial f_2}{\partial x_2}\Delta_2 x_2 & \ldots. & \frac{\partial f_n}{\partial x_2}\Delta_2 x_2 \\ \cdot & \cdot & \cdot & \cdot \\ \frac{\partial f_1}{\partial x_n}\Delta_n x_n & \frac{\partial f_2}{\partial x_n}\Delta_n x_n & \ldots. & \frac{\partial f_n}{\partial x_n}\Delta_n x_n \end{matrix}$$

and the ratio of the determinants of the arrays to be

$$\Sigma\left(\pm\frac{\partial f_1}{\partial x_1}\frac{\partial f_2}{\partial x_2}\cdots\frac{\partial f_n}{\partial x_n}\right).$$

To this there is added "C'est l'expression donnée comme définition par M. Jacobi,"—a remark, however, equally applicable when, as at the outset, the increments of the x's were taken in their most general form. Preparatory to the next example it is pointed out that any n of the variables

$$x_1, x_2, \ldots, x_n, \quad f_1, f_2, \ldots, f_n$$

may have arbitrary values, the other n being then determinable; and that therefore if $n-1$ of them be taken to be invariable, the ratios of the increments of the others may be considered known. We thus see that the two arrays

$$\begin{array}{ccccccccccc} \Delta_1 x_1 & \Delta_1 x_2 & \Delta_1 x_3 & \ldots & \Delta_1 x_n & \Delta_1 f_1 & 0 & 0 & \ldots & 0 \\ 0 & \Delta_2 x_2 & \Delta_2 x_3 & \ldots & \Delta_2 x_n & \Delta_2 f_1 & \Delta_2 f_2 & 0 & \ldots & 0 \\ 0 & 0 & \Delta_3 x_3 & \ldots & \Delta_3 x_n & \Delta_3 f_1 & \Delta_3 f_2 & \Delta_3 f_3 & \ldots & 0 \\ \cdot & \cdot & \cdot & \cdot & \cdot & \cdot & \cdot & \cdot & \cdot & \cdot \\ 0 & 0 & 0 & \ldots & \Delta_n x_n & \Delta_n f_1 & \Delta_n f_2 & \Delta_n f_3 & \ldots & \Delta_n f_n \end{array}$$

of the second example are simultaneously possible, the n independent variables in the case of the first row being $x_1, f_2, f_3, \ldots, f_n$, in the case of the second row $x_1, x_2, f_3, \ldots, f_n$, and so on: and we consequently learn that the functional determinant may be written in the form

$$\left(\frac{\partial f_1}{\partial x_1}\right)\left(\frac{\partial f_2}{\partial x_2}\right)\cdots\left(\frac{\partial f_n}{\partial x_n}\right)$$

on the understanding that the brackets enclosing $\partial f_r/\partial x_r$ imply that f_r is there viewed as a function of $x_1, x_2, \ldots, x_r, f_{r+1}, f_{r+2}, \ldots, f_n$. A third pair of possible arrays is

$$\begin{array}{ccccccccc} \Delta_1 x_1 & 0 & 0 & \ldots. & 0 & \Delta_1 x_{m+1} & \Delta_1 x_{m+2} & \ldots. & \Delta_1 x_n \\ 0 & \Delta_2 x_2 & 0 & \ldots. & 0 & \Delta_2 x_{m+1} & \Delta_2 x_{m+2} & \ldots. & \Delta_2 x_n \\ 0 & 0 & \Delta_3 x_3 & \ldots. & 0 & \Delta_3 x_{m+1} & \Delta_3 x_{m+2} & \ldots. & \Delta_3 x_n \\ \cdot & \cdot & \cdot & \cdot & \cdot & \cdot & \cdot & \cdot & \cdot \\ 0 & 0 & 0 & \ldots. & \Delta_m x_m & \Delta_m x_{m+1} & \Delta_m x_{m+2} & \ldots. & \Delta_m x_n \\ 0 & 0 & 0 & \ldots. & 0 & \Delta_{m+1} x_{m+1} & 0 & \ldots. & 0 \\ 0 & 0 & 0 & \ldots. & 0 & 0 & \Delta_{m+2} x_{m+2} & \ldots. & 0 \\ \cdot & \cdot & \cdot & \cdot & \cdot & \cdot & \cdot & \cdot & \cdot \\ 0 & 0 & 0 & \ldots. & 0 & 0 & 0 & \ldots. & \Delta_n x_n \end{array}$$

and

$$\begin{array}{cccccccccc}
\Delta_1 f_1 & \Delta_1 f_2 & \Delta_1 f_3 & \ldots\ldots & \Delta_1 f_m & 0 & 0 & \ldots\ldots & 0 \\
\Delta_2 f_1 & \Delta_2 f_2 & \Delta_2 f_3 & \ldots\ldots & \Delta_2 f_m & 0 & 0 & \ldots\ldots & 0 \\
\Delta_3 f_1 & \Delta_3 f_2 & \Delta_3 f_3 & \ldots\ldots & \Delta_3 f_m & 0 & 0 & \ldots\ldots & 0 \\
\cdot & \cdot & \cdot & \cdot & \cdot & \cdot & \cdot & \cdot & \cdot \\
\Delta_m f_1 & \Delta_m f_2 & \Delta_m f_3 & \ldots\ldots & \Delta_m f_m & 0 & 0 & \ldots\ldots & 0 \\
\Delta_{m+1} f_1 & \Delta_{m+1} f_2 & \Delta_{m+1} f_3 & \ldots\ldots & \Delta_{m+1} f_m & \Delta_{m+1} f_{m+1} & \Delta_{m+1} f_{m+2} & \ldots\ldots & \Delta_{m+1} f_n \\
\Delta_{m+2} f_1 & \Delta_{m+2} f_2 & \Delta_{m+2} f_3 & \ldots\ldots & \Delta_{m+2} f_m & \Delta_{m+2} f_{m+1} & \Delta_{m+2} f_{m+2} & \ldots\ldots & \Delta_{m+2} f_n \\
\cdot & \cdot & \cdot & \cdot & \cdot & \cdot & \cdot & \cdot & \cdot \\
\Delta_n f_1 & \Delta_n f_2 & \Delta_n f_3 & \ldots\ldots & \Delta_n f_m & \Delta_n f_{m+1} & \Delta_n f_{m+2} & \ldots\ldots & \Delta_n f_n,
\end{array}$$

from which the functional determinant is obtained in the form

$$\Sigma\left[\pm\left(\frac{\partial f_1}{\partial x_1}\right)\left(\frac{\partial f_2}{\partial x_2}\right)\cdots\left(\frac{\partial f_m}{\partial x_m}\right)\right]\cdot\Sigma\left(\pm\frac{\partial f_{m+1}}{\partial x_{m+1}}\frac{\partial f_{m+2}}{\partial x_{m+2}}\cdots\frac{\partial f_n}{\partial x_n}\right)$$

where, from looking as before at corresponding rows of the two arrays, we see that in the first determinant $f_1, f_2, \ldots, f_m$ are to be viewed as functions of $x_1, x_2, \ldots, x_m, f_{m+1}, f_{m+2}, \ldots, f_n$, and in the second determinant $f_{m+1}, f_{m+2}, \ldots, f_n$ are to be viewed as functions of $x_1, x_2, \ldots, x_n$.

The next section is still more interesting, as it concerns the proposition which Jacobi stated incorrectly in his original memoir of 1841 and returned to in 1844. The data according to Bertrand are the usual n functions $f_1, f_2, \ldots, f_n$ each dependent on $x_1, x_2, \ldots, x_n$, with the addition that the said functions when expressed in terms of $x_1, x_2, \ldots, x_n, f_1, f_2, \ldots, f_n$ become $\phi_1, \phi_2, \ldots, \phi_n$: and the problem he sets himself is to find the relation between

$$\Sigma\left(\pm\frac{\partial f_1}{\partial x_1}\frac{\partial f_2}{\partial x_2}\cdots\frac{\partial f_n}{\partial x_n}\right) \quad\text{and}\quad \Sigma\left(\pm\frac{\partial \phi_1}{\partial x_1}\frac{\partial \phi_2}{\partial x_2}\cdots\frac{\partial \phi_n}{\partial x_n}\right),$$

the differentiations in the latter determinant being performed on the understanding that the f's there occurring in the ϕ's are to be viewed as constants. The equations

$$\phi_1 - f_1 = 0, \quad \phi_2 - f_2 = 0, \quad \ldots, \quad \phi_n - f_n = 0,$$

may be held to give implicitly the f's as functions of the x's, and therefore by a previous result

$$\Sigma\left(\pm\frac{\partial f_1}{\partial x_1}\frac{\partial f_2}{\partial x_2}\cdots\frac{\partial f_n}{\partial x_n}\right) = (-1)^n \frac{\Sigma\pm\left(\frac{\partial\phi_1}{\partial x_1}\frac{\partial\phi_2}{\partial x_2}\cdots\frac{\partial\phi_n}{\partial x_n}\right)}{\begin{vmatrix} \frac{\partial\phi_1}{\partial f_1}-1 & \frac{\partial\phi_1}{\partial f_2} & \cdots & \frac{\partial\phi_1}{\partial f_n} \\ \frac{\partial\phi_2}{\partial f_1} & \frac{\partial\phi_2}{\partial f_2}-1 & \cdots & \frac{\partial\phi_2}{\partial f_n} \\ \cdot & \cdot & \cdot & \cdot \\ \frac{\partial\phi_n}{\partial f_1} & \frac{\partial\phi_n}{\partial f_2} & \cdots & \frac{\partial\phi_n}{\partial f_n}-1 \end{vmatrix}}$$

which is the relation desired. If ϕ_1 does not involve f_1, if ϕ_2 does not involve f_1 or f_2, if ϕ_3 does not involve f_1 or f_2 or f_3, and so on,* the determinant in the denominator takes the value $(-1)^n$, and the relation becomes one of equality.

The last section deals with the theorem regarding the change of variables in multiple integrals,—a theorem which in the ten years from Jacobi's memoir to Bertrand's had been discussed by Boole† and Dienger.‡

SPOTTISWOODE, W. (1851, 1853).

[ELEMENTARY THEOREMS RELATING TO DETERMINANTS, ... viii + 63 pp., London. Second edition, as an article in *Crelle's Journ.*, li. pp. 209–271, 328–381.]

Spottiswoode has a special chapter (§ x. pp. 51–57) headed "On Functional Determinants," its contents being a selection of Jacobi's theorems in unimproved form and a reprint of the first three paragraphs of Cayley's paper of 1847. In the second edition (§ ix. pp. 338–343) there is no change, save that the extract from Cayley is left out.

---

* Or if f_n be not involved in ϕ_n, and neither f_n nor f_{n-1} involved in ϕ_{n-1}, and so on.

† *Cambridge Math. Journ.*, iv. (1843), pp. 20-28.

‡ *Archiv d. Math. u. Phys.*, x. (1847), pp. 417-421.

SYLVESTER, J. J. (1853, June).

[On a theory of the syzygetic relations .... *Philos. Transac. R. Soc.* (London), cxliii. pp. 407–548; or *Collected Math. Papers*, i. pp. 429–586.]

In the present connection the only interest of this long and important memoir lies in the fact that Sylvester at page 476 of it uses for the first time the term *Jacobian* and the symbolism $J(f, g)$. His words are "J indicates the Jacobian of the given functions $f, g, \ldots$, meaning thereby the functional determinant of Jacobi."

CAUCHY, A. L. (1853, July).

[Mémoire sur les différentielles et les variations employées comme clefs algébriques. *Comptes rendus .... Acad. des Sci.* (Paris), xxxvii. pp. 38–45, 57–64; or *Œuvres complètes* (1), xii. pp. 46–63.]

The first section of the memoir opens, like his memoir of 1841, with the theorem

$$S(\pm D_x x\, D_y y\, D_z z \ldots)\cdot S(\pm D_x \mathrm{x}\, D_y \mathrm{y}\, D_z \mathrm{z} \ldots) = 1,$$

and then proceeds to the consideration of the arrays

$$\begin{matrix} (aa) & (ab) & (ac) & \ldots \\ (ba) & (bb) & (bc) & \ldots \\ (ca) & (cb) & (cc) & \ldots \\ \ldots & \ldots & \ldots & \ldots , \end{matrix} \qquad \begin{matrix} [aa] & [ab] & [ac] & \ldots \\ [ba] & [bb] & [bc] & \ldots \\ [ca] & [cb] & [cc] & \ldots \\ \ldots & \ldots & \ldots & \ldots , \end{matrix}$$

in which any element (hk) of the first array stands for

$$\begin{vmatrix} D_x h & D_x k \\ D_u h & D_u k \end{vmatrix} + \begin{vmatrix} D_y h & D_y k \\ D_v h & D_v k \end{vmatrix} + \begin{vmatrix} D_z h & D_z k \\ D_w h & D_w k \end{vmatrix} + \ldots\ldots$$

and any element $[hk]$ of the second array stands for

$$\begin{vmatrix} D_h x & D_k x \\ D_h u & D_k u \end{vmatrix} + \begin{vmatrix} D_h y & D_k y \\ D_h v & D_k v \end{vmatrix} + \begin{vmatrix} D_h z & D_k z \\ D_h w & D_k w \end{vmatrix} + \ldots\ldots$$

and $a, b, c, \ldots.$ are $2n$ functions of two sets of n variables $x, y, z, \ldots, u, v, w, \ldots.$ Of course both arrays are recognised to be zero-axial and skew, and under certain conditions it is shown that the elements of the one are expressible in terms of those of the other, and that the product of the determinant of the two arrays is 1.

The language and notation of 'clefs algébriques' are used throughout, but nothing is brought forward as justification for so doing.

DONKIN, W. F. (1854, February).

[On a class of differential equations, including those which occur in dynamical problems, Part I. *Philos. Transac. R. Soc.* (London), cxliv. pp. 71–113.]

It is only the first four pages of Donkin's memoir that concern us, these being introductory and referring to properties of a set of n functions of n variables without any regard to possible connections with dynamics. Drawing attention at the outset to the analogy remarked on by Jacobi and Bertrand, he proposes to signalise it by denoting the determinant of $f_1, f_2, \ldots, f_n$ with respect to $x_1, x_2, \ldots, x_n$ by

$$\frac{\partial(f_1, f_2, \ldots, f_n)}{\partial(x_1, x_2, \ldots, x_n)}.$$

Further, he views the numerator and denominator here as standing for the determinants of Bertrand's arrays of differences, remarking pointedly that the fraction indicated "is a real fraction, provided its numerator and denominator be interpreted in a manner exactly analogous to that in which the numerator and denominator of an ordinary total or partial differential-coefficient are interpreted."

Having thus explained his notation he proceeds to generalise the proposition that

$$\frac{\partial f_r}{\partial x_1}\frac{\partial x_1}{\partial f_s} + \frac{\partial f_r}{\partial x_2}\frac{\partial x_2}{\partial f_s} + \ldots + \frac{\partial f_r}{\partial x_n}\frac{\partial x_n}{\partial f_s} = 1 \text{ or } 0$$

according as r and s are equal or unequal. He recalls Jacobi's theorem (*De determin. funct.* § 11) that if $u_1, u_2, \ldots, u_m$ be functions of $y_1, y_2, \ldots, y_n$, n being greater than m, and the y's be functions of $x_1, x_2, \ldots, x_n$, then

$$\frac{\partial(u_{\alpha_1} u_{\alpha_2} \ldots u_{\alpha_m})}{\partial(x_{\gamma_1} x_{\gamma_2} \ldots x_{\gamma_m})} = \sum_\beta \left\{ \frac{\partial(u_{\alpha_1} u_{\alpha_2} \ldots u_{\alpha_m})}{\partial(y_{\beta_1} y_{\beta_2} \ldots y_{\beta_m})} \cdot \frac{\partial(y_{\beta_1} y_{\beta_2} \ldots y_{\beta_m})}{\partial(x_{\gamma_1} x_{\gamma_2} \ldots x_{\gamma_m})} \right\}$$

where $\alpha_1, \alpha_2, \ldots, \alpha_m$ and $\gamma_1, \gamma_2, \ldots, \gamma_m$ are fixed sets of m integers chosen from $1, 2, \ldots, n$, and $\beta_1, \beta_2, \ldots, \beta_m$ as any such set whatever. Taking then what he considers to be a particular case of this, namely,

$$\frac{\partial(y_{\alpha_1} y_{\alpha_2} \ldots y_{\alpha_m})}{\partial(y_{\gamma_1} y_{\gamma_2} \ldots y_{\gamma_m})} = \sum_\beta \left\{ \frac{\partial(y_{\alpha_1} y_{\alpha_2} \ldots y_{\alpha_m})}{\partial(x_{\beta_1} x_{\beta_2} \ldots x_{\beta_m})} \cdot \frac{\partial(x_{\beta_1} x_{\beta_2} \ldots x_{\beta_m})}{\partial(y_{\gamma_1} y_{\gamma_2} \ldots y_{\gamma_m})} \right\},$$

he points out that if the sets $\alpha_1, \alpha_2, \ldots, \alpha_m$ and $\gamma_1, \gamma_2, \ldots, \gamma_m$ be identical the determinant on the left is equal to 1, and that on the other hand if even one of the γ's be not included among the α's the determinant will have a column of zeros and therefore be zero itself. The generalisation aimed at thus is, that *If* $\mathrm{y}_1, \mathrm{y}_2, \ldots, \mathrm{y}_\mathrm{n}$ *be functions of* $\mathrm{x}_1, \mathrm{x}_2, \ldots, \mathrm{x}_\mathrm{n}$, *and* m *be less than* n, *then*

$$\sum_\beta \left\{ \frac{\partial(y_{\alpha_1} y_{\alpha_2} \ldots y_{\alpha_m})}{\partial(x_{\beta_1} x_{\beta_2} \ldots x_{\beta_m})} \cdot \frac{\partial(x_{\beta_1} x_{\beta_2} \ldots x_{\beta_m})}{\partial(y_{\gamma_1} y_{\gamma_2} \ldots y_{\gamma_m})} \right\} = 1 \text{ or } 0$$

according as the α *set of* m *integers chosen from* 1, 2, . . . , n *is identical or not with the* γ *set.* The illustrative example taken is the case where $n = 2$, and the mode of stating it is that

$$\sum \left\{ \left(\frac{\partial y_p}{\partial x_i} \frac{\partial y_q}{\partial x_j} - \frac{\partial y_p}{\partial x_j} \frac{\partial y_q}{\partial x_i} \right) \left(\frac{\partial x_i}{\partial y_\alpha} \frac{\partial x_j}{\partial y_\beta} - \frac{\partial x_i}{\partial y_\beta} \frac{\partial x_j}{\partial y_\alpha} \right) \right\} = 1 \text{ or } 0$$

according as $\alpha, \beta = p, q$ or not, it being understood that i, j is in succession

$$\begin{array}{llll} 1, 2; & 1, 3; & \ldots; & 1, n \\ & 2, 3; & \ldots; & 2, n \\ & \cdot & \cdot\ \cdot & \cdot \\ & & & n-1, n. \end{array}$$

DONKIN, W. F. (1854, February).

[Demonstration of a theorem of Jacobi's relative to functional determinants. *Cambridge and Dub. Math. Journ.*, ix. pp. 161–163.]

The theorem or identity in question is that of the year 1844. The functions being $u_1, u_2, \ldots, u_n$ and the independent variables $x_1, x_2, \ldots, x_n$, Donkin says that the functional determinant may be represented by

$$\begin{vmatrix} \frac{\partial_1}{\partial x_1} & \frac{\partial_1}{\partial x_2} & \cdots\cdots & \frac{\partial_1}{\partial x_n} \\ \frac{\partial_2}{\partial x_1} & \frac{\partial_2}{\partial x_2} & \cdots\cdots & \frac{\partial_2}{\partial x_n} \\ \cdot & \cdot & \cdot\ \cdot\ \cdot & \cdot \\ \frac{\partial_n}{\partial x_1} & \frac{\partial_n}{\partial x_2} & \cdots\cdots & \frac{\partial_n}{\partial x_n} \end{vmatrix} u_1 u_2 \ldots u_n,$$

it being understood that each symbol of differentiation is operative only upon that one of the functions which has the same suffix as the upper ∂ of the symbol. As a consequence of this he considers that the non-zero member of the identity sought to be established would be

$$\begin{vmatrix} \frac{\partial}{\partial x_1} & \frac{\partial}{\partial x_2} & \cdots\cdots & \frac{\partial}{\partial x_n} \\ \frac{\partial_2}{\partial x_1} & \frac{\partial_2}{\partial x_2} & \cdots\cdots & \frac{\partial_2}{\partial x_n} \\ \cdot & \cdot & \cdot\ \cdot\ \cdot & \cdot \\ \frac{\partial_n}{\partial x_1} & \frac{\partial_n}{\partial x_2} & \cdots\cdots & \frac{\partial_n}{\partial x_n} \end{vmatrix} u_2 u_3 \ldots u_n \qquad \text{(A)}$$

where the upper ∂'s of the first row being now without a suffix are supposed to be no longer restricted in their effect. As, however, the unrestricted symbol $\partial/\partial x_i$ is held to be equivalent to

$$\frac{\partial_2}{\partial x_i} + \frac{\partial_3}{\partial x_i} + \cdots\cdots + \frac{\partial_n}{\partial x_i}$$

the determinant operating on $u_2 u_3 \ldots u_n$ has the first element of each column equal to the sum of all the other elements of the column, and therefore vanishes. The identity is thus thought to be established.

In regard to this so-called demonstration we need only remark in passing (1) that the subject operated on is written in too product-like a form; (2) that an appropriate substitute for it would be $(u_1, u_2, \ldots, u_n)$, this being explained to be such that

$$\frac{\partial_r}{\partial x_s}(u_1, u_2, \ldots, u_n) = \frac{\partial u_r}{\partial x_s},$$

and

$$\frac{\partial}{\partial x_s}(u_1, u_2, \ldots, u_n) = \frac{\partial u_1}{\partial x_s} + \frac{\partial u_2}{\partial x_s} + \ldots + \frac{\partial u_n}{\partial x_s};$$

(3) that the assertion (A) is unsubstantiated, the fact

$$|a_1 b_2 c_3| = a_1|b_2 c_3| - a_2|b_1 c_3| + a_3|b_1 c_2|$$

being nothing more than a suggestion that

$$\begin{vmatrix} \frac{\partial}{\partial x_1} & \frac{\partial}{\partial x_2} & \frac{\partial}{\partial x_3} \\ b_1 & b_2 & b_3 \\ c_1 & c_2 & c_3 \end{vmatrix}$$

may be a suitable *abridged notation* for

$$\frac{\partial}{\partial x_1}|b_2 c_3| - \frac{\partial}{\partial x_2}|b_1 c_3| + \frac{\partial}{\partial x_3}|b_1 c_2|.$$

BRIOSCHI, F. (1854, March).

[LA TEORICA DEI DETERMINANTI, E LE SUE PRINCIPALI APPLICAZIONI. viii+116 pp. Pavia.]

Like Spottiswoode, Brioschi devotes his tenth chapter or section (§ 10) to "determinanti delle funzioni"; but his exposition is much more extensive (pp. 84–106), and, although of course he follows in Jacobi's footsteps, he does so less closely than Spottiswoode.

Thus the fact that the cofactor of $\frac{\partial f_i}{\partial x_k}$ in R is $\mathrm{R}\frac{\partial x_k}{\partial f_i}$, a fact which we may write temporarily in the form

$$\left[\frac{\partial f_i}{\partial x_k}\right] = \mathrm{R}\frac{\partial x_k}{\partial f_i},$$

he obtains by solving $n+1$ sets of equations like

$$\left.\begin{aligned}\frac{\partial x}{\partial f}\frac{\partial f}{\partial x}+\frac{\partial x}{\partial f_1}\frac{\partial f_1}{\partial x}+\ldots\ldots+\frac{\partial x}{\partial f_n}\frac{\partial f_n}{\partial x}&=1\\ \frac{\partial x}{\partial f}\frac{\partial f}{\partial x_1}+\frac{\partial x}{\partial f_1}\frac{\partial f_1}{\partial x_1}+\ldots\ldots+\frac{\partial x}{\partial f_n}\frac{\partial f_n}{\partial x_1}&=0\\ \cdots\cdots\cdots\cdots\cdots&\cdots\\ \frac{\partial x}{\partial f}\frac{\partial f}{\partial x_n}+\frac{\partial x}{\partial f_1}\frac{\partial f_1}{\partial x_n}+\ldots\ldots+\frac{\partial x}{\partial f_n}\frac{\partial f_n}{\partial x_n}&=0\end{aligned}\right\}$$

and by using the same sets of equations after row-by-row multiplication he obtains

$$\sum\left(\pm\frac{\partial x}{\partial f}\frac{\partial x_1}{\partial f_1}\cdots\frac{\partial x_n}{\partial f_n}\right)\cdot\sum\left(\pm\frac{\partial f}{\partial x}\frac{\partial f_1}{\partial x_1}\cdots\frac{\partial f_n}{\partial x_n}\right)=1,$$

$$\text{or, say,}\quad S\cdot R=1.$$

Further, he notes that as a consequence of these two theorems there results

$$\left[\frac{\partial f_i}{\partial x_k}\right]\cdot\left[\frac{\partial x_i}{\partial f_k}\right]=\frac{\partial x_k}{\partial f_i}\cdot\frac{\partial f_k}{\partial x_i},$$

which he might well have generalised by changing the i, k of the second factor of both sides into r, s.

In dealing with the "fundamental lemma" his order of procedure is the reverse of Jacobi's, that is to say, he deduces the form of 1844 from the original form of 1841. Thus, using the latter in regard to S, he has

$$\frac{\partial S}{\partial f_k}=S\cdot\sum_{i=0}^{i=n}\frac{\partial\frac{\partial x_i}{\partial f_k}}{\partial x_i},$$

whence, because of R being the reciprocal of S, he obtains

$$\frac{\partial R}{\partial f_k}+R\cdot\sum_{i=0}^{i=n}\frac{\partial\frac{\partial x_i}{\partial f_k}}{\partial x_i}=0;$$

so that on substituting

$$\sum_{i=0}^{i=n}\frac{\partial R}{\partial x_i}\frac{\partial x_i}{\partial f_k}\quad\text{for}\quad\frac{\partial R}{\partial f_k}$$

there results

$$\sum_{i=0}^{i=n} \frac{\partial\left(\mathrm{R}\frac{\partial x_i}{\partial f_k}\right)}{\partial x_i} = 0,$$

which on further substituting $\left[\frac{\partial f_k}{\partial x_i}\right]$ for $\mathrm{R}\frac{\partial x_i}{\partial f_k}$ becomes the form desired.

The next fresh paragraph (p. 91) appears, although unnecessarily, as an addendum to Jacobi's solution of a set of simultaneous linear equations whose determinant is a functional determinant (*De determ. funct.* § 8). If the square of R be obtained by row-by-row multiplication, and the square of S by column-by-column multiplication it is easily verified that

$$(h^{\text{th}} \text{ row of } \mathrm{S}^2) \times (k^{\text{th}} \text{ column of } \mathrm{R}) = \frac{\partial x_k}{\partial f_h},$$

$$\text{i.e.} \qquad = (k, h)^{\text{th}} \text{ element of S},$$

thus incidentally giving $\mathrm{S}^2\mathrm{R} = \mathrm{S}$ as it should do. From this it is deduced that

$$(h^{\text{th}} \text{ row of } \mathrm{S}^2) \times (k^{\text{th}} \text{ row of } \mathrm{R}^2) = 1 \text{ or } 0$$

according as h and k are equal or unequal,* and that therefore R^2 and S^2 as just defined are in the matter of their primary minors related as R and S have been shown to be.

In the remaining fourteen pages (pp. 92–106) the only matter calling for attention concerns Jacobi's theorem

$$\sum(\pm b b_1^{(1)} \dots b_m^{(m)}) = \mathrm{B}^m \cdot \sum\left(\pm \frac{\partial f}{\partial x} \frac{\partial f_1}{\partial x_1} \cdots \frac{\partial f_{n+m}}{\partial x_{n+m}}\right)$$

where

$$\mathrm{B} = \sum\left(\pm \frac{\partial f}{\partial x} \frac{\partial f_1}{\partial x_1} \cdots \frac{\partial f_{n-1}}{\partial x_{n-1}}\right) \text{ and } b_k^{(i)} = \sum\left(\pm \frac{\partial f}{\partial x} \frac{\partial f_1}{\partial x_1} \cdots \frac{\partial f_{n-1}}{\partial x_{n-1}} \frac{\partial f_{n+i}}{\partial x_{n+k}}\right).$$

From this Brioschi, by taking the f's to be linear functions of the x's, obtains Sylvester's theorem of March 1851 regarding a compound determinant.

* Viewing R and S as matrices of which the conjugates are $\overline{\mathrm{R}}$ and $\overline{\mathrm{S}}$ we have as an equivalent of this

$$\overline{\mathrm{S}}\mathrm{S} \cdot \overline{\mathrm{R}}\mathrm{R} = \overline{\mathrm{S}} \cdot \mathrm{S}\mathrm{R} \cdot \overline{\mathrm{R}}$$
$$= \overline{\mathrm{S}}\,\overline{\mathrm{R}} = 1.$$

PEIRCE, B. (1855).

[A SYSTEM OF ANALYTICAL MECHANICS. (Chap. X. i.: Determinants and Functional Determinants, pp. 172–198.) xl+496 pp., Boston (U.S.A.).]

At the outset of his Tenth Chapter, which deals with the integration of the differential equations of motion, Peirce feels the need for making his reader acquainted with the properties of functional determinants. He accordingly gives as a preparation a brief account (§§ 327–348, pp. 172–183) of determinants in general, and then expounds within the space of sixteen broad-margined pages the main theorems of Jacobi's 'De determinantibus functionalibus.' The treatment of the original is free and masterly, the order being altered with good effect. For example, Jacobi's incorrectly stated proposition is brought forward to occupy the second place, the enunciation being *If either* (i.e. *any one*) *of the given functions contains any of the other functions, these* (*latter*) *functions may be regarded as constant in finding the functional determinant.* There is thence deduced Jacobi's last proposition of all, namely, that expressing the determinant as a single product: and this in turn is used to discuss the connection between the vanishing of the determinant and the interdependence of the functions.

Had Peirce's exposition been less condensed and been published as part of an ordinary text-book of determinants, its value at that time to English-speaking students would have been considerable.

BELLAVITIS, G. (1857, June).

[Sposizione elementare della teorica dei determinanti. *Memorie .... Istituto Veneto* .... vii. pp. 67–144.]

To the subject of a "Determinante formato colle derivate-prime di alquante funzioni di altrettanti variabili" Bellavitis devotes nine and a half pages (pp. 52–61, §§ 65–78), that is to say, about the same as Spottiswoode, though his selection of theorems is not quite the same. In substance he gives nothing fresh. His symbolism for the determinant of $u, v, \ldots$ with respect to $x, y, \ldots$ resembles Cauchy's of 1841, being

$$| D_x u, \ D_y v, \ldots | ;$$

other changes made by him in notation are less satisfactory.

BALTZER, R. (1857).

[THEORIE UND ANWENDUNG DER DETERMINANTEN, mit ..... vi+129 pp. Leipzig, 1857.]

"Die Functionaldeterminante" is the heading of Baltzer's thirteenth chapter or section (§ 13, pp. 61–72). Though the exposition is neither so full nor so fresh as Brioschi's, it has the advantage in arrangement, concision and clearness. Jacobi's last theorem (*De determ. funct.* § 18), expressing the determinant as a single product,

$$\left(\frac{\partial f}{\partial x}\right)\left(\frac{\partial f_1}{\partial x_1}\right)\left(\frac{\partial f_2}{\partial x_2}\right)\cdots\left(\frac{\partial f_n}{\partial x_n}\right),$$

Baltzer makes his first, the proof being readily altered to suit. He then, following Peirce, uses it effectively in dealing with the proposition regarding the vanishing of the determinant. For example, if the determinant vanishes, he can assert that one of the factors of the said product must vanish; and thence step-by-step can infer the vanishing of the succeeding factors including the last,—a result which entails f_n being expressible in terms of the other f's.

A footnote recalls the fact, which we should have noted before this, that Möbius had given in *Crelle's Journ.*, xii. p. 116, in the year 1834, the equation

$$(t_x u_y - t_y u_x)(v_t w_u - v_u w_t)(x_v y_w - x_w y_v) = 1,$$

where t_x stands for $\partial t/\partial x$.

MALMSTEN, C. J. (1858, October).

[Om differential-eqvationers integrering. *K. Svenska Vet.-Akad. Handl.* (Stockholm), iii. No. 2, 94 pp.]

On pp. 9–11 Malmsten enunciates and proves Jacobi's "fundamental lemma" of 1844 without contributing any improvement.

SALMON, G. (1859).

[LESSONS INTRODUCTORY TO THE MODERN HIGHER ALGEBRA. xii + 147 pp., Dublin.]

Salmon gives little, and certainly nothing fresh, on the subject; but his unreserved adoption of Sylvester's word "Jacobian" (§§ 53, 54; p. 37) doubtless helped greatly to spread the usage.

CHAPTER IX.

SKEW DETERMINANTS, FROM 1846 TO 1860.

UNLIKE the special form of the preceding chapter, Skew Determinants received little attention in our first volume, even although in their case the period was extended to 1845 in order to include the whole of Jacobi's work. Unless by implication, indeed, they do not belong to that volume at all, the chapter there assigned to them being really occupied with the related functions afterwards named Pfaffians when the connection between the two came to be recognised. The new form is thus strictly viewable as one of the products of the Cayley-Sylvester period.

CAYLEY, A. (1846).

[Sur quelques propriétés des déterminants gauches. *Crelle's Journ.*, xxxii. pp. 119–123; or *Collected Math. Papers*, i. pp. 332–336.]

This paper, with its author's usual directness, starts at once with a definition, the first words being—

"Je donne le nom de *déterminant gauche* à un déterminant formé par un système de quantités $\lambda_{r,s}$ qui satisfont aux conditions

$$\lambda_{r,s} = -\lambda_{s,r} \quad (r \neq s).$$

J'appelle aussi un tel système, *système gauche*."

So far as can be ascertained, the English equivalent '*skew*,' although it probably was the first of the two in order of thought, did not appear in print until a few years later.

As has been pointed out elsewhere, the title of the paper is quite misleading, the real subject being *the construction of a linear substitution for the transformation of* $x_1^2+x_2^2+x_3^2+\ldots$ *into* $\xi_1^2+\xi_2^2+\xi_3^2+\ldots$. All that can be said in defence of the inaccuracy is that skew determinants are made use of in obtaining the desired substitution. The proper place for giving an account of the contents of the paper is thus under the heading of '*orthogonants*,' if we may so name the *determinants of an orthogonal substitution.*

CAYLEY, A. (1847).

[Sur les déterminants gauches. *Crelle's Journ.*, xxxviii. pp. 93–96; or *Collected Math. Papers*, i. pp. 410–413.]

Here the title and contents agree. At the outset the former definition is repeated, and then for a particular kind of skew determinant, viz., those in which the condition

$$\lambda_{r,s} = -\lambda_{s,r} \qquad (1)$$

is to hold even in the case where s and r are equal, "ou pour lesquels on a

$$\lambda_{r,s} = -\lambda_{s,r} \quad (r \neq s), \qquad \lambda_{r,r} = 0," \qquad (2)$$

the name 'skew symmetric' ("gauche et symétrique") is set apart. The reason for this is evident on the statement of the first theorem, which is to the effect that any skew determinant is expressible in terms of skew symmetric determinants and those elements of the original determinant which are not included in the latter. "En effet," he explains,

"soit Ω le déterminant gauche dont il s'agit, cette fonction peut être présentée sous la forme

$$\Omega = \Omega_0 + \Omega_1\lambda_{11} + \Omega_2\lambda_{22} + \ldots + \Omega_{12}\lambda_{11}\lambda_{22} + \ldots$$

où Ω_0 est ce que devient Ω si $\lambda_{11}, \lambda_{22}, \ldots$ sont réduits à zéro, Ω_1 est ce que devient le coefficient de λ_{11} sous la même condition, et ainsi de suite; c'est à dire, Ω_0 est le déterminant formé par les quantités $\lambda_{r,s}$ en supposant que ces quantités satisfassent aux conditions (2) et en donnant à r, s le valeurs $1, 2, 3, \ldots, n$; Ω_1 est le déterminant formé pareillement en donnant à r, s les valeurs $2, 3, \ldots, n$; Ω_2

s'obtient en donnant à r, s les valeurs 1, 3, . . . , n; et ainsi de suite; cela est aisé de voir si l'on range les quantités $\lambda_{r,s}$ en forme de carré."

At this point a digression is made in order to establish a theorem regarding skew determinants of odd order, and another regarding skew determinants of even order, and thus be enabled to make certain substitutions for the Ω's in the development here announced. Further, as the said substitutions for the Ω's of even order involve the functions dealt with by Jacobi in his paper on the "Pfaffsche Methode,"—functions which Cayley here calls "les fonctions de M. Jacobi," but which at a later date he designated "*Pfaffians*,"—the digression is lengthened by having prefixed to it an account of these functions.

So curious is this account and so likely to be misrepresented by condensation, that the best way of treating it is to reproduce it in the original words.* It stands thus:—

"On obtient ces fonctions (dont je reprends ici la théorie) par les propriétés générales d'un déterminant défini. Car en exprimant par $(1, 2, \ldots, n)$ une fonction quelconque dans laquelle entrent les nombres symboliques $1, 2, \ldots, n$, et par $\pm$ le signe correspondant à une permutation quelconque de ces nombres, la fonction

$$\sum \pm (1\ 2 \ldots n)$$

où $\sum$ désigne la somme de tous les termes qu'on obtient en permutant ces nombres d'une manière quelconque est ce qu'on nomme *Déterminant.* On pourrait encore généraliser cette définition en admettant plusieurs systèmes de nombres $1, 2 \ldots, n$; $1', 2' \ldots, n'$; . . . qui alors devroient être permutés independamment les uns des autres; on obtiendrait de cette manière une infinité d'autres fonctions, mentionnées (T. xxx. p. 7). Dans le cas des déterminants ordinaires, auquel je ne m'arrêterai pas ici, on aura $(1, 2 \ldots n) = \lambda_{\alpha,1}\lambda_{\beta,2} \ldots \lambda_{\kappa,n}$. Pour les cas des fonctions dont il s'agit (les fonctions de M. Jacobi), on supposera n *pair*, et l'on écrira

$$(1\ 2 \ldots n) = \lambda_{1,2}\lambda_{3,4} \ldots \lambda_{n-1,n},$$

où $\lambda_{r,s}$ sont des quantités quelconques qui satisfont aux équations (1). La fonction sera composée d'un nombre $1 . 2 \ldots n$ de termes; mais parmi eux il n'y aura que $1 . 3 \ldots (n-1)$ termes différents qui se trouveront répétés $2^{\frac{1}{2}n}(1 . 2 \ldots \frac{1}{2}n)$ fois, et qu'on obtiendra en permutant cycliquement d'abord les $n-1$ derniers nombres, puis les

* The paper, as it appears in *Crelle's Journal*, is disfigured by misprints, which have not been fully corrected in the *Collected Math. Papers.*

$n-3$ derniers nombres de chaque permutation, et ainsi de suite; le signe étant toujours +. Il pourra être démontré, comme pour les déterminants, que ces fonctions changent de signe en permutant deux quelconques des nombres symboliques, et qu'elles s'evanouissent si deux de ces nombres deviennent identiques. De plus, en exprimant par $[12\ldots n]$ la fonction dont il s'agit, la règle qui vient d'être énoncé, donnera pour la formation de ces fonctions:

$$[1\ 2\ldots n] = \lambda_{12}[3\ 4\ldots n] + \lambda_{13}[4\ldots n,\ 2] + \ldots\ldots\ldots + \lambda_{1n}[2\ 3\ldots n-1].\text{"}$$

Dismissing, as not of present interest, the sentence regarding the generalisation obtained by admitting more than one system of symbolic numbers, we note first of all the peculiar general use of $(1\,2\ldots n)$ for any function the expression of which involves* as suffixes or otherwise the numbers $1, 2, 3, \ldots, n$. Then we are struck with the fact that the use of this along with $\Sigma\pm$ gives a notation for a genus of functions of which determinants, as understood up to the date of the paper, formed a species: thus

$$a_1b_2c_3 + a_2b_3c_1 + a_3b_1c_2 - a_3b_2c_1 - a_2b_1c_3 - a_1b_3c_2$$

is the case of $\Sigma\pm(123)$, where $(123)=a_1b_2c_3$. In the third place we are surprised to find that Cayley seems to propose to extend the meaning of the word *determinant* by transferring the name of the species to the genus, and to call by the name of "ordinary determinants" the functions formerly known as "determinants" merely.

All this is in itself comparatively unimportant, serving perhaps only to recall to us Cauchy's famous paper of 1812, where we have K, the originating term of an alternating function to compare and contrast with Cayley's $(12\ldots n)$, and 'alternating function' to compare and contrast with Cayley's extended meaning of 'determinant.' But what follows by way of second example is very noteworthy, because the originating term taken, viz., $\lambda_{12}\lambda_{34}\ldots\lambda_{n-1,n}$, is one that could not possibly have been used by Cauchy, with whom Σ denoted an operation of a much less simple character than permutation of the integers $1, 2, \ldots, n$.

*Apparently it is meant to be implied that each of the numbers occurs only once in the expression.

Unfortunately the example is not fully exploited.* We are only told that in a certain special case, viz., where the elements are such that rs is always equal to $-sr$, there are only $1.3.5\ldots(2n-1)$ different terms in

$$\sum \pm \lambda_{12}\lambda_{34}\ldots\lambda_{2n-1,2n};$$

---

* Supplying this want we see that in strict accordance with Cayley's definition

$$\begin{array}{rllll}
\Sigma \pm 12{\cdot}34 = & 12{\cdot}34 & +\ 21{\cdot}43 & +\ 31{\cdot}24 & +\ 41{\cdot}32 \\
& -\ 12{\cdot}43 & +\ 23{\cdot}14 & -\ 31{\cdot}42 & +\ 42{\cdot}13 \\
& -\ 13{\cdot}24 & -\ 23{\cdot}41 & -\ 32{\cdot}14 & -\ 42{\cdot}31 \\
& +\ 13{\cdot}42 & -\ 24{\cdot}13 & +\ 32{\cdot}41 & -\ 43{\cdot}12 \\
& +\ 14{\cdot}23 & +\ 24{\cdot}31 & +\ 34{\cdot}12 & +\ 43{\cdot}21 \\
& -\ 14{\cdot}32 & & -\ 34{\cdot}21 & \\
& -\ 21{\cdot}34 & & -\ 41{\cdot}23 &
\end{array}$$

$$\begin{array}{l}
= 2\{\ 12{\cdot}34 - 12{\cdot}43 - 13{\cdot}24 + 13{\cdot}42 \\
\quad + 14{\cdot}23 - 14{\cdot}32 - 21{\cdot}34 + 21{\cdot}43 \\
\quad - 23{\cdot}41 + 24{\cdot}31 - 31{\cdot}42 + 32{\cdot}41\},
\end{array}$$

—a function of twelve variables which is not a determinant in the acceptation either of the present time or of the time preceding Cayley.

It is instructive, in connection with the matter in hand, to note that this function is expressible in terms of four Pfaffians, namely, we have

$$\Sigma \pm 12{\cdot}34 = 2\left\{ \begin{array}{|ccc|} 12 & 13 & 14 \\ & 23 & 24 \\ & & 34 \end{array} - \begin{array}{|ccc|} 12 & 13 & 14 \\ & 32 & 42 \\ & & 43 \end{array} + \begin{array}{|ccc|} 21 & 31 & 41 \\ & 32 & 42 \\ & & 43 \end{array} - \begin{array}{|ccc|} 21 & 31 & 41 \\ & 23 & 24 \\ & & 34 \end{array} \right\};$$

where, be it also observed, the third and fourth Pfaffians are obtainable from the first and second by changing rs in every case into sr, and where, if the condition $rs=-sr$ be introduced, the result is

$$\sum_{rs=-sr} \pm 12{\cdot}34 = 8\cdot \begin{array}{|ccc|} 12 & 13 & 14 \\ & 23 & 24 \\ & & 34 \end{array};$$

so that the Pfaffian on the right may be defined as the eighth part of a certain Cayleyan determinant; or, in Cayley's symbols,

$$[1\,2\,3\,4] = \tfrac{1}{8}\sum_{rs=-sr} \pm 12{\cdot}34,$$

where the 8 is the value of $2^{\frac{1}{2}n}(1.2\ldots.\tfrac{1}{2}n)$ when $n=4$.

Before leaving this it deserves to be noted that when Cayley came in 1889 to re-edit his writings, he appended to this paper a note in which it is stated that part of his purpose was to show "that the definition of a determinant may be so extended as to include within it the Pfaffian" (see *Collected Math. Papers*, i. p. 589).

that the aggregate of these is also got without repetition in a particular way already announced by Jacobi; and that it is this aliquot part of $\Sigma \pm \lambda_{12}\lambda_{34} \ldots \lambda_{2n-1,2n}$ which constitutes 'la fonction de M. Jacobi.' Jacobi's theorem regarding the effect, on the function, of interchanging two indices, is then restated; and a step further is taken in affirming that the function vanishes when two indices are equal. Finally, another law of formation—the recurrent law—is given in the form

$$[12\ldots 2n] = 12[345\ldots 2n]+13[45\ldots 2n,2]+14[5\ldots 2n,2,3]+\ldots$$

which, of course, is in substance not different from Jacobi's

$$\mathrm{R} = a_{1s}\frac{\partial \mathrm{R}}{\partial a_{1s}} + a_{2s}\frac{\partial \mathrm{R}}{\partial a_{2s}} + \ldots .$$

The digression on 'les fonctions de M. Jacobi' being exhausted, Cayley returns to skew symmetric determinants with the requisite material for proving the two theorems above referred to. The first of them, which is not new, is, in later phraseology that "*Any zero-axial skew determinant of odd order vanishes*"; and the second, which is Cayley's own, is that "*Any zero-axial skew determinant of even order is the square of a Pfaffian.*" In both cases the method of proof is the gradational or so-called 'mathematical induction'; and in both cases the main auxiliary theorem used is Cauchy's regarding the expansion of a determinant according to binary products of the elements of a row and the elements of a column.

When n is odd and the elements of the first row and those of the first column are $0,\lambda_{12},\lambda_{13},\ldots,\lambda_{1n}$ and $0,\lambda_{21},\lambda_{31},\ldots,\lambda_{n1}$ respectively, he says it is easy to see that for each term having $\lambda_{1\alpha}\lambda_{\beta 1}$ for a factor, where $\alpha \neq \beta$, there exists an equal term of opposite sign having $\lambda_{1\beta}\lambda_{\alpha 1}$ for a factor; and that therefore, since $\lambda_{1\alpha}\lambda_{\beta 1} = \lambda_{1\beta}\lambda_{\alpha 1}$, these two terms must cancel each other. As for the terms which have $\lambda_{1\alpha}\lambda_{\alpha 1}$ for a factor, the cofactor is a determinant of exactly the same form as the original, but of the order $n-2$; consequently the theorem is seen to hold for any one case if it hold for the case immediately preceding. But for the case where $n=3$, the theorem is self-evident; therefore, "Tout déterminant gauche et symétrique d'un ordre *impair* est zéro."

When n is even, the determinant dealt with is purposely taken more general than one with skew symmetry, although, strange to say, Cayley calls it 'gauche et symétrique,' the elements of the first row and those of the first column being $\lambda_{\alpha\beta}, \lambda_{\alpha 2}, \lambda_{\alpha 3}, \ldots, \lambda_{\alpha n}$ and $\lambda_{\alpha\beta}, \lambda_{2\beta}, \lambda_{3\beta}, \ldots, \lambda_{n\beta}$, and his aim being to prove that such a determinant is equal to the product of two of the functions treated of in the digression, viz., $[\alpha\, 2\, 3 \ldots n]$ and $[\beta\, 2\, 3 \ldots n]$. Developing as in the preceding case, there has this time to be considered the element common to the first row and first column, viz., $\lambda_{\alpha\beta}$, the cofactor of which is seen to be a skew symmetric determinant of odd order $n-1$, and therefore, as has just been shown, is equal to zero. As for the cofactor of $-\lambda_{\alpha\alpha'}\lambda_{\beta'\beta}$, where $\lambda_{\alpha\alpha'}$ is any element of the first row except the first, and $\lambda_{\beta'\beta}$ is any element of the first column except the first, it will be found to be a determinant which Cayley again mistakenly but consistently calls 'gauche et symétrique,' obtained by giving to r all the values $2, 3, \ldots, n$ with the exception of α', and to s all the values $2, 3, \ldots, n$, with the exception of β'. This determinant of the $(n-2)^{\text{th}}$ order is expected to be seen to be of the same kind as that with which we started, and to be temporarily admitted to be equal to

$$[\alpha'+1, \ldots, n, 2, \ldots, \alpha'-1] \cdot [\beta'+1, \ldots, n, 2, \ldots, \beta'-1].$$

The typical term of the expansion will thus be

$$\lambda_{\alpha\alpha'}[\alpha'+1, \ldots, n, 2, \ldots, \alpha'-1] \cdot \lambda_{\beta'\beta}[\beta'+1, \ldots, n, 2, \ldots, \beta'-1];$$

and the sum of all such terms

$$= \left\{\lambda_{\alpha 2}[34 \ldots n] + \lambda_{\alpha 3}[4 \ldots n2] + \ldots + \lambda_{\alpha n}[23 \ldots (n-1)]\right\}$$
$$\cdot \left\{\lambda_{\beta 2}[34 \ldots n] + \lambda_{\beta 3}[4 \ldots n2] + \ldots + \lambda_{\beta n}[23 \ldots (n-1)]\right\}$$

and therefore

$$= [\alpha\, 2\, 3 \ldots . n] \cdot [\beta\, 2\, 3 \ldots n].$$

This means, of course, that if the theorem holds for a determinant of order $n-2$ it will hold for the next succeeding case. But in the simplest case, viz., where $n=2$, it is self-evident that the theorem holds, for the determinant then

$$= \lambda_{\alpha\beta}\lambda_{22} - \lambda_{2\beta}\lambda_{\alpha 2},$$
$$= \lambda_{\beta 2}\lambda_{\alpha 2},$$
$$= [\beta 2]\cdot[\alpha 2];$$

consequently "*Le déterminant gauche et symétrique qu'on obtient en donnant à* r *les valeurs* α,2,3, . . . , n, *et à* s *les valeurs* β,2,3, . . . , n (*où* n *est* pair) *se réduit à*

$$[\alpha\,2\,3\ldots \mathrm{n}]\cdot[\beta\,2\,3\ldots \mathrm{n}];$$

et en particulier, en donnant à r, s *les valeurs* 1,2, . . . , n *ce déterminant se réduit à* $[1\ 2\ 3\ \ldots\ \mathrm{n}]^2$."

Going back now to the expansion of the skew determinant Ω with which the paper opened, and taking for simplicity's sake* $\lambda_{rr} = 1$ in every case, Cayley readily obtains,

for n even,
$$\begin{aligned}\Omega = {} & [123\ldots n]^2\\ & + [34\ldots n]^2 + [24\ldots n]^2 + \ldots\\ & + [56\ldots n]^2 + \ldots\\ & + \ldots\ldots\\ & + 1,\end{aligned}$$

and, for n odd,
$$\begin{aligned}\Omega = {} & [23\ldots n]^2 + [13\ldots n]^2 + \ldots\\ & + [45\ldots n]^2 + \ldots\\ & + \ldots\ldots\\ & + 1.\end{aligned}$$

A special example of each identity is given, namely, the examples in which $n=4$ and 3 respectively. If we make a slight change in the left member, viz., write Ω in Cayley's vertical-line notation (which, by the way, considering the help it would have given, and the fact that it had been introduced six years previously, it is surprising not to find employed in this paper), these examples take the form,—

$$\begin{vmatrix} 1 & \lambda_{12} & \lambda_{13} & \lambda_{14}\\ -\lambda_{12} & 1 & \lambda_{23} & \lambda_{24}\\ -\lambda_{13} & -\lambda_{23} & 1 & \lambda_{34}\\ -\lambda_{14} & -\lambda_{24} & -\lambda_{34} & 1 \end{vmatrix} \quad\text{or}\quad \begin{vmatrix} 1 & 12 & 13 & 14\\ -12 & 1 & 23 & 24\\ -13 & -23 & 1 & 34\\ -14 & -24 & -34 & 1 \end{vmatrix}$$

$$= (\lambda_{12}\lambda_{34} - \lambda_{13}\lambda_{24} + \lambda_{14}\lambda_{23})^2 + \lambda^2_{12} + \lambda^2_{13} + \lambda^2_{14} + \lambda^2_{34} + \lambda^2_{24} + \lambda^2_{23} + 1,$$
$$= [1234]^2 + [12]^2 + [13]^2 + [14]^2 + [34]^2 + [24]^2 + [23]^2 + 1;$$

* And of course without loss of generality, as Cayley might have said.

and

$$\begin{vmatrix} 1 & \lambda_{12} & \lambda_{13} \\ -\lambda_{12} & 1 & \lambda_{23} \\ -\lambda_{13} & -\lambda_{23} & 1 \end{vmatrix} \quad \text{or} \quad \begin{vmatrix} 1 & 12 & 13 \\ -12 & 1 & 23 \\ -13 & -23 & 1 \end{vmatrix}$$

$$= \lambda^2_{23} + \lambda^2_{13} + \lambda^2_{12} + 1$$

$$= [23]^2 + [13]^2 + [12]^2 + 1 .$$

SPOTTISWOODE, W. (1851, 1853).

[ELEMENTARY THEOREMS RELATING TO DETERMINANTS. . . . . . viii+63 pp. London, 1851. Second edition, as an article in *Crelle's Journ.*, li. pp. 209–271, 328–381.]

In this the earliest of modern text-books on Determinants, a special section (§ ix. pp. 46–51; or § vi. pp. 260–266 in second edition) is set apart with the heading "On Skew Determinants." As a matter of fact, however, it is only the latter half of the section which at present concerns us, as the other half deals in reality with Cayley's determinant solution of the problem of orthogonal transformation.

In a sense the mode of treatment is indirect, the general skew determinant being viewed, not as a separate entity, but in its relation to a set of linear equations, the coefficients of which are its elements. The set of equations is

$$\left.\begin{array}{l} (11)x_1 + (12)x_2 + \ldots + (1n)x_n = u_1 \\ (21)x_1 + (22)x_2 + \ldots + (2n)x_n = u_2 \\ \cdot\ \cdot\ \cdot\ \cdot\ \cdot\ \cdot\ \cdot\ \cdot\ \cdot\ \cdot\ \cdot\ \cdot\ \cdot\ \cdot \\ (n1)x_1 + (n2)x_2 + \ldots + (nn)x_n = u_n \end{array}\right\},$$

where it has to be remembered that in every instance $(rr)=0$ and $(rs)+(sr)=0$. The right-hand members of what he calls the "derived" set are $v_1, v_2, \ldots, v_n$; that is to say, there exists simultaneously with the original the set

$$\left.\begin{array}{l} (11)x_1 + (21)x_2 + \ldots + (n1)x_n = v_1 \\ (12)x_1 + (22)x_2 + \ldots + (n2)x_n = v_2 \\ \cdot\ \cdot\ \cdot\ \cdot\ \cdot\ \cdot\ \cdot\ \cdot\ \cdot\ \cdot\ \cdot\ \cdot\ \cdot\ \cdot \\ (1n)x_1 + (2n)x_2 + \ldots + (nn)x_n = v_n \end{array}\right\}$$

whose determinant is got from the determinant of the former set by the change of rows into columns, and may therefore be denominated by the same symbol Δ. Solving the two sets of equations, we have

$$\left.\begin{aligned} x_1\Delta &= [11]u_1 + [12]u_2 + \ldots + [1n]u_n \\ x_2\Delta &= [21]u_1 + [22]u_2 + \ldots + [2n]u_n \\ &\cdots\cdots\cdots\cdots\cdots \\ x_n\Delta &= [n1]u_1 + [n2]u_2 + \ldots + [nn]u_n \end{aligned}\right\}$$

and

$$\left.\begin{aligned} x_1\Delta &= [11]v_1 + [21]v_2 + \ldots + [n1]v_n \\ x_2\Delta &= [12]v_1 + [22]v_2 + \ldots + [n2]v_n \\ &\cdots\cdots\cdots\cdots\cdots \\ x_n\Delta &= [1n]v_1 + [2n]v_2 + \ldots + [nn]v_n \end{aligned}\right\}^{*},$$

where, be it remarked, it would have been much better if in every case the coefficients of u_r and v_r had been interchanged, for then $[rs]$ would have stood for the cofactor of (rs) in Δ. From these by addition and subtraction and by utilising the fact that $u_r + v_r = 0$ † two others are obtained, viz.,

$$\left.\begin{aligned} 2x_1\Delta &= 0 + ([12]-[21])u_2 + \ldots + ([1n]-[n1])u_n \\ 2x_2\Delta &= ([21]-[12])u_1 + 0 + \ldots + ([2n]-[n2])u_n \\ &\cdots\cdots\cdots\cdots\cdots \\ 2x_n\Delta &= ([n1]-[1n])u_1 + ([n2]-[2n])u_2 + \ldots + 0 \end{aligned}\right\}$$

and

$$\left.\begin{aligned} 0 &= 2[11]u_1 + ([12]+[21])u_2 + \ldots + ([1n]+[n1])u_n \\ 0 &= ([21]+[12])u_1 + 2[22]u_2 + \ldots + ([2n]+[n2])u_n \\ &\cdots\cdots\cdots\cdots\cdots \\ 0 &= ([n1]+[1n])u_1 + ([n2]+[2n])u_2 + \ldots + 2[nn]u_n \end{aligned}\right\}.$$

Then follows the very curious sentence—curious, that is to say, from a logical point of view—

* There is herein used the fact, first noted by Rothe in 1800, that the cofactor of rs in any determinant is equal to the cofactor of sr in the conjugate determinant.

† Along with this fact Spottiswoode associates the statements that

$$u_1 + u_2 + \ldots + u_n = 0, \quad v_1 + v_2 + \ldots + v_n = 0,$$

which are manifestly incorrect.

"The comparison of these three systems gives either

$$\left.\begin{array}{cccc} & \Delta = 0 & & \\ * & [12]=]21] & \ldots & [1n]=[n1] \\ [21]=[12] & * & \ldots & [2n]=[n2] \\ \cdot & \cdot & \cdot & \cdot \\ [n1]=[1n] & [n2]=[2n] & \ldots & * \end{array}\right\}$$

or

$$\left.\begin{array}{cccc} [11]=0 & [12]+[21]=0 & \ldots & [1n]+[n1]=0 \\ [21]+[12]=0 & [22]=0 & \ldots & [2n]+[n2]=0 \\ \cdot & \cdot & \cdot & \cdot \\ [n1]+[1n]=0 & [n2]+[2n]=0 & \ldots & [nn]=0 \end{array}\right\};$$

and consequently either a symmetrical skew determinant of an even order or a determinant of an odd order vanishes."

What the first half of the sentence asserts to be proved is the proposition that *If* Δ *be a zero-axial skew determinant, then either*

$$(1)\quad \Delta=0 \quad and \quad [rs]=[sr],$$

$$or\quad (2)\ [rr]=0 \quad and \quad [rs]=-[sr].$$

In this there is evidently included the assertion that *A zero-axial skew determinant either vanishes itself, or all its principal coaxial minors vanish*: but what Spottiswoode finds in it is the much wider assertion that *Either all even-ordered or all odd-ordered zero-axial skew determinants vanish.* If however his accuracy be granted up to this point, there is little objection to the cogency of the next step in the reasoning, which is worded as follows:—

"But since it is found on trial that for $n=1, 3, \ldots,$ Δ vanishes, while for $n=2, 4, \ldots,$ it does not, the following theorems may be enunciated:—

"Theorem XIV. *A symmetrical skew determinant of an odd order in general vanishes, and the system has for its inverse an unsymmetrical skew system.*

"Theorem XV. *A symmetrical skew determinant of an even order does not in general vanish, but the system has for its inverse a symmetrical skew system.*"

The only difficulty to be raised is in regard to the name given to the "inverse system" in the first case. "Unsymmetric skew" is clearly inappropriate when, as we have seen, $[rs]=[sr]$; and it is not improved in the second edition by alteration into "quadratic skew," the fact being that the system is not skew at all, but is symmetric with respect to the principal diagonal, or, in later phraseology, is *axisymmetric.*

The treatment of the next theorem taken up is happier than the foregoing, and is after the outset no less fresh. Taking an even-ordered skew determinant with zeros in the principal diagonal he develops it according to products of an element of the first row and an element of the first column, the result being written in the form

$$\begin{vmatrix} * & 12 & \dots & 1n \\ 21 & * & & 2n \\ \cdot & \cdot & \cdot & \cdot \\ n1 & n2 & & * \end{vmatrix} = (12)^2 \begin{vmatrix} * & 34 & \dots & 3n \\ 43 & * & \dots & 4n \\ \cdot & \cdot & \cdot & \cdot \\ n3 & n4 & \dots & * \end{vmatrix} + 2(12)(13) \begin{vmatrix} 34 & 35 & \dots & 32 \\ * & 45 & \dots & 42 \\ \cdot & \cdot & \cdot & \cdot \\ n4 & n5 & \dots & n2 \end{vmatrix} + \dots$$

where, be it observed, the second typical term on the right has been altered from

$$-\,2(12)(13) \begin{vmatrix} 32 & 34 & \dots & 3n \\ 42 & * & \dots & 4n \\ \cdot & \cdot & \cdot & \cdot \\ n2 & n4 & \dots & * \end{vmatrix}$$

by the translation of the first column to the last place. The determinant in this typical term is then further transformed into the square root of the product of two determinants like that in the term preceding it, the steps of the reasoning being—

$$\begin{vmatrix} 32 & 34 & \dots & 3n \\ 42 & * & \dots & 4n \\ \cdot & \cdot & \cdot & \cdot \\ n2 & n4 & \dots & * \end{vmatrix}^2 = \begin{vmatrix} 23 & 24 & \dots & 2n \\ 43 & * & \dots & 4n \\ \cdot & \cdot & \cdot & \cdot \\ n3 & n4 & \dots & * \end{vmatrix} . \begin{vmatrix} 32 & 34 & \dots & 3n \\ 42 & * & \dots & 4n \\ \cdot & \cdot & \cdot & \cdot \\ n2 & n4 & \dots & * \end{vmatrix},$$

$$= \begin{vmatrix} * & 24 & \dots & 2n \\ 43 & * & \dots & 4n \\ \cdot & \cdot & \cdot & \cdot \\ n3 & n4 & \dots & * \end{vmatrix} . \begin{vmatrix} * & 34 & \dots & 3n \\ 42 & * & \dots & 4n \\ \cdot & \cdot & \cdot & \cdot \\ n2 & n4 & \dots & * \end{vmatrix},$$

the deletion of 23 and 32 in the last step being warranted by the fact that their cofactors are determinants similar to the original but of odd order $n-3$, and therefore have the value zero. The development as thus changed has the form of the square of a polynominal; and consequently by extracting the square root there results

$$\begin{vmatrix} * & 12 & \dots & 1n \\ 21 & * & \dots & 2n \\ . & . & . & . \\ n1 & n2 & \dots & * \end{vmatrix}^{\frac{1}{2}} = 12 \cdot \begin{vmatrix} * & 34 & \dots & 3n \\ 43 & * & \dots & 4n \\ . & . & . & . \\ n3 & n4 & \dots & * \end{vmatrix}^{\frac{1}{2}} + 13 \cdot \begin{vmatrix} * & 45 & \dots & 42 \\ 54 & * & \dots & 52 \\ . & . & . & . \\ 24 & 25 & \dots & * \end{vmatrix}^{\frac{1}{2}} + \dots$$

This, according to the point of view, will be recognised either as Cayley's theorem that an even-ordered skew determinant with zeros in the principal diagonal is a *square*, or as the theorem in Pfaffians formulated by Cayley and which in Jacobi's notation would be written

$$[123 \dots n] = 12[34 \dots n] + 13[45 \dots n2] + 14[56 \dots n23] + \dots$$

The rest of the section or chapter deals with Cayley's extension of this to skew determinants whose principal elements are not zeros, the notation employed being the same.

In the Second Edition, when dealing with Cayley's orthogonal substitution, Spottiswoode gives (p. 261) without proof an important theorem on determinants of the present kind. This will be found formulated under *Orthogonants* (see p. 315 below), and need not be repeated here.

CAYLEY, A. (1851).

[On the theory of permutants. *Camb. and Dub. Math. Journ.*, vii. pp. 40–51; or *Collected Math. Papers*, ii. pp. 16–26.]

By this time the widened definition of a determinant which Cayley had given in his paper of 1847 had been exploited to a certain extent, and had been found profitable both by himself and his fellow-worker Sylvester. The paper we have now come to, however, is the only one of the series that for the present concerns us. In it he implicitly discards his former usage of the word "determinant" in any wider sense than that employed by his predecessors; adopts instead the word "*permutant*" as suggested by Sylvester, and in working out the theory of the general functions under this name assigns to determinants and Pfaffians their proper niches in the new structure, the scheme of classification being

- Permutants
  - (A) (no name)
    - (a) Pfaffians
  - (B) Intermutants (or hyperdeterminants)
    - (b) Commutants
      - (β) Determinants.

CAYLEY, A. (1854).

[Recherches ultérieures sur les déterminants gauches. *Crelle's Journ.*, l. pp. 299–313; or *Collected Math. Papers*, ii. pp. 202–205.]

The development with which this paper of 1847 closes is here recalled and repeated for the case where the skew determinant is of the 5th order and the elements of the diagonal are specialised, the form in which the identity appears being

$$
\begin{aligned}
\overline{12345 \mid 13345} = \; & 11 \cdot 22 \cdot 33 \cdot 44 \cdot 55 \\
& + 11 \cdot 22 \cdot 33 \cdot (45)^2 \\
& + 11 \cdot 22 \cdot 44 \cdot (35)^2 \\
& + 11 \cdot 22 \cdot 55 \cdot (34)^2 \\
& + 11 \cdot 33 \cdot 44 \cdot (25)^2 \\
& + 11 \cdot 33 \cdot 55 \cdot (24)^2 \\
& + 11 \cdot 44 \cdot 55 \cdot (23)^2 \\
& + 22 \cdot 33 \cdot 44 \cdot (15)^2 \\
& + 22 \cdot 33 \cdot 55 \cdot (14)^2 \\
& + 22 \cdot 44 \cdot 55 \cdot (13)^2 \\
& + 33 \cdot 44 \cdot 55 \cdot (12)^2 \\
& + 11 \cdot (2345)^2 \\
& + 22 \cdot (1345)^2 \\
& + 33 \cdot (1245)^2 \\
& + 44 \cdot (1235)^2 \\
& + 55 \cdot (1234)^2,
\end{aligned}
$$

where the symbol on the left stands for the determinant whose elements are 11, 12, . . . , 21, 22, . . . and the peculiarity of skewness is understood but not expressed. Had the specialisation of the elements of the diagonal been as before, the development would clearly have been

$$\begin{gathered}1\\+(45)^2+(35)^2+(34)^2+(25)^2+(24)^2+(23)^2+(15)^2+(14)^2+(13)^2+(12)^2\\+(2345)^2+(1345)^2+(1245)^2+(1235)^2+(1234)^2,\end{gathered}$$

which, if the order be reversed, agrees exactly with the result of putting $n=5$ in the identity towards the end of the paper of 1846. By way of explanation Cayley adds the sentence "Les expressions 12, 1234, etc., à droite sont ici des *Pfaffiens*," which is noteworthy as being the first intimation that he desired "les fonctions de M. Jacobi," as he had formerly called them, to be known by the name of the mathematician whose integration-method had led Jacobi to the discovery of them. The change is easily accounted for by the fact that it was more appropriate to attach Jacobi's name to another class of determinants which were of greater importance and to which Jacobi had given far more attention.

Immediately following this there comes the announcement:—

"J'ai trouvé récemment une formule analogue pour le développement d'un *déterminant gauche* bordé, tel que

$$\overline{\alpha 1234 \mid \beta 1234} = \left|\begin{matrix}\alpha\beta & \alpha 1 & \alpha 2 & \alpha 3 & \alpha 4\\ 1\beta & 11 & 12 & 13 & 14\\ 2\beta & 21 & 22 & 23 & 24\\ 3\beta & 31 & 32 & 33 & 34\\ 4\beta & 41 & 42 & 43 & 44\end{matrix}\right|;$$

Cette formule est:—

$$\overline{\alpha 1234 \mid \beta 1234} = \begin{aligned}&\alpha\beta\cdot 11\cdot 22\cdot 33\cdot 44\\ &\left.\begin{aligned}+\,&\alpha\beta\cdot 12\cdot 12\cdot 33\cdot 44\\ +\,&\alpha\beta\cdot 13\cdot 13\cdot 22\cdot 44\\ +\,&\alpha\beta\cdot 14\cdot 14\cdot 22\cdot 33\\ +\,&\alpha\beta\cdot 23\cdot 23\cdot 11\cdot 44\\ +\,&\alpha\beta\cdot 24\cdot 24\cdot 11\cdot 33\\ +\,&\alpha\beta\cdot 34\cdot 34\cdot 11\cdot 22\end{aligned}\right\}\end{aligned}$$

$$
\left.\begin{array}{l}
+\alpha\beta 1234 \cdot 1234^{*} \\
\left.\begin{array}{l}
+\alpha 1 \cdot \beta 1 \cdot 22 \cdot 33 \cdot 44 \\
+\alpha 2 \cdot \beta 2 \cdot 11 \cdot 33 \cdot 44 \\
+\alpha 3 \cdot \beta 3 \cdot 11 \cdot 22 \cdot 44 \\
+\alpha 4 \cdot \beta 4 \cdot 11 \cdot 22 \cdot 33
\end{array}\right\} \\
\left.\begin{array}{l}
+\alpha 123 \cdot \beta 123 \cdot 44 \\
+\alpha 124 \cdot \beta 124 \cdot 33 \\
+\alpha 134 \cdot \beta 134 \cdot 22 \\
+\alpha 234 \cdot \beta 234 \cdot 11
\end{array}\right\}
\end{array}\right.\text{."}
$$

Naturally enough it is noted by Cayley that the writing of $\alpha=\beta=5$ gives us the less general theorem with which he started; but he does not explain why a third way of arranging the terms of the development is adopted. Stranger still, he does not remark on the fact that by making 11, 22, 33, 44 all vanish there is obtained the identity

$$
\overline{\alpha 1234 \mid \beta 1234}_{rs=-sr,\ rr=0} = \alpha\beta 1234 \cdot 1234,
$$

which is the twin theorem to one given in his previous paper regarding a bordered skew symmetrical determinant of *even* order. It will be remembered, however, that in the statement of this latter theorem, the peculiar narrow use of the word 'bordé' did not occur.

Although what may be called Part Second of the paper (pp. 301, 302) may seem at first sight to concern something else, it really only draws attention to the fact that *the minors* (by which he means those afterwards named *primary* minors) *of a skew determinant are themselves skew, being "gauches ordinaires" when their cofactor in the original determinant is of the form* rr, *and "gauches bordés" when their cofactor is of the form* rs. Considerable space is occupied in verifying by two examples that the same result will be reached whether we apply the theorem of Part First directly to

$$
\overline{123 \ldots n \mid 123 \ldots n}
$$

or to the primary minors in its equivalent

$$
11 \cdot \overline{23 \ldots n \mid 23 \ldots n} - 12 \cdot \overline{23 \ldots n \mid 13 \ldots n} + \ldots\ldots
$$

* A serious misprint in the original is here corrected.

What may be called Part Third (pp. 303–305) is very forbidding, by reason of the defective mode of exposition and of the awkwardness of the notation employed. Probably this accounts for the fact that the interesting theorem which it contains has never emerged until now from its place of sepulture. A portion of it must of necessity be given verbatim, if only for the purpose of preserving historical colour. It commences—

"Je remarque que le nombre des termes du développement (p. 299) du déterminant gauche est toujours une *puissance de* 2, et que de plus, ce nombre se réduit à la moitié, en réduisant à zéro un terme quelconque *aa*. Mais outre cela, le déterminant prend dans cette supposition la forme de déterminant [gauche] d'un ordre inférieur de l'unité. Je considère par exemple le déterminant gauche $\overline{123 \mid 123}$. En y faisant $33 = 0$ et en accentuant, pour y mettre plus de clarté, tous les symboles, on trouve

$$\overline{123 \mid 123'} = 11' \cdot (23')^2 + 22' \cdot (13')^2.$$

De là, en écrivant

$$11 = 13' \cdot 11', \qquad 12 = 11' \cdot 23',$$
$$22 = 13' \cdot 22',$$

on obtient

$$\begin{aligned} \overline{12 \mid 12} &= 11 \cdot 22 + (12)^2, \\ &= 11' \cdot \{22' \cdot (13')^2 + 11' \cdot (23')^2\}, \end{aligned}$$

c'est à dire

$$\overline{12 \mid 12} = 11' \cdot \overline{123 \mid 123'}.$$

On a de même

$$\overline{1234 \mid 1234'} = 11' \cdot 22' \cdot (34')^2 + 11' \cdot 33' \cdot (24')^2 + 22' \cdot 33' \cdot (14')^2 + (1234')^2,$$

et delà, en écrivant

$$\begin{array}{lll} 11 = 14' \cdot 11', & 12 = 11' \cdot 24', & 23 = 1234', \\ 22 = 14' \cdot 22', & 13 = 11' \cdot 34', & \\ 33 = 14' \cdot 33', & & \end{array}$$

on obtient

$$\begin{aligned} \overline{123 \mid 123} &= 11 \cdot 22 \cdot 33 + 11 \cdot (23)^2 + 22 \cdot (31)^2 + 33 \cdot (12)^2, \\ &= 11' \cdot 14' \left\{ \begin{array}{l} 22' \cdot 33' \cdot (14')^2 + (1234')^2 \\ \qquad + 11' \cdot 22' \cdot (34')^2 + 11' \cdot 33' \cdot (24')^2 \end{array} \right\} \end{aligned}$$

c'est à dire

$$\overline{123 \mid 123} = 11' \cdot 14' \cdot \overline{1234 \mid 1234'}."$$

The remainder is devoted to the next two cases, the verification of which, of course, occupies still more space. The theorem thus dealt with may be roughly described as giving *the transformation of a skew determinant, having one zero element in its main diagonal, into a skew determinant of the next lower order;* and in a notation which needs no explanation and which was perfectly familiar to Cayley at the time, the four examples may be written thus:—

$$\begin{vmatrix} 11 & 12 & 13 \\ -12 & 22 & 23 \\ -13 & -23 & . \end{vmatrix} = \begin{vmatrix} 11\cdot13 & 11\cdot23 \\ -11\cdot23 & 22\cdot13 \end{vmatrix} \div 11,$$

$$\begin{vmatrix} 11 & 12 & 13 & 14 \\ -12 & 22 & 23 & 24 \\ -13 & -23 & 33 & 34 \\ -14 & -24 & -34 & . \end{vmatrix} = \begin{vmatrix} 11\cdot14 & 11\cdot24 & 11\cdot34 \\ -11\cdot24 & 22\cdot14 & [1234] \\ -11\cdot34 & -[1234] & 33\cdot14 \end{vmatrix} \div 11\cdot14,$$

$$\begin{vmatrix} 11 & 12 & 13 & 14 & 15 \\ -12 & 22 & 23 & 24 & 25 \\ -13 & -23 & 33 & 34 & 35 \\ -14 & -24 & -34 & 44 & 45 \\ -15 & -25 & -35 & -45 & . \end{vmatrix} = \begin{vmatrix} 11\cdot15 & 11\cdot25 & 11\cdot35 & 11\cdot45 \\ -11\cdot25 & 22\cdot15 & [1235] & [1245] \\ -11\cdot35 & -[1235] & 33\cdot15 & [1345] \\ -11\cdot45 & -[1245] & -[1345] & 44\cdot15 \end{vmatrix} \div 11\cdot(15)^2,$$

$$\begin{vmatrix} 11 & 12 & \ldots & 15 & 16 \\ -12 & 22 & \ldots & 25 & 26 \\ . & . & . & . & . \\ -15 & -25 & \ldots & 55 & 56 \\ -16 & -26 & \ldots & -56 & . \end{vmatrix} = \begin{vmatrix} 11\cdot16 & 11\cdot26 & 11\cdot36 & 11\cdot46 & 11\cdot56 \\ -11\cdot26 & 22\cdot16 & [1236] & [1246] & [1256] \\ -11\cdot36 & -[1236] & 33\cdot16 & [1346] & [1356] \\ -11\cdot46 & -[1246] & -[1346] & 44\cdot16 & [1456] \\ -11\cdot56 & -[1256] & -[1356] & -[1456] & 55\cdot16 \end{vmatrix} \div 11\cdot(16)^3.$$

Of course, this mode of writing does not at once suggest any better mode of proof, but it makes clear the general theorem, which consequently may be enunciated as follows:

"*A skew determinant of the* n^{th} *order which has a zero for the last element of its main diagonal may, if multiplied by* $11\cdot(1\mathrm{n})^{\mathrm{n}-3}$ *be transformed into a skew determinant of the* $(\mathrm{n}-1)^{th}$ *order, which has for its first row the last column of the original determinant multiplied by* 11, *for its main diagonal the main diagonal of the original determinant multiplied by* 1n, *and for the element in every other place* rs *situated between these two lines the Pfaffian* [1rsn]."

The rest of the paper deals with *inverse matrices*, and with the application of them to the problem afterwards known as the *automorphic transformation of a quadric.*

BRIOSCHI, F. (1854).

[LA TEORICA DEI DETERMINANTI, E LE SUE PRINCIPALI APPLICAZIONI. viii+116 pp. Pavia.]

In this, the second text-book, the same importance is given to skew determinants as in Spottiswoode, the first part of the eighth section (pp. 55–72) being devoted to them under the heading "Dei determinanti *gobbi*," which Schellbach translates by *überschlagene.* The arrangement and treatment of the matter, however, are much more logical, zero-axial skew determinants being taken first, then the functions connected with these, namely, Pfaffians, then skew determinants which are not zero-axial, and lastly the use of skew determinants in the consideration of the problem of orthogonal transformation.

The precedence given to determinants which are "gobbi simmetrici" over those which are "puramente gobbi" is explained at the outset by reference to Cayley's theorem regarding the expressibility of the latter in terms of the former, the quite general theorem from which Cayley's immediately follows being carefully enunciated thus:

"Indicando con P_o il determinante nel quale si pongano equali a zero gli elementi principali; e con $({}^m P_{ii})_o$ un determinante minore principale delle' m-esimo ordine del determinante P nel quale siensi annullati gli elementi principali si ha :—

$$P = P_o + \sum_r a_{rr} ({}^1P_{ii})_o + \sum_r \sum_s a_{rr} a_{ss} ({}^2P_{ii})_o + \ . \ . \ . \ + a_{11} a_{22} \ . \ . \ . \ a_{nn}."$$

The proof given of Jacobi's theorem regarding the value of an odd-ordered skew determinant with zeros in the principal diagonal is essentially the same as Cayley's proof (1847), but fuller and clearer. The proof of the corresponding theorem for a determinant of even order resembles Spottiswoode's, the difference lying mainly in the use of the notation of differential-quotients in specifying the minors of the determinant.

Denoting the determinant of even order by P, he starts with the development—

$$P = -a_{1r}^2\frac{\partial^2 P}{\partial a_{1r}\partial a_{r1}} - a_{1s}^2\frac{\partial^2 P}{\partial a_{1s}\partial a_{s1}} \pm 2a_{1r}a_{1s}\frac{\partial^2 P}{\partial a_{1r}\partial a_{s1}} - \cdot \cdot \cdot \cdot \cdot \cdot \cdot \cdot \cdot \cdot \cdot \cdot$$

Then since a previously obtained general identity, originally due to Jacobi, viz.,

$$P\frac{\partial^2 P}{\partial a_{rs}\partial a_{pq}} = \frac{\partial P}{\partial a_{rs}}\cdot\frac{\partial P}{\partial a_{pq}} - \frac{\partial P}{\partial a_{ps}}\cdot\frac{\partial P}{\partial a_{rq}},$$

gives in this special case the identities

$$P\frac{\partial^2 P}{\partial a_{1r}\partial a_{r1}} = \frac{\partial P}{\partial a_{1r}}\cdot\frac{\partial P}{\partial a_{r1}}, \qquad P\frac{\partial^2 P}{\partial a_{1s}\partial a_{s1}} = \frac{\partial P}{\partial a_{1s}}\cdot\frac{\partial P}{\partial a_{s1}},$$

$$P\frac{\partial^2 P}{\partial a_{1r}\partial a_{s1}} = \frac{\partial P}{\partial a_{1r}}\cdot\frac{\partial P}{\partial a_{s1}},$$

because the cofactor, awkwardly denoted by $\partial P/\partial a_{ii}$, of any vanishing element a_{ii} in the principal diagonal is zero in accordance with the preceding theorem of Cayley's. From the first two of these we have

$$P^2\cdot\frac{\partial^2 P}{\partial a_{1r}\partial a_{r1}}\cdot\frac{\partial^2 P}{\partial a_{1s}\partial a_{s1}} = \frac{\partial P}{\partial a_{1r}}\cdot\frac{\partial P}{\partial a_{r1}} \times \frac{\partial P}{\partial a_{1s}}\cdot\frac{\partial P}{\partial a_{s1}},$$

the right side of which can be changed into

$$\left(\frac{\partial P}{\partial a_{1r}}\cdot\frac{\partial P}{\partial a_{s1}}\right)^2$$

by reason of the fact that for a determinant such as P we have in every case

$$\frac{\partial P}{\partial a_{rs}} = -\frac{\partial P}{\partial a_{sr}}.$$

But from the third identity above, by squaring, we obtain on the right the same expression; so that there thus results

$$\left(\frac{\partial P}{\partial a_{1r}\partial a_{s1}}\right)^2 = \frac{\partial^2 P}{\partial a_{1r}\partial a_{r1}}\cdot\frac{\partial^2 P}{\partial a_{1s}\partial a_{s1}},$$

an equation which, in connection with the above development

of P, implies the property that the determinant P is a square ("nella quale equazione trovasi appunto espressa la proprietà che il determinante P è un quadrato").

On looking now to the development with which the demonstration opened, Brioschi is led to an expression for the square in question, viz.:

$$P = \left\{ \pm a_{12}\left(\frac{\partial^2 P}{\partial a_{11}\partial a_{22}}\right)^{\frac{1}{2}} \pm a_{13}\left(\frac{\partial^2 P}{\partial a_{11}\partial a_{33}}\right)^{\frac{1}{2}} \pm \dots \pm a_{1n}\left(\frac{\partial^2 P}{\partial a_{11}\partial a_{nn}}\right)^{\frac{1}{2}}\right\}^2,$$

or, more generally,

$$P = \left\{\textstyle\sum_s \pm a_{rs}\left(\frac{\partial^2 P}{\partial a_{rr}\partial a_{ss}}\right)^{\frac{1}{2}}\right\}^2,$$

where he notes that in every case $a_{rr}=0$ and $\partial^2 P/\partial a_{rr}\partial a_{ss}$, being a determinant of the same kind as P, is a square. The example added is

$$\begin{vmatrix} 0 & a_{12} & a_{13} & a_{14} \\ a_{21} & 0 & a_{23} & a_{24} \\ a_{31} & a_{32} & 0 & a_{34} \\ a_{41} & a_{42} & a_{43} & 0 \end{vmatrix} = \left\{ a_{12}\begin{vmatrix} 0 & a_{34} \\ a_{43} & 0 \end{vmatrix}^{\frac{1}{2}} - a_{13}\begin{vmatrix} 0 & a_{24} \\ a_{42} & 0 \end{vmatrix}^{\frac{1}{2}} + a_{14}\begin{vmatrix} 0 & a_{23} \\ a_{32} & 0 \end{vmatrix}^{\frac{1}{2}}\right\}^2,$$

$$= (a_{12}a_{34} - a_{13}a_{24} + a_{14}a_{23})^2,$$

where the difficulty of the ambiguous sign, although presenting itself more prominently than in the general demonstration, is not referred to.

The new function H, which is the square root of P, is next studied. Differentiating both sides of the equation of relationship Brioschi obtains

$$\frac{\partial P}{\partial a_{rs}} = H\frac{\partial H}{\partial a_{rs}},*$$

where the inconvenience of the differential notation comes out more strikingly than before, the differential-quotient on the left

* Since the left member is what Cayley called a "bordered skew symmetric determinant"; and since, as Jacobi noted, a differential-quotient of H with respect to one of its elements is a function of the same kind as H, we have here unnoticed one half of Cayley's proposition that *a bordered skew symmetric determinant is expressible as the product of two Pfaffians.*

being used conventionally to denote a certain minor of P, and the differentiation on the right being real. By squaring we have

$$\left(\frac{\partial \mathrm{P}}{\partial a_{rs}}\right)^2 = \mathrm{P}\left(\frac{\partial \mathrm{H}}{\partial a_{rs}}\right)^2,$$

and since, as we have seen, it is permissible to substitute

$$\mathrm{P}\frac{\partial^2 \mathrm{P}}{\partial a_{rr}\partial a_{ss}} \text{ for } \left(\frac{\partial \mathrm{P}}{\partial a_{rs}}\right)^2$$

there results

$$\left(\frac{\partial^2 \mathrm{P}}{\partial a_{rr}\partial a_{s}}\right)^{\frac{1}{2}} = \pm\frac{\partial \mathrm{H}}{\partial a_{rs}};$$

so that the expansion for P above obtained may be altered into

$$\mathrm{P} = \left\{\sum_s\left(a_{rs}\frac{\partial \mathrm{H}}{\partial a_{rs}}\right)\right\}^2,$$

from which by extraction of the square root we have

$$\mathrm{H} = \sum_s\left(a_{rs}\frac{\partial \mathrm{H}}{\partial a_{rs}}\right).$$

This will be recognised as a third mode of writing an already well-known result, and, as Brioschi notes, gives a property of the function H similar to a property of determinants ("la quale equazione contiene una proprietà della funzione H analoga ad una nota dei determinanti").

From this he passes to what he calls the characteristic property of H, viz., its change of sign consequent upon the transposition of two indices. Calling H′ what H becomes when r and s are interchanged, he notes that in those terms of H in which the element a_{rs} occurs there can be no other element with the same indices, and that therefore

$$\frac{\partial \mathrm{H}}{\partial a_{rs}} = -\frac{\partial \mathrm{H}'}{\partial a_{rs}}.$$

Then since the same interchange made in P leaves P in reality unaltered,—that is to say, since $\mathrm{H}^2 = \mathrm{H}'^2$,—he obtains

$$\mathrm{H}\frac{\partial \mathrm{H}}{\partial a_{rs}} = \mathrm{H}'\frac{\partial \mathrm{H}'}{\partial a_{rs}};$$

and, it having been shown that the two differential-quotients here appearing are of opposite signs, it follows that so also are H and H′.

Lastly, he passes on to skew determinants in general; and, using the theorem and notation introduced at the outset, he writes Cayley's propositions in the form—

$$n \text{ even,} \qquad \mathrm{P} = \mathrm{P}_o + \sum_r \sum_s a_{rr} a_{ss} ({}^2p_{ii})_o + \ldots + a_{11} a_{22} \ldots a_{nn},$$

$$n \text{ odd,} \qquad \mathrm{P} = \sum_r a_{rr} ({}^1p_{ii})_o + \ldots + a_{11} a_{22} \ldots a_{nn},$$

which, he says, when the principal elements are all unity, become

$$n \text{ even,} \qquad \mathrm{P} = \mathrm{P}_o + \sum_i ({}^2p_{ii})_o + \ldots + 1,$$

$$n \text{ odd,} \qquad \mathrm{P} = \sum_i ({}^1p_{ii})_o + \sum_i ({}^3p_{ii})_o + \ldots + 1,$$

the development now being in each case a sum of squares, as all the minors appearing in it are even-ordered.

BRIOSCHI, F. (1855, March).

[Sur l'analogie entre une classe de déterminants d'ordre pair et les déterminants binaires. *Crelle's Journ.*, lii. pp. 133–141; or *Opere mat.*, v. pp. 511–520. See also *Annali di Sci. mat. e fis.*, vi. pp. 430–432.]

After explaining that his purpose is to generalise a result of Hermite's (*Comptes rendus . . . Acad. des Sci.*, Paris, xl. pp. 249–254) regarding determinants of the fourth order, Brioschi sets out by establishing a necessary lemma regarding determinants of any even order whatever. It is this lemma which is of importance to us in the present connection. Taking the determinant

$$\sum (\pm a_{11} a_{22} \ldots \ldots a_{2m,2m}), \quad \text{or} \quad \mathrm{A} \text{ say},$$

he multiplies it by the equivalent determinant

$$\begin{vmatrix} a_{12} & -a_{11} & a_{14} & -a_{13} & \ldots\ldots & a_{1,2m} & -a_{1,2m-1} \\ a_{22} & -a_{21} & a_{24} & -a_{23} & \ldots\ldots & a_{2,2m} & -a_{2,2m-1} \\ \cdot & \cdot & \cdot & \cdot & \cdot & \cdot & \cdot \\ a_{2m,2} & -a_{2m,1} & a_{2m,4} & -a_{2m,3} & \ldots\ldots & a_{2m,2m} & -a_{2m,2m-1} \end{vmatrix}$$

obtaining the result

$$A^2 = \begin{vmatrix} \cdot & l_{12} & l_{13} & \dots & l_{1,2m} \\ l_{21} & \cdot & l_{23} & \dots & l_{2,2m} \\ l_{31} & l_{32} & \cdot & \dots & l_{3,2m} \\ \cdot & \cdot & \cdot & \cdot & \cdot \\ l_{2m,1} & l_{2m,2} & l_{2m,3} & \dots & \cdot \end{vmatrix}$$

where

$$l_{rs} \equiv a_{r1}a_{s2} - a_{r2}a_{s1} + a_{r3}a_{s4} - a_{r4}a_{s3} + \dots\dots$$

and where therefore

$$l_{rs} = -l_{sr}.$$

Using then Cayley's theorem regarding a determinant which is "gauche symétrique," he concludes that A is expressible as a rational function of the l's. This result he might have put in the form *Any even-ordered determinant is expressible as a Pfaffian*: and at a later date it would have been written

$$\sum(\pm a_{11}a_{22}\dots a_{2m,2m}) = \left| \begin{matrix} l_{12} & l_{13} & l_{14} & \dots & l_{1,2m} \\ & l_{23} & l_{24} & \dots & l_{2,2m} \\ & & \cdot & \cdot & \cdot \\ & & & & l_{2m-1,2m} \end{matrix} \right|.$$

The rest of the paper is occupied with the consideration of the special case where

$$l_{12} = l_{34} = l_{56} = \dots = l_{2m-1,2m},$$

and all the other l's vanish.

BELLAVITIS, G. (1857).

[Sposizione elementare della teorica dei determinanti. *Memorie ... Istituto Veneto* ... vii. pp. 67–144.]

For the determinant which Cayley named "gauche," Bellavitis introduces the term *pseudosimmetrico*, and for "gauche symétrique" he introduces *emisimmetrico* (§ 41). In the matter of notation also he suggests a change, denoting (§ 54) the Pfaffian which is the square root of

$$\begin{vmatrix} 0 & b_a & c_a & d_a \\ a_b & 0 & c_b & d_b \\ a_c & b_c & 0 & d_c \\ a_d & b_d & c_d & 0 \end{vmatrix}$$

by

$$\text{Pf.}\,(a, b, c, d).$$

Nothing else is worth noting save the carefulness of the exposition (§§ 51–54, 59). Part of Cayley's theorem regarding a bordered skew symmetric determinant appears as a theorem regarding a non-coaxial primary minor of a skew symmetric determinant (§ 59).

CAYLEY, A. (1857).

[Théorème sur les déterminants gauches. *Crelle's Journ.*, lv. pp. 277, 278; or *Collected Math. Papers*, iv. pp. 72, 73.]

This is practically a note to rectify the oversight made in the paper of 1854, where, as has been pointed out, he omitted to draw attention to the case in which the skew determinant submitted to the operation of 'bordering' has zeros for the elements of the principal diagonal.

"Un déterminant," he now says, "de cette espèce se réduit toujours au produit de deux *Pfaffiens.* En effet en écrivant dans les exemples $11 = 22 = 33 = 44 = 0$, on obtient:

$$\overline{\alpha 123 \mid \beta 123} = \alpha 123 \cdot \beta 123,$$

$$\overline{\alpha 1234 \mid \beta 1234} = \alpha\beta 1234 \cdot 1234,$$

et de même pour un déterminant gauche et symétrique bordé quelconque, suivant que l'ordre du déterminant est pair ou impair."

To this there is added the suggestive commentary:—

"Je remarque à propos de cela, que dans le cas d'un déterminant d'ordre pair, le terme $\alpha\beta$ est multiplié par un mineur premier lequel (comme déterminant gauche et symétrique d'ordre impair) se réduit à zero; le déterminant ne contient donc pas ce term $\alpha\beta$, et sera par conséquent fonction lineo-linéaire des quantités $\alpha 1$, $\alpha 2$, etc., et 1β, 2β, etc.; de manière qu'on ne saurait être surpris de voir ce déterminant se présenter sous la forme d'un produit de deux facteurs, dont l'un est fonction linéaire de $\alpha 1$, $\alpha 2$, etc., et l'autre fonction linéaire de 1β, 2β, etc. Mais pour un déterminant d'ordre impair, le coefficient du terme $\alpha\beta$ ne se réduit pas à zéro ; en supposant donc que le déterminant puisse s'exprimer comme produit de deux facteurs, il est nécessaire que l'un de ces facteurs soit (comme le déterminant même) fonction linéaire de $\alpha\beta$ et lineo-linéaire de $\alpha 1$, $\alpha 2$, etc., et 1β, 2β, etc.: de cette manière on se rend compte de la différence de la forme des facteurs, qui a lieu dans les deux cas dont il s'agit."

It is finally pointed out that by writing $\beta = \alpha$ we are brought back to

$$\overline{\alpha 123 \mid \alpha 123} = (\alpha 123)^2,$$
$$\overline{\alpha 1234 \mid \alpha 1234} = 0:$$

—"la propriété fondamentale des déterminants gauches et symétriques." There is again, however, an oversight here, for the element $\alpha\alpha$ is taken to be equal to 0, whereas it is only necessarily so in the second case.

BALTZER, R. (1857).

[THEORIE UND ANWENDUNG DER DETERMINANTEN, mit . . . . . vi+129 pp. Leipzig, 1857.]

Following his two predecessors Baltzer also assigned a separate section of his text-book to skew determinants, but without giving them any special designation of his own or even taking over that used by Schellbach. The title of the section (§ 8, pp. 29–34) is thus a little lengthy, viz., "*Determinante eines Systems von Elementen, unter denen die correspondirenden* a_{ik} *und* a_{ki} *entgegengesetzt gleich sind.*"

It must be noted, however, that before this section is reached some theorems which strictly belong to the subject of the section have been already dealt with. These are in the first place (§ 3, 8; p. 12) Jacobi's theorem regarding the vanishing of a zero-axial skew determinant of odd order, and Spottiswoode's theorems regarding conjugate elements of the adjugate or inverse of a zero-axial skew determinant, the mode of proof for all being that used by Jacobi for his own theorem, viz., the multiplication of all rows or all columns by -1, and then comparing the resulting determinant with the original. In the second place

(§ 3, 10; p. 13) we have Brioschi's theorem regarding the differential-quotient of a zero-axial skew determinant of even order, and a suggestive proof of the same which it is desirable to note. It is as follows:—Let the determinant

$$\begin{vmatrix} a_{11} & \dots\dots & a_{1n} \\ \cdot & \cdot\ \cdot\ \cdot\ \cdot & \cdot \\ a_{n1} & \dots\dots & a_{nn} \end{vmatrix}$$

be denoted by Δ, and the cofactor of a_{rs} in Δ by A_{rs}. Then, bearing in mind that Δ is a function of a_{rs} and that a_{sr} is not independent of a_{rs}, we have

$$\begin{aligned} \frac{\partial\Delta}{\partial a_{rs}} &= A_{rs} + A_{sr}\frac{\partial a_{sr}}{\partial a_{rs}}, \\ &= A_{rs} - A_{sr}. \end{aligned}$$

But when n is even we know from Spottiswoode, as above, that $A_{rs} = -A_{sr}$; consequently we have in this case

$$\frac{\partial\Delta}{\partial a_{rs}} = 2A_{rs},$$

as Brioschi affirmed.* In the third place (§ 7, 5; pp. 28, 29) he applies Jacobi's general theorem

$$\begin{vmatrix} A_{rr} & A_{rs} \\ A_{sr} & A_{ss} \end{vmatrix} = \Delta\frac{\partial^2\Delta}{\partial a_{rr}\partial a_{ss}},$$

as Brioschi did, to the case where Δ is zero-axial skew and of odd order to obtain the result

$$A_{rs}^2 = A_{rr}\cdot A_{rs};$$

and he takes the further step of deducing from it the result

$$A_{r1} : A_{r2} : A_{r3} : \dots\dots = \sqrt{A_{11}} : \sqrt{A_{22}} : \sqrt{A_{33}} : \dots$$

* It ought to be noticed also that Baltzer uses the equation

$$\frac{\partial\Delta}{\partial a_{rs}} = A_{rs} - A_{sr}$$

to verify Spottiswoode's theorem for the case where Δ is odd-ordered, the reasoning being that as Δ is then known to be identically zero, so also must $\partial\Delta/\partial a_{rs}$, and that therefore $A_{rs} = A_{sr}$.

thus showing, as he says (1) that the ratios on the left are independent of r, and (2) that, when the sign of one of the roots has been fixed, the others are known ("dass durch das Zeichen einer unter diesen Wurzeln die Zeichen der übrigen Wurzeln bestimmt sind)."

Turning now to the section specially set apart for the consideration of skew determinants, we find that it opens with Cayley's theorem regarding a zero-axial determinant of even order, the requirement being, as here worded, to prove that such a determinant is the square of *a rational integral function of the elements.* The proof is essentially the same as Spottiswoode's and Brioschi's, and differs from Cayley's merely in that it does not begin with a determinant of a more general form than is necessary,—a point which it is desirable to insist upon, as Baltzer ignores the fact, and then does not hesitate to say in a footnote that Cayley's proof "leaves manifold doubts unrelieved." In fact the theorem which Cayley proves is, that *if a zero-axial skew determinant of odd order be 'bordered' the resulting determinant is the product of two Pfaffians*: whereas what the three others prove, is the particular case of this in which the skewness extends to the bordering elements.

The development with which the proof begins Baltzer writes in the form

$$\Delta = a_{11}A_{11} - \sum_{rs} a_{r1}a_{1s}A'_{rs},$$

where A' is the cofactor of a_{rs} in A_{11}, and r and s have the values 2, 3, . . . , n. He then uses the fact that A_{11} is a zero-axial skew determinant of odd order, and that therefore by a preceding result

$$A'_{rs} = A'_{sr} = \sqrt{A'_{rr}A'_{ss}};$$

so that there is obtained

$$\Delta = \sum_{rs} a_{1r}a_{1s}\sqrt{A'_{rr}A'_{ss}};$$

and since in this aggregate the values possible for r are exactly those possible for s, he concludes (without knowing the signs of the terms of the aggregate, be it observed) that it is resolvable into two factors, viz.

$$\Big(\sum_r a_{1r}\sqrt{A'_{rr}}\Big)\Big(\sum_s a_{1s}\sqrt{A'_{ss}}\Big).$$

It is then argued that the two factors are identical even in the signs of their various terms "da durch das Zeichen einer Wurzel die Zeichen der übrigen bestimmt sind"; and that therefore

$$\Delta = \Big(\sum_r a_{1r}\sqrt{A'_{rr}}\Big)^2,$$

$$\text{and} \quad \sqrt{\Delta} = \sum_r a_{1r}\sqrt{A'_{rr}},$$

—an aggregate of $n-1$ terms, since the values to be given to r are $2, 3, \ldots, n$. The next step consists in pointing out that A'_{rr} being a determinant similar to Δ but of order $n-2$, it must follow that $\sqrt{A'_{rr}}$ can in the same way be expressed as an aggregate of $n-3$ terms, and that this process can be continued until the minor under the root-sign is of the 2nd order, when manifestly its value is the square of one of its elements. The final result thus is that $\sqrt{\Delta}$ is expressible as an aggregate of $(n-1)(n-3)\ldots 3.1$ terms, each of which is the product of $\frac{1}{2}n$ elements whose collected suffixes form a permutation of $1, 2, \ldots, n$.

By way of corollary to this it is pointed out that

$$\pm\, a_{12}a_{34}\ldots a_{n-1,n}$$

is one of the terms of the aggregate; and the same is proved by showing that the square of this is a term of Δ, the reasoning being as follows:—Since in every case $a_{rs} = -a_{sr}$ we have

$$(a_{12}a_{34}\ldots a_{n-1,n})^2 = (a_{12}a_{34}\ldots a_{n-1,n})\cdot(-)^{\frac{1}{2}n}(a_{21}a_{43}\ldots a_{n,n-1}),$$

$$\text{and} \therefore \quad = (-)^{\frac{1}{2}n}(a_{12}a_{21}a_{34}a_{43}\ldots a_{n-1,n}a_{n,n-1}),$$

which clearly contains n elements, one from every row and one from every column of Δ, and will therefore be a term of Δ if only we can show that the number of inversions of order in

$$2,1,\ 4,3,\ 6,5,\ \ldots,\ n,\ n-1$$

is $\frac{1}{2}n$, a fact which is self-evident.

Baltzer's proof that the rational integral function H, which is the square root of Δ, changes signs when two suffixes, r and s, are interchanged is a simplification of Brioschi's, the operation and even the notion of differentiation being dispensed with. The function resulting from the change being H′ he concludes like Brioschi that

$$\mathrm{H}^2 = \mathrm{H}'^2;$$

also the aggregate of the terms in H which contain a_{rs} being $a_{rs}\mathrm{B}$, say, he infers as Brioschi does that B cannot be affected by the change, and that therefore $a_{rs}\mathrm{B}$ will be altered into $a_{sr}\mathrm{B}$ or $-a_{rs}\mathrm{B}$. Here, however, he brings the demonstration quickly to a satisfactory end by saying that since some of the terms of H′ are thus seen to differ in sign only from the corresponding terms of H, the equation $\mathrm{H}^2 = \mathrm{H}'^2$ shows all of them must so differ; and this is what was to be proved.

Jacobi's notation for the function H is then introduced, the formal intimation being that $(1,2,3, \ldots, n)$ *is used to denote the aggregate whose first term is* $a_{12}a_{34}, \ldots, a_{n-1,n}$ *and whose square is* Δ. The other value of $\sqrt{\Delta}$ is thus of course representable by $(2,1,3, \ldots, n)$, or $(2,3, \ldots, n,1)$, or .... As this implies also that

$$\sqrt{\mathrm{A}'_{rr}} = \pm(2,3, \ldots, r-1, r+1, \ldots, n)$$

we have now the means, so far as symbolism is concerned, of removing the ambiguity from the various terms of the identity

$$\sqrt{\Delta} = a_{12}\sqrt{\mathrm{A}'_{22}} + a_{13}\sqrt{\mathrm{A}_{33}} + \ldots\ldots + a_{1n}\sqrt{\mathrm{A}'_{nn}}.$$

As for the knowledge necessary to use the symbolism aright, Baltzer's dictum is *that the sign taken to precede*

$$(2,3, \ldots, r-1, r+1, \ldots, n)$$

in substituting for $\sqrt{\mathrm{A}'_{rr}}$ *must be such that the equation*

$$\sqrt{\mathrm{A}'_{rr}} \cdot \sqrt{\mathrm{A}'_{ss}} = \mathrm{A}'_{rs}$$

will be satisfied; and this he proves will take place when the sign-factor of $(2,3, \ldots, r-1, r+1, \ldots, n)$ is $(-1)^r$. By hypothesis, he says, the left-hand side

$$= (-1)^r(2,3,\ldots,r-1,r+1,\ldots,n)\cdot(-1)^s(2,3,\ldots,s-1,s+1,\ldots,n)$$
$$= (-1)^{r+s}(2,3,\ldots,r-1,r+1,\ldots,n)(2,3,\ldots,s-1,s+1,\ldots,n),$$

and therefore by a previous theorem

$$= -(-1)^{r+s}(2,3,\ldots,r-1,r+1,\ldots,n)(3,\ldots,s-1,s+1,\ldots,n,2),$$

the first term of which is

$$-(-1)^{r+s}a_{23}\ldots a_{n-1,n}\cdot a_{34}\ldots a_{n,2},$$

or

$$-(-1)^{r+s}a_{23}a_{34}\ldots a_{n-1,n}a_{n2};$$

and the right-hand side

$$= \text{cofactor of } a_{rs} \text{ in } \begin{vmatrix} a_{22} & a_{23} & \ldots & a_{2n} \\ a_{32} & a_{33} & \ldots & a_{3n} \\ \cdot & \cdot & \cdot & \cdot \\ a_{n2} & a_{n3} & \ldots & a_{nn} \end{vmatrix},$$

$$= (-1)^{r+s}\begin{vmatrix} a_{22} & a_{23} & \ldots\ldots & a_{2,s-1} & a_{2,s+1} & \ldots\ldots & a_{2n} \\ a_{32} & a_{33} & \ldots\ldots & a_{3,s-1} & a_{3,s+1} & \ldots\ldots & a_{3n} \\ \cdot & \cdot & \cdot & \cdot & \cdot & \cdot & \cdot \\ a_{r-1,2} & a_{r-1,3} & \ldots\ldots & a_{r-1,s-1} & a_{r-1,s+1} & \ldots\ldots & a_{r-1,n} \\ a_{r+1,2} & a_{r+1,3} & \ldots\ldots & a_{r+1,s-1} & a_{r+1,s+1} & \ldots\ldots & a_{r+1,n} \\ \cdot & \cdot & \cdot & \cdot & \cdot & \cdot & \cdot \\ a_{n,2} & a_{n,3} & \ldots\ldots & a_{n,s-1} & a_{n,s+1} & \ldots\ldots & a_{n,n} \end{vmatrix},$$

and therefore, on account of the translation of the first column to the last place,

$$= -(-1)^{r+s}\begin{vmatrix} a_{23} & \ldots\ldots & a_{2,s-1} & a_{2,s+1} & \ldots\ldots & a_{2,n} & a_{22} \\ a_{33} & \ldots\ldots & a_{3,s-1} & a_{3,s+1} & \ldots\ldots & a_{3,n} & a_{32} \\ \cdot & \cdot & \cdot & \cdot & \cdot & \cdot & \cdot \\ a_{r-1,3} & \ldots\ldots & a_{r-1,s-1} & a_{r-1,s+1} & \ldots\ldots & a_{r-1,n} & a_{r-1,2} \\ a_{r+1,3} & \ldots\ldots & a_{r+1,s-1} & a_{r+1,s+1} & \ldots\ldots & a_{r+1,n} & a_{r+1,2} \\ \cdot & \cdot & \cdot & \cdot & \cdot & \cdot & \cdot \\ a_{n,3} & \ldots\ldots & a_{n,s-1} & a_{n,s+1} & \ldots\ldots & a_{n,n} & a_{n,2} \end{vmatrix},$$

the first term of which is

$$-(-1)^{r+s}a_{23}a_{34}\ldots a_{n-1,n}a_{n,2},$$

exactly as before.

To the proof no note is appended drawing attention to the fact that the very same result would have been reached by taking

$(-1)^{r-1}$, or indeed $(-1)^{r-t}$, instead of $(-1)^r$ for the sign-factor of $(2,3,\ldots,r-1,r+1,\ldots,n)$.

The very next step taken, in accordance with the above mentioned dictum, is to make the substitution in the right-hand side of the equation

$$\sqrt{\Delta} = a_{12}\sqrt{A'_{22}} + a_{13}\sqrt{A'_{33}} + \ldots + a_{1n}\sqrt{A'_{nn}},$$

the first term being used to decide whether $(1,2,3,\ldots,n)$ or $-(1,2,3,\ldots,n)$ has to be substituted for the left-hand side, and the final result being

$$(1,2,3,\ldots,n) = a_{12}(3,\ldots,n)+a_{13}(4,\ldots,n,2)+\ldots+a_{1n}(2,\ldots,n-1).$$

Since $(3,4\ldots,n)$ is the cofactor of a_{12} in $(1,2,3,\ldots,n)$ and the differential-quotient of the latter with respect to a_{12} is the same, it immediately follows from this that

$$\sqrt{\Delta} = a_{12}\frac{\partial\sqrt{\Delta}}{\partial a_{12}} + a_{13}\frac{\partial\sqrt{\Delta}}{\partial a_{13}} + \ldots + a_{1n}\frac{\partial\sqrt{\Delta}}{\partial a_{1n}}.$$

Baltzer, however, obtains a more general result by going back to the corresponding more general theorem in determinants, viz., the theorem

$$\Delta = a_{r1}A_{r1} + a_{r2}A_{r2} + \ldots + a_{rn}A_{rn},$$

with which he associates

$$0 = a_{r1}A_{s1} + a_{r2}A_{r2} + \ldots + a_{rn}A_{sn};$$

substituting $\sqrt{\Delta}\dfrac{\partial\sqrt{\Delta}}{\partial a_{rs}}$ for A_{rs}; and then dividing both sides by $\sqrt{\Delta}$. In the results,

$$\sqrt{\Delta} = a_{r1}\frac{\partial\sqrt{\Delta}}{\partial a_{r1}} + \ldots + a_{rn}\frac{\partial\sqrt{\Delta}}{\partial a_{rn}},$$

$$0 = a_{r1}\frac{\partial\sqrt{\Delta}}{\partial a_{s1}} + \ldots + a_{rn}\frac{\partial\sqrt{\Delta}}{\partial a_{sn}},$$

it has to be noticed that there is no term in $\partial\sqrt{\Delta}/\partial a_{rr}$.

By comparison of the first of these with the immediately preceding result (the recurring law of development) he deduces the quite general identity regarding the two forms of the cofactor of a_{rs} in $\sqrt{\Delta}$—the identity, that is to say, with which we were inclined to start. His words are—

"Setzt man

$$\begin{aligned}\sqrt{\Delta} &= (r, 1, 2, \ldots, r-1, r+1 \ldots, n) \\ &= a_{r1}(2, \ldots, n) + a_{r2}(3, \ldots, n, 1) + \ldots\end{aligned}$$

so findet man

$$\frac{\partial\sqrt{\Delta}}{\partial a_{rs}} = (s+1, \ldots, n, 1, \ldots, s-1),$$

in welchem Cyclus die Suffixe r and s fehlen."

In regard to this the reader has, of course, to note that $(r,1,2, \ldots, r-1, r+1 \ldots, n)$ being only one of the two values of $\sqrt{\Delta}$, the differential-quotient obtained is also only one of two; in other words, that the result reached is really

$$\partial(r,1,2, \ldots, r-1, r+1, \ldots, n)\big/\partial a_{rs} = (s+1, \ldots, n, 1, \ldots, s-1),$$

where from 1 to $s-1$ and from $s+1$ to n the integers appear in natural order, save that r is omitted.

The remainder of the chapter or section, which contains no new feature, refers to Cayley's expansion of a determinant arranged according to products of elements of the principal diagonal, and the application of this to skew determinants whose diagonal elements are each equal to z.

SCHEIBNER, W. (1859, July).

[Ueber Halbdeterminanten. *Berichte* . . . *Ges. d. Wiss.* (Leipzig): *math.-phys. Cl.*, xi. pp. 151–159.]

This paper is not put forward by its author as containing new matter, being in fact such an exposition of the theory of Pfaffians as would suitably have formed a chapter, and a good one, of a text-book like Brioschi's or Baltzer's.

From the vanishing of a zero-axial skew determinant of odd order Scheibner reaches the already known fact that the product of two of its coaxial primary minors is equal to the square of a non-coaxial primary minor. In a quite fresh manner it is then shown that the square root of this is a rational and integral expression (Pfaffian), whose law of formation is thereafter established. Naturally following on this comes the proof (p. 156) that each of the non-coaxial primary minors is the product of two Pfaffians, the result being written in the form

$$A_{pq} = (p+1, \ldots, 2m, 0, \ldots, p-1)(q+1, \ldots, 2m, 0, \ldots, q-1),$$

where the suffixes of the elements of the original determinant are 0, 1, . . . , $2m$. On a later page (p. 158) it is shown that a similar proposition holds when the original determinant is of even order, namely,

$$A_{pq} = (-1)^p(0,1,2, \ldots, 2m+1)(q+1, \ldots, p-1, p+1, \ldots, q-1).$$

Cayley's theorem regarding a "bordered" skew symmetric determinant thus appears broken up into two parts.

The paper concludes with the suggestions that a skew symmetric determinant should be called a *Wechseldeterminante*, that its square root should be called a *Halbdeterminante*, and that the latter should be denoted by

$$\begin{vmatrix} a_{01} & a_{02} & a_{03} & \cdots & a_{0p} \\ & a_{12} & a_{13} & \cdots & a_{1p} \\ & & a_{23} & \cdots & a_{2p} \\ & & & \cdots & \cdot \\ & & & & a_{p-1,p} \end{vmatrix},$$

an expression which would thus be an alternative for $(0,1,2, \ldots, p)$ and which would vanish for even values of p.

SOUILLART, C. (1860, Sept.).

[Note sur la question 405 et sur une composition de carrés. *Nouv. Annales de Math.*, xix. pp. 320–322.]

Souillart's subject is the skew determinant

$$\begin{vmatrix} a & b & c & d \\ -b & a & -d & c \\ -c & d & a & -b \\ -d & -c & b & a \end{vmatrix},$$

and his observations are (1) that it is equal to

$$(a^2+b^2+c^2+d^2)^2,$$

and (2) that if it be multiplied by the similar determinant which is equal to

$$(p^2+q^2+r^2+s^2)^2$$

the result is a determinant of the same form, whether the multiplication be row-by-row or column-by-column. The object, of course, is to prove Euler's theorem * that the product of two sums of four squares is a sum of four squares.

CAYLEY, A. (1860, Dec.).

[Note on the theory of determinants. *Philos. Magazine*, xxi. pp. 180–185; or *Collected Math. Papers*, v. pp. 45–49.]

After expounding his, or rather Cauchy's last, mode of partitioning the ordinary expansion of a determinant, and giving his own diagrammatic representation of the partition, Cayley applies it to the expansion of a zero-axial skew determinant, showing, of course, that when of odd order it vanishes, and that when of even order it is expressible as a rational integral function of the elements.

TRUDI, N. (1862).

[TEORIA DE' DETERMINANTI, E LORO APPLICAZIONI, di Nicola Trudi. xii+268 pp. Napoli.]

To "determinanti gobbi" Trudi devotes sixteen pages (pp. 78–94) of his text-book, the exposition, which is not a little influenced by Brioschi and Baltzer, being full and simple. There are only one or two points in it worth noting. In the first place, there is his opening proposition that *in any zero-axial skew determinant, conjugate minors, if of even order, are equal, and if of odd order differ only in sign*: this is a slight generalisation of a previously known result. In the second place (p. 80), Jacobi's general theorem

$$\begin{vmatrix} A_{rr} & A_{rs} \\ A_{sr} & A_{ss} \end{vmatrix} = \Delta \times \text{compl. minor of} \begin{vmatrix} a_{rr} & a_{rs} \\ a_{sr} & a_{ss} \end{vmatrix}$$

* *Novi Commentarii Acad. Petropolitanae*, xv. (1770), pp. 75-106. For the conclusion reached see *Nouv. Annales de Math.*, xv. pp. 403–407.

is applied to the case where Δ is a zero-axial skew determinant of *even* order, Δ_{2m} say, and where, therefore, $A_{rr}=0=A_{ss}$ and $A_{sr}=-A_{rs}$, and the said complementary minor is a determinant of the same kind as Δ_{2m} but of the order $2m-2$: and it is thus seen that if Δ_{2m-2} be a square, so also must Δ_{2m}. The use to which this is put is evident.

JANNI, G. (1863).

[Teorica di determinanti simmetrici gobbi. *Giornale di Mat.*, i. pp. 275–278.]

Janni's final result is a troublesome rule for finding the expression whose square is a skew determinant of even order, the line of thought, so far as it goes, being similar to Scheibner's (1859).

CREMONA, L. (1864); D'OVIDIO, TORELLI, MAGNI (1865).

[Quistione 32. *Giornale di Mat.*, ii. p. 62; iii. pp. 5–7, 7–10, 10–14.]

The theorem proposed by Cremona is Spottiswoode's of the year 1853, namely, *if* Δ *be a skew determinant having its diagonal elements* $a_{11}, a_{22}, \ldots, a_{nn}$ *each equal to* z, *then the product of any two rows or any two columns of the adjugate determinant contains* Δ *as a factor, and the determinant of the* n^2 *cofactors equals* Δ^{n-2}. Proofs are given by E. D'Ovidio, G. Torelli, and A. Magni; but the second alone need be attended to here, as the two others are less direct, being connected, as the theorem originally was, with the subject of orthogonal substitution.

Starting with the known result

$$a_{r1}A_{s1}+\ldots+a_{rr}A_{sr}+\ldots+a_{rn}A_{sn} \left.\begin{aligned} &= \Delta \quad \text{when } s=r \\ &= 0 \quad \text{when } s\neq r \end{aligned}\right\},$$

Torelli by subtraction of $2a_{rr}A_{sr}$ and change of signs obtains

$$a_{1r}A_{s1}+a_{2r}A_{s2}+\ldots+a_{rr}A_{sr}+\ldots+a_{nr}A_{sn} \left.\begin{aligned} &= -\Delta+2zA_{sr} \quad \text{when } s=r \\ &= 2zA_{sr} \quad \text{when } s\neq r \end{aligned}\right\}.$$

But $a_{1r}A_{1s}+a_{2r}A_{2s}+\ldots+a_{rr}A_{rs}+\ldots+a_{nr}A_{ns} \begin{array}{l} = \Delta \text{ when } s=r \\ = 0 \text{ when } s\neq r \end{array}\Big\}$,

and thence by addition, whatever s may be,

$$a_{1r}(A_{s1}+A_{1s})+\ldots+a_{nr}(A_{sn}+A_{ns}) = 2zA_{sr}.$$

Writing ω_{sr} for $(A_{sr}+A_{rs})\div 2z$ he thus has the set of n equations in $\omega_{s1}, \omega_{s2}, \ldots, \omega_{sn}$,

$$\left.\begin{array}{l} a_{11}\omega_{s1}+\ldots\ldots+a_{1n}\omega_{sn} = A_{1s} \\ a_{21}\omega_{s1}+\ldots\ldots+a_{2n}\omega_{sn} = A_{2s} \\ \cdot\quad\cdot\quad\cdot\quad\cdot\quad\cdot\quad\cdot\quad\cdot\quad\cdot\quad\cdot\quad\cdot \\ a_{n1}\omega_{s1}+\ldots\ldots+a_{nn}\omega_{sn} = A_{ns} \end{array}\right\},$$

the peculiarity of which is that the right-hand members are cofactors of a column of elements of the determinant formed from the coefficients on the left. The solution thus is

$$\omega_{sr} = \frac{A_{1s}A_{1r}+A_{2s}A_{2r}+\ldots\ldots+A_{ns}A_{nr}}{\Delta},$$

whence

$$\left(A_{1s}, A_{2s}, \ldots, A_{ns} \between A_{1r}, A_{2r}, \ldots, A_{nr}\right) = \frac{A_{sr}+A_{rs}}{2z}\Delta,$$

as desired. Using this n^2 times, he, of course, obtains for the square of the adjugate the expression

$$\Delta^n . \begin{vmatrix} \omega_{11} & \omega_{12} & \ldots & \omega_{1n} \\ \omega_{21} & \omega_{21} & \ldots & \omega_{2n} \\ \cdot & \cdot & \cdot & \cdot \\ \omega_{n1} & \omega_{n2} & \ldots & \omega_{nn} \end{vmatrix},$$

and, it being known otherwise that the square of the adjugate is Δ^{2n-2}, it follows that

$$|\,\omega_{11}\ \omega_{22}\ \ldots\ \omega_{nn}\,| = \Delta^{n-2},$$

which is the other result wanted.

In regard to the elements ω one fact is noted, and is worth noting. Since A_{sr} may be expressed as an aggregate of terms in $z^0, z^1, z^2, \ldots$, namely, say

$$A_{sr} = \Theta_0 + \Theta_1 z + \Theta_2 z^2 + \ldots.$$

and since A_{rs} is got from A_{sr} by altering the signs of all the $(n-1)^2$ elements and then changing $-z$ into z, there results when n is even,

$$A_{sr} + A_{rs} = 2\Theta_1 z + 2\Theta_3 z^3 + \ldots ;$$

in other words, $A_{sr}+A_{rs}$ is then divisible by $2z$.

Two "observations" are added, the first in regard to the case where $z=0$, and the second in regard to an alternative proof of the first part of the foregoing. The latter is interesting in that the expression for $(A_{sr}+A_{rs})\,\Delta$ is not found at once as a whole, but is viewed as consisting of two parts corresponding to $A_{sr}\Delta$ and $A_{rs}\Delta$, the reason being the known existence* of a general theorem of determinants to the effect that if the product of $|a_{11}\ldots a_{nn}|$ and $|b_{11}\ldots b_{nn}|$, obtained in row-by-row fashion, be $|c_{11}\ldots c_{nn}|$, then

$$A_{rs}\cdot|b_{11}\ \ldots\ b_{nn}| = b_{1s}C_{r1} + \ \ldots\ + b_{ns}C_{rn}.$$

This is seen to be immediately applicable on making $|a_{11}\ldots a_{nn}|$ identical with Δ above and the b's identical with the a's; and it, of course, implies that if the product obtained in column-by column fashion be $|c'_{11}\ldots c'_{nn}|$, then

$$A_{sr}\cdot|b_{11}\ \ldots\ b_{nn}| = b_{s1}C'_{1r} + \ \ldots\ + b_{sn}C'_{nr}.$$

Making the said necessary specialisations and noting that the two differently formed axisymmetric products are then identical (in other words, that $C_{rs}=C_{sr}=C'_{rs}=C'_{sr}$), Torelli obtains

$$A_{rs}\cdot\Delta = a_{1s}\cdot\sum_{i=1}^{i=n}A_{ri}A_{1i} + \ \ldots\ + a_{ss}\sum_{i=1}^{i=n}A_{ri}A_{si} + \ \ldots\ + a_{ns}\sum_{i=1}^{i=n}A_{r}\,A_{ni},$$

$$A_{sr}\cdot\Delta = a_{si}\cdot\sum_{i=1}^{i=n}A_{ri}A_{1i} + \ \ldots\ + a_{ss}\sum_{i=1}^{i=n}A_{ri}A_{si} + \ \ldots\ + a_{sn}\sum_{i=1}^{i=n}A_{ri}A_{ni};$$

whence by addition

$$(A_{rs}+A_{sr})\Delta = 2a_{ss}\sum_{i=1}^{i=n}A_{ri}A_{si}.$$

---

* Rubini's *Elementi d'Algebra*, p. 277, is referred to.

CAYLEY, A. (1865, Oct.).

[A supplementary memoir on the theory of matrices. *Philos. Transac. R. Soc.* (London), clvi. pp. 25–35; or *Collected Math. Papers*, v. pp. 438–448.]

The expression of an even-ordered determinant, Δ_{2m}, as a Pfaffian being necessary for the second of the two investigations contained in his paper, Cayley effects the transformation (§§ 15–17) in substantially the same way as that devised by Brioschi ten years previously, the one point of difference being that the form of Δ_{2m} which is employed as a multiplier is got from Δ_{2m} by reversing the order of the columns and then changing the signs of the elements in the last m columns. Thus, Δ_4 being

$$\begin{vmatrix} a & b & c & d \\ e & f & g & h \\ i & j & k & l \\ m & n & o & p \end{vmatrix},$$

the square found for it by row-by-row multiplication is, in Cayley's notation,

	$(d, c, -b, -a)$	$(h, g, -f, -e)$	$(l, k, -j, -i)$	$(p, o, -n, -m)$
(a, b, c, d)	”	”	”	”
(e, f, g, h)	”	”	”	”
(i, j, k, l)	”	”	”	”
(m, n, o, p)	”	”	”	”

which is readily seen to be zero-axial skew.

Another expression is, of course, got by treating the conjugate of Δ_4 in the same manner.

Brioschi's paper of 1855 is not referred to.

HORNER, J. (1865, Oct.).

[Notes on determinants. *Quart. Journ. of Math.*, viii. pp. 157–162.]

The second of Horner's three notes consists of a fresh proof that a zero-axial skew determinant of even order, Δ_{2m} say, is the square of a rational function of the elements.

Δ_{2m} multiplied by the square of the product of the non-zero elements of the first row is evidently equal to a zero-axial skew determinant of the same order, Δ'_{2m} say, having 0, 1, 1, . . . , 1 for its first row. But by performing in order the operations which we may conveniently specify by

$$\begin{array}{llll} \text{row}_{2m} - \text{row}_{2m-1}, & \text{row}_{2m-1} - \text{row}_{2m-2}, & \ldots & \text{row}_3 - \text{row}_2, \\ \text{col}_{2m} - \text{col}_{2m-1}, & \text{col}_{2m-1} - \text{col}_{2m-2}, & \ldots & \text{col}_3 - \text{col}_2, \end{array}$$

it is seen that for Δ'_{2m} we may substitute a zero-axial skew determinant of the next lower even order, Δ_{2m-2} say. The factor thus shown to connect Δ_{2m} and Δ_{2m-2} being a square, the little that needs to be added is evident.

CHAPTER X.

ORTHOGONANTS, FROM 1841 TO 1860.

NOTWITHSTANDING the generalisations made by Jacobi and Cauchy, the special case with which the whole theory originated continued from time to time to attract attention. In 1843 William Thomson, afterwards known as Lord Kelvin, published under the signature "T." in the *Cambridge Math. Journ.*, iii. pp. 247–248, a short note in which he proved the detached theorem that if $l_1, m_1, n_1, l_2, \ldots$ be nine quantities such that

$$\begin{aligned} l_1^2 + m_1^2 + n_1^2 &= 1, & l_1l_2 + m_1m_2 + n_1n_2 &= 0, \\ l_2^2 + m_2^2 + n_2^2 &= 1, & l_2l_3 + m_2m_3 + n_2n_3 &= 0, \\ l_3^2 + m_3^2 + n_3^2 &= 1, & l_3l_1 + m_3m_1 + n_3n_1 &= 0, \end{aligned}$$

then it follows that

$$\begin{aligned} l_1^2 + l_2^2 + l_3^2 &= 1, & l_1m_1 + l_2m_2 + l_3m_3 &= 0, \\ m_1^2 + m_2^2 + m_3^2 &= 1, & m_1n_1 + m_2n_2 + m_3n_3 &= 0, \\ n_1^2 + n_2^2 + n_3^2 &= 1, & n_1l_1 + n_2l_2 + n_3l_3 &= 0. \end{aligned}$$

This led to a short paper by A. Göpel in the *Archiv d. Math. u. Phys.*, iv. (1843), pp. 244–246. The subject was again taken up in 1848 by L. Schläfli in the *Mitteilungen d. naturf. Ges. in Bern*, Nos. 112, 113, pp. 27–33,* and in 1850 by V.-A. Lebesgue in the *Nouv. Annales de Math.*, ix. pp. 46–51. Details of these papers need not be given. We may take the opportunity

* Published also in *Archiv d. Math. u. Phys.*, xiii. pp. 276-281.

to note, however, that after the appearance of Cayley's paper on matrices in 1857 the known general theorem embracing the one just mentioned might have been briefly formulated by saying that—*If* $MM' = 1$, *where* M *is any square matrix and* M′ *its conjugate, then also* $M'M = 1$.

KUMMER, E. E. (1843).

[Bemerkungen über die cubische Gleichung, durch welche die Haupt-Axen der Flächen zweiten Grades bestimmt werden. *Crelle's Journ.*, xxvi. pp. 268-272.]

To prove the reality of all the roots of the equation mentioned in the title of his paper—a problem first solved by Lagrange in 1773—Kummer sought to show that the expression for the product of their squared differences was inherently positive. This he succeeded in doing by transforming the said expression into a sum of squares, the result being reached by proceeding from particular to general, and by a combined process of guess and test. The equation being

$$\begin{vmatrix} a-x & h & g \\ h & b-x & f \\ g & f & c-x \end{vmatrix} = 0;$$

or, say,
$$x^3 - Px^2 + Qx - R = 0,$$

the expression referred to is*

$$P^2Q^2 - 4P^3R + 18PQR - 4Q^3 - 27R^2;$$

and Kummer's equivalent for it is

$$\begin{aligned} &15 \mathop{\sum}^{\circ} \Big[gh(b-c) + f(g^2-h^2)\Big]^2 \\ &+ \mathop{\sum}^{\circ} \Big[2(b-c)(c-a)h + (2c-a-b)fg + (2h^2-f^2-g^2)h\Big]^2 \\ &+ \quad \Big[(b-c)(c-a)(a-b) + (b-c)f^2 + (c-a)g^2 + (a-b)h^2\Big]^2, \end{aligned}$$

* *I.e.* $-\frac{1}{3}$ of what afterwards came to be called the *discriminant* of
$$x^3 - Px^2y + Qxy^2 - Ry^3.$$

where we use $\overset{\circ}{\Sigma}$ to indicate the summing of the expressions obtained by performing simultaneously the cyclical substitutions

$$\begin{pmatrix} a & b & c \\ b & c & a \end{pmatrix}, \quad \begin{pmatrix} f & g & h \\ g & h & f \end{pmatrix}.$$

JACOBI, C. G. J. (1844, March).

[Sulla condizione di ugualianza di due radici dell' equazione cubica, dalla quale dipendono gli assi principali di una superficie del second' ordine. *Giornale Arcadico*, xcix. pp. 3–11; or *Crelle's Journ.*, xxx. pp. 46–50; or *Gesammelte Werke*, i. pp. 271–276.]

By using A, B, ... for $bc-f^2$, $ca-g^2$, ... Jacobi first puts Kummer's sum of squares in a neater form, namely,

$$15\overset{\circ}{\sum}(g\mathrm{H}-h\mathrm{G})^2 + \overset{\circ}{\sum}(b\mathrm{F}-f\mathrm{B}+c\mathrm{F}-f\mathrm{C}-2a\mathrm{F}+2f\mathrm{A})^2 \\ + (b\mathrm{C}-c\mathrm{B}+c\mathrm{A}-a\mathrm{C}+a\mathrm{B}-b\mathrm{A})^2,$$

and he then gives a lengthy but thorough verification of its accuracy.

From the fundamental identity

$$ax^2+by^2+cz^2+2fyz+2gzx+2hxy \\ = \mathrm{L}(a_1x+\beta_1y+\gamma_1z)^2+\mathrm{M}(a_2x+\beta_2y+\gamma_2z)^2+\mathrm{N}(a_3x+\beta_3y+\gamma_3z)^2,$$

where L, M, N are, as in his paper of 1827, the roots whose reality is to be established, he obtains at once

$$\begin{aligned} a &= \mathrm{L}a_1^2+\mathrm{M}a_2^2+\mathrm{N}a_3^2, & f &= \mathrm{L}\beta_1\gamma_1+\mathrm{M}\beta_2\gamma_2+\mathrm{N}\beta_3\gamma_3, \\ b &= \mathrm{L}\beta_1^2+\mathrm{M}\beta_2^2+\mathrm{N}\beta_3^2, & h &= \mathrm{L}\gamma_1a_1+\mathrm{M}\gamma_2a_2+\mathrm{N}\gamma_3a_3, \\ c &= \mathrm{L}\gamma_1^2+\mathrm{M}\gamma_2^2+\mathrm{N}\gamma_3^2, & g &= \mathrm{L}a_1\beta_1+\mathrm{M}a_2\beta_2+\mathrm{N}a_3\beta_3, \end{aligned}$$

and thence derives

$$\begin{aligned} \mathrm{A} &= \mathrm{MN}a_1^2+\mathrm{NL}a_2^2+\mathrm{LM}a_3^2, & \mathrm{F} &= \mathrm{MN}\beta_1\gamma_1+\mathrm{NL}\beta_2\gamma_2+\mathrm{LM}\beta_3\gamma_3 \\ \mathrm{B} &= \mathrm{MN}\beta_1^2+\mathrm{NL}\beta_2^2+\mathrm{LM}\beta_3^2, & \mathrm{G} &= \mathrm{MN}\gamma_1a_1+\mathrm{NL}\gamma_2a_2+\mathrm{LM}\gamma_3a_3, \\ \mathrm{C} &= \mathrm{MN}\gamma_1^2+\mathrm{NL}\gamma_2^2+\mathrm{LM}\gamma_3^2, & \mathrm{H} &= \mathrm{MN}a_1\beta_1+\mathrm{NL}a_2\beta_2+\mathrm{LM}a_3\beta_3. \end{aligned}$$

From these it can be shown with more or less trouble* that

$$\begin{aligned} gH-hG &= \Pi\cdot\alpha_1\alpha_2\alpha_3, \\ bF-fB+cF-fC-2aF+2fA &= \Pi\cdot(\alpha_1\beta_2\beta_3+\alpha_2\beta_3\beta_1+\alpha_3\beta_1\beta_2 \\ &\qquad -\alpha_1\gamma_2\gamma_3-\alpha_2\gamma_3\gamma_1-\alpha_3\gamma_1\gamma_2), \\ bC-cB+cA-aC+aB-bA &= \Pi\cdot(\alpha_1\beta_2\gamma_3+\alpha_2\beta_3\gamma_1+\alpha_3\beta_1\gamma_2 \\ &\qquad +\alpha_1\beta_3\gamma_2+\alpha_2\beta_1\gamma_3+\alpha_3\beta_2\gamma_1), \end{aligned}$$

where Π stands for $(L-M)(M-N)(N-L)$. Kummer's sum of squares is thus made to take the form

$$\Pi^2\cdot\Big[15\mathring{\sum}(\alpha_1\alpha_2\alpha_3)^2+\sum\big\{\alpha_1(\beta_2\beta_3-\gamma_2\gamma_3)+\alpha_2(\beta_3\beta_1-\gamma_3\gamma_1)+\alpha_3(\beta_1\beta_2-\gamma_1\gamma_2)\big\}^2 \\ +\big\{\alpha_1(\beta_2\gamma_3+\beta_3\gamma_2)+\alpha_2(\beta_3\gamma_1+\beta_1\gamma_3)+\alpha_3(\beta_1\gamma_2+\beta_2\gamma_1)\big\}^2\Big],$$

where we use $\mathring{\Sigma}$ to indicate the sum of a set of terms produced by the cyclical substitution $\alpha\to\beta$, $\beta\to\gamma$, $\gamma\to\alpha$. After this the cofactor of Π^2 is shown with seeming ease to be 1, and the desired result is reached.

A knowledge of the relationships existing between the elements of the orthogonant $|\alpha_1\beta_2\gamma_3|$ is, of course, a constant requirement throughout the demonstration; and to two of these relationships special attention is drawn by Jacobi himself. The first is

$$\begin{aligned} &2\{\alpha_1^2\alpha_2^2\alpha_3^2+\beta_1^2\beta_2^2\beta_3^2+\gamma_1^2\gamma_2^2\gamma_3^2\} \\ &\quad= \alpha_1\beta_2\gamma_3\cdot\alpha_2\beta_3\gamma_1 + \alpha_2\beta_3\gamma_1\cdot\alpha_3\beta_1\gamma_2 + \alpha_3\beta_1\gamma_2\cdot\alpha_1\beta_2\gamma_3 \\ &\qquad +\alpha_1\beta_3\gamma_2\cdot\alpha_2\beta_1\gamma_3 + \alpha_2\beta_1\gamma_3\cdot\alpha_3\beta_2\gamma_1 + \alpha_3\beta_2\gamma_1\cdot\alpha_1\beta_3\gamma_2, \end{aligned}$$

and the second is

$$\alpha_1^2\alpha_2^2\alpha_3^2 + \beta_1^2\beta_2^2\beta_3^2 + \gamma_1^2\gamma_2^2\gamma_3^2 = \alpha_1^2\beta_1^2\gamma_1^2 + \alpha_2^2\beta_2^2\gamma_2^2 + \alpha_3^2\beta_3^2\gamma_3^2,$$

*The modern reader would do well to use Binet's theorem regarding the determinant which is viewable as the product of two rectangular arrays. Thus

$$A = \begin{vmatrix} b & f \\ f & c \end{vmatrix} = \begin{Vmatrix} L\beta_1 & M\beta_2 & N\beta_3 \\ L\gamma_1 & M\gamma_2 & N\gamma_3 \end{Vmatrix}\cdot\begin{Vmatrix} \beta_1 & \beta_2 & \beta_3 \\ \gamma_1 & \gamma_2 & \gamma_3 \end{Vmatrix},$$

$$= LM\,|\beta_1\gamma_2|^2+MN\,|\beta_2\gamma_3|^2+NL\,|\beta_1\gamma_3|^2 = LM\alpha_3^2+MN\alpha_1^2+NL\alpha_2^2$$

and

$$\begin{vmatrix} g & h \\ G & H \end{vmatrix} = \begin{vmatrix} L & M & N \\ MN & NL & LM \end{vmatrix}\cdot\begin{Vmatrix} \gamma_1\alpha_1 & \gamma_2\alpha_2 & \gamma_3\alpha_3 \\ \alpha_1\beta_1 & \alpha_2\beta_2 & \alpha_3\beta_3 \end{Vmatrix}$$

$$= N(L^2-M^2)\cdot\alpha_1\alpha_2\,|\gamma_1\beta_2| + \dots\dots$$

and so forth.

the latter's existence being due to the fact that the right hand member of the former is not altered by the interchange

$$\begin{pmatrix} \alpha_2 & \alpha_3 & \beta_3 \\ \beta_1 & \gamma_1 & \gamma_2 \end{pmatrix}.$$

BORCHARDT, C. W. (1845, January).

[Neue Eigenschaft der Gleichung, mit deren Hülfe man die seculären Störungen der Planeten bestimmt. *Crelle's Journ.*, xxx. pp. 38–45; or, in an extended form, *Journ.* (*de Liouville*) *de Math.*, xii. pp. 50–67; or *Werke*, pp. 3–13.]

Borchardt's "new property" is the naturally desirable generalisation of Kummer's identity. In his mode of designating the equation * he does not follow Kummer and Jacobi, but goes back to Cauchy (1829), the implied reference being to Laplace's *Mécanique Céleste*, partie i., livre ii., § 56 (1799).

The set of equations, from which by elimination there is obtained the equation referred to in the title, being

$$\left.\begin{aligned} gx_1 &= a_{11}x_1 + a_{12}x_2 + \ldots + a_{1n}x_n \\ gx_2 &= a_{21}x_1 + a_{22}x_2 + \ldots + a_{2n}x_n \\ &\ldots\ldots\ldots\ldots\ldots \\ gx_n &= a_{n1}x_1 + a_{n2}x_2 + \ldots + a_{nn}x_n \end{aligned}\right\},$$

where $a_{ik}=a_{ki}$, Borchardt multiplies the two sides of each equation of the set by g, and then on the right-hand side substitutes for $gx_1, gx_2, \ldots, gx_n$ their equivalents as given. There thus results the new set

$$\left.\begin{aligned} g^2x_1 &= a_{11}^{(2)}x_1 + a_{12}^{(2)}x_2 + \ldots + a_{1n}^{(2)}x_n \\ g^2x_2 &= a_{21}^{(2)}x_1 + a_{22}^{(2)}x_2 + \ldots + a_{2n}^{(2)}x_n \\ &\ldots\ldots\ldots\ldots\ldots \\ g^2x_n &= a_{n1}^{(2)}x_1 + a_{n2}^{(2)}x_2 + \ldots + a_{nn}^{(2)}x_n \end{aligned}\right\},$$

where $a_{ik}^{(2)} = a_{ki}^{(2)} = \sum_{s=1}^{s=n} a_{is}a_{sk}$. Repeating this operation he finds

* The most appropriate designation would seem to be "Lagrange's determinantal equation," because of this mathematician's early (1773) and successful investigation of the cubic. See his *Œuvres complètes*, iii. pp. 600–603.

generally that

$$\left.\begin{array}{rcl} g^m x_1 &=& a_{11}^{(m)} x_1 + a_{12}^{(m)} x_2 + \ldots + a_{1n}^{(m)} x_n \\ g^m x_2 &=& a_{21}^{(m)} x_1 + a_{22}^{(m)} x_2 + \ldots + a_{2n}^{(m)} x_n \\ \cdot & \cdot & \cdot \quad \cdot \quad \cdot \quad \cdot \quad \cdot \quad \cdot \quad \cdot \quad \cdot \\ g^m x_n &=& a_{n1}^{(m)} x_1 + a_{n2}^{(m)} x_2 + \ldots + a_{nn}^{(m)} x_n \end{array}\right\}$$

where

$$a_{ik}^{(m)} = a_{ki}^{(m)} = \sum_{s_1=1}^{s_1=n} \sum_{s_2=1}^{s_2=n} \cdots \sum_{s_{m-1}=1}^{s_{m-1}=n} a_{is_1} a_{s_1 s_2} a_{s_2 s_3} \ldots a_{s_{m-1} k}.$$

In the next place, $g_1, g_2, \ldots, g_n$ being the roots of the resultant of the initial set of equations, it is readily seen, from the expression for the said resultant when arranged according to descending powers of g, that

$$g_1 + g_2 + \ldots + g_n = a_{11} + a_{22} + \ldots + a_{nn}.$$

Similarly, by considering the resultant of the second set of equations we learn that

$$g_1^2 + g_2^2 + \ldots + g_n^2 = a_{11}^{(2)} + a_{22}^{(2)} + \ldots + a_{nn}^{(2)},$$

and generally that

$$g_1^m + g_2^m + \ldots + g_n^m = a_{11}^{(m)} + a_{22}^{(m)} + \ldots + a_{nn}^{(m)}.$$

Consequently, if we use s_m to stand for the sum of the mth powers of the g's we have

$$s_m = \sum_{i=1}^{i=n} a_{ii}^{(m)}.$$

In the third place the difference-product of the g's being

$$\sum(\pm g_1^0 g_2^1 g_3^2 \ldots g_n^{n-1}),$$

Borchardt has only to use the multiplication-theorem of determinants to obtain as an equivalent for the product of the squared differences the determinant of the system

$$\begin{array}{lllll} s_0 & s_1 & s_2 & \ldots & s_{n-1} \\ s_1 & s_2 & s_3 & \ldots & s_n \\ s_2 & s_3 & s_4 & \ldots & s_{n+1} \\ \cdot & \cdot & \cdot & \cdot & \cdot \\ s_{n-1} & s_n & s_{n+1} & \ldots & s_{2n-2}. \end{array}$$

It is this last determinant, therefore, which he has to aim at expressing as a sum of squares.

The process devised by him for doing so is very interesting. Returning to the original set of equations and the sets derived therefrom, he takes the μth set and multiplies both sides of each equation by g^ν and then on the right-hand side substitutes for $g^\nu x_1, g^\nu x_2, \ldots, g^\nu x_n$ their equivalents as obtainable from the νth set. A comparison of the results with the equations of the $(\mu+\nu)^{\text{th}}$ set, he says, gives the noteworthy result

$$a_{ik}^{(\mu+\nu)} = \sum_{s=1}^{s=n} a_{si}^{(\mu)} a_{sk}^{(\nu)},$$

or

$$a_{ik}^{(m)} = \sum_{s=1}^{s=n} a_{si}^{(r)} a_{sk}^{(m-r)}.$$

This includes, of course, the recurrent law of formation

$$a_{ik}^{(m)} = \sum_{s=1}^{s=n} a_{si} a_{sk}^{(m-1)},$$

if we remember that by implication $a_{si}^{(1)}$ must be viewed to be the same as a_{si}. The variety of expressions for $a_{ik}^{(m)}$ which the identity gives makes possible a like variety for

$$a_{11}^{(m)} + a_{22}^{(m)} + \ldots + a_{nn}^{(m)},$$

that is, for s_m. We may, in fact, put as an equivalent for s_m any one of the $m-1$ expressions got from

$$\sum_{i=1}^{i=n} \sum_{s=1}^{s=n} a_{si}^{(r)} a_{si}^{(m-r)}$$

by taking $r=1, 2, \ldots, m-1$. We may even obtain an mth equivalent by making $r=0$ if we agree to consider $a_{si}^{(0)}=0$ or 1 according as s is different from i or the same as i: in other words, if we agree to place before the original set of equations the set

$$\left.\begin{aligned} g^0 x_1 &= 1x_1 + 0x_2 + \ldots + 0x_n \\ g^0 x_2 &= 0x_1 + 1x_2 + \ldots + 0x_n \\ &\cdots\cdots\cdots\cdots \\ g^0 x_n &= 0x_1 + 0\cdot x_2 + \ldots + 1x_n \end{aligned}\right\}$$

the truth of which is incontestable. As a consequence the array of s's above given may be replaced by

$$\begin{array}{ccccc}
\sum\sum a_{ik}^{(0)}a_{ik}^{(0)} & \sum\sum a_{ik}^{(0)}a_{ik}^{(1)} & \sum\sum a_{ik}^{(0)}a_{ik}^{(2)} & \ldots & \sum\sum a_{ik}^{(0)}a_{ik}^{(n-1)} \\
\sum\sum a_{ik}^{(1)}a_{ik}^{(0)} & \sum\sum a_{ik}^{(1)}a_{ik}^{(1)} & \sum\sum a_{ik}^{(1)}a_{ik}^{(2)} & \ldots & \sum\sum a_{ik}^{(1)}a_{ik}^{(n-1)} \\
\sum\sum a_{ik}^{(2)}a_{ik}^{(0)} & \sum\sum a_{ik}^{(2)}a_{ik}^{(1)} & \sum\sum a_{ik}^{(2)}a_{ik}^{(2)} & \ldots & \sum\sum a_{ik}^{(2)}a_{ik}^{(n-1)} \\
\cdot & \cdot & \cdot & \cdot & \cdot \\
\sum\sum a_{ik}^{(n-1)}a_{ik}^{(0)} & \sum\sum a_{ik}^{(n-1)}a_{ik}^{(1)} & \sum\sum a_{ik}^{(n-1)}a_{ik}^{(2)} & \ldots & \sum\sum a_{ik}^{(n-1)}a_{ik}^{(n-1)},
\end{array}$$

where s_2, for example, is represented in the 1st, 2nd, 3rd rows by

$$\sum\sum a_{ik}^{(0)}a_{ik}^{(2)}, \quad \sum\sum a_{ik}^{(1)}a_{ik}^{(1)}, \quad \sum\sum a_{ik}^{(2)}a_{ik}^{(0)}$$

respectively, and where by reason of the range of the two Σ's each element is the sum of n^2 binary products. Any said element may thus be represented as the product of two rows of n^2 elements each, and a little examination shows that only n rows of the latter kind are necessary for the representation of all. In other words, the array of s's can be represented by the product obtained by multiplying the array

$$\begin{array}{ccccccccccc}
a_{11}^{(0)} & a_{12}^{(0)} & \ldots & a_{1n}^{(0)} & a_{21}^{(0)} & a_{22}^{(0)} & \ldots & a_{2n}^{(0)} & a_{31}^{(0)} & \ldots & a_{nn}^{(0)} \\
a_{11}^{(1)} & a_{12}^{(1)} & \ldots & a_{1n}^{(1)} & a_{21}^{(1)} & a_{22}^{(1)} & \ldots & a_{2n}^{(1)} & a_{31}^{(1)} & \ldots & a_{nn}^{(1)} \\
\cdot & \cdot & \cdot & \cdot & \cdot & \cdot & \cdot & \cdot & \cdot & \cdot & \cdot \\
a_{11}^{(n-1)} & a_{12}^{(n-1)} & \ldots & a_{1n}^{(n-1)} & a_{21}^{(n-1)} & a_{22}^{(n-1)} & \ldots & a_{2n}^{(n-1)} & a_{31}^{(n-1)} & \ldots & a_{nn}^{(n-1)}
\end{array}$$

by itself, and therefore is, by Binet's theorem, expressible as a sum of squares.

By way of illustration, Borchardt takes the case where $n=3$. The product of the squared differences of the roots is then, in later notation,

$$\left\| \begin{array}{ccccccccc}
1 & . & . & . & 1 & . & . & . & 1 \\
a & h & g & h & b & f & g & f & c \\
r_1r_1 & r_1r_2 & r_1r_3 & r_2r_1 & r_2r_2 & r_2r_3 & r_3r_1 & r_3r_2 & r_3r_3
\end{array} \right\|^2$$

where $r_\alpha r_\beta$ means the product of the αth and β^{th} rows of

$$\begin{vmatrix} a & h & g \\ h & b & f \\ g & f & c \end{vmatrix}.$$

By performing on the 3-by-9 array the operation

$$\text{row}_3 - (a+b+c)\,\text{row}_2 + (ab+bc+ca-f^2-g^2-h^2)\,\text{row}_1$$

there is obtained

$$\begin{Vmatrix} 1 & . & . & . & 1 & . & . & . & 1 \\ a & h & g & h & b & f & g & f & c \\ \mathrm{A} & \mathrm{H} & \mathrm{G} & \mathrm{H} & \mathrm{B} & \mathrm{F} & \mathrm{G} & \mathrm{F} & \mathrm{C} \end{Vmatrix}^2$$

which is readily shown to be equal to Kummer's sum of squares.

It is a little curious that Borchardt nowhere draws attention to the fact that the determinant of the coefficients in the right-hand members of his m^{th} set of equations is the mth power of the determinant of the corresponding coefficients of the original set.

JACOBI, C. G. J. (1845, August).

[Ueber ein leichtes Verfahren die in der Theorie der Säcularstörungen vorkommenden Gleichungen numerisch aufzulösen. *Crelle's Journ.*, xxx. pp. 51–94; or *Gesammelte Werke*, i. pp. 227–270; or *Nouv. Annales de Math.*, x. pp. 258–265.]

This long memoir being intended for astronomical mathematicians and computers, there is little of it that concerns us except two of the introductory sections (§§ 2, 3, pp. 52–56); and even these need not detain us, as they are in effect but a well-constructed abstract of Cauchy's paper of 1829, the starting-point being the set of $n+1$ equations

$$\left.\begin{array}{l} (a_{11}-\theta)x_1 + a_{12}x_2 + \ldots + a_{1n}x_n = 0 \\ a_{21}x_1 + (a_{22}-\theta)x_2 + \ldots + a_{2n}x_n = 0 \\ \ldots\ldots\ldots\ldots\ldots\ldots\ldots\ldots \\ a_{n1}x_1 + a_{n2}x_2 + \ldots + (a_{nn}-\theta)x_n = 0 \\ x_1^2 + x_2^2 + \ldots + x_n^2 = 1 \end{array}\right\}$$

considered without any regard to the mode in which they may have originated.

CAYLEY, A. (1846).

[Sur quelques propriétés des déterminants gauches. *Crelle's Journ.*, xxxii. pp. 119–123; or *Collected Math. Papers*, i. pp. 332–336.]

There is clear evidence that Rodrigues' paper of 1840 made a strong impression upon Cayley. In a paper published in 1843* he introduces his subject by speaking of Rodrigues as having "given some very elegant formulæ for determining the position of two sets of rectangular axes with respect to each other, employing rational functions of three quantities only"; and he proceeds at once to demonstrate these formulæ as a necessary preliminary to the essential part of his paper. In another paper published in 1845,† the first part of which deals with a quaternion identity, he makes the important observation that a set of nine coefficients which occur in the identity is precisely the same as the set of nine given in Rodrigues' transformation; and he adds, "It would be an interesting question to account *à priori* for the appearance of these coefficients here." We are thus not wholly unprepared for a communication from Cayley himself on the subject of the construction of a linear substitution for the transformation of $x_1^2+x_2^2+\ \ldots$ into $\xi_1^2+\xi_2^2+\ \ldots$. The following is his procedure, four variables being used in place of his n.

With unity and any six quantities whatever there is first formed the square array

$$\begin{matrix} 1 & l_{12} & l_{13} & l_{14} \\ -l_{12} & 1 & l_{23} & l_{24} \\ -l_{13} & -l_{23} & 1 & l_{34} \\ -l_{14} & -l_{24} & -l_{34} & 1, \end{matrix} \qquad \text{or say} \qquad \begin{matrix} l_{11} & l_{12} & l_{13} & l_{14} \\ l_{21} & l_{22} & l_{23} & l_{24} \\ l_{31} & l_{32} & l_{33} & l_{34} \\ l_{41} & l_{42} & l_{43} & l_{44} \end{matrix}$$

---

* Cayley, A., "On the motion of rotation of a solid body." *Cambridge Math. Journ.*, iii. pp. 224–232; or *Collected Math. Papers*, i. pp. 28–35.

† Cayley, A., "On certain results relating to quaternions." *Philos. Magazine*, xxvi. pp. 141–145; or *Collected Math. Papers*, i. pp. 123–126.

remembering that $l_{rr}=1$ and $l_{rs}=-l_{sr}$. Then taking a new set of four variables θ_1, θ_2, θ_3, θ_4, and using for their coefficients the quantities in the square array, firstly as disposed in rows, and secondly as disposed in columns, he puts

$$\left.\begin{array}{l} l_{11}\theta_1 + l_{12}\theta_2 + l_{13}\theta_3 + l_{14}\theta_4 = x_1 \\ l_{21}\theta_1 + l_{22}\theta_2 + l_{23}\theta_3 + l_{24}\theta_4 = x_2 \\ l_{31}\theta_1 + l_{32}\theta_2 + l_{33}\theta_3 + l_{34}\theta_4 = x_3 \\ l_{41}\theta_1 + l_{42}\theta_2 + l_{43}\theta_3 + l_{44}\theta_4 = x_4 \end{array}\right\}$$

and

$$\left.\begin{array}{l} l_{11}\theta_1 + l_{21}\theta_2 + l_{31}\theta_3 + l_{41}\theta_4 = \xi_1 \\ l_{12}\theta_1 + l_{22}\theta_2 + l_{32}\theta_3 + l_{42}\theta_4 = \xi_2 \\ l_{13}\theta_1 + l_{23}\theta_2 + l_{33}\theta_3 + l_{43}\theta_4 = \xi_3 \\ l_{14}\theta_1 + l_{24}\theta_2 + l_{34}\theta_3 + l_{44}\theta_4 = \xi_4 \end{array}\right\},$$

thereby ensuring that

$$x_1^2+x_2^2+x_3^2+\ldots=\xi_1^2+\xi_2^2+\xi_3^3\ldots$$

Solving the two sets of equations separately for each of the θ's and equating the results, he next obtains

$$\left.\begin{array}{l} L_{11}x_1 + L_{21}x_2 + L_{31}x_3 + L_{41}x_4 = L_{11}\xi_1 + L_{12}\xi_2 + L_{13}\xi_3 + L_{14}\xi_4 \\ L_{12}x_1 + L_{22}x_2 + L_{32}x_3 + L_{42}x_4 = L_{21}\xi_1 + L_{22}\xi_2 + L_{23}\xi_3 + L_{24}\xi_4 \\ L_{13}x_1 + L_{23}x_2 + L_{33}x_3 + L_{43}x_4 = L_{31}\xi_1 + L_{32}\xi_2 + L_{33}\xi_3 + L_{34}\xi_4 \\ L_{14}x_1 + L_{24}x_2 + L_{34}x_3 + L_{44}x_4 = L_{41}\xi_1 + L_{42}\xi_2 + L_{43}\xi_3 + L_{44}\xi_4 \end{array}\right\},$$

where L_{rs} is used for the cofactor of l_{rs} in the determinant (Δ say) of the initial array. It only then remains to obtain from this the x's in terms of the ξ's, or the ξ's in terms of the x's. This Cayley does by using as multipliers, in the former case the elements of any row of the original array, and in the latter case the elements of any column. Thus, multiplying by l_{11}, l_{12}, l_{13}, l_{14} respectively and adding, he obtains

$$\Delta x_1 = (2l_{11}L_{11}-\Delta)\xi_1 + 2l_{11}L_{12}\xi_2 + 2l_{11}L_{13}\xi_3+2l_{11}L_{14}\xi_4,$$

the full substitution being

$$\left.\begin{array}{llll} x_1 = \left(\frac{2L_{11}}{\Delta}-1\right)\xi_1+ & \frac{2L_{12}}{\Delta}\xi_2+ & \frac{2L_{13}}{\Delta}\xi_3+ & \frac{2L_{14}}{\Delta}\xi_4 \\ x_2 = \frac{2L_{21}}{\Delta}\xi_1+ & \left(\frac{2L_{22}}{\Delta}-1\right)\xi_2+ & \frac{2L_{23}}{\Delta}\xi_3+ & \frac{2L_{24}}{\Delta}\xi_4 \\ x_3 = \frac{2L_{31}}{\Delta}\xi_1+ & \frac{2L_{32}}{\Delta}\xi_2+ & \left(\frac{2L_{33}}{\Delta}-1\right)\xi_3+ & \frac{2L_{34}}{\Delta}\xi_4 \\ x_4 = \frac{2L_{41}}{\Delta}\xi_1+ & \frac{2L_{42}}{\Delta}\xi_2+ & \frac{2L_{43}}{\Delta}\xi_3+ & \left(\frac{2L_{44}}{\Delta}-1\right)\xi_4 \end{array}\right\}.$$

We may add, that had the relation of the reverse substitution to this not been already known it would have been evident from the set of equations which here produce both. The result reached is that *the* n^2 *coefficients* $a_{11}, \ldots, a_{nn}$ *for the transformation of rectangular co-ordinates can be expressed rationally in terms of* $\frac{1}{2}n(n-1)$ *arbitrary quantities* l_{rs} *satisfying the conditions* $l_{rs} = -l_{sr}$, $l_{rr}=1$ *by forming the determinant* $|l_{11}l_{22}\ldots l_{nn}|$, *or* Δ *say, and thereafter the adjugate determinant* $|L_{11}L_{22}\ldots L_{nn}|$, *and taking*

$$a_{rs} = \frac{2L_{rs}}{\Delta}, \quad a_{rr} = \frac{2L_{rr}}{\Delta}-1.$$

By way of illustration Cayley works out the cases where $n=3$ and where $n=4$. For $n=3$ he begins with three quantities

$$\begin{array}{cc} \nu & -\mu \\ & \lambda \end{array}$$

and obtains the substitution-coefficients

$$\begin{array}{ccc} \frac{1+\lambda^2-\mu^2-\nu^2}{1+\lambda^2+\mu^2+\nu^2} & \frac{2(\lambda\mu+\nu)}{1+\lambda^2+\mu^2+\nu^2} & \frac{2(\nu\lambda-\mu)}{1+\lambda^2+\mu^2+\nu^2} \\ \frac{2(\lambda\mu-\nu)}{1+\lambda^2+\mu^2+\nu^2} & \frac{1+\mu^2-\nu^2-\lambda^2}{1+\lambda^2+\mu^2+\nu^2} & \frac{2(\mu\nu+\lambda)}{1+\lambda^2+\mu^2+\nu^2} \\ \frac{2(\nu\lambda+\mu)}{1+\lambda^2+\mu^2+\nu^2} & \frac{2(\mu\nu-\lambda)}{1+\lambda^2+\mu^2+\nu^2} & \frac{1+\nu^2-\lambda^2-\mu^2}{1+\lambda^2+\mu^2+\nu^2}, \end{array}$$

remarking, in passing, on Rodrigues' introduction of them (but on this point see Euler's memoir of 1770) and on their connection with the theory of quaternions. For $n=4$ he begins with the six arbitrary quantities

$$\begin{matrix} a & b & c \\ & -h & g \\ & & -f \end{matrix}$$

and obtains for the substitution-coefficients the following quantities all divided by Δ:

$$\begin{matrix} \Delta-2(a^2+b^2+c^2+\theta^2) & 2(f\theta+a+bh-cg) & 2(g\theta+b+cf-ah) & 2(h\theta+c+ag-bf) \\ 2(-f\theta-a+bh-cg) & \Delta-2(g^2+h^2+a^2+\theta^2) & 2(-c\theta-h+fg-ab) & 2(b\theta+g+hf-ca) \\ 2(-g\theta-b+cf-ah) & 2(c\theta+h+fg-ab) & \Delta-2(h^2+f^2+b^2+\theta^2) & 2(-a\theta-f+gh-bc) \\ 2(-h\theta-c+ag-bf) & 2(-b\theta-g+hf-ca) & 2(a\theta+f+gh-bc) & \Delta-2(f^2+g^2+c^2+\theta^2) \end{matrix}$$

where $\theta=af+bg+ch$ and $\Delta=1+a^2+b^2+c^2+f^2+g^2+h^2+\theta^2$.

Before leaving Cayley's very interesting paper it should be noted that the essential part of it is contained in the first few lines, where in effect he says that *if we put*

$$\left.\begin{matrix} \theta_1+\lambda\theta_2+\mu\theta_3+\ldots=x_1 \\ -\lambda\theta_1+\theta_2+\nu\theta_3+\ldots=x_2 \\ -\mu\theta_1-\nu\theta_2+\theta_3+\ldots=x_3 \\ \ldots\ldots\ldots\ldots \end{matrix}\right\} \text{ and } \left\{\begin{matrix} \theta_1-\lambda\theta_2-\mu\theta_3-\ldots=\xi_1 \\ \lambda\theta_1+\theta_2-\nu\theta_3-\ldots=\xi_2 \\ \mu\theta_1+\nu\theta_2+\theta_3-\ldots=\xi_3 \\ \ldots\ldots\ldots\ldots \end{matrix}\right.$$

then $x_1, x_2, x_3, \ldots$ *and* $\xi_1, \xi_2, \xi_3, \ldots$ *are orthogonally related, the coefficients of the linear substitutions connecting them being rational functions of* $\lambda, \mu, \nu, \ldots$ The rest of the paper is taken up with the finding of these coefficients, that is to say, with the elimination of $\theta_1, \theta_2, \theta_3, \ldots$ and the expression of each of the remaining variables as a linear function of all the variables of the set to which this variable does not belong.

HERMITE, C. (1849, January).

[Sur une question relative à la théorie des nombres. *Journ.* (*de Liouville*) *de Math.*, xiv. pp. 21–30; or *Œuvres* i. pp. 265–273.]

The problem here solved has only a distant connection with our subject. What is given is a set of mutually prime integers forming the first column of a determinant, and the requirement is to find all the other elements so that the square of the determinant may be 1.

SPOTTISWOODE, W. (1851).

[ELEMENTARY THEOREMS RELATING TO DETERMINANTS. . . . . .
viii + 63 pp., London.]

Following Cayley, Spottiswoode places the construction of an orthogonal substitution at the opening of his section (§ 9) on skew determinants. The mode of treatment differs from Cayley's in being verificatory rather than investigative. Starting with the two sets of equations

$$\left.\begin{array}{c} l_{11}\theta_1 + l_{12}\theta_2 + l_{13}\theta_3 + l_{14}\theta_4 = x_1 \\ \cdot \quad \cdot \quad \cdot \quad \cdot \quad \cdot \quad \cdot \quad \cdot \quad \cdot \quad \cdot \quad \cdot \quad \cdot \end{array}\right\}$$

and

$$\left.\begin{array}{c} l_{11}\theta_1 + l_{21}\theta_2 + l_{31}\theta_3 + l_{41}\theta_4 = \xi_1 \\ \cdot \quad \cdot \quad \cdot \quad \cdot \quad \cdot \quad \cdot \quad \cdot \quad \cdot \quad \cdot \quad \cdot \quad \cdot \end{array}\right\}$$

he does not seek to ascertain therefrom the linear substitutions connecting the x's with the ξ's, but bringing forward the coefficients of these substitutions as found by Cayley, namely,

$$\left.\begin{array}{cccc} \dfrac{2L_{11}}{\Delta} - 1 & \dfrac{2L_{12}}{\Delta} & \dfrac{2L_{13}}{\Delta} & \dfrac{2L_{14}}{\Delta} \\ \cdot \quad \cdot \quad \cdot & \cdot \quad \cdot \quad \cdot & \cdot \quad \cdot \quad \cdot & \cdot \quad \cdot \quad \cdot \end{array}\right\}$$

he affirms that by using as multipliers, along with the first set of equations, the elements of the various columns of this array in succession we shall have

$$\left.\begin{array}{c} \left(\dfrac{2L_{11}}{\Delta} - 1\right)x_1 + \dfrac{2L_{21}}{\Delta}x_2 + \dfrac{2L_{31}}{\Delta}x_3 + \dfrac{2L_{41}}{\Delta}x_4 = \xi_1 \\ \cdot \quad \cdot \quad \cdot \quad \cdot \quad \cdot \quad \cdot \quad \cdot \quad \cdot \quad \cdot \quad \cdot \quad \cdot \quad \cdot \quad \cdot \quad \cdot \quad \cdot \end{array}\right\},$$

and that by using along with the second set of equations the elements of the various *rows* of the array in succession we shall have

$$\left.\begin{array}{c} \left(\dfrac{2L_{11}}{\Delta} - 1\right)\xi_1 + \dfrac{2L_{12}}{\Delta}\xi_2 + \dfrac{2L_{13}}{\Delta}\xi_3 + \dfrac{2L_{14}}{\Delta}\xi_4 = x_1 \\ \cdot \quad \cdot \quad \cdot \quad \cdot \quad \cdot \quad \cdot \quad \cdot \quad \cdot \quad \cdot \quad \cdot \quad \cdot \quad \cdot \quad \cdot \quad \cdot \quad \cdot \end{array}\right\}.$$

At the close of a preceding section (§ 6) he devotes three pages (pp. 35–37) to an investigation of the conditions under which

Lagrange's determinantal equation shall have all its roots positive. The result is not so interesting in connection with our present subject as a theorem made use of in the process of attaining it, namely:—*If we have given the set of equations*

$$\left.\begin{array}{l} a_{11}x_1 + a_{12}x_2 + \ldots + a_{1n}x_n = \theta x_1 \\ a_{21}x_1 + a_{22}x_2 + \ldots + a_{2n}x_n = \theta x_2 \\ \cdot \quad \cdot \quad \cdot \quad \cdot \quad \cdot \quad \cdot \quad \cdot \quad \cdot \quad \cdot \quad \cdot \\ a_{n1}x_1 + a_{n2}x_2 + \ldots + a_{nn}x_n = \theta x_n \end{array}\right\}$$

where $a_{rs} = a_{sr}$, *and if we put* Δ *for* $|\,a_{11}a_{22}\ldots a_{nn}\,|$, *then*

$$\left.\begin{array}{l} A_{11}x_1 + A_{21}x_2 + \ldots + A_{n1}x_n = \dfrac{\Delta}{\theta}x_1 \\ A_{12}x_1 + A_{22}x_2 + \ldots + A_{n2}x_n = \dfrac{\Delta}{\theta}x_2 \\ \cdot \quad \cdot \quad \cdot \quad \cdot \quad \cdot \quad \cdot \quad \cdot \quad \cdot \quad \cdot \quad \cdot \\ A_{1n}x_1 + A_{2n}x_2 + \ldots + A_{nn}x_n = \dfrac{\Delta}{\theta}x_n \end{array}\right\}.$$

No proof of this is given, but one is readily got by using the elements of the rth column of the adjugate of Δ as multipliers in connection with the given set of equations and performing addition, when there results the rth equation of the required set.*

*It should be noted that the theorem holds when Δ is any determinant whatever. Further, there is implied in it another of at least equal importance, namely:—*If* Δ *stand for* $|\,a_{11}a_{22}\ldots a_{nn}\,|$, *the equation whose roots are* Δ *times the reciprocals of the roots of the equation*

$$\begin{vmatrix} a_{11}-\theta & a_{12} & \ldots & a_{1n} \\ a_{21} & a_{22}-\theta & \ldots & a_{2n} \\ \cdot & \cdot & \cdot & \cdot \\ a_{n1} & a_{n2} & \ldots & a_{nn}-\theta \end{vmatrix} = 0$$

is

$$\begin{vmatrix} A_{11}-\Theta & A_{12} & \ldots & A_{1n} \\ A_{21} & A_{22}-\Theta & \ldots & A_{2n} \\ \cdot & \cdot & \cdot & \cdot \\ A_{n1} & A_{n2} & \ldots & A_{nn}-\Theta \end{vmatrix} = 0.$$

An independent proof of this is readily obtained by substituting Δ/Θ for θ in the original equation, expanding the determinant in a series arranged according to descending powers of Δ/Θ, using Θ^n/Δ as a multiplier, substituting $A_{11}, A_{12}\ldots$ for their equivalents, and returning to the determinant form.

HESSE, O. (1851, April).

[Ueber die Eigenschaften der linearen Substitutionen, durch welche eine homogene ganze Function zweiten Grades, welche nur die Quadrate von vier Variabeln enthält, in eine Function von derselben Form transformirt wird. *Crelle's Journ.*, xlv. pp. 93–101; or *Werke*, pp. 307–317.]

Starting with the supposition that the substitution

$$y_k = a_{k1}x_1 + a_{k2}x_2 + \ldots + a_{kn}x_n \Big\}_{k=1}^{k=n}$$

makes

$$b_1y_1^2 + b_2y_2^2 + \ldots + b_ny_n^2 = a_1x_1^2 + a_2x_2^2 + \ldots + a_nx_n^2,$$

Hesse obtains by differentiation with respect to x_1, x_2, x_3, x_4, the reverse substitution

$$a_kx_k = a_{1k}b_1y_1 + a_{2k}b_2y_2 + \ldots + a_{nk}b_ny_n \Big\}_{k=1}^{k=n};$$

and having thus found that the latter substitution will make

$$a_1x_1^2 + a_2x_2^2 + \ldots + a_nx_n^2 = b_1y_1^2 + b_2y_2^2 + \ldots + b_ny_n^2$$

he is able by putting η_k for a_kx_k and ξ_k for b_ky_k to say that the substitution

$$\eta_k = a_{1k}\xi_1 + a_{2k}\xi_2 + \ldots + a_{nk}\xi_k \Big\}_{k=1}^{k=n}$$

will make

$$\frac{1}{a_1}\eta_1^2 + \frac{1}{a_2}\eta_2^2 + \ldots + \frac{1}{a_n}\eta_n^2 = \frac{1}{b_1}\xi_1^2 + \frac{1}{b_2}\xi_2^2 + \ldots + \frac{1}{b_n}\xi_n^2.$$

The result is the theorem that—*If the substitution*

$$y_k = a_{k1}x_1 + a_{k2}x_2 + \ldots + a_{kn}x_n \Big\}_{k=1}^{k=n}$$

changes

$$b_1y_1^2 + b_2y_2^2 + \ldots + b_ny_n^2 \quad \textit{into} \quad a_1x_1^2 + a_2x_2^2 + \ldots + a_na_n^2$$

the conjugate substitution will change

$$\frac{1}{a_1}y_1^2 + \frac{1}{a_2}y_2^2 + \ldots + \frac{1}{a_n}y_n^2 \quad \textit{into} \quad \frac{1}{b_1}x_1^2 + \frac{1}{b_2}x_2^2 + \ldots + \frac{1}{b_n}x_n^2.$$

The rest of the paper is occupied with theorems which hold only in the case of four variables.

SYLVESTER, J. J. (1852, July).

[A demonstration of the theorem that every homogeneous quadratic polynomial is reducible by real orthogonal substitutions to the form of a sum of positive and negative squares. *Philos. Magazine* (4), iv. pp. 138–142; or *Collected Math. Papers*, i. pp. 378–381.]

The terms "orthogonal transformation" and "orthogonal substitution" date from the year 1852, the former appearing in a paper of Sylvester's published in the February part of the *Cambridge and Dub. Math. Journ.* (see vol. vii. p. 57), and the latter in the title of the paper now reached. In the former paper, too, the word "unimodular," as applied to a transformation, is first used (see p. 52), the meaning being that the modulus—that is to say, the determinant of the coefficients of transformation—is then unity.

As has been already noted * when dealing with axisymmetric determinants, this opens with the proposition that when $a_{rs} = a_{sr}$,

$$\begin{vmatrix} a_{11}+x & a_{12} & \dots & a_{1n} \\ a_{21} & a_{22}+x & \dots & a_{2n} \\ \cdot & \cdot & \cdot & \cdot \\ a_{n1} & a_{n2} & \dots & a_{nn}+x \end{vmatrix} \cdot \begin{vmatrix} a_{11}-x & a_{12} & \dots & a_{1n} \\ a_{21} & a_{22}-x & \dots & a_{2n} \\ \cdot & \cdot & \cdot & \cdot \\ a_{n1} & a_{n2} & \dots & a_{nn} \end{vmatrix}$$

$$= \begin{vmatrix} q_{11}-x^2 & q_{12} & \dots & q_{1n} \\ q_{21} & q_{22}-x^2 & \dots & q_{2n} \\ \cdot & \cdot & \cdot & \cdot \\ q_{n1} & q_{n2} & \dots & q_{nn}-x^2 \end{vmatrix}$$

where $$q_{rs} = (a_{r1}a_{r2}\dots a_{rn} \between a_{1s}a_{2s}\dots a_{ns}),$$

and where therefore

$$|q_{11}q_{22}\dots q_{nn}| = |a_{11}a_{22}\dots a_{nn}|^2.$$

It is then pointed out that the last determinant multiplied by $(-1)^n$ is expressible in the form

$$(x^2)^n - Q_1(x^2)^{n-1} + Q_2(x^2)^{n-2} - \dots\dots;$$

* On verifying this, see also the account of the related paper published in the *Nouv. Annales de Math.* for November 1852.

that Q_1, Q_2, ... can be shown to be sums of squares; that consequently the values of x^2 in the equation

$$(x^2)^n - Q_1(x^2)^{n-1} + Q_2(x^2)^{n-2} - \dots = 0$$

are all positive; and therefore, finally, that the values of x in the equation

$$\begin{vmatrix} a_{11}-x & a_{12} & \dots & a_{1n} \\ a_{21} & a_{22}-x & \dots & a_{2n} \\ \cdot & \cdot & \cdot & \cdot \\ a_{n1} & a_{n2} & \dots & a_{nn}-x \end{vmatrix} = 0$$

are all real.*

The remainder of the paper deals with the "Law of Inertia for Quadratic Forms," this law being "that by whatever linear substitutions, orthogonal or otherwise, a given polynomial is reduced to the form $\Sigma A_1\xi_1^2$, the number of positive and negative coefficients is invariable."

LAMÉ, G. (1852).

[LEÇONS SUR LA THÉORIE MATHÉMATIQUE DE L'ELASTICITÉ DES CORPS SOLIDES. xvi+336 pp., Paris.]

While discussing (§§ 18–22) the axes of the ellipsoid of elasticity Lamé gives in substance the theorem that *if* $|\alpha_1\beta_2\gamma_3|$ *be an orthogonant, and the ordinary multiplication-theorem produce the identity*

$$\begin{vmatrix} \alpha_1 & \beta_1 & \gamma_1 \\ \alpha_2 & \beta_2 & \gamma_2 \\ \alpha_3 & \beta_3 & \gamma_3 \end{vmatrix} \cdot \begin{vmatrix} a & f & e \\ f & b & d \\ e & d & c \end{vmatrix} \cdot \begin{vmatrix} \alpha_1 & \alpha_2 & \alpha_3 \\ \beta_1 & \beta_2 & \beta_3 \\ \gamma_1 & \gamma_2 & \gamma_3 \end{vmatrix} = \begin{vmatrix} P_1 & Q_3 & Q_2 \\ Q_3 & P_2 & Q_1 \\ Q_2 & Q_1 & P_3 \end{vmatrix},$$

then

$$P_1+P_2+P_3 = a+b+c,$$

$$\begin{vmatrix} P_1 & Q_3 \\ Q_3 & P_2 \end{vmatrix} + \begin{vmatrix} P_2 & Q_1 \\ Q_1 & P_3 \end{vmatrix} + \begin{vmatrix} P_1 & Q_2 \\ Q_2 & P_3 \end{vmatrix} = \begin{vmatrix} a & f \\ f & b \end{vmatrix} + \begin{vmatrix} b & d \\ d & c \end{vmatrix} + \begin{vmatrix} a & e \\ e & c \end{vmatrix},$$

* This proof, for the case where $n=3$, is given free of determinants by Grunert in the *Archiv d. Math. u. Phys.*, xxix. (1857), pp. 442–446.

and of course

$$\begin{vmatrix} P_1 & Q_3 & Q_2 \\ Q_3 & P_2 & Q_1 \\ Q_2 & Q_1 & P_3 \end{vmatrix} = \begin{vmatrix} a & f & e \\ f & b & d \\ e & d & c \end{vmatrix}.$$

No determinant notation, however, is used, nor are determinants spoken of.

HERMITE, C. (1853, May).

[Sur la théorie des formes quadratiques ternaires indéfinies. *Crelle's Journ.*, xlvii. pp. 307–312; or *Œuvres*, i. pp. 193–199.]

[Remarques sur un mémoire de M. Cayley relatif aux déterminants gauches. *Cambridge and Dub. Math. Journ.*, ix. pp. 63–67; or *Œuvres*, i. pp. 290–295.]

In his paper of 1846 Cayley, as we have seen, gave a general solution of the problem of the transformation of $x_1^2+x_2^2+\ldots$ into $\xi_1^2+\xi_2^2+\ldots$ by means of a linear substitution. Hermite now faces a more general problem, namely, "la transformation en elle-même d'une forme quadratique *quelconque*," a problem which in itself is rather outside our subject, but which, by reason of the important modification made in the initial step of the solution, deserves attention.

The quadric being $f(x_1, x_2, \ldots)$, the problem is to find the most general linear substitution which will transform

$$f(x_1, x_2, \ldots) \text{ into } f(\xi_1, \xi_2, \ldots);$$

and Hermite having before him Cayley's expressions, in the simpler case, for the x's and ξ's in terms of an intermediary set of variables, and observing that any member of the intermediary set is the arithmetic mean of the corresponding members of the two given sets, begins by imagining merely "que les quantités x et ξ soient exprimés par des indéterminées auxiliaires θ, de sorte qu'on ait en général

$$x_r+\xi_r = 2\theta_r."$$

There is thus obtained

$$f(x_1, x_2, \ldots) = f(2\theta_1 - \xi_1, 2\theta_2 - \xi_2, \ldots),$$
$$= 4f(\theta_1, \theta_2, \ldots) - 2\left(\xi_1\frac{\partial f}{\partial\theta_1} + \xi_2\frac{\partial f}{\partial\theta_2} + \ldots\right) + f(\xi_1, \xi_2, \ldots),$$

so that in order to have $f(x_1, x_2, \ldots) = f(\xi_1, \xi_2, \ldots)$ it is seen to be necessary that

$$\xi_1\frac{\partial f}{\partial\theta_1} + \xi_2\frac{\partial f}{\partial\theta_2} + \ldots = 2f(\theta_1, \theta_2, \ldots).$$

Now this condition is manifestly satisfied by putting $\xi_r = \theta_r$, but "la manière la plus générale de la vérifier en exprimant les quantités ξ en θ sera de faire

$$\xi_r = \theta_r + \tfrac{1}{2}\sum_{s=1}^{s=n}\lambda_{rs}\frac{\partial f}{\partial\theta_s},$$

les indéterminées λ étant assujetées à la condition $\lambda_{rs} = -\lambda_{sr}$." This of course implies that

$$x_r = \theta_r - \tfrac{1}{2}\sum_{s=1}^{s=n}\lambda_{rs}\frac{\partial f}{\partial\theta_s};$$

and there have thus been obtained in their general form the two sets of equations with which Cayley started in his special case.

For those who may wish to pursue the subject of "automorphic transformation" farther than these papers of Hermite's we may note that the actual expression of the x's in terms of the ξ's was given by Cayley in a paper dated 24th May 1854,* and that he extended his result to a *bipartite* quadric function in a paper dated 10th December, 1857.†

Another problem, which in the early history of orthogonants we have seen to be of interest, namely, the simultaneous trans-

* Cayley, A., "Sur la transformation d'une fonction quadratique en elle-même par des substitutions linéaires," *Crelle's Journ.*, l. pp. 288–299; or *Collected Math. Papers*, ii. pp. 192–201. See also Brioschi in *Annali di Sci. mat. e fis.*, v. pp. 201–206.

† Cayley, A., "A Memoir on the Automorphic Linear Transformation of a Bipartite Quadric Function," *Philos. Transac. R. Soc.* (London), cxlviii. pp. 39–46; or *Collected Math. Papers*, ii. pp. 497–505.

formation of two quadrics, Cayley also dealt with, the first time in 1849 and the second in 1857.*

SYLVESTER, J. J. (1853).

[The algebraical theory of the secular-inequality determinantive equation generalised. *Philos. Magazine*, vi. pp. 214–216; or *Collected Math. Papers*, i. pp. 634–636.]

The fundamental theorem here is that if

$$X_1 = ax+a, \quad X_2 = \begin{vmatrix} ax+a & bx+\beta \\ bx+\beta & cx+\gamma \end{vmatrix}, \quad X_3 = \begin{vmatrix} ax+a & bx+\beta & dx+\delta \\ bx+\beta & cx+\gamma & ex+\epsilon \\ dx+\delta & ex+\epsilon & fx+\phi \end{vmatrix}, \ldots$$

and the coefficients of the highest powers of x in $X_1, X_2, X_3, \ldots$ have all the same sign, then the roots of X_i will be all real and will lie respectively in the intervals comprised between $+\infty$, the successive descending roots of X_{i-1}, and $-\infty$. The mode of proof is Cauchy's (1829).

SPOTTISWOODE, W. (1853, August).

[Elementary theorems relating to determinants. Second edition, rewritten and much enlarged by the author. *Crelle's Journ.*, li. pp. 209–271, 328–381.]

In trying to insert in his second edition an alternative process for establishing Cayley's result of 1846, Spottiswoode is very unfortunate. The place selected by him is immediately after the sentence defining *skew*, and therefore immediately preceding the former process; but in making the insertion (p. 260) the predicate of the important sentence in question has suffered excision, along with a very necessary explanation regarding the diagonal elements of the initial determinant. Further, at the utmost all that is established is the fact that the determinant of Cayley's substitution is equal to $+1$. Such neglect, however, can well be overlooked in view of certain deductions which he

* *Cambridge and Dublin Math. Journ.*, iv. pp. 47-50; and *Quart. Journ. of Math.*, ii. pp. 192–195; or *Collected Math. Papers*, i. pp. 428-431, and iii. pp. 129-131.

records, and which he says can be made from Cayley's result. These may be enunciated in more modern form as follows:—

If $|a_{11}\ a_{22}\ \ldots\ a_{nn}|$ *or* Δ *be a unit-axial skew determinant,* $|A_{11}\ A_{22}\ \ldots\ A_{nn}|$ *its adjugate, and* $|\omega_{11}\ \omega_{22}\ \ldots\ \omega_{nn}|$ *Cayley's orthogonant formed therefrom, then*

$$\left.\begin{aligned} A_{1r}^2 + A_{2r}^2 + \ldots + A_{nr}^2 &= A_{rr}\,.\,\Delta, \\ A_{1r}A_{1s} + A_{2r}A_{2s} + \ldots + A_{nr}A_{ns} &= \tfrac{1}{2}(A_{rs}+A_{sr})\,.\,\Delta \end{aligned}\right\} \qquad (\alpha)$$

and

$$\left.\begin{aligned} a_{1r}\omega_{1r} + a_{2r}\omega_{2r} + \ldots + a_{nr}\omega_{nr} &= a_{rr} \\ a_{1r}\omega_{1s} + a_{2r}\omega_{2s} + \ldots + a_{nr}\omega_{ns} &= a_{rs} \end{aligned}\right\} \qquad (\beta)$$

The former, (α), belongs strictly to the theory of skew determinants, as has already been mentioned in the proper place.

CAYLEY, A. (1853, November).

[On the homographic transformation of a surface of the second order into itself. *Philos. Magazine*, vi. pp. 326–333; or *Collected Math. Papers*, ii. pp. 105–112.]

Here Cayley recalculates the general orthogonant of the 4th order, taking note in passing of the related identity

$$\begin{aligned} &(-ax-by-cz+w)^2 \\ &+(x+\nu y-\mu z+aw)^2+(-\nu x+y+\lambda z+bw)^2+(\mu x-\lambda y+z+cw)^2 \\ &= x^2+y^2+z^2+w^2+(ax+by+cz)^2 \\ &\qquad +(\nu y-\mu z+aw)^2+(-\nu x+\lambda z+bw)^2+(\mu x-\lambda y+cw)^2. \end{aligned}$$

We may add that if the last eight squares be subtracted from both sides of this there remains on the left-hand side a quadric having a zero-axial discriminant.

BRIOSCHI, F. (1854, March).

[La Teorica dei Determinanti, e le sue principali Applicazioni. viii+116 pp., Pavia.]

In Brioschi's text-book, the paragraphs dealing with a "sostituzione ortogonale" are somewhat scattered, most of them appearing among the applications (pp. 24–26, 47–51, 62–69).

The first deserving of notice (p. 49) concerns the product $QP\bar{Q}$, where P and Q are determinants of the same order and $\bar{Q}$ is the conjugate of Q. Viewing the product as $Q\cdot(P\bar{Q})$ Brioschi first uses a result of Cauchy's to express any m-line minor of $Q\cdot(P\bar{Q})$ in terms of m-line minors of Q and $P\bar{Q}$: then for the said m-line minors of $P\bar{Q}$ he substitutes with the same assistance expressions involving m-line minors of P and Q: there is thus obtained for any m-line minor of $QP\bar{Q}$ an expression involving only m-line minors of P and Q. The result may be put in the form

$$(QP\bar{Q})^{(m)}_{\nu,s} = \sum_r \left[Q^{(m)}_{\nu,r}\left\{P^{(m)}_{r1}Q^{(m)}_{s1} + P^{(m)}_{r2}Q^{(m)}_{s2} + \ldots + P^{(m)}_{rz}Q^{(m)}_{sz}\right\}\right],$$

if z be put for $n(n-1)\ldots(n-m+1)/1.2.\ldots m$, and if generally we use $A^{(m)}_{sr}$ to stand for an m-line minor of an n-line determinant A, the rows of A taken to form $A^{(m)}_{rs}$ being those whose numbers constitute the rth combination of m of the integers 1,2, . . . , n, and the columns those whose numbers constitute the sth like combination. Putting $\nu=s$ we obtain the expression of an m-line *coaxial* minor of $QP\bar{Q}$, and thence for the sum of all such minors the expression

$$\sum_r \sum_s \left[Q^{(m)}_{sr}\left\{P^{(m)}_{r1}Q^{(m)}_{s1} + P^{(m)}_{r2}Q^{(m)}_{s2} + \ldots + P^{(m)}_{rz}Q^{(m)}_{sz}\right\}\right],$$

which changes into

$$\sum_r \left\{P^{(m)}_{r1}M^{(m)}_{r1} + P^{(m)}_{r2}M^{(m)}_{r2} + \ldots + P^{(m)}_{rz}M^{(m)}_{rz}\right\},$$

if M be the determinant which equals Q^2. Specialising still further by making Q the determinant of an orthogonal substitution so that

$$M^{(m)}_{rr} = 1 \quad \text{and} \quad M^{(m)}_{rs} = 0,$$

Brioschi finally obtains the important "formula nota"

$$\sum_s (QP\bar{Q})^{(m)}_{ss} = \sum_s P^{(m)}_{ss},$$

which we may express in words for ourselves thus:—*If* Q *be an orthogonant and* P *any other determinant of the same order, then the sum of the* m-*line coaxial minors of* $QP\bar{Q}$ *is the same as the sum of the* m-*line coaxial minors of* P.

The other paragraph requiring notice concerns the determinant arising from Cayley's of 1846 by subtracting 1 from each diagonal element. The value of this is shown (p. 65) to be 0 when n is odd, and $2^n\Delta_0/\Delta$ when n is even, Δ being the basic determinant, and Δ_0 what Δ becomes on making all its diagonal elements zero. The result is easily reached on multiplying the given determinant by Δ and showing that the product is $(-1)^n 2^n \Delta_0$.

BRIOSCHI, F. (1854, August).

[Note sur un théorème relatif aux déterminants gauches. *Journ. (de Liouville) de Math.*, xix. pp. 253–256; or in the French translation of his *Teorica dei Determinanti*, pp. 144–147; or *Opere mat.*, v. pp. 161–164.]

Brioschi's subject is really the equation

$$\begin{vmatrix} \omega_{11}-x & \omega_{12} & \dots & \omega_{1n} \\ \omega_{21} & \omega_{22}-x & \dots & \omega_{2n} \\ \cdot & \cdot & \cdot & \cdot \\ \omega_{n1} & \omega_{n2} & \dots & \omega_{nn}-x \end{vmatrix} = 0,$$

in which the left-hand member is the determinant of Cayley's orthogonal substitution with $-x$ affixed to each diagonal element. He notes at once, of course, that if the basic determinant be $|a_{11}a_{22}\dots a_{nn}|$, or Δ say, the equation may be changed into

$$\begin{vmatrix} A_{11}-y & A_{12} & \dots & A_{1n} \\ A_{21} & A_{22}-y & \dots & A_{2n} \\ \cdot & \cdot & \cdot & \cdot \\ A_{n1} & A_{n2} & \dots & A_{nn}-y \end{vmatrix} = 0,$$

where y is put for $\frac{1}{2}(1+x)\Delta$. A further transformation is then effected by multiplying both sides by Δ and putting z for $1-\Delta/y$, the result being

$$\begin{vmatrix} z & a_{21} & \dots & a_{n1} \\ a_{12} & z & \dots & a_{n2} \\ \cdot & \cdot & \cdot & \cdot \\ a_{1n} & a_{2n} & \dots & z \end{vmatrix} = 0.$$

Using Cayley's expansion (1847) for the determinant on the left, it is seen that when n is odd the equation resolves itself into $z=0$ and an equation in z^2 with positive coefficients, and that when n is even it is already of the latter form. All values of z^2 thus obtainable must be negative, and consequently all the values of z save the value 0 must be imaginary and must occur in pairs whose sum is zero. But as

$$z = 1 - \frac{\Delta}{\frac{1}{2}(1+x)\Delta} = \frac{x-1}{x+1},$$

and

$$\therefore \quad x = \frac{1+z}{1-z},$$

it is clear that for every pair of values of z that differ only in sign there must be a pair of values of x that are reciprocals. The theorem reached by Brioschi we may thus enunciate for ourselves as follows:—*The roots of the equation*

$$\begin{vmatrix} \omega_{11}-x & \omega_{12} & \cdots & \omega_{1n} \\ \omega_{21} & \omega_{22}-x & \cdots & \omega_{2n} \\ \cdot & \cdot & \cdot & \cdot \\ \omega_{n1} & \omega_{n2} & \cdots & \omega_{nn}-x \end{vmatrix} = 0,$$

where $|\omega_{11}\,\omega_{22}\cdots\omega_{nn}|$ *is Cayley's orthogonant, are arrangeable in pairs of reciprocal imaginaries, save when* n *is odd, in which case there is the single real root* 1.

When instead of the ω's we take the coefficients of the substitution which transforms a *general* quadric into itself, the words "reciprocal imaginaries" need to be changed into "reciprocals." This generalisation Brioschi published a month or two sooner (see *Annali di Sci. mat. e fis.*, v. pp. 201–206).

BRUNO, F. FAÀ DI (1854, September).

[Note sur un théorème de M. Brioschi. *Journ. (de Liouville) de Math.*, xix. p. 304.]

On multiplying both sides of Brioschi's equation (1854, August) by $|\omega_{11}\,\omega_{22}\cdots\omega_{nn}|$ and dividing by $(-x)^n$ an equation is obtained which differs from the original simply in having x^{-1} for x. The portion of the theorem which concerns "reciprocity" Bruno thus readily establishes.

CAUCHY, A. L. (1857, Feb.).

[Sur les fonctions quadratiques et homogènes de plusieurs variables. *Comptes rendus . . . Acad. des Sci.* (Paris), xliv, pp. 361–370, 416; or *Œuvres complètes* (1), xii. pp. 421–432. 444–445.]

The second section of this bears the title "Sur l'équation qui détermine les maxima et minima d'une fonction réelle quadratique et homogène de plusieurs variables dont les carrés donne pour somme l'unité," and at once recalls the important memoir of 1829. The subject is the same, and any additional result obtained is quite unimportant. Further, the mode of treatment is not essentially different, the language and notation of 'clefs anastrophiques' being for some obscure reason substituted for those of 'sommes alternées.'

We have only to add, as being well worthy of note in passing, that this was Cauchy's last contribution to the literature of our subject, his first and greatest, and probably the greatest of all, having been made so long before as forty-five years. Three months after the last was presented to the Academy he was dead.

BALTZER, R. (1857).

[Theorie und Anwendungen der Determinanten, mit . . . . . vi+129 pp. Leipzig.]

Baltzer devotes a whole section (§ 15) of seventeen pages (pp. 80–96) to the subject of "Die lineare, insbesondere die orthogonale Substitutionen." The section, like its fellows, is noteworthy, not for freshness of matter, but for good arrangement, clearness and compactness.

In treating of Cayley's orthogonant (§ 15, 6) he takes l, not 1, as the constant element of the basic determinant: and, when in the course of the proof he obtains the two values for each of Cayley's θ's, he does not equate them, but uses with each of them Hermite's observation

$$x_i + \xi = 2l\theta_i,$$

thus reaching the elements

$$\frac{2l\mathrm{L}_{rr}}{\Delta} - 1, \quad \frac{2l\mathrm{L}_{rs}}{\Delta},$$

of the desired substitutions without more trouble. On the other hand, he fails to note that Cayley's θ's are so introduced as to ensure from the outset the equality of $x_1^2+x_2^2+\ldots$ and $\xi_1^2+\xi_2^2+\ldots$, and thus he is led to prove propositions already established (§ 15, 5).

Brioschi's equation of August 1854 being denoted (§ 15, 9) by $f(x)=0$, he multiplies $f(x)$ by $f(-x)$, and obtains for $f(x)\cdot f(-x)/x^n$ a skew determinant having each diagonal element equal to $1/x-x$. This determinant being therefore expressible as a sum of squares when n is even, and as $1/x-x$ times a sum of squares when n is odd, the part of Brioschi's proposition which asserts the unreality of the roots follows by a *reductio ad absurdum*.

SALMON, G. (1859).

[LESSONS INTRODUCTORY TO THE MODERN HIGHER ALGEBRA, . . . xii+147 pp., Dublin.]

In Salmon's treatment of the subject (§§ 118, 139, 142, 156–7, 163–4) only two points call for remark. In the first place, "orthogonal transformation" with him is not as with his predecessors a transformation which merely changes

$$x^2+y^2+z^2+\ldots \quad \text{into} \quad \xi^2+\eta^2+\zeta^2+\ldots,$$

but one which at the same time changes

$$ax^2+by^2+cz^2+\ldots+2fyz+2gzx+2hxy+\ldots \quad \text{into} \quad \mathrm{A}\xi^2+\mathrm{B}\eta^2+\mathrm{C}\zeta^2+\ldots$$

In the second place, he has a fresh mode of arriving at the equation for determining A, B, C, . . . Calling the four quadrics just mentioned V, V′, U, U′, he forms the discriminant of $\mathrm{U}-\lambda\mathrm{V}$, and asserts that the coefficient of all the several powers of λ in it must be invariants, and that, therefore, if the said discriminant be put equal to 0 and the equation so obtained be solved for λ, the roots resulting must be identical with the roots of the equation

$$\text{Discrim. } (\mathrm{U}'-\lambda\mathrm{V}') = 0;$$

in other words, that we must have identically

$$\begin{vmatrix} a-\lambda & h & g & \dots \\ h & b-\lambda & f & \dots \\ g & f & c-\lambda & \dots \\ \cdot & \cdot & \cdot & \cdot \end{vmatrix} = \begin{vmatrix} \mathrm{A}-\lambda & 0 & 0 & \dots \\ 0 & \mathrm{B}-\lambda & 0 & \dots \\ 0 & 0 & \mathrm{C}-\lambda & \dots \\ \cdot & \cdot & \cdot & \cdot \end{vmatrix};$$

so that A, B, C, ... are the values of λ in the equation

$$\text{Discrim. } (\mathrm{U}-\lambda\mathrm{V}) = 0.$$

HESSE, O. (1859, October).

[Neue Eigenschaften der linearen Substitutionen welche gegebene homogene Functionen des zweiten Grades in andere transformiren die nur die Quadrate der Variabeln enthalten. *Crelle's Journ.*, lvii. pp. 175–182; or *Werke*, pp. 489–496.]

Hesse's object is that of Kummer (1843), Jacobi (1844, March), and Borchardt (1845, January), namely, to prove the reality of the roots of Lagrange's determinantal equation by showing that the product of their squared differences is essentially positive.

Taking the linear substitution

$$\xi_k = a_{k1}x_1 + a_{k2}x_2 + \dots + a_{kn}x_n \Big\}_{k=1}^{k=n}$$

we readily see that $\xi_1\xi_2 \dots \xi_n$ is expressible as a sum of terms of the form $\mathrm{C}x_1^{e_1}x_2^{e_2} \dots x_n^{e_n}$, where $e_1+e_2+\dots+e_n=n$ and C is an integral function of a's—a result which Hesse writes

$$\xi_1\xi_2 \dots \xi_n = \sum \mathrm{A}_{e_1e_2\dots e_n}x_1^{e_1}x_2^{e_2} \dots x_n^{e_n},$$

the coefficient of any term being denoted by an A with n suffixes identical with the n exponents of the x's. Now let us suppose the substitution to be orthogonal, in which case we know that

$$x_k = a_{1k}\xi_1 + a_{2k}\xi_2 + \dots + a_{nk}\xi_n \Big\}_{k=1}^{k=n};$$

and let us thereby transform $\sum \mathrm{A}_{e_1e_2\dots e_n}x_1^{e_1}x_2^{e_2} \dots x_n^{e_n}$ so as to have it again in terms of the ξ's. In doing this Hesse pays attention only to the term in $\xi_1\xi_2 \ldots \xi_n$, making the assertion that *the coefficient of* $\xi_1\xi_2 \ldots \xi_n$ *in* $x_1^{e_1}x_2^{e_2} \ldots x_n^{e_n}$ *is either the same as the coefficient of* $x_1^{e_1}x_2^{e_2} \ldots x_n^{e_n}$ *in* $\xi_1\xi_2 \ldots \xi_n$ *or differs from the latter coefficient by a merely arithmetical multiplier.* From this it follows that the coefficient of $\xi_1\xi_2 \ldots \xi_n$ in any term $A_{e_1e_2 \ldots e_n}x_1^{e_1}x_2^{e_2} \ldots x_n^{e_n}$ is a merely arithmetical multiple of $A^2_{e_1e_2 \ldots e_n}$; and, if the multiplier in question be denoted by $\Theta_{e_1e_2 \ldots e_n}$, there results from the equatement of coefficients

$$1 = \sum \Theta_{e_1e_2 \ldots e_n} A^2_{e_1e_2 \ldots e_n}.$$

Next, let us suppose in addition that our substitution transforms an n-ary quadric

$$f_1(x_1, x_2, \ldots, x_n) \quad \text{into} \quad g_1\xi_1^2 + g_2\xi_2^2 + \ldots + g_n\xi_n^2,$$

a step which, as we know, introduces the quantities whose reality is in question. In regard to them Hesse first recalls Jacobi's proof (1833) that they are such that

$$g_1^2\xi_1^2 + g_2^2\xi_2^2 + \ldots + g_n^2\xi_n^2, \quad g_1^3\xi_1^2 + g_2^3\xi_2^2 + \ldots + g_n^3\xi_n^2, \quad \ldots.$$

are also expressible as homogeneous quadric functions of the x's, and that the coefficients of these quadrics are rational integral functions of the coefficients of the original quadric f_1. It is seen to be not inappropriate therefore to use

$$f_p(x_1, x_2, \ldots, x_n) \quad \text{for} \quad g_1^p\xi_1^2 + g_2^p\xi_2^2 + \ldots + g_n^p\xi_n^2$$

and to denote the partial differential-quotient of $f_p(x_1, x_2, \ldots, x_n)$ with respect to x_k by $f'_p(x_k)$, thus giving

$$\tfrac{1}{2}f'_p(x_k) = \alpha_{k1}g_1^p\xi_1 + \alpha_{k2}g_2^p\xi_2 + \ldots + \alpha_{kn}g_n^p\xi_n.$$

The next step is the deduction of an important result from the consideration of the determinant

$$\begin{vmatrix} x_1 & x_2 & \ldots & x_n \\ \tfrac{1}{2}f'_1(x_1) & \tfrac{1}{2}f'_1(x_2) & \ldots & \tfrac{1}{2}f'_1(x_n) \\ \tfrac{1}{2}f'_2(x_1) & \tfrac{1}{2}f'_2(x_2) & \ldots & \tfrac{1}{2}f'_2(x_n) \\ \cdot & \cdot & \cdot & \cdot \\ \tfrac{1}{2}f'_{n-1}(x_1) & \tfrac{1}{2}f'_{n-1}(x_2) & \ldots & \tfrac{1}{2}f'_{n-1}(x_n) \end{vmatrix}, \text{ or } \Delta \text{ say.}$$

Each element being linear in the x's, the determinant is of the nth degree in those variables, and therefore we may put

$$\Delta = \sum \mathrm{B}_{e_1e_2\ldots e_n} x_1^{e_1} x_2^{e_2} \ldots x_n^{e_n}.$$

On the other hand, if we substitute for each element its expression in terms of the ξ's, the result is manifestly a product-determinant, and we learn that

$$\Delta = \begin{vmatrix} a_{11} & a_{12} & a_{13} & \ldots & a_{1n} \\ a_{21} & a_{22} & a_{23} & \ldots & a_{2n} \\ \cdot & \cdot & \cdot & \cdot & \cdot \\ a_{n1} & a_{n2} & a_{n3} & \ldots & a_{nn} \end{vmatrix} \cdot \begin{vmatrix} \xi_1 & g_1\xi_1 & g_1^2\xi_1 & \ldots & g_1^{n-1}\xi_1 \\ \xi_2 & g_2\xi_2 & g_2^2\xi_2 & \ldots & g_2^{n-1}\xi_2 \\ \cdot & \cdot & \cdot & \cdot & \cdot \\ \xi_n & g_n\xi_n & g_n^2\xi_n & \ldots & g_n^{n-1}\xi_n \end{vmatrix}$$

$$= (\pm 1)\cdot | g_1^0 g_2^1 \ldots g_n^{n-1} | \cdot \xi_1\xi_2 \ldots \xi_n.$$

Equating these two values and substituting the expression found at the outset for $\xi_1\xi_2 \ldots \xi_n$ we obtain

$$\sum \mathrm{B}_{e_1e_2\ldots e_n} x_1^{e_1} x_2^{e_2} \ldots x_n^{e_n} = (\pm 1)\cdot | g_1^0 g_2^1 \ldots g_n^{n-1} | \cdot \sum \mathrm{A}_{e_1e_2\ldots e_n} x_1^{e_1} x_2^{e_2} \ldots x_n^{e_n},$$

and thus see that

$$\mathrm{B}_{e_1e_2\ldots e_n} = (\pm 1)\cdot | g_1^0 g_2^1 \ldots g_n^{n-1} | \cdot \mathrm{A}_{e_1e_2\ldots e_n},$$

as Jacobi had shown in 1845 in the case of $n=3$.

With the help of this, Hesse's first result at once becomes

$$| g_1^0 g_2^1 \ldots g_n^{n-1} |^2 = \sum \Theta_{e_1e_2\ldots e_n} \mathrm{B}^2_{e_1e_2\ldots e_n},$$

and the desired end is reached.

CHAPTER XI.

PERSYMMETRIC DETERMINANTS, FROM 1841 TO 1860.

As has already been pointed out (*History*, i. pp. 485–487*), the special form of determinant named "persymmetric" in 1853 by Sylvester came first to light in 1835 in a paper of Jacobi's on the elimination of the unknown from two equations of the nth degree, the fact being that the adjugate of Bezout's condensed eliminant —in other words, the adjugate of the determinant resulting from Bezout's "abridged method" of elimination—is there shown to be such that the elements of it whose place-numbers have the same sum are equal.

The essentials of the proof are easily made clear if we accept the fact that from the equations

$$\left.\begin{aligned} a_1x + a_2y + a_3z &= 0 \\ b_1x + b_2y + b_3z &= 0 \\ c_1x + c_2y + c_3z &= 0 \end{aligned}\right\}$$

it can be shown for non-zero values of x, y, z that

$$\begin{aligned} x : y : z &:: \mathrm{A}_1 : \mathrm{A}_2 : \mathrm{A}_3 \\ &:: \mathrm{B}_1 : \mathrm{B}_2 : \mathrm{B}_3 \\ &:: \mathrm{C}_1 : \mathrm{C}_2 : \mathrm{C}_3 . \end{aligned}$$

This is something more than what Jacobi had then occasion to use, but in 1841 the portion of it which holds when there is one equation fewer was stated by him in all its generality in § 7 of

* The 7th and 8th lines of p. 486 have unfortunately been transposed by the printer. Also, in the first determinant of the footnote on the same page the first b_1 should be b_0.

the *De Formatione* ... Specialising from it in two directions, namely, (1) by taking x, x^2, x^3 instead of x, y, z, and (2) by taking the determinant of the coefficients to be axisymmetric, we can assert that if the equations

$$\left.\begin{aligned} ax + fx^2 + ex^3 &= 0 \\ fx + bx^2 + dx^3 &= 0 \\ ex + dx^2 + cx^3 &= 0 \end{aligned}\right\}$$

hold for a non-zero value of x, then

$$\begin{aligned} x : x^2 : x^3 &:: \mathrm{A} : \mathrm{F} : \mathrm{E} \\ &:: \mathrm{F} : \mathrm{B} : \mathrm{D} \\ &:: \mathrm{E} : \mathrm{D} : \mathrm{C}; \end{aligned}$$

and it will follow that $\mathrm{B} = \mathrm{E}$, and that

$$x : x^2 : x^3 : x^4 : x^5 :: \mathrm{A} : \mathrm{F} : \mathrm{E} : \mathrm{D} : \mathrm{C}.$$

Now the set of equations from which Bezout's condensed eliminant is derived is of the very special type here posited: consequently it is seen that the adjugate of the said eliminant must be persymmetric, and that its different elements form an equirational progression whose common multiplier is the root common to the original pair of equations.

ROSENHAIN, G. (1844).

[Exercitationes analyticæ in theorema Abelianum de integralibus functionum algebraicarum. *Crelle's Journ.*, xxviii. pp. 249–278.]

What concerns our subject here is a digression (§§ 5–10, pp. 263–278) on the elimination of the unknown from two equations of the nth degree. The first two sections (pp. 263–268) are little else than a reproduction of part of Jacobi's paper of 1835 dealing with Bezout's so-called "abridged method," and the remainder contains a discussion of other methods. In subject, therefore, the digression resembles Cauchy's paper of 1840.

At this point we have to recall the fact already reported,

that in Borchardt's paper of 1845 (January) a determinant of the special form we are now considering appeared as an expression for the square of the difference-product, and that a generalisation of this result was given by Cayley the year following. These two papers as well as four others dealt with under Alternants should be kept in view in reading the present chapter. The full list is—

1845	Borchardt, C. W.,	p. 159.	1854	Brioschi, F.,	p. 172.
1846	Cayley, A.,	p. 162.	1857	Bellavitis, G.,	p. 181.
1854	Joachimsthal, F.,	p. 169.	1847	Baltzer, R.	p. 183.

JACOBI, C. G. J. (1845, August).

[Ueber die Darstellung einer Reihe gegebner Werthe durch eine gebrochne rationale Function. *Crelle's Journ.*, xxx. pp. 127–156; or *Gesammelte Werke*, iii. pp. 479–511.]

The subject here dealt with by Jacobi is that first considered by Cauchy in the fifth note to the *Analyse Algébrique* of 1821, namely, the extension of Lagrange's interpolation-formula, or the finding of a function u of the form $\mathrm{N}(x)/\mathrm{M}(x)$ which shall have the values $u_1, u_2, \ldots, u_{n+m+1}$ when x has the values $x_1, x_2, \ldots, x_{n+m+1}$, it being understood that N and M are respectively of the nth and m^{th} degrees in x.

The given $n+m+1$ equations

$$u_1\mathrm{M}(x_1) = \mathrm{N}(x_1), \quad u_2\mathrm{M}(x_2) = \mathrm{N}(x_2), \quad \ldots\ldots$$

are first used to eliminate the $n+1$ coefficients of $\mathrm{N}(x)$, and thereby obtain m equations for the determination of the ratios of the coefficients of $\mathrm{M}(x)$. This is interestingly accomplished by using the multipliers $x_1^p/f'(x_1)$, $x_2^p/f'(x_2)$, ..., where $f(x)=(x-x_1)(x-x_2)\ldots\ldots(x-x_{n+m+1})$, then performing addition, and finally utilising a known theorem regarding "partial fractions." The result is that for any one value of p we have

$$\sum_{i=1}^{i=n+m+1}\frac{u_i x_i^p \mathrm{M}(x_i)}{f'(x_i)} = \sum_{i=1}^{i=n+m+1}\frac{x_i^p \mathrm{N}(x_i)}{f'(x_i)};$$

and that therefore when p has any one of the values 0, 1, 2, ... $m-1$, we have

$$\sum_{i=1}^{i=n+m+1} \frac{u_i x_i^p \mathrm{M}(x_i)}{f'(x_i)} = 0.$$

By putting

$$v_p \quad \text{for} \quad \frac{x_1^p u_1}{f'(x_1)} + \frac{x_2^p u_2}{f'(x_2)} + \ldots + \frac{x_{n+m+1}^p u_{n+m+1}}{f'(x_{n+m+1})}$$

and

$$a + a_1 x + a_2 x^2 + \ldots + a_m x^m \quad \text{for} \quad \mathrm{M}(x)$$

these last m equations become

$$\left.\begin{array}{l} v_0 a + v_1 a_1 + v_2 a_2 + \ldots + v_m a_m = 0 \\ v_1 a + v_2 a_1 + v_3 a_2 + \ldots + v_{m+1} a_m = 0 \\ \cdot \quad \cdot \quad \cdot \quad \cdot \quad \cdot \quad \cdot \quad \cdot \quad \cdot \quad \cdot \quad \cdot \\ v_{m-1} a + v_m a_1 + v_{m+1} a_2 + \ldots + v_{2m-1} a_m = 0 \end{array}\right\}$$

whence for $\mathrm{M}(x)$ there is obtained the expression *

$$\begin{vmatrix} 1 & x & x^2 & \ldots & x_m \\ v_0 & v_1 & v_2 & \ldots & v_m \\ v_1 & v_2 & v_3 & \ldots & v_{m+1} \\ \cdot & \cdot & \cdot & \cdot & \cdot \\ v_{m-1} & v_m & v_{m+1} & \ldots & v_{2m-1} \end{vmatrix},$$

* By making the observation that the v's are neatly expressible as determinants the whole matter may be put much more simply. Thus, taking the case where $u=(\beta_0+\beta_1 x)/(a_0+a_1 x+a_2 x^2)$, we see at a glance that

$$\begin{vmatrix} 1 & x_1 & x_1^2 & x_1^p(\beta_0+\beta_1 x_1) \\ 1 & x_2 & x_2^2 & x_2^p(\beta_0+\beta_1 x_2) \\ 1 & x_3 & x_3^2 & x_3^p(\beta_0+\beta_1 x_3) \\ 1 & x_4 & x_4^2 & x_4^p(\beta_0+\beta_1 x_4) \end{vmatrix} = 0 \quad \text{when } p = 0 \text{ or } 1,$$

and therefore from the data that

$$\begin{vmatrix} 1 & x_1 & x_1^2 & u_1 x_1^p(a_0+a_1 x_1+a_2 x_1^2) \\ 1 & x_2 & x_2^2 & u_2 x_2^p(a_0+a_1 x_2+a_2 x_2^2) \\ 1 & x_3 & x_3^2 & u_3 x_3^p(a_0+a_1 x_3+a_2 x_3^2) \\ 1 & x_4 & x_4^2 & u_4 x_4^p(a_0+a_1 x_4+a_2 x_4^2) \end{vmatrix} = 0 \quad \text{when } p = 0 \text{ or } 1,$$

or, what is the same thing, that

$$\begin{vmatrix} 1 & x_1 & x_1^2 & u_1 x_1^p \\ 1 & x_2 & x_2^2 & u_2 x_2^p \\ 1 & x_3 & x_3^2 & u_3 x_3^p \\ 1 & x_4 & x_4^2 & u_4 x_4^p \end{vmatrix} a_0 + \begin{vmatrix} 1 & x_1 & x_1^2 & u_1 x_1^{p+1} \\ 1 & x_2 & x_2^2 & u_2 x_2^{p+1} \\ 1 & x_3 & x_3^2 & u_3 x_3^{p+1} \\ 1 & x_4 & x_4^2 & u_4 x_4^{p+1} \end{vmatrix} a_1 + \begin{vmatrix} 1 & x_1 & x_1^2 & u_1 x_1^{p+2} \\ 1 & x_2 & x_2^2 & u_2 x_2^{p+2} \\ 1 & x_3 & x_3^2 & u_3 x_3^{p+2} \\ 1 & x_4 & x_4^2 & u_4 x_4^{p+2} \end{vmatrix} a_2 = 0.$$

or, by further putting $w = v_{p+1} - xv_p$,

$$\begin{vmatrix} w_0 & w_1 & \dots & w_{m-1} \\ w_1 & w_2 & \dots & w_m \\ \cdot & \cdot & \cdot & \cdot \\ w_{m-1} & w_m & \dots & w_{2m-2} \end{vmatrix}.$$

After finding other forms for $\mathrm{M}(x)$, and varying (§ 2) the mode of finding them, Jacobi proceeds (§ 3, pp. 140–146) to deal with $\mathrm{N}(x)$, first remarking, of course, that the one function is immediately determinable from the other, because the problem of representing $u_1, u_2, \dots$ by $\mathrm{N}(x)/\mathrm{M}(x)$ is the same as the problem of representing $u_1^{-1}, u_2^{-1}, \dots$ by $\mathrm{M}(x)/\mathrm{N}(x)$. Instead of utilising this, however, he takes from the theory of "partial fractions" the result

$$-\frac{\mathrm{N}(x)}{f(x)} = \sum_{i=1}^{i=n+m+1} \frac{\mathrm{N}(x_i)}{(x_i - x)f'(x_i)},$$

whence follows

$$-\frac{\mathrm{N}(x)}{f(x)} = \sum_{i=1}^{i=n+m+1} \frac{u_i \mathrm{M}(x_i)}{(x_i - x)f'(x_i)};$$

so that if we put

$$\mathrm{R}_p \quad \text{for} \quad \sum_{i=1}^{i=n+m+1} \frac{x_i^p u_i}{(x_i - x)f'(x_i)},$$

we have

$$-\frac{\mathrm{N}(x)}{f(x)} = a\mathrm{R}_0 + a_1\mathrm{R}_1 + \dots + a_m\mathrm{R}_m,$$

From these two equations on solving for $a_0 : a_1 : a_2$ and substituting in $a_0 + a_1x + a_2x^2$ we obtain

$$\mathrm{M}(x) = \begin{vmatrix} 1 & x & x^2 \\ v_0 & v_1 & v_2 \\ v_1 & v_2 & v_3 \end{vmatrix}$$

where $v^p = | x_1^0\, x_2^1\, x_3^2\, x_4^p u_4 |$, or

$$\mathrm{M}(x) = \begin{vmatrix} \omega_0 & \omega_1 \\ \omega_1 & \omega_2 \end{vmatrix}$$

where $\omega_p = v_{p+1} - xv_p = | x_1^0\, x_2^1\, x_3^2\, x_4^p u_4(x_4 - x) |$.

and therefore, by substituting the already found values of $a:a_1:a_2:\ldots:a_m$,

$$-\frac{\mathrm{N}(x)}{f(x)} = \begin{vmatrix} \mathrm{R}_0 & \mathrm{R}_1 & \ldots & \mathrm{R}_m \\ v_0 & v_1 & \ldots & v_m \\ v_1 & v_2 & \ldots & v_{m+1} \\ \cdot & \cdot & \cdot & \cdot \\ v_{m-1} & v_m & \ldots & v_{2m-1} \end{vmatrix}.$$

As, however, $x\mathrm{R}_p + v_p = \mathrm{R}_{p+1}$, we can change the elements of the second row here into $\mathrm{R}_1, \mathrm{R}_2, \ldots, \mathrm{R}_{m+1}$, and then the elements of the third row into $\mathrm{R}_2, \mathrm{R}_3, \ldots, \mathrm{R}_{m+2}$, and so on, thus arriving at a determinant of the same special form as in the case of $\mathrm{M}(x)$.*

Combining the two results, Jacobi is thus led to the theorem that

$$-\frac{1}{f(x)}\cdot\frac{\mathrm{N}(x)}{\mathrm{M}(x)} = \begin{vmatrix} \mathrm{R}_0 & \mathrm{R}_1 & \ldots & \mathrm{R}_m \\ \mathrm{R}_1 & \mathrm{R}_2 & \ldots & \mathrm{R}_{m+1} \\ \mathrm{R}_2 & \mathrm{R}_3 & \ldots & \mathrm{R}_{m+2} \\ \cdot & \cdot & \cdot & \cdot \\ \mathrm{R}_m & \mathrm{R}_{m+1} & \ldots & \mathrm{R}_{2m} \end{vmatrix} \div \begin{vmatrix} w_0 & w_1 & \ldots & w_{m-1} \\ w_1 & w_2 & \ldots & w_m \\ \cdot & \cdot & \cdot & \cdot \\ w_{m-1} & w_m & \ldots & w_{2m-1} \end{vmatrix},$$

—a result not easily verifiable by giving x one of its $n+m+1$ values.

---

* Continuing the case of the previous footnote we should prefer to begin with

$$\frac{\mathrm{N}(x)}{f(x)}\,|\,x_1^0 x_2^1 x_3^2 x_4^3\,| = \begin{vmatrix} 1 & x_1 & x_1^2 & \mathrm{N}(x_1)/(x-x_1) \\ 1 & x_2 & x_2^2 & \mathrm{N}(x_2)/(x-x_2) \\ 1 & x_3 & x_3^2 & \mathrm{N}(x_3)/(x-x_3) \\ 1 & x_4 & x_4^2 & \mathrm{N}(x_4)/(x-x_4) \end{vmatrix},$$

and then proceeding exactly as before we should arrive at

$$\frac{\mathrm{N}(x)}{f(x)}\,|\,x_1^0 x_2^1 x_3^2 x_4^3\,| = \begin{vmatrix} \rho_0 & \rho_1 & \rho_2 \\ \rho_1 & \rho_2 & \rho_3 \\ \rho_2 & \rho_3 & \rho_4 \end{vmatrix},$$

where $\rho_p = |\,x_1^0\ x_2^1\ x_3^2\ x_4^p u_4/(x-x_4)\,|$.

The function sought would then be

$$\frac{(x-x_1)(x-x_2)(x-x_3)(x-x_4)\begin{vmatrix} \rho_0 & \rho_1 & \rho_2 \\ \rho_1 & \rho_2 & \rho_3 \\ \rho_2 & \rho_3 & \rho_4 \end{vmatrix}}{|\,x_1^0 x_2^1 x_3^2 x_4^3\,|\cdot\begin{vmatrix} \omega_0 & \omega_1 \\ \omega_1 & \omega_2 \end{vmatrix}}.$$

For ourselves we may add that the theorem becomes still more interesting when it is pointed out that, by reason of the identity

$$|y_1^0 y_2^1 \ldots y_{n-1}^{n-2} \mathrm{Y}_n| \div |y_1^0 y_2^1 \ldots y_{n-1}^{n-2} y_n^{n-1}| = \frac{\mathrm{Y}_1}{\phi'(y_1)} + \frac{\mathrm{Y}_2}{\phi'(y_2)} + \ldots + \frac{\mathrm{Y}_n}{\phi'(y_n)}$$

where $\phi(y) = (y-y_1)(y-y_2) \ldots (y-y_n)$, the R's like the w's are all expressible as determinants of the order $n+m+1$, that these determinants in both cases belong to the special type known as alternants, and that R_p differs from w_p in the last column only;—in fact, that

$$\mathrm{R}_p = \begin{vmatrix} 1 & x_1 & x_1^2 & \ldots & x_1^{n+m-1} & x_1^p u_1/(x_1-x) \\ 1 & x_2 & x_2^2 & \ldots & x_2^{n+m-1} & x_2^p u_2/(x_2-x) \\ 1 & x_3 & x_3^2 & \ldots & x_3^{n+m-1} & x_3^p u_3/(x_3-x) \\ \cdot & \cdot & \cdot & \cdot & \cdot & \cdot \end{vmatrix} \div \zeta^{\frac{1}{2}},$$

$$w_p = \begin{vmatrix} 1 & x_1 & x_1^2 & \ldots & x_1^{n+m-1} & x_1^p u_1(x_1-x) \\ 1 & x_2 & x_2^2 & \ldots & x_2^{n+m-1} & x_2^p u_2(x_2-x) \\ 1 & x_3 & x_3^2 & \ldots & x_3^{n+m-1} & x_3^p u_3(x_3-x) \\ \cdot & \cdot & \cdot & \cdot & \cdot & \cdot \end{vmatrix} \div \zeta^{\frac{1}{2}},$$

where $\zeta^{\frac{1}{2}}$ is the difference-product of $x_1, x_2, \ldots, x_{n+m+1}$.

BORCHARDT, C. W. (1847, February).

[Développements sur l'équation à l'aide de laquelle on détermine les inégalités séculaires du mouvement des planètes. *Journ. (de Liouville) de Math.*, xii. pp. 50–67; *Gesammelte Werke*, pp. 15–30.]

The new section of this paper, which is an extension of Borchardt's of 1845 (January), is the third (pp. 54–60), and explains at length how, for the purpose of ascertaining the total number of real roots of the equation of the nth degree $f(x)=0$, the coefficients of highest powers in the series of Sturm's functions $f(x)$, $f_1(x)$, $f_2(x)$, ... may be replaced, according to Sylvester, by

$$1, \quad n, \quad \sum(x_2-x_1)^2, \quad \sum(x_2-x_1)^2(x_3-x_1)^2(x_2-x_1)^2, \quad \ldots.$$

where $x_1, x_2, \ldots$ are the roots, and therefore by

$$1, \quad s_0, \quad \begin{vmatrix} s_0 & s_1 \\ s_1 & s_2 \end{vmatrix}, \quad \begin{vmatrix} s_0 & s_1 & s_2 \\ s_1 & s_2 & s_3 \\ s_2 & s_3 & s_4 \end{vmatrix}, \quad \ldots .$$

where $s_r = x_1^r + x_2^r + \ldots + x_n^r$. All this, however, is practically implied in Cayley's paper of 1846 (August).*

SYLVESTER, J. J. (1851, May).

[ESSAY ON CANONICAL FORMS: Supplement to a "Sketch of a Memoir on Elimination, Transformation, and Canonical Forms," 36 pp., London. Or *Collected Math. Papers*, i. pp. 203–216.]

In giving a preliminary notice of his general method for reducing odd-degreed functions to their canonical form, Sylvester says he based his method on the proposition that every one of the n-line minor determinants of the array

$$\begin{matrix} T_1 & T_2 & T_3 & \ldots & T_{n+1} \\ T_2 & T_3 & T_4 & \ldots & T_{n+2} \\ T_3 & T_4 & T_5 & \ldots & T_{n+3} \\ \cdot & \cdot & \cdot & \cdot & \cdot \\ T_n & T_{n+1} & T_{n+2} & \ldots & T_{2n} \end{matrix}$$

vanishes if

$$T_i = a_1^{r-i} b_1^{s+1} + a_2^{r-i} b_2^{s+i} + \ldots + a_{n-1}^{r-i} b_{n-1}^{s+i}.$$

This, which he hastily calls "a beautiful and striking theorem," and which he generalises in Note B of an Appendix, arises from the simple fact that each determinant is the product of two zeros, T_i being

$$(a_1^{r-i}, a_2^{r-i}, \ldots, a_{n-1}^{r-i}, 0 \between b_1^{s+i}, b_2^{s+i}, \ldots, b_{n-1}^{s+1}, 0).$$

It is of more importance, therefore, to recall that it was in

* The proposition Borchardt is concerned with is of course that *The equation* f(x) = 0 *has as many pairs of imaginary roots as there are changes of sign in any one of the three series mentioned.*

this year that Sylvester made the fruitful observation, already chronicled,* that the persymmetric determinants

$$ac - b^2, \quad ace + 2bcd - ae^2 - bd^2 - c^3 , \quad \ldots .$$

are expressible as "commutants," or rather that these special determinants could be represented in the umbral notation by using umbræ not wholly unconnected with one another. Thus, while

$$\begin{vmatrix} 0 & 1 & 2 \\ 0 & 1 & 2 \end{vmatrix} \text{ stands for the general determinant } \begin{vmatrix} 00 & 01 & 02 \\ 10 & 11 & 12 \\ 20 & 21 & 22 \end{vmatrix}$$

so long as the umbræ are understood to be entirely independent, it might also be used to stand for the special determinant

$$\begin{vmatrix} 00 & 01 & 02 \\ 01 & 02 & 03 \\ 02 & 03 & 04 \end{vmatrix}$$

if some mark were added to indicate that in the development 01 is to be put for 10, 02 for 20 or 11, 03 for 12 or 21, and 04 for 22.

SYLVESTER, J. J. (1851, October).

[On a remarkable discovery in the theory of canonical forms and of hyperdeterminants. *Philos. Magazine,* ii. pp. 391–410; or *Collected Math. Papers,* i. pp. 265–283.]

The consideration of the problem of the canonisation of the binary quintic led Sylvester to the more general problem of determining the p's and q's in

$$(p_1x + q_1y)^{2n+1} + (p_2x + q_2y)^{2n+1} + \ldots . + (p_{n+1}x + q_{n+1}y)^{2n+1}$$

so as to make this expression identical with

$$a_0x^{2n+1} + (2n+1)a_1x^{2n}y^1 + \tfrac{1}{2}(2n+1)2na_2x^{2n+1}y^2 + \ldots . + a_{2n+1}y^{2n+1}.$$

* See above, pp. 68, ...

This is at once seen to depend on the solution of the peculiar set of $2n+2$ equations

$$\left.\begin{array}{lllll}\pi_1 & +\ \pi_2 & +\ \ldots\ldots + & \pi_{n+1} & = a_0 \\ \pi_1\lambda_1 & +\ \pi_2\lambda_2 & +\ \ldots\ldots + & \pi_{n+1}\lambda_{n+1} & = a_1 \\ \pi_1\lambda_1^2 & +\ \pi_2\lambda_2^2 & +\ \ldots\ldots + & \pi_{n+1}\lambda_{n+1}^2 & = a_2 \\ \cdot\ \cdot\ \cdot & \cdot\ \cdot\ \cdot & \cdot\ \cdot\ \cdot & \cdot\ \cdot\ \cdot & \cdot\ \cdot \\ \pi_1\lambda_1^{2n+1} & +\ \pi_2\lambda_2^{2n+1} & +\ \ldots\ldots + & \pi_{n+1}\lambda_{n+1}^{2n+1} & = a_{2n+1}\end{array}\right\}$$

where the new unknowns $\pi_1, \pi_2, \ldots\ldots, \pi_{n+1}, \lambda_1, \lambda_2, \ldots\ldots, \lambda_{n+1}$ are introduced merely for shortness' sake, namely

$$\pi_r \text{ for } p_r^{2n+1} \quad \text{and} \quad \lambda_r \text{ for } q_r \div p_r .$$

Taking $n+2$ consecutive equations beginning with the first, and eliminating the π's, there is obtained

$$\begin{vmatrix} 1 & 1 & \ldots\ldots & 1 & a_0 \\ \lambda_1 & \lambda_2 & \ldots\ldots & \lambda_{n+1} & a_1 \\ \lambda_1^2 & \lambda_2^2 & \ldots\ldots & \lambda_{n+1}^2 & a_2 \\ \cdot & \cdot & \cdot & \cdot & \cdot \\ \lambda_1^{n+1} & \lambda_2^{n+1} & \ldots\ldots & \lambda_{n+1}^{n+1} & a_{n+1} \end{vmatrix} = 0,$$

which, if division by the difference-product of the λ's be effected, gives

$$a_{n+1} - a_n \sum\lambda_1 + a_{n-1}\sum\lambda_1\lambda_2 - \ldots\ldots = 0.$$

A similar result is evidently reached by taking *any* $n+2$ consecutive equations, so that altogether we shall have

$$\left.\begin{array}{l} a_{n+1} - a_n \sum\lambda_1 + a_{n-1}\sum\lambda_1\lambda_2 - \ldots\ldots = 0 \\ a_{n+2} - a_{n+1}\sum\lambda_1 + a_n \sum\lambda_1\lambda_2 - \ldots\ldots = 0 \\ a_{n+3} - a_{n+2}\sum\lambda_1 + a_{n+1}\sum\lambda_1\lambda_2 - \ldots\ldots = 0 \\ \cdot\ \cdot\ \cdot\ \cdot\ \cdot\ \cdot\ \cdot\ \cdot\ \cdot\ \cdot\ \cdot\ \cdot \\ a_{2n+1} - a_{2n}\sum\lambda_1 + a_{2n-1}\sum\lambda_1\lambda_2 - \ldots\ldots = 0 \end{array}\right\}$$

—that is to say, a set of $n+1$ equations in the $n+1$ unknowns $\Sigma\lambda_1, \Sigma\lambda_1\lambda_2, \ldots\ldots\ldots, \lambda_1\lambda_2\ldots\lambda_{n+1}$, the solution of which is

$$\frac{1}{A_0} = \frac{\sum\lambda_1}{A_1} = \frac{\sum\lambda_1\lambda_2}{A_2} = \ldots\ldots$$

where A_r is the determinant whose array is got by deleting the $(r+1)^{\text{th}}$ column from the array

$$\begin{array}{ccccc} a_{n+1} & a_n & a_{n-1} & \ldots & a_0 \\ a_{n+2} & a_{n+1} & a_n & \ldots & a_1 \\ \cdot & \cdot & \cdot & \cdot & \cdot \\ a_{2n+1} & a_{2n} & a_{2n-1} & \ldots & a_n. \end{array}$$

From this it follows that the λ's are the roots of the equation

$$A_0\lambda^{n+1} - A_1\lambda^n + A_2\lambda^{n-1} - \ldots = 0,$$

$$i.e. \quad \begin{vmatrix} \lambda^{n+1} & \lambda^n & \lambda^{n-1} & \ldots\ldots & \lambda^0 \\ a_{n+1} & a_n & a_{n-1} & \ldots\ldots & a_0 \\ a_{n+2} & a_{n+1} & a_n & \ldots\ldots & a_1 \\ \cdot & \cdot & \cdot & \cdot & \cdot \\ a_{2n+1} & a_{2n} & a_{2n-1} & \ldots\ldots & a_n \end{vmatrix} = 0,$$

$$i.e. \quad \begin{vmatrix} a_{n+1} - a_n\lambda & a_n - a_{n-1}\lambda & \ldots\ldots & a_1 - a_0\lambda \\ a_{n+2} - a_{n+1}\lambda & a_{n+1} - a_n\lambda & \ldots\ldots & a_2 - a_1\lambda \\ \cdot & \cdot & \cdot & \cdot \\ a_{2n+1} - a_{2n}\lambda & a_{2n} - a_{2n-1}\lambda & \ldots\ldots & a_{n+1} - a_n\lambda \end{vmatrix} = 0.$$

On substituting in the first $n+1$ equations of the original set the values of $\lambda_1, \lambda_2, \ldots, \lambda_{n+1}$ thus found, the values of $\pi_1, \pi_2, \ldots, \pi_{n+1}$ are obtainable from a set of linear equations of the type associated with the name of Lagrange.

The latter part of this procedure is not given by Sylvester, who on reaching the set of equations in $\Sigma\lambda_1$, $\Sigma\lambda_1\lambda_2$, ... suddenly draws the seemingly irrelevant conclusion "that

$$(x+\lambda_1 y)(x+\lambda_2 y)\ldots\ldots(x+\lambda_{n+1}y)$$

is a constant multiple of the determinant

$$\begin{vmatrix} x^{n+1} & -x^n y & x^{n+1}y^2 & \ldots\ldots \\ a_{n+1} & a_n & a_{n-1} & \ldots\ldots \\ a_{n+2} & a_{n+1} & a_n & \ldots\ldots \\ \cdot & \cdot & \cdot & \cdot \\ a_{2n+1} & a_{2n} & a_{2n-1} & \ldots\ldots \end{vmatrix}\text{''}, \quad \text{or } \Delta, \text{ say.}$$

As a matter of fact $(p_1x+q_1y)(p_2x+q_2y)\ \ldots\ldots\ (p_{n+1}x+q_{n+1}y)$

$$= p_1p_2\ldots p_{n+1}(x+\lambda_1y)(x+\lambda_2y)\ \ldots\ldots\ (x+\lambda_{n+1}y),$$

$$= p_1p_2\ldots p_{n+1}\Big(x^{n+1} + \sum\lambda_1\cdot x^ny + \sum\lambda_1\lambda_2\cdot x^{n-1}y^2 + \ldots\ldots\Big),$$

$$= \frac{p_1p_2\ldots p_{n+1}}{\mathrm{A}_0}(\mathrm{A}_0x^{n+1}+\mathrm{A}_1x^ny+\mathrm{A}_2x^{n-1}y^2+\ldots),$$

$$= \frac{p_1p_2\ldots p_{n+1}}{\mathrm{A}_0}\Delta,$$

$$= \frac{p_1p_2\ldots p_{n+1}}{\mathrm{A}_0}\begin{vmatrix} a_{n+1}y+a_nx & a_ny+a_{n-1}x & \ldots\ldots \\ a_{n+2}y+a_{n+1}x & a_{n+1}y+a_nx & \ldots\ldots \\ \cdot\ \cdot\ \cdot & \cdot\ \cdot\ \cdot & \cdot\ \cdot\ \cdot \\ a_{2n+1}y+a_{2n}x & a_{2n}y+a_{2n-1}x & \ldots\ldots \end{vmatrix},$$

from which we see (1) the point which Sylvester wished to make, namely, that p_1x+q_1y, p_2x+q_2y , ... being viewed as the original unknowns, it is important to know that their values are multiples of the linear factors of Δ, and (2) that

$$\Delta=\mathrm{A}_0(x+\lambda_1y)(x+\lambda_2y)\ \ldots\ldots$$

Of course the conclusion drawn is that the transformation of a binary $(2n+1)$-ic into the sum of $n+1$ powers depends on the solution of a determinantal equation of the $(n+1)^{\text{th}}$ degree. As examples, the quintic and septimic are taken, the latter mainly for the purpose of drawing attention to the fact that the conditions of "catalecticism," that is, of $(a, b, \ldots, h \between x, y)^7$ being expressible in the form of the sum of three seventh powers—instead of four, as the general rule provides—require that the cofactors of the elements of the first row of the determinant

$$\begin{vmatrix} y^4 & -y^3x & y^2x^2 & -yx^3 & x^4 \\ a & b & c & d & e \\ b & c & d & e & f \\ c & d & e & f & g \\ d & e & f & g & h \end{vmatrix}$$

must all vanish, or, what by the homaloidal law is the same thing, that *two* of them vanish.

The analogous problem for even-degreed functions is next taken up, a beginning being made with the transformation of the quartic $(a, b, \ldots, e \between x, y)^4$ into the form

$$(p_1x+q_1y)^4 + (p_2x+q_2y)^4 + 6\epsilon(p_1x+q_1y)^2(p_2x+q_2y)^2.$$

On putting

$$q_1 = p_1\lambda_1, \quad p_2 = q_2\lambda_2, \quad \epsilon p_1^2p_2^2 = \mu, \quad \lambda_1+\lambda_2 = s_1, \quad \lambda_1\lambda_2 = s_2$$

there is obtained by equatement of like powers of x and y

$$\left.\begin{aligned} a &= p_1^4 + p_2^4 + 6\mu \\ b &= p_1^4\lambda_1 + p_2^4\lambda_2 + 3\mu s_1 \\ c &= p_1^4\lambda_1^2 + p_2^4\lambda_2^2 + \mu s_1^2 + 2\mu s_2 \\ d &= p_1^4\lambda_1^3 + p_2^4\lambda_2^3 + 3\mu s_1 s_2 \\ e &= p_1^4\lambda_1^4 + p_2^4\lambda_2^4 + 6\mu s_2^2 \end{aligned}\right\},$$

and from these by operations which lead to the elimination of p_1^4, p_2^4 from every consecutive triad of equations

$$\left.\begin{aligned} as_2 - bs_1 + c - \mu(8s_2-2s_1^2) &= 0 \\ bs_2 - cs_1 + d - \mu(4s_2-s_1^2)s_1 &= 0 \\ cs_2 - ds_1 + e - \mu(8s_2-2s_1^2)s_2 &= 0 \end{aligned}\right\}.$$

or, if we put ν for $-\mu(8s_2-2s_1^2)$,

$$\left.\begin{aligned} as_2 - bs_1 + (c+\nu) &= 0 \\ bs_2 - (c-\tfrac{1}{2}\nu)s_1 + d &= 0 \\ (c+\nu)s_2 - ds_1 + e &= 0 \end{aligned}\right\},$$

From the resulting cubic equation

$$\begin{vmatrix} a & b & c+\nu \\ b & c-\frac{1}{2}\nu & d \\ c+\nu & d & e \end{vmatrix} = 0$$

ν can be determined, and thence in backward order s_1, s_2; μ; λ_1, λ_2; p_1, p_2; q_1, q_2; m.

In passing, note is taken of the fact that the said cubic when arranged according to powers of ν is

$$\nu^3 - (ae-4bd+3c^2)\nu + 2\begin{vmatrix} a & b & c \\ b & c & d \\ c & d & e \end{vmatrix} = 0,$$

and that $ae - 4bd + 3c^2$ and the determinant here appearing are the two invariants * of the quartic under investigation.

The reduction of the octavic $(a_0, a_1, \ldots, a_8 \between x, y)^8$ to the form

$$u_1^8 + u_2^8 + u_3^8 + u_4^8 + 70\epsilon u_1^2 u_2^2 u_3^2 u_4^2,$$

where $u_r = p_r x + q_r y$, is shown in similar fashion to depend on the solution of the quintic equation

$$\begin{vmatrix} a_0 & a_1 & a_2 & a_3 & a_4 - \nu \\ a_1 & a_2 & a_3 & a_4 + \frac{1}{4}\nu & a_5 \\ a_2 & a_3 & a_4 - \frac{1}{6}\nu & a_5 & a_6 \\ a_3 & a_4 + \frac{1}{4}\nu & a_5 & a_6 & a_7 \\ a_4 - \nu & a_5 & a_6 & a_7 & a_8 \end{vmatrix} = 0,$$

where

$$\nu = 72\epsilon p_1^2 p_2^2 p_3^2 p_4^2 \mathrm{I} \quad \text{and} \quad \mathrm{I} = s_4 - \tfrac{1}{4} s_1 s_3 + \tfrac{1}{12} s_2^2,$$

I being the quadratic invariant of

$$x^4 + s_1 x^3 y + s_2 x^2 y^2 + s_3 x y^3 + s_4 y^4 \quad \text{or} \quad (x + \lambda_1 y)(x + \lambda_2 y)(x + \lambda_3 y)(x + \lambda_4 y).$$

The fact that the coefficients of ν^3, ν^2, ν^1, ν^0 are invariants of the octavic is insisted on, and generalisations are effected for functions of the degree $4m$ and the degree $4m + 2$.

Further, it is pointed out that when the said even-degreed functions after transformation are without the last (or unique) term,—that is to say, are in Sylvester's phraseology "meio-catalectic,"—the last of the series of invariants must vanish: for example, the condition that $(a_0, a_1, \ldots, a_6 \between x, y)^6$ may be expressible as the sum of three sixth powers is

$$\begin{vmatrix} a_0 & a_1 & a_2 & a_3 \\ a_1 & a_2 & a_3 & a_4 \\ a_2 & a_3 & a_4 & a_5 \\ a_3 & a_4 & a_5 & a_6 \end{vmatrix} = 0.$$

This, of course, may be proved independently, but is seen to be a conclusion from putting $\epsilon = 0$ in the foregoing.

* The term "invariant" is first used in this paper.

SYLVESTER, J. J. (1852, April).

[On the principles of the calculus of forms. *Cambridge and Dub. Math. Journ.*, vii. pp. 52–97, 179–217; or *Collected Math. Papers*, i. pp. 284–327, 328–363.]

Here the same subjects and the same special determinants are dealt with as in the preceding; and the determinant whose vanishing has been seen to be the condition for "meio-catalecticism" is denominated (p. 62) the *catalecticant** of the even-degreed function in question, while the determinant whose resolution into linear factors furnishes Sylvester's canonical form of an odd-degreed function is called the *canonizant*† of the said function. As the former is an invariant of its function, so the latter is a covariant.

BRUNO, F. FAÀ DI (1852, May).

[Démonstration d'un théorème relatif à la réduction des fonctions homogènes à deux lettres à leur forme canonique. *Journ.* (*de Liouville*) *de Math.* . . . xvii. pp. 193–201.]

The subject of the whole of this paper is simply the solution of the set of equations dealt with in Sylvester's paper of 1851 (October). The process is lengthy and uninviting, the sole point of interest being that the equation in λ comes out in the form

$$\begin{vmatrix} a_0\lambda - a_1 & a_1\lambda - a_2 & \dots\dots & a_n\lambda - a_{n+1} \\ a_0\lambda^2 - a_2 & a_1\lambda^2 - a_3 & \dots\dots & a_n\lambda^2 - a_{n+2} \\ \cdot & \cdot & \cdot & \cdot \\ a_0\lambda^{n+1} - a_{n+1} & a_1\lambda^{n+1} - a_{n+2} & \dots\dots & a_n\lambda^{n+1} - a_{2n+1} \end{vmatrix} = 0,$$

where the determinant is easily shown to be the same as one of Sylvester's forms by diminishing each row in order, beginning with the last, by λ times the row immediately preceding.

* "Meicatalecticizant," Sylvester truly says, would have been the more correct word, but even he took alarm sometimes.

† The name would have been equally appropriate for the determinants of the preceding paper which have ν in their diagonal.

CHIO, F. (1853, June).

[Mémoire sur les fonctions connues sous le nom de résultantes ou de déterminans. 32 pp., Turin.]

The second part (pp. 23–32) of Chio's memoir, which is headed "Exemples," mainly concerns Sylvester's set of equations of 1851 (October). His procedure is much more interesting than Faà di Bruno's. Using any multipliers A_0, A_1, ... with the first $n+2$ equations he obtains by addition

$$\begin{aligned} & x_0(A_0+A_1\lambda_0+A_2\lambda_0^2+\ \dots\ +A_{n+1}\lambda_0^{n+1}) \\ + & x_1(A_0+A_1\lambda_1+A_2\lambda_1^2+\ \dots\ +A_{n+1}\lambda_1^{n+1}) \\ + & \dots\dots\dots\dots\dots \\ + & x_n(A_0+A_1\lambda_n+A_2\lambda_n^2+\ \dots\ +A_{n+1}\lambda_n^{n+1}) = A_0a_0+A_1a_1+\dots+A_{n+1}a_{n+1}; \end{aligned}$$

and, the ratios of A_0, A_1, ... being supposed to be determined so as to make the coefficients of $x_0, x_1, \dots, x_n$ vanish, there results

$$A_0a_0 + A_1a_1 + \dots + A_{n+1}a_{n+1} = 0.$$

If each succeeding set of $n+2$ consecutive equations be treated in the same manner, it will be found that the *same* multipliers will make the coefficients of the x's vanish in every case: consequently there is obtained

$$\begin{aligned} A_0a_1 + A_1a_2 \quad + \dots + A_{n+1}a_{n+2} &= 0, \\ A_0a_2 + A_1a_3 \quad + \dots + A_{n+1}a_{n+3} &= 0, \\ \dots\dots\dots\dots\dots \\ A_0a_n + A_1a_{n+1} + \dots + A_{n+1}a_{2n+1} &= 0. \end{aligned}$$

This derived set of $n+1$ equations suffices to give the values of the ratios of A_0, A_1, ..., A_{n+1} in terms of the a's, and the substitution of the said values in

$$A_0 + A_1\lambda + A_2\lambda^2 + \dots + A_{n+1}\lambda^{n+1} = 0$$

gives the equation for the determination of the λ's.

It is not noted by the author that having $n+2$ equations linear and homogeneous in the A's he could at once deduce

$$\begin{vmatrix} 1 & \lambda & \lambda^2 & \ldots\ldots & \lambda^{n+1} \\ a_0 & a_1 & a_2 & \ldots\ldots & a_{n+1} \\ a_1 & a_2 & a_3 & \ldots\ldots & a_{n+2} \\ \cdot & \cdot & \cdot & \cdot\;\cdot\;\cdot\;\cdot & \cdot \\ a_n & a_{n+1} & a_{n+2} & \ldots\ldots & a_{2n+1} \end{vmatrix} = 0.$$

The other forms of the equation, however, he gives full attention to.

SYLVESTER, J. J. (1853, June).

[On a theory of the syzygetic relations of two rational integral functions, ... *Philos. Transac. R. Soc.* (London). cxliii. pp. 407–548; or *Collected Math. Papers*, i. pp. 429–586.]

When dealing in art. 7 with Bezout's condensed eliminant of two equations of the nth degree, Sylvester illustrates by the case of $n=5$, that is to say, where the equations are

$$\left.\begin{array}{l} a_0x^5+a_1x^4+a_2x^3+a_3x^2+a_4x+a_5 = 0 \\ b_0x^5+b_1x^4+b_2x^3+b_3x^2+b_4x+b_5 = 0 \end{array}\right\},$$

pointing out that the eliminant may be constructed by first forming the array

$$\begin{array}{ccccc} |a_0b_1| & |a_0b_2| & |a_0b_3| & |a_0b_4| & |a_0b_5| \\ |a_0b_2| & |a_0b_3| & |a_0b_4| & |a_0b_5| & |a_1b_5| \\ |a_0b_3| & |a_0b_4| & |a_0b_5| & |a_1b_5| & |a_2b_5| \\ |a_0b_4| & |a_0b_5| & |a_1b_5| & |a_2b_5| & |a_3b_5| \\ |a_0b_5| & |a_1b_5| & |a_2b_5| & |a_3b_5| & |a_4b_5| \end{array}$$

and then, as it were, superposing the array

$$\begin{array}{ccc} |a_1b_2| & |a_1b_3| & |a_1b_4| \\ |a_1b_3| & |a_1b_4| & |a_2b_4| \\ |a_1b_4| & |a_2b_4| & |a_3b_4| \end{array}$$

and next the array

$$|a_2b_3|.$$

In regard to these arrays he says in a footnote (p. 424), "A square arrangement having this kind of symmetry, namely, such as obtains in the so-called Pythagorean addition-table as distinguished from that which obtains in the multiplication-table, may be universally called *persymmetric*." This is apparently the first use of the word.

SPOTTISWOODE, W. (1853, August).

[Elementary theorems relating to determinants. Second edition, ... *Crelle's Journ.*, li. pp. 209–271, 328–381.]

Just as Spottiswoode viewed an axisymmetric determinant as the determinant of an n-ary quadric, so he closely associated a persymmetric determinant with an even-ordered binary quantic. Taking, for example, the binary quartic which Cayley would a year later have denoted by $(a_0, a_1, \ldots, a_4 \between x, y)^4$, namely,

$$a_0x^4 + 4a_1x^3y + 6a_2x^2y^2 + 4a_3xy^3 + a_4y^4,$$

Spottiswoode writes it in the form *

$$\begin{aligned} &(a_0x^2+2a_1xy+a_2y^2)x^2 \\ +2&(a_1x^2+2a_2xy+a_3y^2)xy \\ +\;&(a_2x^2+2a_3xy+a_4y^2)y^2; \end{aligned}$$

* A preferable form, because making the "catalecticant" still more prominent, is

$$\begin{array}{ccc|l} x^2 & 2xy & y^2 & \\ \hline a_0 & a_1 & a_2 & x^2 \\ a_1 & a_2 & a_3 & 2xy \\ a_2 & a_3 & a_4 & y^2. \end{array}$$

Similarly an odd-degreed function may be represented so as to bring the "canonizant" into prominence: for example $(a, b, \ldots, f \between x, y)^5$ may be written $(ax+by, bx+cy, \ldots, ex+fy \between x, y)^4$, or

$$\begin{array}{ccc|l} x^2 & 2xy & y^2 & \\ \hline ax+by & bx+cy & cx+dy & x^2 \\ bx+cy & cx+dy & dx+ey & 2xy \\ cx+dy & dx+ey & ex+fy & y^2. \end{array}$$

and calling it U points out that

$$\left.\begin{aligned}\frac{\partial^2 U}{\partial x^2} &= 12(a_0x^2+2a_1xy+a_2y^2)\\ \frac{\partial^2 U}{\partial x\,\partial y} &= 12(a_1x^2+2a_2xy+a_3y^2)\\ \frac{\partial^2 U}{\partial y^2} &= 12(a_2x^2+2a_3xy+a_4y^2)\end{aligned}\right\},$$

and thus like Sylvester concludes that the evanescence of

$$\begin{vmatrix} a_0 & a_1 & a_2 \\ a_1 & a_2 & a_3 \\ a_2 & a_3 & a_4 \end{vmatrix}$$

is the condition that the second differential-quotients of U shall simultaneously vanish, or, say, that we shall have

$$\frac{\partial^2 U}{\partial(x,y)^2} = 0.$$

BRIOSCHI, F. (1854, March).

[LA TEORICA DEI DETERMINANTI, E LE SUE PRINCIPALI APPLICAZIONI. viii+116 pp. Pavia.]

Denoting by s_r the sum of the rth powers of the roots of the equation

$$a_n \quad + a_{n-1}x + \ldots + a_1x^{n-1} + x^n \quad = 0,$$

Brioschi recalls the n known relations

$$\begin{aligned} a_ns_0 &+ a_{n-1}s_1 + \ldots + a_1s_{n-1} + s_n &= 0\\ a_ns_1 &+ a_{n-1}s_2 + \ldots + a_1s_n + s_{n+1} &= 0\\ &\cdots\cdots\cdots\cdots\cdots\cdots\\ a_ns_{n-1} &+ a_{n-1}s_n + \ldots + a_1s_{2n-2} + s_{2n-1} &= 0,\end{aligned}$$

and thus derives by elimination

$$\begin{vmatrix} s_0 & s_1 & \ldots & s_n \\ s_1 & s_2 & \ldots & s_{n+1} \\ \cdot & \cdot & \cdot & \cdot \\ s_{n-1} & s_n & \ldots & s_{2n-1} \\ 1 & x & \ldots & x^n \end{vmatrix} = 0,$$

and therefore

$$\begin{vmatrix} s_1 - s_0 x & s_2 - s_1 x & \dots & s_n - s_{n-1} x \\ s_2 - s_1 x & s_3 - s_2 x & \dots & s_{n+1} - s_n x \\ \cdot & \cdot & \cdot & \cdot \\ s_n - s_{n-1} x & s_{n+1} - s_n x & \dots & s_{2n-1} - s_{2n-2} x \end{vmatrix} = 0.$$

Further, he points out that if the last determinant be denoted by V_n, and the cofactor of its last element by V_{n-1}, and so on, then V_n being axisymmetric it follows from Cauchy's theorem of 1829 that $V_n, V_{n-1}, \dots, V_1$, 1 possess the characteristic property of Sturm's remainders.

It is not noted that the set of n relations used gives each of the a's in terms of the s's, and that substitution in the original equation then gives

$$\begin{vmatrix} s_0 & s_1 & \dots & s_n \\ s_1 & s_2 & \dots & s_{n+1} \\ \cdot & \cdot & \cdot & \cdot \\ s_{n-1} & s_n & \dots & s_{2n-1} \\ 1 & x & \dots & x^n \end{vmatrix} = (x^n + a_1 x^{n-1} + \dots + a_n) \begin{vmatrix} s_0 & s_1 & \dots & s_{n-1} \\ s_1 & s_2 & \dots & s_n \\ \cdot & \cdot & \cdot & \cdot \\ s_{n-1} & s_n & \dots & s_{2n-2} \end{vmatrix},$$

as may be otherwise seen.

BRIOSCHI, F. (1854, February).

[Sur les fonctions de Sturm. *Nouv. Annales de Math.*, xiii. pp. 71–80; or *Opere mat.*, v. pp. 89–97.]

Brioschi in effect here recalls that if $f, f_1, f_2, \dots$ be the series of Sturm's functions originating in the consideration of the equation

$$x^n + a_1 x^{n-1} + a_2 x^{n-2} + \dots + a^n = 0, \text{ or, say, } f(x) = 0,$$

and $q_1, q_2, \dots$ be the linear functions of x which are the quotients obtained in the process of finding $f_2, f_3, \dots$ then

(1) $f = q_1 f_1 - f_2,\ f_1 = q_2 f_2 - f_3,\ \dots,\ f_{r-2} = q_{r-1} f_{r-1} - f_r;$

(2) f_r is of the $(n-r)^{\text{th}}$ degree in x;

(3) from (1) $\dfrac{f_1}{f} = \dfrac{1}{q_1 -}\,\dfrac{1}{q_2 -}\,\dfrac{1}{q_3 -}\, \ldots\, ;$

(4) the successive convergents $\dfrac{1}{q_1}$, $\dfrac{q_2}{q_1 q_2 - 1}$, ... to this continued fraction being N_1/D_1, N_2/D_2, ...

$$N_r = \begin{vmatrix} q_2 & 1 & . & \ldots & . \\ 1 & q_3 & 1 & \ldots & . \\ . & 1 & q_4 & \ldots & . \\ . & . & . & \ldots & . \\ . & . & . & \ldots & q_r \end{vmatrix}, \quad D_r = \begin{vmatrix} q_1 & 1 & . & \ldots & . \\ 1 & q_2 & 1 & \ldots & . \\ . & 1 & q_3 & \ldots & . \\ . & . & . & \ldots & . \\ . & . & . & \ldots & q_r \end{vmatrix}.$$

(5) from (1) after eliminating $f_2, f_3, \ldots, f_{r-1}$ by repeated substitution or otherwise

$$f_r = D_{r-1} f_1 - N_{r-1} f;$$

(6) from (1) after eliminating $f_1, f_2, \ldots, f_{r-2}$

$$f = D_{r-1} f_{r-1} - D_{r-2} f_r;$$

With these facts before him he seeks to find expressions for f_r, D_r, N_r, or, say, for the coefficients in

$$\begin{array}{l} A_{r,1}x^{n-r} + A_{r,2}x^{n-r-1} + \ldots, \\ B_{r,1}x^{r} \quad + B_{r,2}x^{r-1} \quad + \ldots, \\ C_{r,1}x^{r-1} + C_{r,2}x^{r-2} \quad + \ldots, \end{array}$$

failing to note that, Cayley having in 1846 found such an expression for Sylvester's substitute for f_r, the annexure of a known multiplier to Cayley's result would have given him the most important of the three expressions sought.

In the first place he deduces from (4) that the coefficient of the highest power of x in D_{r-1} is always $\dfrac{1}{n}$ of the coefficient of the highest power of x in N_{r-1}, because $q_1 = \dfrac{1}{n}x + \dfrac{a_1}{n^2}$; and from (6), by equating coefficients of x^n, that the coefficient of the highest power of x in f_{r-1} is the reciprocal of the highest power of x in D_{r-1}: in other words, that

$$C_{r,1} = nB_{r,1} = n/A_{r,1}.$$

In the next place, denoting the roots of the given equation by $x_1, x_2, \ldots, x_n$ he has from the theory of "partial fractions"

$$\sum_{i=1}^{i=n}\frac{f_r(x_i)}{f_1(x_i)} = 0, \quad \sum_{i=1}^{i=n}\frac{x_i f_r(x_i)}{f_1(x_i)} = 0, \quad \sum_{i=1}^{i=n}\frac{x_i^2 f_r(x_i)}{f_1(x_i)} = 0,$$

$$\ldots\ldots, \quad \sum_{i=1}^{i=n}\frac{x_i^{r-2} f_r(x_i)}{f_1(x_i)} = 0, \quad \sum_{i=1}^{i=n}\frac{x_i^{r-1} f_r(x_i)}{f_1(x_i)} = A_{r,1};$$

and therefore from (5)

$$\sum_{i=1}^{i=n} D_{r-1}(x_i) = 0, \quad \sum_{i=1}^{i=n} x_i D_{r-1}(x_i) = 0, \quad \sum_{i=1}^{i=n} x_i^2 D_{r-1}(x_i) = 0,$$

$$\ldots\ldots, \quad \sum_{i=1}^{i=n} x_i^{r-2} D_{r-1}(x_i) = 0, \quad \sum_{i=1}^{i=n} x_i^{r-1} D_{r-1}(x_i) = A_{r,1};$$

and consequently on putting

$$B_{r-1,1}x^{r-1} + B_{r-1,2}x^{r-2} + \ldots + B_{r-1,r} \quad \text{for} \quad D_{r-1}(x)$$

$$\text{and} \quad s_m \quad \text{for} \quad x_1^m + x_2^m + \ldots + x_n^m$$

there results

$$\left.\begin{array}{lllll} B_{r-1,r}s_0 & + B_{r-1,r-1}s_1 & + \ldots + B_{r-1,1}s_{r-1} & = 0 \\ B_{r-1,r}s_1 & + B_{r-1,r-1}s_2 & + \ldots + B_{r-1,1}s_r & = 0 \\ \ldots & \ldots & \ldots & \ldots \\ B_{r-1,r}s_{r-2} & + B_{r-1,r-1}s_{r-1} & + \ldots + B_{r-1,1}s_{2r-3} & = 0 \\ B_{r-1,r}s_{r-1} & + B_{r-1,r-1}s_r & + \ldots + B_{r-1,1}s_{2r-2} & = A_{r,1} \end{array}\right\}.$$

The solution of this set of equations gives the B's in terms of $A_{r,1}$ and the s's, so that the finding of $A_{r,1}$ in terms of the s's is the next desideratum. This with the help of the relation $A_{r,1}\,B_{r,1}=1$ is easily obtained; for from the set of equations it is seen that Δ_r being written for the persymmetric determinant of s's

$$B_{r-1,1} = \frac{A_{r,1}\Delta_{r-1}}{\Delta_r}$$

and therefore

$$A_{r,1} = \frac{\Delta_r}{\Delta_{r-1}}\cdot\frac{1}{A_{r-1,1}};$$

whence,

$$\left.\begin{aligned} &\text{for } r \text{ even} \quad \mathrm{A}_{r,1} = \left(\frac{\Delta_2\Delta_4 \ldots \Delta_{r-2}}{\Delta_1\Delta_3 \ldots \Delta_{r-1}}\right)^2 \Delta_r \\ \text{and} \\ &\text{for } r \text{ odd} \quad \mathrm{A}_{r,1} = \left(\frac{\Delta_1\Delta_3 \ldots \Delta_{r-2}}{\Delta_2\Delta_4 \ldots \Delta_{r-1}}\right)^2 \Delta_r \end{aligned}\right\},$$

—a result in agreement, as far as it goes, with Sturm's of 1842, Sturm's non-determinant p_r being the equivalent of Brioschi's Δ_r.

The obtaining of Δ_r in terms of the coefficients of $f(x)$ is next illustrated by changing Δ_4 into the form *

$$\begin{vmatrix} 1 & . & . & . & . & . \\ . & 1 & . & . & . & . \\ . & . & s_0 & s_1 & s_2 & s_4 \\ . & s_0 & s_1 & s_2 & s_3 & s_4 \\ s_0 & s_1 & s_2 & s_3 & s_4 & s_5 \\ s_1 & s_2 & s_3 & s_4 & s_5 & s_6 \end{vmatrix}$$

and performing operations which we may denote by

$$\begin{aligned} &\mathrm{col}_6 + a_1\mathrm{col}_5 + a_2\mathrm{col}_4 + \ldots\ldots + a_5\mathrm{col}_1, \\ &\mathrm{col}_5 + a_1\mathrm{col}_4 + \ldots\ldots + a_4\mathrm{col}_1, \\ &\ldots\ldots\ldots\ldots\ldots\ldots\ldots\ldots \\ &\mathrm{col}_2 + a_1\mathrm{col}_1, \end{aligned}$$

---

* Brioschi (and afterwards Baltzer, § 12, 9) would have done better to change into the similar form of the *seventh* order, for then the result would have been

$$\Delta_4 = \begin{vmatrix} 1 & a_1 & a_2 & a_3 & a_4 & a_5 & a_6 \\ . & 1 & a_1 & a_2 & a_3 & a_4 & a_5 \\ . & . & 1 & a_1 & a_2 & a_3 & a_4 \\ . & . & . & n & (n-1)a_1 & (n-2)a_2 & (n-3)a_3 \\ . & . & n & (n-1)a_1 & (n-2)a_2 & (n-3)a_3 & (n-4)a_4 \\ . & n & (n-1)a_1 & (n-2)a_2 & (n-3)a_3 & (n-4)a_4 & (n-5)a_5 \\ n & (n-1)a_1 & (n-2)a_2 & (n-3)a_3 & (n-4)a_4 & (n-5)a_5 & (n-6)a_6 \end{vmatrix},$$

in which the determinant on the right has a simpler law of formation than Brioschi's and yet is readily reducible to the latter, and which, as we see on putting $n=4$, has the further merit of showing that Δ_n equals the dialytic eliminant of $f(x)=0, f'(x)=0$.

the result being

$$\Delta_4 = \begin{vmatrix} a_1 & 2a_2 & 3a_3 & 4a_4 & 5a_5 & 6a_6 \\ 1 & a_1 & a_2 & a_3 & a_4 & a_5 \\ \cdot & 1 & a_1 & a_2 & a_3 & a_4 \\ \cdot & \cdot & n & (n-1)a_1 & (n-2)a_2 & (n-3)a_3 \\ \cdot & n & (n-1)a_1 & (n-2)a_2 & (n-3)a_3 & (n-4)a_4 \\ n & (n-1)a_1 & (n-2)a_2 & (n-3)a_3 & (n-4)a_4 & (n-5)a_5 \end{vmatrix}.$$

To obtain the desired expression for $D_{r-1}(x)$ Brioschi takes the set of equations in the x's together with the equation from which the set was derived, and eliminates the B's, the result in the case of $r=3$ being

$$\begin{vmatrix} 1 & x & x^2 & -D_2 \\ s_0 & s_1 & s_2 & \cdot \\ s_1 & s_2 & s_3 & \cdot \\ s_2 & s_3 & s_4 & A_{3,1} \end{vmatrix} = 0,$$

whence of course he deduces for D_2 the expression

$$\frac{A_{3,1}}{\Delta_3}\begin{vmatrix} 1 & x & x^2 \\ s_0 & s_1 & s_2 \\ s_1 & s_2 & s_3 \end{vmatrix} \quad \text{or} \quad \left(\frac{\Delta_1}{\Delta_2}\right)^2 \begin{vmatrix} 1 & x & x^2 \\ s_0 & s_1 & s_2 \\ s_1 & s_2 & s_3 \end{vmatrix}$$

and the alternative form

$$\frac{n^2}{\begin{vmatrix} a_1 & 2a_2 \\ n & (n-1)a_1 \end{vmatrix}^2} \cdot \begin{vmatrix} a_1 & 2a_2 & 3a_3 \\ n & (n-1)a_1 & (n-2)a_2 \\ 1 & x+a_1 & x^2+a_1x+a_2 \end{vmatrix}.$$

The process of finding f_r is quite similar to this but much more troublesome, the equation taken along with the set of equations in the s's preparatory for elimination being

$$\frac{f_r(x)}{f(x)} = B_{r-1,r}u_0 + B_{r-1,r-1}u_1 + \ldots + B_{r-1,1}u_{r-1}$$

where $$u_m = \frac{x_1^m}{x-x_1} + \frac{x_2^m}{x-x_2} + \ldots + \frac{x_n^m}{x-x_n}.$$

The two previous steps necessary to reach this are

$$\frac{f_r(x)}{f(x)} = \sum_{i=1}^{i=n} \frac{f_r(x_i)}{(x-x_i)f_1(x_i)} = \sum_{i=1}^{i=n} \frac{D_{r-1}(x_i)}{x-x_i};$$

and the result of the elimination is

$$\frac{f_r(x)}{f(x)} = \frac{A_r}{\Delta_r} \begin{vmatrix} s_0 & s_1 & \dots & s_{r-1} \\ s_1 & s_2 & \dots & s_r \\ \cdot & \cdot & \cdot\cdot\cdot & \cdot \\ s_{r-2} & s_{r-1} & \dots & s_{2r-3} \\ u_0 & u_1 & \dots & u_{r-1} \end{vmatrix}$$

which in the case of $r=4$ is changed by means of the substitutions

$$u_m = xu_{m-1} - s_{m-1}, \quad u_0 = f_1(x) \div f(x),$$

into

$$f_4(x) = \left(\frac{\Delta_2}{\Delta_1\Delta_3}\right)^2 \begin{vmatrix} a_1 & 2a_2 & 3a_3 & 4a_4 & 5a_5 \\ 1 & a_1 & a_2 & a_3 & a_4 \\ \cdot & n & (n-1)a_1 & (n-2)a_2 & (n-3)a_3 \\ n & (n-1)a_1 & (n-2)a_2 & (n-3)a_3 & (n-4)a_4 \\ \cdot & f_1(x) & Z_1 & Z_2 & Z_3 \end{vmatrix}$$

where

$$\begin{aligned} Z_1 &= (x+a_1)f_1(x) - n\,.f(x), \\ Z_2 &= (x^2+a_1x+a_2)f_1(x) - [nx+(n-1)a_1]f(x), \\ Z_3 &= (x^3+a_1x^2+a_2x+a_3)f_1(x) - [nx^2+(n-1)a_1x+(n-2)a_2]f(x). \end{aligned}$$

The worthlessness of this in itself is apparent as soon as we note the presence of $f_1(x)$ and $f(x)$: when, however, the determinant is partitioned into two, one having $f_1(x)$ for a factor and the other $f(x)$, and the result compared with $f_4 = D_3f_1 - N_3f$, we obtain for N_3 an expression similar to that for D_3.

BRIOSCHI, F. (1854, August).

[Intorno ad alcune questioni d' algebra superiore. *Annali di Sci. mat. e fis.*, v. pp. 301–312; or French translation of Brioschi's *Teorica dei Determinanti*, pp. 151–170; or *Opere mat.*, i. pp. 127–142.]

The questions referred to are much the same as those of his paper on Sturm's functions (1854, February), the first function

$$a_0x^n+a_1x^{n-1}+\ldots+a_n \quad \text{or} \quad a_0(x-x_1)(x-x_2)\ldots(x-x_n) \quad \text{or} \quad f(x)$$

being, however, no longer connected with the second

$$b_0x^m+b_1x^{m-1}+\ldots+b_m \quad \text{or} \quad \phi(x) \quad \text{where} \quad m<n.$$

Putting

$$\mathrm{S}_r \quad \text{for} \quad \frac{x_1^r\phi(x_1)}{f'(x_1)}+\frac{x_2^r\phi(x_2)}{f'(x_2)}+\ldots+\frac{x_n^r\phi(x_n)}{f'(x_n)}$$

he first proves the interesting theorem

$$a_0\mathrm{S}_r+a_1\mathrm{S}_{r-1}+\ldots+a_r\mathrm{S}_0 = b_{r+m-n+1} \quad (r<n).$$

Then temporarily denoting

$$\left\{\phi(x_r)\div f'(x_r)\right\}^{\frac{1}{2}} \quad \text{by} \quad \mathrm{A}_r$$

he squares in two ways the determinant

$$\begin{vmatrix} \mathrm{A}_1 & \mathrm{A}_2 & \ldots & \mathrm{A}_n \\ \mathrm{A}_1x_1 & \mathrm{A}_2x_2 & \ldots & \mathrm{A}_nx_n \\ \cdot & \cdot & \cdot & \cdot \\ \mathrm{A}_1x_1^{n-1} & \mathrm{A}_2x_2^{n-1} & \ldots & \mathrm{A}_nx_n^{n-1} \end{vmatrix},$$

thus obtaining

$$\begin{vmatrix} \mathrm{S}_0 & \mathrm{S}_1 & \ldots & \mathrm{S}_{n-1} \\ \mathrm{S}_1 & \mathrm{S}_2 & \ldots & \mathrm{S}_n \\ \cdot & \cdot & \cdot & \cdot \\ \mathrm{S}_{n-1} & \mathrm{S}_n & \ldots & \mathrm{S}_{2n-2} \end{vmatrix} = \mathrm{A}_1^2\mathrm{A}_2^2\ldots\mathrm{A}_n^2 \begin{vmatrix} s_0 & s_1 & \ldots & s_{n-1} \\ s_1 & s_2 & \ldots & s_n \\ \cdot & \cdot & \cdot & \cdot \\ s_{n-1} & s_2 & \ldots & s_{2n-2} \end{vmatrix},$$

from which, on putting $(-1)^{\frac{1}{2}n(n-1)}f'(x_1).f'(x_2)\ldots f'(x_n)$ for the determinant on the right * and substituting for the A's, he deduces

$$\begin{vmatrix} S_0 & S_1 & \dots & S_{n-1} \\ S_1 & S_2 & \dots & S_n \\ \cdot & \cdot & \cdot & \cdot \\ S_{n-1} & S_n & \dots & S_{2n-2} \end{vmatrix} = (-1)^{n(n-1)}\phi(x_1).\phi(x_2)\dots\phi(x_n).$$

This result, be it noted, is not given in the original paper, but appears first in Combescure's translation (1856), which contains six pages (pp. 153–159) more than the original. Brioschi does not point out its significance in connection with Euler's first form of the resultant of $f(x)=0$, $\phi(x)=0$.

The remainder of the paper is of little interest in the present connection.

BRIOSCHI, F. (1855, January).

[Sur les questions 241 et 141. *Nouv. Annales de Math.*, xiv. pp. 20–24; or *Opere mat.*, v. pp. 107–111.]

If for all positive integral values of r and s we have

$$A_{r+s} = a_1 A_{r+s-1} + a_2 A_{r+s-2} + \dots + a_s A_r,$$

—in other words, if this last be a "recurrence-formula,"—it is readily seen that the last column of the persymmetric determinant

$$\begin{vmatrix} A_r & A_{r+1} & \dots & A_{r+s-1} \\ A_{r+1} & A_{r+2} & \dots & A_{r+s} \\ \cdot & \cdot & \cdot & \cdot \\ A_{r+s-1} & A_{r+s} & \dots & A_{r+2s-2} \end{vmatrix}, \quad \text{or } \Delta_{rs} \text{ say,}$$

may be legitimately changed into

$$a_s A_{r-1}, \ a_s A_r, \ \dots, \ a_s A_{r+s-2}$$

so that there is deducible

$$\Delta_{r,s} = (-1)^{s-1} a_s \Delta_{r-1,s},$$

* Brioschi unfortunately neglects the sign-factor. See *History*, i. p. 345, where the footnote might have made mention of the fact that the identity there spoken of as used by Jacobi had already appeared in one of Cauchy's own memoirs of the year 1813. (See *Journ. de l'éc. polyt.*, x. cah. 17, p. 485.)

and thence

$$\Delta_{r,s} = (-1)^{r(s-1)} a_s^r \Delta_{0,s},$$

thus implying that $\Delta_{r,s} / (-1)^{r(s-1)} a_s^r$ is independent of r,—a result suggested to Brioschi by an old proposition of Euler's which is referred to in our chapter on Continuants.

CAYLEY, A. (1856, March).

[A third memoir on quantics. *Philos. Transac. R. Soc.* (London), cxlvi. pp. 627–647; or *Collected Math. Papers*, ii. pp. 310–335.]

Among Cayley's tables of invariants there naturally appear the catalecticants of the binary quartic, sextic, and octavic; so that we have from him the final expansions of the persymmetric determinants of the 3rd, 4th, 5th orders. Of canonizants only that of the quintic is given. The four results are those which he numbers 10, 34, 35, 16. The first and last we need not reproduce. The second is

$$\begin{vmatrix} a & b & c & d \\ b & c & d & e \\ c & d & e & f \\ d & e & f & g \end{vmatrix} = \begin{cases} aceg - acf^2 - ad^2g + 2adef - ae^3 - b^2eg + b^2f^2 \\ +2bcdg - 2bcef - 2bd^2f + 2bde^2 - c^3g + 2c^2df \\ +c^2e^2 - 3cd^2e + d^4, \end{cases}$$

where the terms are arranged, as with Cayley, in alphabetical order. The third, altered in form, is

$$\begin{vmatrix} a & b & c & d & e \\ b & c & d & e & f \\ c & d & e & f & g \\ d & e & f & g & h \\ e & f & g & h & i \end{vmatrix} = \begin{aligned} & e^5 - e^3(ai + 3cg + 4df) \\ & \quad + e^2\{2(afh + bdi) + (ag^2 + c^2i) + 4(bfg + cdh) + 3(cf^2 + d^2g)\} \\ & \quad - e\{(ach^2 + b^2gi) + 2(adgh + bcfi) + 3(af^2g + cd^2i) \\ & \qquad + 4(bdg^2 + c^2fh) + 2(bf^3 + d^3h) - acgi \\ & \qquad - 2adfi - b^2h^2 - 2bcgh - 2c^2g^2 + 2cdfg + 3d^2f^2\} \\ & \quad + \{-(acf^2i + ad^2gi) + 2(acfgh + bcdgi) - (acg^3 + c^3gi) \\ & \qquad + (ad^2h^2 + b^2f^2i) - 2(adf^2h + bd^2fi) + 2(adfg^2 + c^2dfi) \\ & \qquad + (af^4 + d^4i) - 2(b^2fgh + bcdh^2) + (b^2g^3 + c^3h^2) \\ & \qquad + 2(bcf^2h + bd^2gh) - 2(bcfg^2 + c^2dgh) + 2(bdf^2g + cd^2fh) \\ & \qquad + (c^2f^2g + cd^2g^2) - 2(cdf^3 + d^3fg)\}. \end{aligned}$$

It is the term, L, independent of λ in the almost persymmetric determinant which Cayley * on Sylvester's suggestion calls the *lamdaic*, namely,

$$\begin{vmatrix} a & b & c & d & e-12\lambda \\ b & c & d & e+3\lambda & f \\ c & d & e-2\lambda & f & g \\ d & e+3\lambda & f & g & h \\ e-12\lambda & f & g & h & i \end{vmatrix},$$

and which, if I, J, K be the other invariants of the octavic, is equal to

$$-\ 2592\lambda^5 + 18\mathrm{I}\lambda^3 + 3\mathrm{J}\lambda^2 + 2\mathrm{K}\lambda + \mathrm{L}.$$

The expansions of I, J, K are those numbered 39, 43, 44 in Cayley's collection.

SCHEIBNER, W. (1856, May).

[Ueber die Auflösung eines gewissen Gleichungsystems. *Berichte .... Ges. d. Wiss.* (Leipzig): *math.-phys. Cl.*, viii. pp. 65–76.]

This is still another attempt to deal with Sylvester's set of equations of 1851 (October); but any interest which the process of solution possesses is unconnected with determinants.

BRUNO, F. FAÀ DI (1856, August).

[Sopra i resti di Sturm. *Annali di Sci. mat. e fis.*, vii. pp. 313–317.]

Beginning with two unrelated functions, P, Q, of the nth and $(n-1)^{\text{th}}$ degrees, Bruno gives an expression for any one of the Sturmian series of functions thence derived, the coefficients of x

* Cayley, A. Mémoire sur la forme canonique des fonctions binaires. *Crelle's Journ.*, liv. pp. 48–58, 292; or *Collected Math. Papers*, iv. pp. 43–52. If "lamdaic" be not used as a noun, "lamdaic canonizant" would be better than "lamdaic determinant."

in this expression being determinants which resemble in outward form those of Cayley's analogous expression of 1846 (August), but which have for elements the coefficients of $x^{-1}, x^{-2}, \ldots.$ in Q/P. He says that the expression "salvo qualche modificazione" was found by Cauchy, but gives no reference.

BELLAVITIS, G. (1857, June).

[Sposizione elementare della teorica dei determinanti. *Memorie . . . Istituto Veneto* . . . . vii. pp. 67–144.]

Although the persymmetric determinant

$$\begin{vmatrix} s_0 & s_1 & s_2 & \ldots. \\ s_1 & s_2 & s_3 & \ldots. \\ s_2 & s_3 & s_4 & \ldots. \\ \cdot & \cdot & \cdot & \cdot \; \cdot \; \cdot \end{vmatrix}$$

where s_r is the sum of the rth powers of the roots of the equation $x^n + a_1 x^{n-1} + \ldots + a_n = 0$ has repeatedly come before us, it has always been with the understanding that no two of the roots were equal, and the order of the determinant has never been greater than the nth. Bellavitis takes up the subject (§§ 45–50) with these conditions removed. He affirms that *when the number of rows exceeds the number of different roots, the persymmetric determinant*

$$\begin{vmatrix} s_{0+r} & s_{1+r} & s_{2+r} & \ldots. \\ s_{1+r} & s_{2+r} & s_{3+r} & \ldots. \\ s_{2+r} & s_{3+r} & s_{4+r} & \ldots. \\ \cdot & \cdot & \cdot & \cdot \; \cdot \; \cdot \end{vmatrix}$$

vanishes, and when the numbers are the same the determinant is a multiple of the square of the difference-product of the said roots. By way of proof of the second part of the proposition the special case is taken where the roots are a, a, a, b, b, c, and where, since

$$s_0 = 6, \quad s_1 = 3a + 2b + c, \quad s_2 = 3a^2 + 2b^2 + c^2, \quad \ldots\ldots$$

we have

$$\begin{vmatrix} s_5 & s_6 & s_7 \\ s_6 & s_7 & s_8 \\ s_7 & s_8 & s_9 \end{vmatrix} = \begin{vmatrix} 3 & 2 & 1 \\ 3a & 2b & c \\ 3a^2 & 2b^2 & c^2 \end{vmatrix} \cdot \begin{vmatrix} a^5 & b^5 & c^5 \\ a^6 & b^6 & c^6 \\ a^7 & b^7 & c^7 \end{vmatrix},$$

$$= 6\,|\,a^0b^1c^2\,| \cdot a^5b^5c^5\,|\,a^0b^1c^2\,|,$$

$$= 6a^5b^5c^5\,|\,a^0b^1c^2\,|^2.$$

The other part of the proposition rests on the statement that a similar procedure leads in that case to factors having at least one column of zeros.

The case where the number of rows is *less* than the number of different roots is not considered; but the first part of the proposition is used to obtain the modification which it is possible to make in the relation

$$s_{n+r} + a_1 s_{n+r-1} + a_2 s_{n+r-2} + \ldots + a_n s_r = 0$$

when the roots cease to be all different.

ROUCHÉ, E. (1858, Dec.).

[Sur les fonctions X_n de Legendre. *Comptes rendus ... Acad. des Sci.* (Paris), xlvii. pp. 917–921.]

X_n being the nth differential-quotient of $(x^2-1)^n$ Rouché first proves that every rational integral function of x of the nth degree $x^n + a_1x^{n-1} + \ldots + a_{n-1}x + a_n$, or V_n say, which for all integral values of $k < n$ satisfies the equation

$$\int_{-1}^{+1} x^k V_n dx = 0$$

does not differ from X_n save by a constant multiplier. To determine V_n,—in other words, to determine its coefficients $a_1, a_2, \ldots, a_n$,—there is thus available the set of n equations

$$\int_{-1}^{+1} V_n dx = 0, \quad \int_{-1}^{+1} xV_n dx = 0, \quad \ldots, \quad \int_{-1}^{+1} x^{n-1}V_n dx = 0,$$

which if we put i_r for $\frac{1}{2}\int_{-1}^{+1} x^r dx$ take the form

$$\left.\begin{array}{l} a_n i_0 + a_{n-1} i_1 + \dots + a_1 i_{n-1} + i_n = 0 \\ a_n i_1 + a_{n-1} i_2 + \dots + a_1 i_n + i_{n+1} = 0 \\ a_n i_2 + a_{n-1} i_3 + \dots + a_1 i_{n+1} + i_{n+2} = 0 \\ \dots\dots\dots\dots\dots\dots \\ a_n i_{n-1} + a_{n-1} i_n + \dots + a_1 i_{2n-2} + i_{2n-1} = 0 \end{array}\right\}.$$

Solving for $a_1, a_2, a_3, \dots, a_n$ and substituting in V_n Rouché of course obtains

$$V_n = \begin{vmatrix} i_0 & i_1 & i_2 & \dots & i_{n-1} & i_n \\ i_1 & i_2 & i_3 & \dots & i_n & i_{n+1} \\ i_2 & i_3 & i_4 & \dots & i_{n+1} & i_{n+2} \\ \dots & \dots & \dots & \dots & \dots & \dots \\ i_{n-1} & i_n & i_{n+1} & \dots & i_{2n-2} & i_{2n-1} \\ 1 & x & x^2 & \dots & x^{n-1} & x^n \end{vmatrix} \div \varpi,$$

where ϖ stands for the cofactor of x^n in the determinant, and being independent of x may be left out of account; and, since i_r is $1/(r+1)$ or 0 according as r is even or odd, it follows that $X_1, X_2, X_3, \dots$ are proportional to

$$\begin{vmatrix} 1 & 0 \\ 1 & x \end{vmatrix}, \quad \begin{vmatrix} 1 & 0 & \frac{1}{3} \\ 0 & \frac{1}{3} & 0 \\ 1 & x & x^2 \end{vmatrix}, \quad \begin{vmatrix} 1 & 0 & \frac{1}{2} & 0 \\ 0 & \frac{1}{3} & 0 & \frac{1}{5} \\ \frac{1}{3} & 0 & \frac{1}{5} & 0 \\ 1 & x & x^2 & x^3 \end{vmatrix}, \quad \dots$$

respectively.

By reason of the peculiar distribution of the zero-elements these determinants are changeable into

$$x, \quad \frac{1}{3}\begin{vmatrix} 1 & \frac{1}{3} \\ 1 & x \end{vmatrix}, \quad \begin{vmatrix} 1 & \frac{1}{3} \\ \frac{1}{3} & \frac{1}{5} \end{vmatrix} \cdot \begin{vmatrix} \frac{1}{3} & \frac{1}{5} \\ x & x^3 \end{vmatrix}, \quad \begin{vmatrix} \frac{1}{3} & \frac{1}{5} \\ \frac{1}{5} & \frac{1}{7} \end{vmatrix} \cdot \begin{vmatrix} 1 & \frac{1}{3} & \frac{1}{5} \\ \frac{1}{3} & \frac{1}{5} & \frac{1}{7} \\ 1 & x^2 & x^4 \end{vmatrix}, \quad \dots$$

and the result may consequently be put in an alternative form, namely,

$$X_{2n} = \begin{vmatrix} 1 & \frac{1}{3} & \frac{1}{5} & \cdots & \frac{1}{2n+1} \\ \frac{1}{3} & \frac{1}{5} & \frac{1}{7} & \cdots & \frac{1}{2n+3} \\ \frac{1}{5} & \frac{1}{7} & \frac{1}{9} & \cdots & \frac{1}{2n+5} \\ \cdot & \cdot & \cdot & \cdot & \cdot \\ \frac{1}{2n-1} & \frac{1}{2n+1} & \frac{1}{2n+3} & \cdots & \frac{1}{4n-1} \\ 1 & x^2 & x^4 & \cdots & x^{2n} \end{vmatrix} \times \text{ a constant,}$$

and

$$X_{2n+1} = \begin{vmatrix} \frac{1}{3} & \frac{1}{5} & \frac{1}{7} & \cdots & \frac{1}{2n+3} \\ \frac{1}{5} & \frac{1}{7} & \frac{1}{9} & \cdots & \frac{1}{2n+5} \\ \frac{1}{7} & \frac{1}{9} & \frac{1}{11} & \cdots & \frac{1}{2n+7} \\ \cdot & \cdot & \cdot & \cdot & \cdot \\ \frac{1}{2n+1} & \frac{1}{2n+3} & \frac{1}{2n+5} & \cdots & \frac{1}{4n+1} \\ x & x^3 & x^5 & \cdots & x^{2n+1} \end{vmatrix} \times \text{ a constant.}$$

Lastly it is pointed out that by throwing X_n into the form

$$\begin{vmatrix} i_1 - i_0 x & i_2 - i_1 x & \cdots & i_n - i_{n-1} x \\ i_2 - i_1 x & i_3 - i_2 x & \cdots & i_{n+1} - i_n x \\ \cdot & \cdot & \cdot & \cdot \\ i_n - i_{n-1} x & i_{n+1} - i_n x & \cdots & i_{2n-1} - i_{2n-2} x \end{vmatrix} \times \text{ a constant}$$

it is possible in the manner of Cauchy (1829) and Sylvester (1853) to draw conclusions regarding the situation and character of the roots of the equation $X_n = 0$.

SALMON, G. (1859).

[LESSONS INTRODUCTORY TO THE MODERN HIGHER ALGEBRA, ... xii+147 pp., Dublin.]

In Salmon's treatment of the foregoing subjects (p. 14, §§ 119–126, 162–165, p. 146) there are several points of freshness. His proof, for example, that *if*

$$(a,b,c,d,e,f\between x,y)^5 = (l_1x+m_1y)^5 + (l_2x+m_2y)^5 + (l_3x+m_3y)^5,$$

the persymmetric form of the canonizant is equal to

$$|l_1m_2|^2\,|l_1m_3|^2\,|l_2m_3|^2\cdot(l_1x+m_1y)(l_2x+m_2y)(l_3x+m_3y),$$

is accomplished (§ 119) by using four differentiations to show that

$$\begin{aligned} ax+by &= l_1^4(l_1x+m_1y) + l_2^4(l_2x+m_2y) + l_3^4(l_3x+m_3y) \\ &= l_1^4u + l_2^4v + l_3^4w, \text{ say}, \\ bx+cy &= l_1^3m_1u + l_2^3m_2v + l_3^3m_3w, \\ &\cdots\cdots\cdots \end{aligned}$$

substituting these trinomials for $ax+by$, $bx+cy$, . . . in the canonizant, and then examining * the twenty-seven determinants into which the latter can be partitioned. He also gives (§ 121) a new mode of arriving at the canonizant in the form

$$\left(\begin{Vmatrix} a & b & c & d \\ b & c & d & e \\ c & d & e & f \end{Vmatrix}\between x,y\right)^3.$$

In the course of a short "Note on Commutants" he suggests (p. 146) that the two rows of umbræ

$$\begin{pmatrix} 0 & 1 & 2 \\ 0 & 1 & 2 \end{pmatrix} \text{ should stand for } \begin{vmatrix} a_0 & a_1 & a_2 \\ a_1 & a_2 & a_3 \\ a_2 & a_3 & a_4 \end{vmatrix},$$

in which case the three suffixes of any term of the developed determinant are got by *adding* the first row of umbræ to a permutation of the second row.

* It is better to note at this stage that the determinant is the product of

$$\begin{vmatrix} l_1^2u & l_2^2v & l_3^2w \\ l_1m_1u & l_2m_2v & l_3m_3w \\ m_1^2u & m_2^2v & m_3^2w \end{vmatrix} \text{ and } \begin{vmatrix} l_1^2 & l_2^2 & l_3^2 \\ l_1m_1 & l_2m_2 & l_3m_3 \\ m_1^2 & m_2^2 & m_3^2 \end{vmatrix}$$

and therefore is equal to $|l_1^2\ l_2m_2\ m_3^2|^2\,.\,uvw$.

CHAPTER XII.

BIGRADIENTS, UP TO 1860.

As we have already pointed out (*History*, i. p. 487), bigradients were first brought to light by Sylvester in 1840 in the paper in which he made known his so-called "dialytic" method of eliminating the unknown from two equations of the same or different degrees. Shortly afterwards Richelot and Cauchy recalled attention to Euler's and Bezout's method of 1764, as giving substantially the same result as Sylvester's, the fact being that the determinant obtained by Sylvester differs from that obtainable in the other case merely by being its conjugate. The details of these papers and of others related to them have already been given.

CAYLEY, A. (1844).

[Note sur deux formules données par MM. Eisenstein et Hesse. *Crelle's Journ.*, xxix. pp. 54–57; or *Collected Math. Papers*, i. pp. 113–116.]

Although Eisenstein's property * of the discriminant

$$a^2d^2 - 3b^2c^2 + 4ac^3 + 4b^3d - 6abcd, \quad \text{or } \Delta \text{ say,}$$

of the binary cubic $ax^3 + 3bx^2y + 3cxy^2 + dy^3$,—namely, the property that

$$A^2D^2 - 3B^2C^2 + 4AC^3 + 4B^3D - 6ABCD$$
$$= (a^2d^2 - 3b^2c^2 + 4ac^3 + 4b^3d - 6abcd)^3$$

* *Crelle's Journ.*, xxvii. pp. 105-106, 319–321.

when

$$\mathrm{A,\ B,\ C,\ D} = -\frac{1}{2}\frac{\partial\Delta}{\partial d},\quad \frac{1}{6}\frac{\partial\Delta}{\partial c},\quad -\frac{1}{6}\frac{\partial\Delta}{\partial b},\quad \frac{1}{2}\frac{\partial\Delta}{\partial a},$$

—can be expressed in the form

$$\begin{vmatrix} \mathrm{A} & 2\mathrm{B} & \mathrm{C} & \cdot \\ \cdot & \mathrm{A} & 2\mathrm{B} & \mathrm{C} \\ \mathrm{B} & 2\mathrm{C} & \mathrm{D} & \cdot \\ \cdot & \mathrm{B} & 2\mathrm{C} & \mathrm{D} \end{vmatrix} = \begin{vmatrix} a & 2b & c & \cdot \\ \cdot & a & 2b & c \\ b & 2c & d & \cdot \\ \cdot & b & 2c & d \end{vmatrix}^3,$$

it is not as a relation between two four-line determinants that it has been studied.

Cayley in effect says that if we wish to find substitutes A, B, C, . . . for $a, b, c, \ldots$ so that

$$\phi(\mathrm{A, B, C}, \ldots) = \{\phi(a, b, c, \ldots)\}^p,$$

we must (1) find a quantic u of which $\phi(a, b, c, \ldots)$ is an invariant; (2) express $\phi(a, b, c, \ldots)$ as a determinant of the same number of lines as u has facients; and (3) transform u into U by a linear substitution of which the said determinant is the modulus. The coefficients of U will then be the substitutes required.

For example, Δ being an invariant of the binary cubic and being expressible in the form

$$\begin{vmatrix} bc - ad & 2(c^2-bd) \\ 2(b^2-ac) & bc - ad \end{vmatrix},$$

we should have to transform the said cubic by the substitution

$$\left.\begin{aligned} x &= (bc-ad)\xi + 2(c^2-bd)\eta \\ y &= 2(b^2-ac)\xi + (bc-ad)\eta \end{aligned}\right\},$$

and the discriminant of the new cubic thus obtained being Δ multiplied by a power of the modulus must be a power of Δ. Unfortunately, in this case it would be Δ^7, whereas in Eisenstein's case the power-index is 3. Instead of the binary cubic, therefore, Cayley takes the binary trilinear

$$ax_1y_1z_1 + bx_1y_1z_2 + cx_1y_2z_1 + dx_1y_2z_2 + ex_2y_1z_1 + fx_2y_1z_2 + gx_2y_2z_1 + hx_2y_2z_2,$$

of which a generalisation of Δ, namely,

$$\left.\begin{array}{l} a^2h^2 + b^2g^2 + c^2f^2 + d^2e^2 + 4adfg + 4bceh \\ -\ 2ahbg - 2ahcf - 2ahde - 2bgcf - 2bgde - 2cfde \end{array}\right\} \text{ or Q say,}$$

is an invariant; expresses Q as a two-line determinant

$$\begin{vmatrix} ah - bg - cf + de & -2(eh-fg) \\ -2(ad-bc) & ah - bg - cf + de \end{vmatrix}$$

makes the substitution

$$\left.\begin{array}{l} x_1 = (ah-bg-cf+de)\xi_1 - 2(eh-fg)\xi_2 \\ y_1 = \qquad -2(ad-bc)\xi_1 + (ah-bg-cf+de)\xi_2 \end{array}\right\};$$

and, as the multiplier connecting the new Q and the old is now the *second* power of the modulus, he obtains what was wanted.*

The substitutes found turn out to be

$$\frac{1}{2}\frac{\partial Q}{\partial a}, \quad \frac{1}{2}\frac{\partial Q}{\partial b}, \quad \frac{1}{2}\frac{\partial Q}{\partial c}, \quad \ldots .$$

but no explanation of this is vouchsafed.

Eisenstein's case is the degeneration reached by putting

$$\begin{array}{l} \quad a,\ b,\ c,\ d,\ e,\ f,\ g,\ h \\ = a,\ b,\ b,\ c,\ b,\ c,\ c,\ d. \end{array}$$

Had the two other sets of variables been at the same time transformed with the same determinant for modulus, we should have had the new Q equal to Q^7.

BOOLE, G. (1844, June).

[Notes on linear transformation. *Cambridge Math. Journ.*, iv. pp. 167–171.]

The third note being devoted to the proof of his well-known theorem of 1841 regarding what afterwards came to be known

---

* This short paper of Cayley's teems with misprints, both in the original and in the *Collected Math. Papers.*

as 'the invariance of the discriminant,' the fourth gives at full length and correctly the discriminant of

$$ax^4+4bx^3y+6cx^2y^2+4dxy^3+ey^4,$$

—that is to say, the eliminant of

$$\left.\begin{array}{l} ax^3 + 3bx^2y + 3cxy^2 + dy^3 = 0 \\ bx^3 + 3cx^2y + 3dxy^2 + ey^3 = 0 \end{array}\right\}.$$

HEILERMANN, [H.] (1845).

[Ueber die Verwandlung der Reihen in Kettenbrüche. *Crelle's Journ.*, xxxiii. pp. 174–188.]

The determinant which here appears for the first time is different from but resembles those to which Sylvester's dialytic method of elimination leads, being exemplified for the 4th and 5th orders by

$$\begin{vmatrix} a_3 & a_2 & b_2 & b_3 \\ a_2 & a_1 & b_1 & b_2 \\ a_1 & a_0 & b_0 & b_1 \\ a_0 & \cdot & \cdot & b_0 \end{vmatrix}, \quad \begin{vmatrix} a_4 & a_3 & a_2 & b_3 & b_4 \\ a_3 & a_2 & a_1 & b_2 & b_3 \\ a_2 & a_1 & a_0 & b_1 & b_2 \\ a_1 & a_0 & \cdot & b_0 & b_1 \\ a_0 & \cdot & \cdot & \cdot & b_0 \end{vmatrix}.$$

Calling these Δ_3, Δ_4 Heilermann writes his main result in the form

$$\frac{a_0 + a_1x + \ldots + a_nx^n}{b_0 + b_1x + \ldots + b_mx^m} = \frac{\Delta_0}{b_0} - \frac{\Delta_1 x}{\Delta_0} - \frac{\Delta_2 x}{\Delta_1} - \frac{\Delta_0\Delta_3 x}{\Delta_2} - \frac{\Delta_1\Delta_4 x}{\Delta_3 - \ldots}$$

the end on the right being

$$-\frac{\Delta_{2n-3}\Delta_{2n}x}{\Delta_{2n-1}} \quad \text{or} \quad -\frac{\Delta_{2m-4}\Delta_{2m-1}}{\Delta_{2m-2}}$$

according as $m <$ or $> n$.

CAYLEY, A. (1848, August).

[Nouvelles recherches sur les fonctions de M. Sturm. *Journ.* (*de Liouville*) *de Math.*, xiii. pp. 269–274; or *Collected Math. Papers*, i. pp. 392–396.]

Recalling his former paper on the same subject in *Liouville's Journal*, xi. (1846), pp. 297–299, where Sturm's functions had been expressed in terms of sums of powers of the roots of the original function, he intimates now the discovery of more simple expressions in terms of the *coefficients* of the said function. At the same time he draws attention to the fact that his result may be viewed as unconnected with Sturm's division-process, and it is in this general light that he prefers to state it. Beginning with two functions V and V′ of the nth degree, namely,

$$ax^n + bx^{n-1} + \ldots \qquad a'x^n + b'x^{n-1} + \ldots$$

and forming therefrom the series of functions

$$\begin{vmatrix} \mathrm{V} & \mathrm{V}' \\ a & a' \end{vmatrix}, \quad \begin{vmatrix} x\mathrm{V} & \mathrm{V} & x\mathrm{V}' & \mathrm{V}' \\ a & . & a' & . \\ b & a & b' & a' \\ c & b & c' & b' \end{vmatrix}, \quad \begin{vmatrix} x^2\mathrm{V} & x\mathrm{V} & \mathrm{V} & x^2\mathrm{V}' & x\mathrm{V}' & \mathrm{V}' \\ a & . & . & a' & . & . \\ b & a & . & b' & a' & . \\ c & b & a & c' & b' & a' \\ d & c & b & d' & c' & b' \\ e & d & c & e' & d' & c' \end{vmatrix},$$

etc., which he denotes by $-\mathrm{F}_1, \mathrm{F}_2, -\mathrm{F}_3, \ldots$, he affirms that there is a homogeneous linear relation connecting every consecutive three of the latter functions,* namely,

$$\mathrm{P}_1^2\mathrm{F}_3 + (x\mathrm{P}_1\mathrm{P}_2 + \mathrm{P}_1\mathrm{P}_2' + \mathrm{P}_1'\mathrm{P}_2)\mathrm{F}_2 + \mathrm{P}_2^2\mathrm{F}_1 = 0,$$
$$\mathrm{P}_2^2\mathrm{F}_4 + (x\mathrm{P}_2\mathrm{P}_3 + \mathrm{P}_2\mathrm{P}_3' + \mathrm{P}_2'\mathrm{P}_3)\mathrm{F}_3 + \mathrm{P}_3^2\mathrm{F}_2 = 0,$$
$$\cdots\cdots\cdots\cdots$$

* The signs require verification.

where by $P_1, P_2, P_3, \ldots.$ are meant the determinants

$$\begin{vmatrix} a & a' \\ b & b' \end{vmatrix}, \quad \begin{vmatrix} a & . & a' & . \\ b & a & b' & a' \\ c & b & c' & b' \\ d & c & d' & c' \end{vmatrix}, \quad \begin{vmatrix} a & . & . & a' & . & . \\ b & a & . & b' & a' & . \\ c & b & a & c' & b' & a' \\ d & c & b & d' & c' & b' \\ e & d & c & e' & d' & c' \\ f & e & d & f' & e' & d' \end{vmatrix}, \ldots.$$

and by $P'_1, P'_2, P'_3, \ldots.$ the determinants got from $P_1, P_2, P_3, \ldots.$ by altering the last rows into

$$c\ c'; \quad e\ d\ e'\ d'; \quad g\ f\ e\ g'\ f'\ e'; \quad \ldots.$$

No proof is given of the relations; indeed, after pointing out that they involve the proposition that the first and last of three consecutive functions are of opposite sign for every value of x that makes the intermediate function vanish, Cayley adds: "Je n'ai pas encore réussi à démontrer dans toute la généralité l'équation identique d'où dépend cette propriété."

The case which brings him into closer contact with Sturm, namely, where V' is the differential-quotient of V, is dealt with in some detail.

HEILERMANN, [H.] (1852, December).

[Independente Berechnung der Sturm'schen Reste. *Crelle's Journ.*, xlviii. pp. 190–206.]

The subject of this paper is of course closely connected with that of the author's previous work (1845). Like Cayley, he begins with two functions that are unrelated, and subsequently passes to the special case where the one is the derivate of the other; but, unlike Cayley, he makes the said functions of different degrees, namely,

$$c_{00}x^n \ + c_{10}x^{n-1} + c_{20}x^{n-2} + \ \ldots. \ + c_{n0},$$
$$c_{01}x^{n-1} + c_{11}x^{n-2} + c_{21}x^{n-3} + \ \ldots. \ + c_{n-1,1}.$$

Following the ordinary division-process for expressing the ratio of the second function to the first as a continued fraction of the form

$$\frac{p_0}{x+}\frac{p_1}{1+}\frac{p_2}{x+}\frac{p_3}{1+}\cdots$$

and denoting the remainders in order by

$$\frac{1}{c_{01}}\left\{c_{02}x^{n-1}+c_{12}x^{n-2}+c_{22}x^{n-3}+\ldots\right\},$$

$$\frac{1}{c_{02}}\left\{c_{03}x^{n-2}+c_{13}x^{n-3}+c_{23}x^{n-4}+\ldots\right\},$$

$$\frac{1}{c_{01}c_{03}}\left\{c_{04}x^{n-3}+c_{14}x^{n-4}+c_{24}x^{n-5}+\ldots\right\},$$

$$\cdots\cdots\cdots\cdots$$

he finds that

$$p_0=\frac{c_{01}}{c_{00}},\quad p_{r+1}=\frac{c_{0,r+2}}{c_{0,r}c_{0,r+1}},\quad p_{2n-1}=\frac{c_{1,2n-2}}{c_{0,2n-2}}.$$

The second suffix of any one of the new c's is seen to indicate the remainder-function to which the c belongs, and the first suffix the position of the c in that remainder. To obtain expressions for these in terms of the original two sets of c's it is taken for granted, and with reason, that as a result of the process we have generally

$$c_{r,s}=c_{0,s-1}c_{r+1,s-2}-c_{0,s-2}c_{r+1,s-1}=\begin{vmatrix} c_{0,s-1} & c_{0,s-2} \\ c_{r+1,s-1} & c_{r+1,s-2} \end{vmatrix}. \tag{1}$$

By using this twice upon itself, so as to lower the second suffixes of the first column, there is found

$$c_{r,s}=\begin{vmatrix} c_{0,s-2} & c_{0,s-3} & \cdot \\ c_{1,s-2} & c_{1,s-3} & c_{0,s-2} \\ c_{r+2,s-2} & c_{r+2,s-3} & c_{r+1,s-2} \end{vmatrix}, \tag{2}$$

where the second suffixes are now $s-2$, $s-3$. A page is then occupied in ridding (2) in the same way of the elements which have $s-3$ for a suffix, the result being

$$c_{r,s} = c_{0,s-3} \begin{vmatrix} c_{0,s-3} & c_{0,s-4} & \cdot & \cdot \\ c_{1,s-3} & c_{1,s-4} & c_{0,s-3} & c_{0,s-4} \\ c_{2,s-3} & c_{2,s-4} & c_{1,s-3} & c_{1,s-4} \\ c_{r+3,s-3} & c_{r+3,s-4} & c_{r+2,s-3} & c_{r+2,s-4} \end{vmatrix} \qquad (3)$$

With increasing tediousness a five-line determinant is reached having elements with $s-4$ and $s-5$ for second suffixes, a six-line determinant having elements with $s-5$ and $s-6$ for second suffixes, and so on. The form of the determinant of the $(2q+2)^{\text{th}}$ order is thus deduced, the factor preceding it being said to be

$$(c_{0,s-3}c_{0,s-4})(c_{0,s-5}c_{0,s-6})^2(c_{0,s-7}c_{0,s-8})^3 \ldots (c_{0,s-2q+1}c_{0,s-2q})^{q-1}(c_{0,s-2q-1})^q.$$

Sturm's division-process, in which each remainder is of a lower degree than the remainder preceding it, and for which, therefore, the corresponding continued fraction has each partial denominator a linear function of x, has close relationship with the above division-process, because the continued fraction already obtained is identical with*

$$\cfrac{p_0}{x+p_1} - \cfrac{p_1p_2}{x+p_2+p_3} - \cfrac{p_3p_4}{x+p_4+p_5} - \cdots$$

Expressions for the remainders corresponding to this continued fraction are thus readily obtainable, and a section (§ 3) is devoted to finding simplified substitutes for them. The remaining section concerns the strictly Sturmian case, where one of the original functions is the derivate of the other.

BRUNO, F. FAÀ DI (1855, July).

[Sulle funzioni simmetriche delle radici di un' equazione. *Annali di Sci. mat. e fis.*, vi. pp. 412–419.]

As evidence of the value of a certain theorem Bruno adduces the ease with which the expansion of the resultant of a pair of

* This identity Heilermann published again separately in 1860 (see *Zeitschrift f. Math. u. Phys.*, v. pp. 262–263). It is included, however, in a result given by Stern in 1833 (see *Crelle's Journ.*, x. p. 156); and a still more general identity will be found in the *Proceed. Edinburgh Math. Soc.*, xxiii. p. 37.

equations may be calculated, and he prints at full length the resultants $R_{2,2}$, $R_{3,3}$, $R_{4,4}$, that is to say, the final expansions of

$$\begin{vmatrix} \cdot & a & b & c \\ a & b & c & \cdot \\ \cdot & p & q & r \\ p & q & r & \cdot \end{vmatrix}, \text{ etc.,}$$

the arrangement of the terms being such as to make evident the fact that each resultant is unaltered by reversing the order of the two sets of coefficients of which it is a function: for example:—*

$$R_{2,2} = (a^2r^2+c^2p^2) - (abqr+bcpq) - 2acpr + acq^2 + b^2pr.$$

CAYLEY, A. (1856).

[A second memoir on quantics. *Philos. Transac. R. Soc.* (London), cxlvi. pp. 101–126; or *Collected Math. Papers*, ii. pp. 250–275.]

In addition to previously calculated discriminants Cayley gives (No. 26) the discriminant of the binary quintic,—that is to say, the eliminant of

$$\left.\begin{aligned} ax^4 + 4bx^3y + 6cx^2y^2 + 4dxy^3 + ey^4 &= 0 \\ bx^4 + 4cx^3y + 6dx^2y^2 + 4exy^3 + fy^4 &= 0 \end{aligned}\right\}\dagger$$

CAYLEY, A. (1856, December).

[Memoir on the resultant of a system of two equations. *Philos. Transac. R. Soc.* (London), cxlvii. pp. 703–715; or *Collected Math. Papers*, ii. pp. 440–453.]

As the resultant, $R_{3,2}$ say, of the pair of equations

$$ax^3+bx^2y+cxy^2+dy^3 = 0, \quad px^2+qxy+ry^2 = 0,$$

is homogeneous and of the 3rd degree in the coefficients of the second equation, and at the same time homogeneous and of the 2nd degree in the coefficients of the first equation, Cayley

* The expression for $R_{4,4}$ is full of inaccuracies.

† In the reprint, p. 288, the coefficient of b^4cef^2 should be -1920, not $+1920$.

seeks a convenient form of representation in which this double homogeneity will be prominent. What he obtains is *

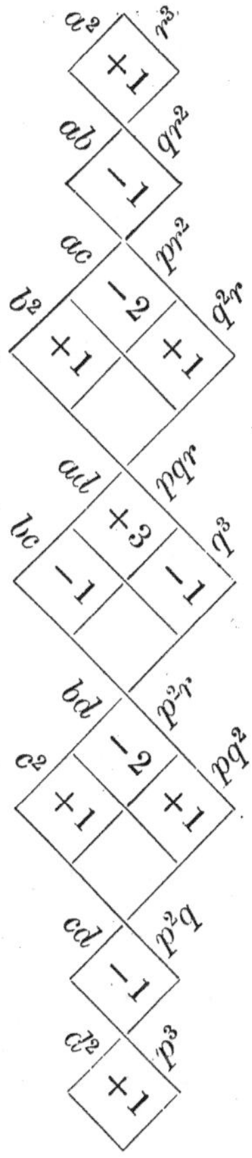

where the first square represents a^2r^3, the second $-abqr^2$, the third $-2acpr^2+b^2pr^2+acq^2r$, and so on. The result is reached in two ways, the second being by developing the dialytic eliminant

$$\begin{vmatrix} \cdot & a & b & c & d \\ a & b & c & d & \cdot \\ \cdot & \cdot & p & q & r \\ \cdot & p & q & r & \cdot \\ p & q & r & \cdot & \cdot \end{vmatrix}.$$

* Three misprints being corrected.

The two-line minors of the first two rows of this determinant being denoted by 12, 13, . . . , 45, and the three-line minors of the remaining rows by 123, 124, . . . , 345, it is seen that

$$R_{3,2} = \left. \begin{array}{l} 12\cdot345 - 13\cdot245 + 14\cdot235 - 15\cdot234 \\ \qquad\qquad + 23\cdot145 - 24\cdot135 + 25\cdot134 \\ \qquad\qquad\qquad\qquad + 34\cdot125 - 35\cdot124 \\ \qquad\qquad\qquad\qquad\qquad\qquad - 45\cdot123 \end{array} \right\} ;$$

and it will be found that

$$12\cdot345, \quad -13\cdot245, \quad 14\cdot235 + 23\cdot145, \quad -(15\cdot234 + 24\cdot135), \quad \ldots$$

correspond to the 1st, 2nd, 3rd, 4th, . . . squares of Cayley's expression.

The paper closes with the six resultants $R_{2,2}$, $R_{3,2}$, $R_{4,2}$, $R_{3,3}$, $R_{4,3}$, $R_{4,4}$ printed each in the new form as a chain of squares;* they occupy four quarto pages.

ZEIPEL, V. v. (1858, June).

[Demonstration of a theorem of Mr. Cayley's in relation to Sturm's functions. *Quart. Journ. of Math.*, iii. pp. 108–117, or *Nouv. Annales de Math.*, xix. pp. 220–224.]

The theorem referred to is that of August 1848. Zeipel's proof (pp. 108–114) is lengthy and unattractive, and scarcely warrants reproduction. The remaining pages are occupied with the curious identities

$$\begin{vmatrix} P_{r-1} & P_r & . \\ P'_{r-1} & P'_r & P_r \\ P''_{r-1} & P''_r & P'_r \end{vmatrix} = -P^2_{r-1}P_{r+1}, \qquad \begin{vmatrix} P_{r-1} & P_r & . \\ P'_{r-1} & P'_r & P_r \\ P'''_{r-1} & P'''_r & P''_r \end{vmatrix} = -P^2_{r-1}P'_{r+1},$$

and the corresponding identities in which the left-hand members are of a higher odd order than the third.

* In the expression for $R_{4,4}$ there is at least one misprint, namely, a^2c^2 for a^2b^2 outside the third square of the chain.

HESSE, O. (1858, October).

[Il determinante di Sylvester ed il risultante di Eulero. *Annali di Mat.* . . . ii. pp. 5–8; or *Werke*, 475–480.]

The determinant referred to is that to which Sylvester was led in 1840 by his so-called "dialytic" method, and which, as we have already seen, Hesse himself arrived at in 1843; and the resultant coupled with it is Euler's of 1748, which takes the form of a product of differences of the roots of the two given equations. Both forms, as well as others, are treated of by Cauchy in his paper of 1840, already dealt with.

The equations being

$$\left.\begin{aligned} a_3x^3 + a_2x^2 + a_1x^1 + a_0 &= 0 \\ b_2x^2 + b_1x^1 + b_0 &= 0 \end{aligned}\right\} \text{ or say } \left\{\begin{aligned} \phi(x) &= 0 \\ \psi(x) &= 0, \end{aligned}\right.$$

Hesse multiplies Sylvester's eliminant by $s_0s_2 - s_1^2$, the squared difference-product of the roots β_1, β_2 of the second equation, with the result

$$\begin{vmatrix} a_0 & a_1 & a_2 & a_3 & \cdot \\ \cdot & a_0 & a_1 & a_2 & a_3 \\ b_0 & b_1 & b_2 & \cdot & \cdot \\ \cdot & b_0 & b_1 & b_2 & \cdot \\ \cdot & \cdot & b_0 & b_1 & b_2 \end{vmatrix} \cdot \begin{vmatrix} s_0 & s_1 & s_2 & s_3 & s_4 \\ s_1 & s_2 & s_3 & s_4 & s_5 \\ \cdot & \cdot & 1 & \cdot & \cdot \\ \cdot & \cdot & \cdot & 1 & \cdot \\ \cdot & \cdot & \cdot & \cdot & 1 \end{vmatrix}$$

$$= \begin{vmatrix} a_0s_0+a_1s_1+\ldots+a_3s_3 & a_0s_1+a_1s_2+\ldots+a_3s_4 & a_2 & a_3 & \cdot \\ a_0s_1+a_1s_2+\ldots+a_3s_4 & a_0s_2+a_1s_3+\ldots+a_3s_5 & a_1 & a_2 & a_3 \\ b_0s_0+b_1s_1+b_2s_2 & b_0s_1+b_1s_2+b_2s_3 & b_2 & \cdot & \cdot \\ b_0s_1+b_1s_2+b_2s_3 & b_0s_2+b_1s_3+b_2s_4 & b_1 & b_2 & \cdot \\ b_0s_2+b_1s_3+b_2s_4 & b_0s_3+b_1s_4+b_2s_5 & b_0 & b_1 & b_2 \end{vmatrix},$$

where $s_p = \beta_1^p + \beta_2^p$, and where, therefore, the elements in the places 31, 32, 41, 42, 51, 52 of the right-hand member all vanish. There is thus obtained, if we denote Sylvester's eliminant by S,

$$S \cdot \begin{vmatrix} s_0 & s_1 \\ s_1 & s_2 \end{vmatrix} = b_2^3 \begin{vmatrix} a_0s_0+a_1s_1+\ldots+a_3s_3 & a_0s_1+a_1s_2+\ldots+a_3s_4 \\ a_0s_1+a_1s_2+\ldots+a_3s_4 & a_0s_2+a_1s_3+\ldots+a_3s_5 \end{vmatrix}.$$

But the determinant on the right is resolvable into

$$\begin{vmatrix} 1 & 1 \\ \beta_1 & \beta_2 \end{vmatrix} \cdot \begin{vmatrix} a_0+a_1\beta_1+a_2\beta_1^2+a_3\beta_1^3 & a_0+a_1\beta_2+a_2\beta_2^2+a_3\beta_2^3 \\ a_0\beta_1+a_1\beta_1^2+a_2\beta_1^3+a_3\beta_1^4 & a_0\beta_2+a_1\beta_2^2+a_2\beta_2^3+a_3\beta_2^4 \end{vmatrix},$$

that is, into

$$\begin{vmatrix} 1 & 1 \\ \beta_1 & \beta_2 \end{vmatrix} \cdot \begin{vmatrix} \phi(\beta_1) & \phi(\beta_2) \\ \beta_1\phi(\beta_1) & \beta_2\phi(\beta_2) \end{vmatrix}$$

and therefore into

$$\begin{vmatrix} 1 & 1 \\ \beta_1 & \beta_2 \end{vmatrix}^2 \cdot \phi(\beta_1) \cdot \phi(\beta_2).$$

We thus have

$$S = b_2^3 \phi(\beta_1)\phi(\beta_2),$$

and, consequently, if a_1, a_2, a_3 be the roots of the first given equation,

$$S = b_2^3 a_3^2 (\beta_1 - a_1)(\beta_1 - a_2)(\beta_1 - a_3)$$
$$(\beta_2 - a_1)(\beta_2 - a_2)(\beta_2 - a_3),$$

which is what was to be shown, the cofactor of $b_2^3 a_3^2$ being Euler's product of differences.

ZEIPEL, V. v. (1859, April).

[Undersökningar i högre algebran, jemte några deraf beroende theoremer i determinant-theorien. *K. Svenska Vet.-Akad. Handl.* (Stockholm), iii. No. 4, 32 pp.]

This simply written and fully detailed memoir is an extension of the author's paper of the preceding year. It consists of five chapters, of which the fifth contains the freshest matter.

The first, and much the longest (pp. 3–16), is occupied in § 1 (pp. 3–11) with the application of the Sturmian division-process to

$$a_0x^m + a_1x^{m-1} + a_2x^{m-2} + \ldots + a_m, \quad \text{or } F(x),$$
$$b_0x^{m-1} + b_1x^{m-2} + \ldots + b_{m-1}, \quad \text{or } f(x),$$

with the object of obtaining Cayley's expressions of August 1848 for the remainders (R_r). In this he is successful, and he adds

Brioschi's of 1854 for comparison. In § 2 (pp. 12–15) he works out for himself a third form, which, in the case of $m=4$, gives

$$R_1 = \begin{vmatrix} a_0 & b_0 & \cdot & \cdot & \cdot \\ a_1 & b_1 & b_0 & \cdot & \cdot \\ a_2 & b_2 & b_1 & -1 & \cdot \\ a_3 & b_3 & b_2 & x & -1 \\ a_4 & \cdot & b_3 & \cdot & x \end{vmatrix}, \quad R_2 = \begin{vmatrix} a_0 & \cdot & b_0 & \cdot & \cdot & \cdot \\ a_1 & a_0 & b_1 & b_0 & \cdot & \cdot \\ a_2 & a_1 & b_2 & b_1 & b_0 & \cdot \\ a_3 & a_2 & b_3 & b_2 & b_1 & \cdot \\ a_4 & a_3 & \cdot & b_3 & b_2 & -1 \\ \cdot & a_4 & \cdot & \cdot & b_3 & x \end{vmatrix}, \quad R_3 = \ldots$$

—expressions which take a more familiar appearance when developed and arranged according to descending powers of x.

He then passes to the case where $f(x)=F'(x)$, his second chapter being occupied with a short but very complete historical sketch of Sturm's functions.

In the third chapter the allied subject of the common roots of two equations is taken up, and to illustrate the advantage of his own procedure over Lagrange's and Brioschi's he takes the equations

$$a_0x^3 + a_1x^2 + a_2x + a_3 = 0,$$
$$b_0x^3 + b_1x^2 + b_2x + b_3 = 0,$$

and (1) supposing them to have *one* common root gives

$$\begin{vmatrix} a_0 & \cdot & b_0 & \cdot \\ a_1 & a_0 & b_1 & b_0 \\ a_2 & a_1 & b_2 & b_1 \\ a_3 & a_2 & b_3 & b_2 \end{vmatrix} x + \begin{vmatrix} a_0 & \cdot & b_0 & \cdot \\ a_1 & a_0 & b_1 & b_0 \\ a_2 & a_1 & b_2 & b_1 \\ \cdot & a_3 & \cdot & b_3 \end{vmatrix} = 0$$

as the equation for determining it; (2) gives the vanishing of the two determinants in this equation as the conditions requisite for the existence of *two* common roots; and (3) gives

$$\begin{vmatrix} a_0 & b_0 \\ a_1 & b_1 \end{vmatrix} x^2 + \begin{vmatrix} a_0 & b_0 \\ a_2 & b_2 \end{vmatrix} x + \begin{vmatrix} a_0 & b_0 \\ a_3 & b_3 \end{vmatrix} = 0$$

as the equation for determining the said two. In this connection Sylvester's paper of 1840 might well have been referred to (see *History*, i. pp. 236–238).

The fourth chapter is headed "Relation between any three

remainders," but its main subject is Cayley's relation of August 1848, and the obtaining of others like it.

Lastly, a chapter is devoted to the bigradient compound determinants which he (Zeipel) drew attention to in his paper of 1858, and which when of the $(2r-1)^{\text{th}}$ order are equal to

$$\mathrm{P}_n^{n-2}\mathrm{P}_{n+r} \quad \text{and} \quad \mathrm{P}_n^{n-2}\mathrm{P}_{n+r}^{(r)}$$

respectively. Even to-day these excite a real interest, and a purely determinantal proof of the identities is much to be desired.

BRUNO, F. FAÀ DI (1859).

[THÉORIE GÉNÉRALE DE L'ÉLIMINATION. Par le Chevalier François Faà di Bruno. . . . x+224 pp. Paris.]

In his section (pp. 32–40) dealing with the dialytic eliminant, Bruno, besides reprinting $\mathrm{R}_{3,3}$, $\mathrm{R}_{4,4}$, gives the full expansion of the discriminant of the equation

$$ax^4 + 4bx^3 + 6cx^2 + 4dx + e = 0$$

and of the discriminant of

$$ax^5 + 5bx^4 + 10cx^3 + 10dx^2 + 5ex + f = 0 .$$

In the printing of the latter discriminant, however, there are at least seven mistakes. In the next section (pp. 40–46) he seeks to improve on what we have called Cayley's "chain of squares" by combining the last square with the first, the second from the end with the second from the beginning, and so on. For example, his expression for $\mathrm{R}_{3,3}$, that is to say, for

$$(a^3s^3-d^3p^3) + (-a^2brs^2+cd^2p^2q) + \{2(-a^2cqs^2+bd^2p^2r) + \ldots\} + \ldots$$

is

$$\begin{array}{c|c|} & \begin{smallmatrix} s^3 \\ d^3 \end{smallmatrix} \\ \hline \begin{smallmatrix} a^3 \\ p^3 \end{smallmatrix} & \pm 1 \\ \hline \end{array} + \begin{array}{c|c|} & \begin{smallmatrix} rs^2 \\ cd^2 \end{smallmatrix} \\ \hline \begin{smallmatrix} a^2b \\ p^2q \end{smallmatrix} & \mp 1 \\ \hline \end{array} + \begin{array}{c|c|c|} & \begin{smallmatrix} qs^2 \\ bd^2 \end{smallmatrix} & \begin{smallmatrix} r^2s \\ c^2d \end{smallmatrix} \\ \hline \begin{smallmatrix} a^2c \\ p^2r \end{smallmatrix} & \mp 2 & \pm 1 \\ \hline \begin{smallmatrix} ab^2 \\ pq^2 \end{smallmatrix} & \pm 1 & \\ \hline \end{array} + \begin{array}{c|c|c|c|} & \begin{smallmatrix} ps^2 \\ ad^2 \end{smallmatrix} & \begin{smallmatrix} qrs \\ bcd \end{smallmatrix} & \begin{smallmatrix} r^3 \\ c^3 \end{smallmatrix} \\ \hline \begin{smallmatrix} a^2d \\ p^2s \end{smallmatrix} & \mp 3 & \pm 3 & \mp 1 \\ \hline \begin{smallmatrix} abc \\ pqr \end{smallmatrix} & \pm 3 & \mp 1 & \\ \hline \begin{smallmatrix} b^3 \\ q^3 \end{smallmatrix} & \mp 1 & & \\ \hline \end{array} + \begin{array}{c|c|c|c|} & \begin{smallmatrix} prs \\ acd \end{smallmatrix} & \begin{smallmatrix} q^2s \\ b^2d \end{smallmatrix} & \begin{smallmatrix} qr^2 \\ bc^2 \end{smallmatrix} \\ \hline \begin{smallmatrix} abd \\ pqs \end{smallmatrix} & \mp 1 & \mp 2 & \pm 1 \\ \hline \begin{smallmatrix} ac^2 \\ pr^2 \end{smallmatrix} & \mp 2 & \pm 1 & \\ \hline \begin{smallmatrix} b^2c \\ q^2r \end{smallmatrix} & \pm 1 & & \\ \hline \end{array} ,$$

a marked improvement on which would be

$$\begin{array}{c} s^3 \\ p^3 \\ \begin{array}{c} a^3 \\ d^3 \end{array}\boxed{\pm 1} \end{array} + \begin{array}{c} rs^2 \\ p^2q \\ \begin{array}{c} a^2b \\ cd^2 \end{array}\boxed{\mp 1} \end{array} + \ldots .$$

the second term in each binomial being derived from the first term by the change of $a, b, c, d, p, q, r, s,$ into $d, c, b, a, s, r, q, p,$ —that is to say, by the interchange

$$\begin{pmatrix} a & b & p & q \\ d & c & s & r \end{pmatrix}.$$

Further, he improves upon Cayley's squares by so transposing their rows, where necessary, as to bring about axisymmetry.* Lastly, he tries (pp. 43–46) to justify Cayley's rule for calculating the coefficients placed inside any square of the chain.

SALMON, G. (1859).

[Lessons introductory to the Modern Higher Algebra. ... xii+147 pp. Dublin.]

Having stated Euler's and Bezout's method of 1764 for eliminating the unknown from two equations (see Cauchy's account of it, *History*, i. pp. 241–242), and having previously defined such eliminant as the expression whose vanishing is the condition for the two equations having a common root, Salmon naturally thinks (p. 32, § 47) of extending the method to find the conditions for the equations having *two* common roots. The equations being

$$\left.\begin{aligned} ax^4 + bx^3 + cx^2 + dx + e &= 0 \\ \alpha x^3 + \beta x^2 + \delta x + \epsilon &= 0 \end{aligned}\right\} \quad \text{or say} \quad \left\{\begin{aligned} \phi(x) &= 0 \\ \psi(x) &= 0, \end{aligned}\right.$$

he asserts that if $\phi(x)$ and $\psi(x)$ have two linear factors in common the result of multiplying $\phi(x)$ by the remaining linear factor of $\psi(x)$ must be the same as the result of multiplying

* The expression for $R_{4,4}$, though now given more accurately than before, is still disfigured by at least ten misprints.

$\psi(x)$ by the remaining quadratic factor of $\phi(x)$. Taking, therefore, these remaining factors to be

$$\mathrm{A}x + \mathrm{B} \quad \text{and} \quad \mathrm{A}'x^2 + \mathrm{B}'x + \mathrm{C}',$$

and performing the multiplications, he is entitled to equate the coefficients of like powers in the two products. He thus obtains

$$\left.\begin{array}{l} \mathrm{A}a - \mathrm{A}'a = 0 \\ \mathrm{A}b + \mathrm{B}a - \mathrm{A}'\beta - \mathrm{B}'a = 0 \\ \mathrm{A}c + \mathrm{B}b - \mathrm{A}'\gamma - \mathrm{B}'\beta - \mathrm{C}'a = 0 \\ \mathrm{A}d + \mathrm{B}c - \mathrm{A}'\delta - \mathrm{B}'\gamma - \mathrm{C}'\beta = 0 \\ \mathrm{A}e + \mathrm{B}d - \mathrm{B}'\delta - \mathrm{C}'\gamma = 0 \\ \mathrm{B}e - \mathrm{C}'\delta = 0 \end{array}\right\},$$

from which the deduction is

$$\begin{Vmatrix} a & b & c & d & e & \cdot \\ \cdot & a & b & c & d & e \\ a & \beta & \gamma & \delta & \cdot & \cdot \\ \cdot & a & \beta & \gamma & \delta & \cdot \\ \cdot & \cdot & a & \beta & \gamma & \delta \end{Vmatrix} = 0.$$

BORCHARDT, C. W. (1859, November).

[Vergleichung zweier Formen der Eliminations-Resultante. *Crelle's Journ.*, lvii. pp. 183–186; or *Gesammelte Werke*, pp. 145–150.]

The problem here is exactly the same as Hesse's of the previous year. Instead, however, of multiplying S by $\zeta(\beta_1, \beta_2)$, he preferably multiplies $\zeta^{\frac{1}{2}}(\beta_1, \beta_2, a_1, a_2, a_3)$ by S, thus obtaining

$$\begin{vmatrix} 1 & \beta_1 & \beta_1^2 & \beta_1^3 & \beta_1^4 \\ 1 & \beta_2 & \beta_2^2 & \beta_2^3 & \beta_2^4 \\ 1 & a_1 & a_1^2 & a_1^3 & a_1^4 \\ 1 & a_2 & a_2^2 & a_2^3 & a_2^4 \\ 1 & a_3 & a_3^2 & a_3^3 & a_3^4 \end{vmatrix} \cdot \begin{vmatrix} a_0 & a_1 & a_2 & a_3 & \cdot \\ \cdot & a_0 & a_1 & a_2 & a_3 \\ b_0 & b_1 & b_2 & \cdot & \cdot \\ \cdot & b_0 & b_1 & b_2 & \cdot \\ \cdot & \cdot & b_0 & b_1 & b_2 \end{vmatrix}$$

$$= \begin{vmatrix} \phi(\beta_1) & \beta_1\phi(\beta_1) & \cdot & \cdot & \cdot \\ \phi(\beta_2) & \beta_2\phi(\beta_2) & \cdot & \cdot & \cdot \\ \cdot & \cdot & \psi(a_1) & a_1\psi(a_1) & a_1^2\psi(a_1) \\ \cdot & \cdot & \psi(a_3) & a_2\psi(a_2) & a_2^2\psi(a_2) \\ \cdot & \cdot & \psi(a_3) & a_3\psi(a_3) & a_3^2\psi(a_3) \end{vmatrix}.$$

Now, E being Euler's product of differences, the first determinant on the left is resolvable into

$$\zeta^{\frac{1}{2}}(\beta_1, \beta_2) \,.\, \zeta^{\frac{1}{2}}(a_1, a_2, a_3) \,.\, \mathrm{E},$$

as was first observed by Rosenhain in 1845 (Sept.); and the determinant on the right is resolvable into

$$\left\{\zeta^{\frac{1}{2}}(\beta_1, \beta_2) \,.\, \phi(\beta_1) \,.\, \phi(\beta_2)\right\}\left\{\zeta^{\frac{1}{2}}(a_1, a_2, a_3) \,.\, \psi(a_1) \,.\, \psi(a_2) \,.\, \psi(a_3)\right\}.$$

We thus have

$$\begin{aligned} \mathrm{ES} &= \left\{\phi(\beta_1) \,.\, \phi(\beta_2)\right\}\left\{\psi(a_1) \,.\, \psi(a_2) \,.\, \psi(a_3)\right\}, \\ &= a_3{}^2\mathrm{E} \,.\, b_2{}^3\mathrm{E}, \end{aligned}$$

and

$$\therefore\quad \mathrm{S} = a_3{}^2 b_2{}^3\mathrm{E}, \text{ as before.}$$

CHAPTER XIII.

HESSIANS, UP TO 1860.

SPECIAL cases of the determinant

$$\begin{vmatrix} \frac{\partial^2 u}{\partial x^2} & \frac{\partial^2 u}{\partial x \partial y} & \frac{\partial^2 u}{\partial x \partial z} & \cdots \\ \frac{\partial^2 u}{\partial y \partial x} & \frac{\partial^2 u}{\partial y^2} & \frac{\partial^2 u}{\partial y \partial z} & \cdots \\ \frac{\partial^2 u}{\partial z \partial x} & \frac{\partial^2 u}{\partial z \partial y} & \frac{\partial^2 u}{\partial z^2} & \cdots \\ \cdot & \cdot & \cdot & \cdots \end{vmatrix},$$

where u is a function of $x, y, z, \ldots$, may well have appeared at a very early date in the history of determinants. The case where $u = ax^2 + 2bxy + cy^2$ may be viewed as traceable to Lagrange (1773), and the case where $u = ax^2 + by^2 + cz^2 + 2dyz + 2ezx + 2fxy$ to Gauss (1801); but it is certain that in those cases the elements of the determinants were not looked on as second differential-quotients of u. The general conception first occurred to Hesse in the year 1843.

HESSE, O. (1844, January).

[Ueber die Elimination der Variabeln aus drei algebraischen Gleichungen vom zweiten Grade mit zwei Variabeln. *Crelle's Journal*, xxviii. pp. 68–96; or *Werke*, pp. 89–122.]

In § 15 (p. 83) Hesse passes from the direct subject of his paper to the special case in which the three functions f_1, f_2, f_3 are the first differential-quotients of the homogeneous function of the third degree

$$\Sigma a_{\kappa,\lambda,\mu} x_\kappa x_\lambda x_\mu\,, \quad \text{or } f \text{ say,}$$

where each of the suffixes κ, λ, μ may be 1 or 2 or 3. The determinant, afterwards called the *Jacobian*, of f_1, f_2, f_3 he says may in that case be styled "the determinant of f." This expression at once recalls that used by Gauss in 1801, namely, "determinant of a form of the second degree," the determinant of

$$ax^2 + 2bxy + cy^2,$$

according to Gauss, being $b^2 - ac$, and the determinant of

$$a_1x^2 + a_2y^2 + a_3z^2 + 2b_1yz + 2b_2zx + 2b_3xy$$

being $a_1b_1^2 + a_2b_2^2 + a_3b_3^2 - a_1a_2a_3 - 2b_1b_2b_3$. The two usages, when Hesse's is restricted to the second degree, are not so far apart: for, according to Hesse, the determinants of the same two forms are

$$\begin{vmatrix} 2a & 2b \\ 2b & 2c \end{vmatrix}, \qquad \begin{vmatrix} 2a_1 & 2b_3 & 2b_2 \\ 2b_3 & 2a_2 & 2b_1 \\ 2b_2 & 2b_1 & 2a_3 \end{vmatrix},$$

i.e. $\qquad -2^2(b^2 - ac), \quad -2^3(a_1b_1^2 + a_2b_2^2 + \ldots\ldots).$

The elements of Hesse's "determinant of f" being evidently the second differential-quotients of f, those in conjugate places must be equal—that is to say, the determinant is axisymmetric.

The first result enunciated is (p. 85)—*Die Determinante der Determinante einer gegebenen homogenen Function dritten Grades von drei Variabeln ist gleich der Summe der gegebenen Function und ihre Determinante, jede mit einem passenden constanten Factor multiplicirt.* In symbols at a later date this would have been written

$$\mathrm{H}\{\mathrm{H}(u_{33})\} = cu_{33} + c'\mathrm{H}(u_{33}).$$

Following thereupon is a theorem of like type

$$\mathrm{H}\{c_1u_{33} + c_2\mathrm{H}(u_{33})\} = c_3u_{33} + c_4\mathrm{H}(u_{33}),$$

and this is used to solve the equation

$$\mathrm{H}(u'_{33}) = u_{33},$$

where u_{33} and u'_{33} stand for ternary cubics, and u'_{33} is the unknown.

The effect of linear transformation, so strikingly brought to the front by Boole three years before, is then (§ 19) entered on, f being no longer a ternary cubic, but any function whatever of $x_1, x_2, \ldots, x_n$, and supposed to be expressed also as a function of the variables $y_1, y_2, \ldots, y_n$ by means of the equations

$$\left.\begin{array}{l} a_1^{(1)}x_1 + a_1^{(2)}x_2 + \ldots . + a_1^{(n)}x_n = y_1 \\ a_2^{(1)}x_1 + a_2^{(2)}x_2 + \ldots . + a_2^{(n)}x_n = y_2 \\ \ldots \ldots \ldots \ldots \ldots \ldots \\ a_n^{(1)}x_1 + a_n^{(2)}x_2 + \ldots . + a_n^{(n)}x_n = y_n \end{array}\right\} .$$

Denoting the determinant of f when viewed as a function of the x's by ϕ, and when viewed as a function of the y's by ϕ', Hesse affirms that

$$\phi = r^2\phi',$$

where r is the determinant formed from the coefficients of the x's in the transforming equations. His proof is essentially that still followed; that is to say, he recalls that from the multiplication theorem we have

$$\sum \pm u_1^{(1)}u_2^{(2)} \ldots u_n^{(n)} = \sum \pm a_1^{(1)}a_2^{(2)} \ldots a_n^{(n)} \cdot \sum \pm w_1^{(1)}w_2^{(2)} \ldots w_n^{(n)},$$

if $\qquad u_\kappa^{(\lambda)} = a_1^{(\lambda)}w_1^{(\kappa)} + a_2^{(\lambda)}w_2^{(\kappa)} + \ldots + a_n^{(\lambda)}w_n^{(\kappa)};$

and

$$\sum \pm u_1^{(1)}u_2^{(2)} \ldots u_n^{(n)} = r^2 \cdot \sum \pm v_1^{(1)}v_2^{(2)} \ldots v_n^{(n)},$$

if in addition $\qquad u_\kappa^{(\lambda)} = a_1^{(\lambda)}v_1^{(\kappa)} + a_2^{(\lambda)}v_2^{(\kappa)} + \ldots + a_n^{(\lambda)}v_n^{(\kappa)};$

and he then merely asserts that application to the case where

$$u_\kappa^{(\lambda)} = \frac{\partial^2 f}{\partial x_\kappa \partial x_\lambda}, \qquad w_\lambda^{(\kappa)} = \frac{\partial\left(\dfrac{\partial f}{\partial x_\kappa}\right)}{\partial y_\lambda}, \qquad v_\kappa^{(\lambda)} = \frac{\partial^2 f}{\partial y_\kappa \partial y_\lambda}$$

accomplishes the desired aim. The result, as stated in later phraseology, is that "the Hessian is a covariant." The case where $f = ax^2 + 2bxy + cy^2$ was given by Lagrange in 1773, and the case $f = ax^2 + by^2 + cz^2 + 2dyz + 2ezx + 2fxy$ by Gauss in 1801.

The ternary cubic u_{33} is next returned to and shown to be transformable by a linear substitution into the form

$$y_1^3 + y_2^3 + y_3^3 + 6\pi y_1 y_2 y_3$$

and to be such that constants c_1, c_2 are determinable which make

$$c_1 u_{33} + c_2 \mathrm{H}(u_{33})$$

resolvable into linear factors.

CAYLEY, A. (1845, early).

[Note sur deux formules données par MM. Eisenstein et Hesse. *Crelle's Journ.*, xxix. pp. 54–57; or *Collected Math. Papers*, i. pp. 113–116.]

Cayley, who had, like others, been attracted by Boole's epoch-making paper on Linear Transformations, and was about to publish his own first paper on the subject (*Camb. and Dubl. Math. Journ.*, i. pp. 104–122), was naturally interested in that part of Hesse's paper which concerned the "determinant of *f*." He consequently wrote the note we have now reached, for the purpose of adding to Hesse's results and of extending an identity of Eisenstein's not distantly related to the same subject.

The "équation remarquable" of Hesse's which he starts with he writes in the form

$$\nabla(\mathrm{U}+a\nabla\mathrm{U}) = \mathrm{AU}+\mathrm{B}\nabla\mathrm{U},$$

noting that its author had not given the values of the coefficients A, B, "ce qui parait être très difficile à effectuer." He then announces the analogous theorem: "Soit U une fonction homogène et de l'ordre ν des deux variables x, y, et ∇U la déterminante

$$\frac{\partial^2\mathrm{U}}{\partial x^2}\cdot\frac{\partial^2\mathrm{U}}{\partial y^2} - \left(\frac{\partial^2\mathrm{U}}{\partial x\,\partial y}\right)^2,$$

l'on a

$$(\nu-2)(\nu-3)\cdot\nabla(\mathrm{U}+a\nabla\mathrm{U}) = \left\{-\nu(\nu-1)(\nu-3)^2 a\mathrm{J} + \nu(\nu-1)(2\nu-5)^2 a^2\mathrm{I}\right\}\mathrm{U}$$
$$+\left\{(\nu-2)(\nu-3)^3 + (\nu-2)(\nu-3)(2\nu-5)a^2\mathrm{J}\right\}\nabla\mathrm{U}.$$

En représentant par i, j, k, l, m les coefficients différentiels du quatrième ordre de U, on a

$$\begin{aligned} \mathrm{I} &= ikm - il^2 - mj^2 - k^3 + 2jkl, \\ \mathrm{J} &= 4jl - 3k^2 - mi, \end{aligned}$$

de manière que I, J sont des fonctions de x, y des ordres $3(\nu-4)$ et $2(\nu-4)$ respectivement." To this he adds the remarkable fact, that if the binary quartic

$$i\xi^4 + 4j\xi^3\eta + 6k\xi^2\eta^2 + 4l\xi\eta^3 + m\eta^4,$$

where i, j, k, l, m are now any quantities independent of ξ, η, be transformed by the substitution

$$\left.\begin{aligned} \xi &= \lambda\xi' + \mu\eta' \\ \eta &= \lambda'\xi' + \mu'\eta' \end{aligned}\right\}$$

and I and J thus become I′ and J′, then

$$\mathrm{I}' = (\lambda\mu' - \lambda'\mu)^6\mathrm{I}, \qquad \mathrm{J}' = (\lambda\mu' - \lambda'\mu)^4\mathrm{J}.$$

In other words, he makes known for the first time the two "invariants" of a binary quartic, and notes the curious fact that expressions of exactly the same form occur in his equivalent for $\nabla(\mathrm{U}_{2,\nu} + a\nabla\mathrm{U}_{2,\nu})$.

Another remark is equally suggestive, namely, that simpler results might be reached if U were taken a homogeneous function in x', y' as well as in x, y, and $\nabla\mathrm{U}$ were defined as

$$\frac{\partial^2\mathrm{U}}{\partial x\,\partial x'} \cdot \frac{\partial^2\mathrm{U}}{\partial x\,\partial y'} - \frac{\partial^2\mathrm{U}}{\partial y\,\partial y'} \cdot \frac{\partial^2\mathrm{U}}{\partial y\,\partial x'}.$$

For example, U being a quadric in both sets of variables, namely,

$$\begin{aligned} \mathrm{U} = &\quad\ x_1{}^2(\mathrm{A}x^2 + 2\mathrm{B}xy + \mathrm{C}y^2) \\ &+ 2x_1y_1(\mathrm{A}'x^2 + 2\mathrm{B}'xy + \mathrm{C}'y^2) \\ &+\quad y_1{}^2(\mathrm{A}''x^2 + 2\mathrm{B}''xy + \mathrm{C}''y^2), \end{aligned}$$

or in later notation

$$\mathrm{U} = \begin{array}{ccc|l} x^2 & 2xy & y^2 & \\ \hline \mathrm{A} & \mathrm{B} & \mathrm{C} & x_1{}^2 \\ \mathrm{A}' & \mathrm{B}' & \mathrm{C}' & 2x_1y_1 \\ \mathrm{A}'' & \mathrm{B}'' & \mathrm{C}'' & y_1{}^2, \end{array}$$

then we should have*

$$\nabla\nabla U = 2^{10}\begin{vmatrix} A & B & C \\ A' & B' & C' \\ A'' & B'' & C'' \end{vmatrix} U.$$

In connection with the first of these results of Cayley's the reader should note that on putting $\nu=4$ we obtain

$$\nabla(U+\alpha\nabla U) = (-6\alpha J+54\alpha^2 I)U + (1+3\alpha^2 J)\nabla U,$$

or $$\nabla(\alpha U+\beta\nabla U) = (-6\alpha\beta J+54\beta^2 I)U + (\alpha^2+3\beta^2 J)\nabla U,$$

and $\therefore$ $$\nabla\nabla U = 54\, I\cdot U+3\, J\cdot\nabla U.$$

Further, if the particular form of U be

$$ax^2+4bx^3y+6cx^2y^2+4dxy^3+ey^4,$$

this gives

$$\nabla\nabla U = 12^3(432\, I\cdot U-J\cdot\nabla U)$$

where $$I = ace+2bcd-ad^2-eb^2-c^3,$$

and $$J = ae+3c^2-4bd.$$

In his famous first paper (February 1845) "On the Theory of Linear Transformations," of which this is merely an offshoot, Cayley states that the invariance of I had been communicated to him by Boole, along with the still more interesting fact that Boole's invariant (*i.e.* the discriminant) is equal to J^3-27I^2.

CAYLEY, A. (1846).

[On homogeneous functions of the third order with three variables. *Cambridge and Dub. Math. Journ.*, i. pp. 97–104; or *Collected Math. Papers*, i. pp. 230–233.]

One of such functions, U say, being taken in the form

$$ax^3 + by^3 + cz^3 + 3iy^2z + 3jz^2x + 3kx^2y$$
$$+ 3i_1yz^2 + 3j_1zx^2 + 3k_1xy^2 + 6lxyz,$$

Cayley calculates ∇U, and writing it in the form

$$Ax^3 + By^3 + Cz^3 + 3Iy^2z + \ldots\ldots$$

gives the values† of A, B, C, 3I, ... in terms of $a, b, c, i, \ldots\ldots$

* Instead of 2^{10} we find in the original 2^6, and in the *Collected Math. Papers* 2^8.

† In the value of $3K_1$ there is printed $-2bi_1l$ instead of $-2bj_1l$.

CAYLEY, A. (1847).

[Note sur les hyperdéterminants. *Crelle's Journal*, xxxiv. pp. 148–152; or *Collected Math. Papers*, i. pp. 352–355.]

As already noted, the second section of this short paper concerns what would, a few years later, have been called "the Hessian of the discriminant

$$6abcd+3b^2c^2-a^2d^2-4ac^3-4b^3d$$

of the binary cubic." The result, which is rather inelegantly verified, is that the said Hessian is a numerical multiple of the square of the discriminant.

HESSE, O. (1847, August).

[Ueber Curven dritter Classe und Curven dritter Ordnung. *Crelle's Journ.*, xxxviii. pp. 241–256; or *Werke*, pp. 193-210.]

Any homogeneous function of the mth degree in the variables x_1, x_2, x_3, being denoted by u, its first differential-quotients by u_1, u_2, u_3, and its second differential-quotients by u_{11}, u_{12}, . . . , there is obtained from Euler

$$\left.\begin{aligned} u_{11}x_1 + u_{12}x_2 + u_{13}x_3 &= (m-1)u_1 \\ u_{21}x_1 + u_{22}x_2 + u_{23}x_3 &= (m-1)u_2 \\ u_{31}x_1 + u_{32}a_2 + u_{33}x_3 &= (m-1)u_3 \end{aligned}\right\},$$

and thence, on solving,

$$\left.\begin{aligned} \frac{\Delta}{m-1}x_1 &= \mathrm{U}_{11}u_1+\mathrm{U}_{12}u_2+\mathrm{U}_{13}u_3 \\ \frac{\Delta}{m-1}x_2 &= \mathrm{U}_{21}u_1+\mathrm{U}_{22}u_2+\mathrm{U}_{23}u_3 \\ \frac{\Delta}{m-1}x_3 &= \mathrm{U}_{31}u_1+\mathrm{U}_{32}u_2+\mathrm{U}_{33}u_3 \end{aligned}\right\},$$

where, evidently, Δ is used for Hesse's determinant of u, and U_{rs} for the cofactor of u_{rs} in Δ. Using in connection with the latter three equations the multipliers u_1, u_2, u_3, and adding, Hesse derives the interesting result

$$\frac{m}{m-1}u\Delta = \mathrm{U}_{11}u_1{}^2+\mathrm{U}_{22}u_2{}^2+\mathrm{U}_{33}u_3{}^2+2\mathrm{U}_{23}u_2u_3+2\mathrm{U}_{31}u_3u_1+2\mathrm{U}_{12}u_1u_2,$$

and this by a process of differentiation leads to six results of the type

$$u_{12}u_1u_3+u_{13}u_1u_2-u_{23}u_1{}^2-u_{11}u_2u_3 = \frac{m}{m-1}\mathrm{U}_{23}u - \frac{x_2x_3}{(m-1)^2}\Delta.$$

The rest of the paper is geometrical.

HESSE, O. (1849, January).

[Transformation einer beliebigen gegebenen homogenen Function 4ten Grades von zwei Variabeln. . . . . *Crelle's Journ.*, xli. pp. 243–263; or *Werke*, pp. 223–246.]

A binary quartic u_{24} being the only other homogeneous integral function whose determinant, in Hesse's sense, is of the same degree as the function, there was naturally an inclination to make a study of its properties in the same fashion as has been followed with the ternary cubic. Analogous results are reached, such, for example, as the theorem that *it is possible to determine constants* c_1, c_2 *so that*

$$c_1u_{24}+c_2u\mathrm{H}(_{24})$$

may be an exact square. Most of the matter, however, more directly concerns the quartic than its so-called determinant.

ARONHOLD, S. (1849, July).

[Zur Theorie der homogenen Functionen dritten Grades von drei Variabeln. *Crelle's Journ.*, xxxix. pp. 140–159.]

This is an inspiration from, and a striking development of, the latter part of Hesse's paper of the year 1844, and like that paper may be said to concern itself more with the ternary cubic than with the so-called determinant of that function. In regard to the latter, however, there is one very noteworthy result: for, just as Cayley in 1845 established the two invariants I and J of a binary quartic u_{24}, and used them for the expression of $\mathrm{H}\left\{a\cdot u_{24}+b\cdot \mathrm{H}(u_{24})\right\}$ in the form $\mathrm{A}\cdot u_{24}+\mathrm{B}\cdot \mathrm{H}(u_{24})$, so Aronhold here announces the two invariants S and T of a ternary cubic, and gives the similar expression for

$$\mathrm{H}\left\{a\cdot u_{33}+b\cdot \mathrm{H}(u_{33})\right\}.$$

Obtained from this by putting $a=0$ is the result

$$\mathrm{H}\left\{\mathrm{H}(u_{33})\right\} = 3\mathrm{S}^3\cdot u_{33}-2\mathrm{T}\cdot \mathrm{H}(u_{33})$$

—the longed-for definite form of Hesse's theorem of the year 1844.

We may also note in passing that the result of eliminating x, y, z from the equations

$$\frac{\partial u_{33}}{\partial x} = 0, \quad \frac{\partial u_{33}}{\partial y} = 0, \quad \frac{\partial u_{33}}{\partial z} = 0$$

is expressed in terms of S and T, namely,

$$\mathrm{T}^2-\mathrm{S}^3 = 0;$$

or, in later phraseology, that the discriminant of u_{33} is $\mathrm{T}^2-\mathrm{S}^3$.

HESSE AND JACOBI (1849, December).

[Auszug zweier Schreiben des Prof. Hesse an den Herrn Prof. Jacobi und eines Schreibens des Prof. Jacobi an Herrn Prof. Hesse. *Crelle's Journ.*, xl. pp. 316–318.]

Hesse having communicated to Jacobi a theorem regarding a homogeneous function of three variables, Jacobi sent back a proof showing that the theorem held in the case of n variables. The function being noted by u, and being of the mth degree in the variables $x_1, x_2, \ldots, x_n$, Jacobi, like Hesse himself in his paper of 1847 (August), obtains the equations

$$\left.\begin{array}{l} x_1 u_{11} + x_2 u_{21} + \ldots + x_n u_{n1} = (m-1)u_1 \\ x_1 u_{12} + x_2 u_{22} + \ldots + x_n u_{n2} = (m-1)u_2 \\ \ldots\ldots\ldots\ldots\ldots\ldots\ldots\ldots \\ x_1 u_{1n} + x_2 u_{2n} + \ldots + x_n n_{nn} = (m-1)u_n \end{array}\right\},$$

and thence

$$x_i\Delta = (m-1)\left\{U_{i1}u_1 + U_{i2}u_2 + \ldots + U_{in}u_n\right\}. \qquad (\alpha)$$

Differentiating both sides of this with respect to x_k, we have

$$\begin{aligned} x_i\frac{\partial\Delta}{\partial x_k} &= (m-1)\left\{U_{i1}u_{k1} + U_{i2}u_{k2} + \ldots + U_{in}u_{kn}\right\} \\ &\quad + (m-1)\left\{u_1\frac{\partial U_{i1}}{\partial x_k} + u_2\frac{\partial U_{i2}}{\partial x_k} + \ldots + u_n\frac{\partial U_{in}}{\partial x_k}\right\} \\ &= (m-1)\left\{u_1\frac{\partial U_{i1}}{\partial x_k} + u_2\frac{\partial U_{i2}}{\partial x_k} + \ldots + u_n\frac{\partial U_{in}}{\partial x_k}\right\} \qquad (\beta) \end{aligned}$$

if k be different from i. A second differentiation, but this time with respect to x_l, gives

$$\begin{aligned} x_i\frac{\partial^2\Delta}{\partial x_l\partial x_k} &= (m-1)\left\{u_1\frac{\partial^2 U_{i1}}{\partial x_l\partial x_k} + u_2\frac{\partial^2 U_{i2}}{\partial x_l\partial x_k} + \ldots + u_n\frac{\partial^2 U_{in}}{\partial x_l\partial x_k}\right\} \\ &\quad + (m-1)\left\{u_{l1}\frac{\partial U_{i1}}{\partial x_k} + u_{l2}\frac{\partial U_{i2}}{\partial x_k} + \ldots + u_{ln}\frac{\partial U_{in}}{\partial x_k}\right\}. \end{aligned}$$

But on the supposition that l is different from i we have

$$u_{l1}U_{i1} + u_{l2}U_{i2} + \ldots + u_{ln}U_{in} = 0,$$

and therefore by differentiation with respect to x_k

$$\begin{aligned} &u_{l1}\frac{\partial U_{i1}}{\partial x_k} + u_{l2}\frac{\partial U_{i2}}{\partial x_k} + \ldots + u_{ln}\frac{\partial U_{in}}{\partial x_k} \\ &\quad + u_{kl1}U_{i1} + u_{kl2}U_{i2} + \ldots + u_{kln}U_{in} = 0. \end{aligned}$$

Consequently by substitution

$$\begin{aligned} x_i\frac{\partial^2\Delta}{\partial x_l\partial x_k} &= (m-1)\left\{u_1\frac{\partial^2 U_{i1}}{\partial x_l\partial x_k} + u_2\frac{\partial U_{i2}}{\partial x_l\partial x_k} + \ldots + u_n\frac{\partial^2 U_{in}}{\partial x_l\partial x_k}\right\} \\ &\quad -(m-1)\left\{u_{kl1}U_{i1} + u_{kl2}U_{i2} + \ldots + u_{kln}U_{in}\right\}, \qquad (\gamma) \end{aligned}$$

where l and k are each different from i.

If now special values of $x_1, x_2, \ldots x_n$ make $u_1, u_2, \ldots, u_n$ all vanish, then, Jacobi says, we shall also have

$$\Delta = 0, \quad \frac{\partial\Delta}{\partial x_k} = 0, \quad U_{rs} = Nx_rx_s$$

for all values of r and s *; and consequently from (γ)

$$x_i\frac{\partial^2\Delta}{\partial x_l\partial x_k} = -(m-1)\left\{Nx_ix_1\frac{\partial u_{kl}}{\partial x_1} + Nx_ix_2\frac{\partial u_{kl}}{\partial x_2} + \dots\right\}$$

$$= -(m-1)Nx_i\left\{x_1\frac{\partial u_{kl}}{\partial x_1} + x_2\frac{\partial u_{kl}}{\partial x_2} + \dots\right\}$$

$$= -(m-1)(m-2)Nx_iu_{kl},$$

whence

$$\frac{\partial^2\Delta}{\partial x_l\partial x_k} = -(m-1)(m-2)N\cdot\frac{\partial^2 u}{\partial x_l\partial x_k}\cdot$$

This result we may formally enunciate as follows:—*If the first differential-quotients of a homogeneous rational integral function all vanish, the elements of the Hessian of the function are proportional to the elements of the Hessian of the Hessian.*

TERQUEM, O. (1851, March).

[Note sur les déterminants. *Nouv. Annales de Math.*, x. pp. 124–131.]

This is an elementary exposition of Hesse's determinant, with simple illustrations from algebraic geometry, the property of "covariance" being made prominent. A curious distinction is made between what are called the "first" and "second" determinants: for example,

$$4ac - b^2$$

---

* The first two of these results, which follow from (α) and (β), there is no pressing reason for mentioning: it would have been equally pertinent to note that $u=0$. The third result Jacobi probably obtained (see *Crelle's Journal*, xv. p. 304) by taking in every possible way $n-1$ of the initiatory set of equations and deducing

$$\begin{aligned} x_1 : x_2 : \dots : x_n &= U_{11} : U_{21} : \dots : U_{n1}, \\ &= U_{12} : U_{22} : \dots : U_{n2}, \\ &\dots\dots\dots\dots \\ &= U_{1n} : U_{2n} : \dots : U_{nn}. \end{aligned}$$

This implies that $\qquad U_{rs} : U_{rs'} = x_s : x_{s'}$

and $\qquad U_{s'r} : U_{s'r'} = x_r : x_{r'}$;

and from these, by reason of the equality of $U_{rs'}$ and $U_{s'r}$ there is got by multiplication

$$U_{rs} : U_{s'r'} = x_sx_r : x_{s'}x_{r'}.$$

is styled "le *premier déterminant* de la fonction hexanôme du second degré à deux variables," and

$$8acf + 4bde - 2ae^2 - 2cd^2 - 2fb^2$$

the "second déterminant de la même fonction rendue homogène et ternaire." The former is, in fact, the determinant of

$$ax^2 + bxy + cy^2 + dxz + eyz + fz^2$$

or $$ax^2 + bxy + cy^2 + dx + ey + f$$

with respect to x, y; and the latter the determinant with respect to x, y, z.

HESSE, O. (1851, March).

[Ueber die Bedingung, unter welcher eine homogene ganze Function von n unabhängigen Variabeln durch lineare Substitutionen von n andern unabhängigen Variabeln auf eine homogene Function sich zurück führen lässt, die eine Variabel weniger enthält. *Crelle's Journ.*, xlii. pp. 117–124; or *Werke*, pp. 289–296.]

Hesse here returns to the subject of § 19 of his original paper, calling $\Sigma \pm u_{11}u_{22} \ldots u_{nn}$ or Δ the determinant of u with respect to the variables $x_1, x_2, \ldots x_n$, and $\Sigma \pm u^{11}u^{22} \ldots u^{nn}$ or ∇ the determinant of u with respect to $y_1, y_2, \ldots, y_n$, and proving once more his theorem that

$$\nabla = r^2 \Delta.$$

He then supposes that in the result of the transformation y_n does not appear, and says that as this implies that $u^{1n}, u^{2n}, \ldots, u^{nn}$ all vanish, it follows that $\nabla = 0$, and that therefore from the said theorem Δ also must vanish. There is thus obtained the result that, "*Wenn eine homogene ganze Function der* n *unabhängigen Variabeln* $\mathrm{x}_1, \mathrm{x}_2, \ldots, \mathrm{x}_n$, *durch*

$$\mathrm{x}_k = \mathrm{a}_1^k \mathrm{y}_1 + \mathrm{a}_2^k \mathrm{y}_2 + \ldots + \mathrm{a}_n^k \mathrm{y}_n$$

in eine Function der Variabeln $\mathrm{y}_1, \mathrm{y}_2, \ldots, \mathrm{y}_n$ *übergeht, in welcher eine dieser Variabeln fehlt, so ist die Determinante dieser Function in Rücksicht auf die Variabeln* $\mathrm{x}_1, \mathrm{x}_2, \ldots, \mathrm{x}_n$, *identisch gleich* 0."

The rest of the paper is occupied with the converse theorem; but as the author himself came to be dissatisfied with his attempt at a proof and returned to the subject seven years later, it need not be entered on here.

SYLVESTER, J. J. (1851, April).

[Sketch of a memoir on elimination, transformation, and canonical forms. *Cambridge and Dub. Math. Journ.*, vi. pp. 186–200; or *Collected Math. Papers*, i. pp. 184–197.]

The expression "determinant of a function" or, more definitely, "determinant of a function in respect to certain variables" occurs repeatedly in Sylvester's writings of the year 1850, the accompanying notation being*

$$\square(u);$$

for example, when dealing with ternary quadrics U and V, expressions like

$$\underset{\lambda\mu}{\boxed{}}\ \underset{xyz}{\boxed{}}\ (\lambda \mathrm{U}+\mu \mathrm{V})$$

are in constant use by him. It is clear, however, that the determinant which he had in mind was not Hesse's, but that which the year following he named the "discriminant." †

The interest of the present paper lies in the fact that, amid much other matter, not only are the said two determinants clearly defined and distinguished, but are shown to be viewable as having a common parentage, being indeed two extreme members of a family group. In the first place, the determinant of any homogeneous integral function is incidentally defined as the resultant of the first partial differential coefficients of the function, when drawing attention to Boole's proposition (1843) that the said determinant "is unaltered by any linear transformation of the variables, except so far as regards the introduction of

---

* $\square$ was used by Cayley in 1846 as the symbol of hyperdeterminant derivation. See *Collected Math. Papers*, i. p. 97.

† See *Philos. Magazine*, ii. (1851), p. 406, and *Cambridge and Dub. Math. Journ.*, vii. (1852), p. 52; or Sylvester's *Collected Math. Papers*, i. pp. 280, 284.

a power of the modulus of transformation." It is spoken of later in the paper as the "common constant determinant" or the "ordinary determinant" of the function, the word *discriminant* not being proposed until a later date in the same year. In the second place, there is brought into notice in connection with any homogeneous integral function $\phi(x, y, \ldots, z)$ of the nth degree the family of functions

$$\left(\xi\frac{\partial}{\partial x}+\eta\frac{\partial}{\partial y}+\ \ldots\ +\zeta\frac{\partial}{\partial z}\right)^r\phi(x, y, \ldots, z),$$

where r has the values $1, 2, \ldots, n$. Corresponding to these there is a family of determinants (*i.e.* discriminants), namely

$$\underset{\xi,\,\eta,\,\ldots}{\boxed{}}\left(\xi\frac{\partial}{\partial x}+\eta\frac{\partial}{\partial y}+\ \ldots\ +\zeta\frac{\partial}{\partial z}\right)^r\phi(x, y, \ldots, z),$$

where $r=2, 3, \ldots, n$, the first being according to Sylvester the "Hessian" or "First Boolian" determinant* of ϕ, and the last the "Final Boolian" or "ordinary determinant" of ϕ. The reader is left in the former case to reconcile the new definition with Hesse's own definition, and in the latter case to observe that

$$\left(\xi\frac{\partial}{\partial x}+\eta\frac{\partial}{\partial y}+\ \ldots\ +\zeta\frac{\partial}{\partial z}\right)^n\phi(x, y, \ldots, z) = \phi(\xi, \eta, \ldots, \zeta).$$

The notation used for the Hessian of ϕ is $\mathrm{H}(\phi)$: by "second Hessian" he says he means "Hessian of the Hessian"; by "post-Hessian" the determinant of the function got by taking $r=3$; and similarly for "præter-post-Hessian"!

We may at once remark that much of this nomenclature had a very short life, being supplanted by other coinages made by Sylvester himself. The functions

$$\left(\xi\frac{\partial}{\partial x}+\eta\frac{\partial}{\partial y}+\ldots\right)^r\phi(x,\ y,\ \ldots)$$

---

* On p. 194 he says the Hessian of $\mathrm{F}(x, y)$ is "the determinant *of the determinant*, in respect of ξ and η, of

$$\left(\xi\frac{\partial}{\partial x}+\eta\frac{\partial}{\partial y}\right)^2\mathrm{F}(x, y)\text{"}$$

—an error which is repeated in the *Collected Math. Papers*.

he soon named the *emanants* of ϕ: and thus the Hessian, post-Hessian, præter-post-Hessian, and other determinants forming the "Hessian (or Boolian) Scale" became known as the discriminants of the emanants of ϕ. To the first member of the scale, however, the word "Hessian" became permanently attached, although Sylvester's mode of defining it as the "discriminant of the quadratic (or second) emanant"* did not spread. It was introduced by Salmon into the first edition of his *Higher Plane Curves* (see p. 72) about a year after Sylvester's first use of it, and met with rapid acceptance.

SALMON, G. (1852).

[A Treatise on the Higher Plane Curves. . . . . xii+316 pp. Dublin.]

In sect. ix. (pp. 181–195) Salmon deals with the "General Equation of the Third Degree" on the lines of Aronhold's paper of 1849, and the Hessian naturally comes in for attention. The ternary cubic U which constitutes the non-zero side of the equation he writes in the form

$$a_1x^3+b_2y^3+c_3z^3+3a_2x^2y+3b_3y^2z+3c_1z^2x \\ +3a_3x^2z+3b_1y^2x+3c_2z^2y+6dxyz,$$

and following Cayley gives its Hessian† as

$$\mathrm{a}_1x^3+\mathrm{b}_2y^3+\mathrm{c}_3z^3+3\mathrm{a}_2x^2y+\ldots,$$

where

$$\begin{aligned} \mathrm{a}_1 &= a_1d^2+b_1a_3^2+c_1a_2^2-a_1b_1c_1-2da_2a_3, \\ \mathrm{b}_2 &= b_2d^2+c_2b_1^2+a_2b_3^2-a_2b_2c_2-2db_3b_1, \\ \mathrm{c}_3 &= c_3d^2+a_3c_2^2+b_3c_1^2-a_3b_3c_3-2dc_1c_2, \end{aligned}$$

* See *Philos. Magazine*, v. p. 122; or *Collected Math. Papers*, i. p. 591.

† It should be noted that this is $\frac{1}{216}$ of the Hessian as defined, that a_1 is expressible as a three-line determinant, that $3\mathrm{a}_2$ and $3\mathrm{a}_3$ are expressible as the sum of two such determinants, that 6d is expressible as the sum of three such determinants, and that the performance of the circular substitutions

$$a_1,\ b_2,\ c_3{=}b_2,\ c_3,\ a_1, \quad a_2,\ b_3,\ c_1{=}b_3,\ c_1,\ a_2, \quad a_3,\ b_1,\ c_2{=}b_1,\ c_2,\ a_3,$$

on the expressions for a_1, $3\mathrm{a}_2$, $3\mathrm{a}_3$, gives us six other of the expressions.

$$
\begin{aligned}
3\mathrm{a}_2 &= c_2a_2{}^2 + a_2c_1b_1 - 2a_2a_3b_3 - a_2d^2 + 2a_1b_3d + b_2a_3{}^2 - a_1c_1b_2 - a_1b_1c_2,\\
3\mathrm{b}_3 &= a_3b_3{}^2 + b_3a_2c_2 - 2b_3b_1c_1 - b_3d^2 + 2b_2c_1d + c_3b_1{}^2 - b_2a_2c_3 - b_2c_2a_3,\\
3\mathrm{c}_1 &= b_1c_1{}^2 + c_1b_3a_3 - 2c_1c_2a_2 - c_1d^2 + 2c_3a_2d + a_1c_2{}^2 - c_3b_3a_1 - c_3a_3b_1,\\
3\mathrm{a}_3 &= b_3a_3{}^2 + a_3b_1c_1 - 2a_3a_2c_2 - a_3d^2 + 2a_1c_2d + c_3a_2{}^2 - a_1b_1c_3 - a_1c_1b_3,\\
3\mathrm{b}_1 &= c_1b_1{}^2 + b_1c_2a_2 - 2b_1b_3a_3 - b_1d^2 + 2b_2a_3d + a_1b_3{}^2 - b_2c_2a_1 - b_2a_2c_1,\\
3\mathrm{c}_2 &= a_2c_2{}^2 + c_2a_3b_3 - 2c_2c_1b_1 - c_2d^2 + 2c_3b_1d + b_2c_1{}^2 - c_3a_3b_2 - c_3b_3a_2,\\
6\mathrm{d} &= -2d^3 + 2d(b_1c_1+c_2a_2+a_3b_3) + (a_1b_3c_2+b_2c_1a_3+c_3a_2b_1) - a_1b_2c_3\\
&\qquad - 3(a_2b_3c_1+a_3b_1c_2).
\end{aligned}
$$

The invariants S and T are also printed in full, viz.

$$
\begin{aligned}
\mathrm{S} = {} & d^4 - 2d^2(b_1c_1+c_2a_2+a_3b_3) + 3d(a_2b_3c_1+a_3b_1c_2) - d\cdot a_1b_2c_3\\
&+ d(a_1b_3c_2+b_2c_1a_3+c_3a_2b_1) - (b_1c_1\cdot c_2a_2+c_2a_2\cdot a_3b_3 + a_3b_3\cdot b_1c_1)\\
&+ (b_1{}^2c_1{}^2+c_2{}^2a_2{}^2+a_3{}^2b_3{}^2) - (a_1b_2\cdot c_1c_2+b_2c_3\cdot a_2a_3+c_3a_1\cdot b_3b_1)\\
&+ (b_3c_3a_2{}^2+c_1a_1b_3{}^2+a_2b_2c_1{}^2+b_2c_2a_3{}^2+c_3a_3b_1{}^2+a_1b_1c_2{}^2),
\end{aligned}
$$

$$\mathrm{T} = -8d^6 + 24d^4(b_1c_1+c_2a_2+a_3b_3) - \dots\dots$$

As these differ from Aronhold's by numerical factors, we are prepared to find corresponding differences in the expressions for the Hessian of the Hessian and for the discriminant, namely,

$$4\mathrm{S}^2\cdot\mathrm{U} - \mathrm{T}\cdot\mathrm{H(U)} \quad\text{and}\quad \mathrm{T}^2 - 64\mathrm{S}^3$$

respectively.

BRIOSCHI, F. (1852, August).

[Sur les déterminants des formes quadratiques. *Nouv. Annales de Math.*, xi. pp. 307–311; or *Opere mat.* v. pp. 81–85.]

After an introduction of two pages on determinants in general, the determinant of a quadratic form is defined as the determinant whose elements are the second differential-quotients of the form, the editor adding in a footnote the words, "c'est le déterminant *hessien* des Anglais." Starting then from the known fact that if $a_1a_2 - b_1^2 = 0$

$$a_1x_1^2 + a_2x_2^2 + 2b_1x_1x_2 = \frac{1}{a_1}(a_1x_1+b_1x_2)^2,$$

Brioschi states that similarly, if the determinants of

$$a_1x_1^2 + a_2x_2^2 + a_3x_3^2 + 2b_1x_1x_2 + 2b_2x_1x_3 + 2c_1x_2x_3\,,$$
$$a_1x_1^2 + a_2x_2^2 + 2b_1x_1x_2\,,$$
$$a_1x_1^2 + a_3x_3^2 + 2b_2x_1x_3\,,$$

all vanish, the ternary quadric is equal to

$$\frac{1}{a_1}(a_1x_1+b_1x_2+b_2x_3)^2\,;$$

and if the determinants of

$$\left.\begin{array}{r} a_1x_1^2 + a_2x_2^2 + a_3x_3^2 + a_4x_4^2 + 2b_1x_1x_2 + 2b_2x_1x_3 + 2b_3x_1x_4 \\ + 2c_1x_2x_3 + 2c_2x_2x_4 + 2d_1x_3x_4 \end{array}\right\},$$
$$a_1x_1^2 + a_2x_2^2 + a_3x_3^2 + 2b_1x_1x_2 + 2b_2x_1x_3 + 2c_1x_2x_3\,,$$
$$a_1x_1^2 + a_2x_2^2 + a_4x_4^2 + 2b_1x_1x_2 + 2b_3x_1x_4 + 2c_2x_2x_4\,,$$
$$a_1x_1^2 + a_2x_2^2 + 2b_1x_1x_2\,,$$
$$a_1x_1^2 + a_3x_3^2 + 2b_2x_1x_3\,,$$
$$a_1x_1^2 + a_4x_4^2 + 2b_3x_1x_4\,,$$

all vanish, the quaternary quadric is equal to

$$\frac{1}{a_1}(a_1x_1+b_1x_2+b_2x_3+b_3x_4)^2\,;$$

and so on generally. An alternative set of conditions is referred to, and is exemplified by the case of the ternary quadric, where the vanishing of $a_1c_1-b_1b_2$ is substituted for the vanishing of

$$a_1a_2a_3 + 2b_1b_2c_1 - a_1c_1^2 - a_2b_2^2 - a_3b_1^2\,,$$

the latter being equal to

$$\left\{(a_1a_2-b_1^2)(a_1a_3-b_2^2) - (a_1c_1-b_1b_2)^2\right\} \div a_1\,.$$

SYLVESTER, J. J. (1853).

[On the conditions necessary and sufficient to be satisfied in order that a function of any number of variables may be linearly equivalent to a function of any less number of variables. *Philos. Magazine*, v. pp. 119–126; or *Collected Math. Papers*, i. pp. 587–594.]

The title at once suggests a connection with Hesse's converse theorem of 1851 (March): the investigation, however, proceeds on totally different lines, and only concerns us because of the doubt thrown on the truth of the said theorem by Sylvester's assertion that the Hessian "is really foreign to the nature" of the question under discussion.

SPOTTISWOODE, W. (1853, August).

[Elementary theorems relating to determinants; second edition, rewritten and much enlarged by the author. *Crelle's Journ.*, li. pp. 209–271, 328–381.]

The latter portion (pp. 343–350) of his chapter (§ ix.) "On Functional Determinants" Spottiswoode devotes to one or two theorems connected with Hessians, and mainly to Hesse's theorem of the year 1849 (December): his proof of the latter, however, is not an improvement on Jacobi's. One of his notations for the Hessian, U being the function and $x, y, z, \ldots$ the variables, is

$$\left\{\begin{matrix} \frac{\partial}{\partial x} & \frac{\partial}{\partial y} & \frac{\partial}{\partial z} & \cdots \\ \frac{\partial}{\partial x} & \frac{\partial}{\partial y} & \frac{\partial}{\partial z} & \cdots \end{matrix}\right\} \mathrm{U},$$

suggested, doubtless, by Sylvester's general umbral notation. Further, he uses the word "Hessian" in a geometrical sense, namely, for the locus represented by

$$\mathrm{H}(\mathrm{U}) = 0.$$

SALMON, G. (1854, Feb.).

[Exercises in the hyperdeterminant calculus. *Cambridge and Dub. Math. Journ.*, ix. pp. 19–33.]

Of these exercises, which usefully served as an exposition of Cayley's so-called hyperdeterminant method of deriving invariants and covariants, it is the seventh (pp. 24–25) which here concerns us, the subject being the calculation of the Hessian of the Hessian of a binary quantic. In this method the symbol $\overline{12}^2$, primarily introduced to stand for the operator

$$\begin{vmatrix} \dfrac{\partial}{\partial x_1} & \dfrac{\partial}{\partial y_1} \\ \dfrac{\partial}{\partial x_2} & \dfrac{\partial}{\partial y_2} \end{vmatrix}^2,$$

comes by the adoption of additional conventions to represent the Hessian of a binary quantic: and similar considerations lead to the Hessian of the Hessian being denoted by

$$(\overline{13}+\overline{14}+\overline{23}+\overline{24})^2\,\overline{12}^2\cdot\overline{34}^2.$$

Adopting this expression Salmon succeeds in transposing it so as to give Cayley's binomial expression of the year 1845. The proof is reproduced with improvements in his *Modern Higher Algebra* of 1859 (§§ 170, 171: pp. 140–141), where also the special case of the binary quartic is referred to (§ 135, pp. 103–104), and Hesse's analogous theorem regarding the ternary cubic (§§ 143, 145: pp. 112–114).

BRIOSCHI, F. (1854).

[Solutions des questions 285, 286. *Nouv. Annales de Math.*, xiii. pp. 402–409.]

The theorems which had been set for proof were geometrical theorems due to Hesse, and as a foundation on which to base them and others Brioschi establishes a general result regarding Hessians. This with only slight departures from the original may be formally enunciated as follows:—*If* u *be a homogeneous integral function of the* m^{th} *degree in* r+1 *variables* $\mathrm{x}_0, \mathrm{x}_1, \mathrm{x}_2, \ldots, \mathrm{x}_\mathrm{r}$, *the Hessian of which with respect to those variables is* $\mathrm{H}_{\mathrm{r}+1}$ *and with respect to the variables* $\mathrm{x}_1, \mathrm{x}_2, \ldots, \mathrm{x}_\mathrm{r}$ *is* H_r, *then the determinant which is the result of bordering* H_r *by prefixing*

$$0,\ \frac{\partial \mathrm{u}}{\partial \mathrm{x}_1},\ \frac{\partial \mathrm{u}}{\partial \mathrm{x}_2},\ \ldots,\ \frac{\partial \mathrm{u}}{\partial \mathrm{x}_\mathrm{r}}$$

as a first row and as a first column is equal to

$$(-1)^{\mathrm{r}+1}\frac{\mathrm{m}}{\mathrm{m}-1}\mathrm{uH}_\mathrm{r} + \frac{\mathrm{x}_0{}^2}{(\mathrm{m}-1)^2}\mathrm{H}_{\mathrm{r}+1}.$$

The bordered Hessian, B say, being equal to

$$\frac{1}{m-1}\begin{vmatrix} \cdot & u_1 & u_2 & \dots & u_r \\ (m-1)u_1 & u_{11} & u_{12} & \dots & u_{1r} \\ (m-1)u_2 & u_{21} & u_{22} & \dots & u_{2r} \\ \cdot & \cdot & \cdot & \cdot & \cdot \\ (m-1)u_r & u_{r1} & u_{r2} & \dots & u_{rr} \end{vmatrix}$$

and Euler's theorem regarding the differentiating of homogeneous functions giving

$$\left.\begin{aligned} mu &= x_0u_0 + x_1u_1 + \dots + x_ru_r \\ (m-1)u_0 &= x_0u_{00} + x_1u_{01} + \dots + x_ru_{0r} \\ (m-1)u_1 &= x_0u_{10} + x_1u_{11} + \dots + x_ru_{1r} \\ &\dots\dots\dots\dots \\ (m-1)u_r &= x_0u_{r0} + x_1u_{r1} + \dots + x_ru_{rr} \end{aligned}\right\},$$

there is obtained

$$\mathrm{B} = \frac{1}{m-1}\begin{vmatrix} x_0u_0 - mu & u_1 & u_2 & \dots & u_r \\ x_0u_{10} & u_{11} & u_{12} & \dots & u_{1r} \\ x_0u_{20} & u_{21} & u_{22} & \dots & u_{2r} \\ \cdot & \cdot & \cdot & \cdot & \cdot \\ x_0u_{r0} & u_{r1} & u_{r2} & \dots & u_{rr} \end{vmatrix}$$

$$= -\frac{m}{m-1}u\begin{vmatrix} u_{11} & u_{12} & \dots & u_{1r} \\ u_{21} & u_{22} & \dots & u_{2r} \\ \cdot & \cdot & \cdot & \cdot \\ u_{r1} & u_{r2} & \dots & u_{rr} \end{vmatrix} + \frac{x_0}{m-1}\begin{vmatrix} u_0 & u_1 & \dots & u_r \\ u_{10} & u_{11} & \dots & u_{1r} \\ u_{20} & u_{21} & \dots & u_{2r} \\ \cdot & \cdot & \cdot & \cdot \\ u_{r0} & u_{r1} & \dots & u_{rr} \end{vmatrix}.$$

By similar treatment, however, the second determinant on the right of this

$$= \frac{1}{m-1}\begin{vmatrix} (m-1)u_0 & (m-1)u_1 & \dots & (m-1)u_r \\ u_{10} & u_{11} & \dots & u_{1r} \\ u_{20} & u_{21} & \dots & u_{2r} \\ \cdot & \cdot & \cdot & \cdot \\ u_{r0} & u_{r1} & \dots & u_{rr} \end{vmatrix},$$

$$= \frac{x_0}{m-1}\begin{vmatrix} u_{00} & u_{01} & \dots & u_{0r} \\ u_{10} & u_{11} & \dots & u_{1r} \\ \cdot & \cdot & \cdot & \cdot \\ u_{r0} & u_{r1} & \dots & u_{rr} \end{vmatrix},$$

so that finally we have

$$B = -\frac{m}{m-1}uH_r + \frac{x_0^{\ 2}}{(m-1)^2}H_{r+1},$$

as desired.

As a corollary it is noted that if H_{r+1} vanishes identically, then

$$(m-1)B + muH_r = 0,$$

i.e.

$$\begin{vmatrix} mu & u_1 & \dots & u_r \\ (m-1)u_1 & u_{11} & \dots & u_{r1} \\ (m-1)u_2 & u_{21} & \dots & u_{2r} \\ \cdot & \cdot & \cdot & \cdot \\ (m-1)u_r & u_{r1} & \dots & u_{rr} \end{vmatrix} = 0,$$

"et l'équation

$$u(x_1, x_2, \dots, x_r) = 0$$

est elle-même homogène"—a sort of converse of Euler's theorem above referred to.

BRIOSCHI, F. (1854, March).

[LA TEORICA DEI DETERMINANTI, E LE SUE PRINCIPALI APPLICAZIONI. viii+116 pp. Pavia.]

The last section (§ 11, pp. 106–116) of Brioschi's text-book is headed "Del determinante di Hesse." Opening with the definition of "l'Hessiano," it gives a clear and orderly exposition of a goodly number of the main theorems up till then discovered, with geometrical applications.*

Separated altogether, however, from these is a demonstration (pp. 20, 21) which strictly belongs to this section. Recognising that Hesse's expression (1847, August) for the product of u and its Hessian Δ is in reality obtained by eliminating x_1, x_2, x_3 from four equations, Brioschi performs this elimination openly, with the result:

* The reason given for the deduction $U_{rs} = Nx_r x_s$, which occurs in his presentation of Jacobi's proof of the year 1849, is disappointing.

$$0 = \begin{vmatrix} \frac{m}{m-1}u & u_1 & u_2 & \dots & u_n \\ u_1 & u_{11} & u_{12} & \dots & u_{1n} \\ \cdot & \cdot & \cdot & \cdot & \cdot \\ u_n & u_{n1} & u_{n2} & \dots & u_{nn} \end{vmatrix}:$$

$$= \frac{m}{m-1}u\Delta + \begin{vmatrix} \cdot & u_1 & u_2 & \dots & u_n \\ u_1 & u_{11} & u_{12} & \dots & u_{1n} \\ u_2 & u_{21} & u_{22} & \dots & u_{2n} \\ \cdot & \cdot & \cdot & \cdot & \cdot \\ u_n & u_{n1} & u_{n2} & \dots & u_{nn} \end{vmatrix}.$$

CAYLEY, A. (1856).

[A second memoir on quantics. A third memoir on quantics. *Philos. Transac. R. Soc.* (London), cxlvi. pp. 101–126, pp. 627–647; or *Collected Math. Papers*, ii. pp. 250–275, pp. 310–332.]

In his tables of invariants and covariants Cayley gives the Hessian of the binary quartic, binary quintic, binary sextic, binary octavic and ternary cubic. The results are those numbered 9, 15, 33, 42, 61.*

BELLAVITIS, G. (1857, June).

[Sposizione elementare della teorica dei determinanti. *Memorie . . . Istituto Veneto . . . .* vii. pp. 67–144.]

Bellavitis (§§ 79, 80) denotes "l'*Hessiano* delle funzione ϕ" by

$$| D_x D_x \ D_y D_y \ \dots\dots | \ \phi,$$

calling it also "il determinante delle derivate-seconde." He confines himself to three of the main theorems. Hesse's theorem of 1851 (March) he amplifies, his enunciation being:—*If* u, *a homogeneous integral function of the variables* $x_1, x_2, \dots, x_n$, *be transformed by means of the substitution*

$$x_k = a_1^{(k)}y_1 + a_2^{(k)}y_2 + \dots + a_n^k y_n$$

*In the third column of this last in the *Collected Math. Papers*, *cij* and *fkl* should be *cfj* and *gil*, and in the eighth column *ach* should be *ack*.

into v, *and one of the new variables, say* y_1, *be absent from* v, *then* (1) *the Hessian* H *of* u *must vanish identically,* (2) *the cofactor of the elements of any row of* H *must be proportional to the coefficients of* y_1 *in the substitution,* (3) *the product of the first differential-quotients of* u *by the said column of coefficients is equal to* 0. The third of these Bellavitis reaches very easily, because generally we have

$$\frac{\partial v}{\partial y_r} = \frac{\partial u}{\partial x_1}\cdot\frac{\partial x_1}{\partial y_r} + \frac{\partial u}{\partial x_2}\cdot\frac{\partial x_2}{\partial y_r} + \ldots + \frac{\partial u}{\partial x_n}\cdot\frac{\partial x_n}{\partial y_r},$$

and therefore when $r=1$

$$0 = \frac{\partial u}{\partial x_1}a_1^{(1)} + \frac{\partial u}{\partial x_2}a_1^{(2)} + \ldots + \frac{\partial u}{\partial x_n}a_1^{(n)}$$

or

$$0 = u_1 a_1^{(1)} + u_2 a_1^{(2)} + \ldots + u_n a_1^{n}. \qquad (\pi)$$

As regards the second he notes that on account of the vanishing of H we have in the first place

$$U_{11} : U_{12} : \ldots : U_{1n} = U_{s1} : U_{s2} : \ldots : U_{sn};$$

and in the second place * the set of equations

$$\left.\begin{array}{l} u_1 U_{11} + u_2 U_{12} + \ldots + u_n U_{1n} = 0 \\ u_1 U_{21} + u_2 U_{22} + \ldots + u_n U_{2n} = 0 \\ \cdot\ \cdot\ \cdot\ \cdot\ \cdot\ \cdot\ \cdot\ \cdot\ \cdot\ \cdot\ \cdot\ \cdot\ \cdot\ \cdot \end{array}\right\},$$

from which there is the evident deduction that the said set reduces to a single equation: the identity of this equation with (π) is then assumed.

Hesse's converse theorem he treats with a wise caution, deducing as before from the vanishing of H the existence of a single equation of the form

$$\alpha\frac{\partial u}{\partial x_1} + \beta\frac{\partial u}{\partial x_2} + \ldots = 0,$$

but then adding, "ma rimane da dimostrare che le $\alpha, \beta, \ldots$ sieno quantità costanti."

Lastly, he notes that if there be *two* such equations with constant coefficients, the function is transformable into one with two fewer variables, and all the primary minors of H vanish.

* See (a) in Jacobi's proof of 1849.

BALTZER, R. (1857).

[THEORIE UND ANWENDUNGEN DER DETERMINANTEN, mit . . . . . vi+129 pp. Leipzig, 1857.]

In Hesse's converse theorem of 1851 (March) Baltzer (§ 13, 3) wisely substitutes for $\Delta = 0$ the condition

$$c_1 u_1 + c_2 u_2 + \ldots + c_n u_n = 0$$

(which by a property of Jacobians implies $\Delta = 0$), his proof being that the substitution

$$x_k = b_{k1} y_1 + b_{k2} y_2 + \ldots + b_{k,n-1} y_{n-1} + c_k y_n$$

will then give $\partial u / \partial y_n = 0$, for

$$\begin{aligned} \frac{\partial u}{\partial y_n} &= \frac{\partial u}{\partial x_1}\frac{\partial x_1}{\partial y_n} + \frac{\partial u}{\partial x_2}\frac{\partial x_2}{\partial y_n} + \ldots + \frac{\partial u}{\partial x_n}\frac{\partial x_n}{\partial y_n} \\ &= u_1 c_1 + u_2 c_2 + \ldots + u_n c_n . \end{aligned}$$

In the second place, from the same $n+1$ equations, namely

$$\left.\begin{array}{l} -(m-1)\dfrac{mu}{m-1} + u_1 x_1 + \ldots + u_n x_n = 0 \\ -(m-1)u_1 + u_{11} x_1 + \ldots + u_{n1} x_n = 0 \\ \cdot\quad\cdot\quad\cdot\quad\cdot\quad\cdot\quad\cdot\quad\cdot\quad\cdot\quad\cdot\quad\cdot \\ -(m-1)u_n + u_{1n} x_1 + \ldots + u_{nn} x_n = 0 \end{array}\right\},$$

he obtains (§ 14, 4)

$$\left.\begin{aligned} -(m-1) : x_1 : x_2 : \ldots : x_n &= \Delta : V_1 : V_2 : \ldots : V_n \\ &= V_1 : V_{11} : V_{21} : \ldots : V_{n1} \\ &= \cdot\quad\cdot\quad\cdot\quad\cdot\quad\cdot\quad\cdot\quad\cdot \\ &= V_n : V_{1n} : V_{2n} : \ldots : V_{nn} \end{aligned}\right\},$$

where the V's are the cofactors of the corresponding u's in the resultant of the set of equations. The first and $(r+1)^{\text{th}}$ lines imply respectively

$$-(m-1) : x_r = \Delta : V_r ,$$

$$\text{and} \qquad -(m-1) : x_s = V_r : V_{sr} ,$$

the former of which gives

$$V_r = -\frac{x_r}{m-1}\Delta , \tag{a}$$

and the two together

$$V_{sr} = \frac{x_s x_r}{(m-1)^2}\Delta . \qquad (\beta)$$

The result (a) is essentially the same as the first result reached in Jacobi's proof of 1849 (December)—a proof which Baltzer restates (§ 14, 7, 8) without noting the fact; and (β) is essentially the same as the second of the two results given in Hesse's paper of 1847 (August), it being noted, however, that the case of this where $r=s$ had been established by Hesse in 1844 (see *Crelle's Journal*, xxviii. p. 103, lines 1 and 2).

HESSE, O. (1858).

[Zur Theorie der ganzen homogenen Functionen. *Crelle's Journ.*, lvi. pp. 263–269; or *Werke*, pp. 481–488.]

The first part of Hesse's attempted proof of his converse theorem of 1851 (March) was to show that the vanishing of "the determinant of u" led to the establishment of a linear relation connecting the first differential coefficients of u. In this there was an oversight, which Brioschi repeated, but which Bellavitis and Baltzer, from the course followed by them, must have been conscious of. Accordingly, Hesse now returns to the subject, the one object of his six-page paper being "diesen Lehrsatz strenger zu begründen." He first clears the ground a little by setting aside the case where one, and therefore all, of the primary minors of Δ vanish, merely stating that a linear substitution is then possible which will transform u into a function with *two* variables less than before. He then sets himself to supply the want which Bellavitis had drawn attention to, the result being a lengthy (pp. 265–268) and still unconvincing argument.

CHAPTER XIV.

CIRCULANTS, UP TO 1860.

So far as mathematical writers have as yet noted, a set of equations of the type

$$\left.\begin{array}{l} a_1x_1 + a_2x_2 + \ldots + a_nx_n \quad = u_1 \\ a_nx_1 + a_1x_2 + \ldots + a_{n-1}x_n = u_2 \\ a_{n-1}x_1 + a_nx_2 + \ldots + a_{n-2}x_n = u_3 \\ \cdot \quad \cdot \quad \cdot \quad \cdot \quad \cdot \quad \cdot \quad \cdot \quad \cdot \quad \cdot \quad \cdot \\ a_2x_1 + a_3x_2 + \ldots + a_1x_n \quad = u_n \end{array}\right\}$$

had not made its appearance in mathematical work prior to the year 1846: and it is almost absolutely certain that before that year the *determinant* of such a set had never been considered. It is not at all unlikely, however, that the expression

$$a^3+b^3+c^3-3abc$$

which is the case of the determinant for n equal to 3 had more than once turned up in other connections, and that its divisibility by $a+b+c$ had been noted: but of this, too, there is no record.

CATALAN, E. (1846).

[Recherches sur les déterminants. *Bull. de l'Acad. roy. . . . . . de Belgique*, xiii. pp. 534–555.]

As has been already explained, about half of Catalan's paper is occupied with an elementary exposition of known properties of determinants and with the establishment of a fresh theorem of his own, which in his notation might have been written in the form

$$\text{dét.}\,(A_1+A_2+\ldots+A_n,\ A_1-A_2,\ A_2-A_3,\ \ldots,\ A_{n-1}-A_n)$$
$$=(-1)^{n-1}\cdot n\cdot\text{dét.}\,(A_1,\ A_2,\ \ldots,\ A_n),$$

and which is to the effect that *If from a determinant* Δ *of the* n^{th} *order we form another* Δ', *such that the first row of* Δ' *is the sum of all the rows of* Δ, *and every other row of* Δ' *is got by subtracting the corresponding row of* Δ *from the row preceding it in* Δ, *then*

$$\Delta'=(-1)^{n-1}n\Delta.*$$

Strange to say, almost all the examples given in illustration of this theorem (of § 13) are of the special form distinguished at a later date by the name "circulant," and consequently fall now to be considered. He says (§ 17):—

"Afin de sortir de ces généralités, considérons les équations

$$\left.\begin{array}{r} -x_1+x_2+x_3+\ \ldots\ +x_n = u_1 \\ x_1-x_2+x_3+\ \ldots\ +x_n = u_2 \\ \ldots\ldots\ldots\ldots \\ x_1+x_2+x_3+\ \ldots\ -x_n = u_n \end{array}\right\}.$$

Pour obtenir le déterminant Δ, je remplace d'abord les équations données par les suivantes :

$$\left.\begin{array}{r} (n-2)x_1+(n-2)x_2+\ \ldots\ +(n-2)x_n = u_1+u_2+\ \ldots\ +u_n \\ -2x_1+2x_2 = u_1-u_2 \\ -2x_2+2x_3 = u_2-u_3 \\ \ldots\ldots\ldots \\ -2x_{n-1}+2x_n = u_{n-1}-u_n \end{array}\right\}.$$

D'après ce qui précède, le déterminant Δ' du nouveau système sera $(-1)^{n-1}n\Delta$. Mais, d'un autre côté, en comparant Δ' au déterminant Δ'' du système

$$\left.\begin{array}{r} x_1+x_2+\ \ldots\ +x_3 = \ldots \\ -x_1\ \ +x_2 = \ldots \\ -x_2\ \ +x_3 = \ldots \\ \ldots\ldots\ldots \\ -x_{n-1}+x_n = \ldots \end{array}\right\}$$

* This result is reached in a way different from Catalan's by performing on Δ' the operation

$$\text{row}_1 + (n-1)\,\text{row}_2 + (n-2)\,\text{row}_3 + \ldots + \text{row}_n,$$

separating out the factor n, and then showing that the resulting determinant is $(-1)^{n-1}\Delta$.

on a $\Delta' = (n-2)\,2^{n-1}\Delta''$. Enfin, d'après le n° 13, et en observant que les quantités $A_1 - A_2$, $A_2 - A_3$, . . . ont ici changé de signe

$$\Delta'' = n.$$

On déduit, de ces diverses formules

$$\Delta = (n-2)(-2)^{n-1}."$$

This result, which at a later date would have been written

$$C(-1, 1, 1, \ldots, 1) = (n-2)(-2)^{n-1},$$

and which, we may point out in passing, could also be reached by the operations

$$\begin{aligned} &\text{row}_1 + \text{row}_2 + \ldots + \text{row}_n, \\ &\text{removal of factor } n-2, \\ &\text{row}_n - \text{row}_{n-1},\ \text{row}_{n-1} - \text{row}_{n-2},\ \ldots . \end{aligned}$$

is then attempted to be generalised (§ 18) by withdrawing the restriction as to the number of negative units in a row. The reasoning, however, seems to have been incautiously conducted, the extension arrived at being

$$C(-1, -1, \ldots, 1, 1)_{p,n-p} = (n-2p)(-2)^{n-1},$$

where the number of consecutive negative units in the first row is p, and the number of positive units $n-p$.

Catalan then passes (§ 19) to the consideration of the similar circulant whose first row consists of p consecutive positive units followed by $n-p$ zeros, separating the investigation into two parts, (1) the case where p and n have a common factor other than unity, (2) where they are mutually prime. In the former case he shows that the equations which have the circulant in question for determinant are "indéterminées ou incompatibles"; in the latter case he shows that the equations are determinate. He thereupon goes on to supplement the information in the second case by proving that the circulant is equal to p: he omits, however, any similar proof that in the first case the circulant is zero.*

* Catalan proposed a question on this subject in *Nouv. Annales de Math.* xv. (1856), p. 257, and returned again to it in *Nouv. Corresp. Math.* iv. (1868), p. 78.

Lastly, he attacks the general circulant, or, as he calls it, "le déterminant du système

$$\begin{matrix} a_1 & a_2 & a_3 & \dots & a_n \\ a_2 & a_3 & a_4 & \dots & a_1 \\ a_3 & a_4 & a_5 & \dots & a_2 \\ \cdot & \cdot & \cdot & \cdot & \cdot \\ a_n & a_1 & a_2 & \dots & a_{n-1}. \end{matrix}$$"

The procedure, however, is rather perverse, the theorem of § 13 being forced into service. This gives

$$\Delta = (-1)^{n-1}\frac{1}{n}\Delta',$$

where Δ' is the determinant of the system

$$\begin{matrix} s & a_1-a_2 & a_2-a_3 & \dots\dots & a_{n-1}-a_n \\ s & a_2-a_3 & a_3-a_4 & \dots\dots & a_n\ \ -a_1 \\ s & a_3-a_4 & a_4-a_5 & \dots\dots & a_1\ \ -a_2 \\ \cdot & \cdot & \cdot & \cdot & \cdot \\ s & a_n-a_1 & a_1-a_2 & \dots\dots & a_{n-2}-a_{n-1}, \end{matrix}$$

after which Δ'/s is partitioned into determinants with monomial elements, and certain more or less evident reductions made. The result is "Le déterminant du système proposé s'obtiendra en multipliant $a_1 + a_2 + \dots + a_n$ (i.e. s) par le déterminant du système

$$\begin{matrix} a_1-a_2 & a_2-a_3 & \dots\dots & a_{n-1}-a_n \\ a_2-a_3 & a_3-a_4 & \dots\dots & a_n\ \ -a_1 \\ \cdot & \cdot & \cdot & \cdot \\ a_{n-1}-a_n & a_n-a_1 & \dots\dots & a_{n-3}-a_{n-2}, \end{matrix}$$"

a theorem which afterwards came to be written in the form

$$\begin{aligned} C(a_1, a_2, \dots, a_n) = {} & (a_1+a_2+\dots+a_n) \\ & \cdot P(a_1-a_2, \dots, a_{n-1}-a_n, a_n-a_1, \dots, a_{n-3}-a_{n-2}), \end{aligned}$$

the symbol $P(x, y, z, w, v)$ being used to stand for the "persymmetric" determinant

$$\begin{vmatrix} x & y & z \\ y & z & w \\ z & w & v \end{vmatrix}.$$

BERTRAND, J. (1850).

[TRAITÉ ELÉMENTAIRE D'ALGÈBRE, avec un grand nombre d'exercices: par Joseph Bertrand. 407 pp. Paris.]

On page 25 the student is asked to prove that if

$$\begin{aligned} A &= bc' + cb' + aa', \\ B &= ab' + ba' + cc', \\ C &= ac' + ca' + bb', \end{aligned}$$

then

$$\begin{aligned} A+B+C &= (a+b+c)(a'+b'+c'), \\ A^2+B^2+C^2 - AB-AC-BC &= (a^2+b^2+c^2-ab-ac-bc) \\ &\quad \cdot(a'^2+b'^2+c'^2-a'b'-a'c'-b'c'), \\ A^3+B^3+C^3 - 3ABC &= (a^3+b^3+c^3-3abc) \\ &\quad \cdot(a'^3+b'^3+c'^3-3a'b'c'). \end{aligned}$$

These results may possibly have been got by the multiplication of two circulants of the third order, but as against this it has to be noted that determinants are not referred to in the book.

SPOTTISWOODE, W. (1853).

[Elementary theorems relating to determinants. Rewritten and much enlarged by the author. *Crelle's Journ.*, li. pp. 209–271, 328–381.]

In the section (§ xi.) which did not appear in the first edition, and which bears the title "Miscellaneous instances of determinants," the following is given (p. 375), being the fourth of the said instances:—

"Let 1, i_1, i_2, . . . , i_n be the $n+1$ roots of the equation

$$x^{n+1} - 1 = 0,$$

then, whatever be the values of A, A_1, A_2, . . . , A_n

$$\begin{vmatrix} A & A_1 & \ldots & A_n \\ A_1 & A_2 & \ldots & A_1 \\ \cdot & \cdot & \cdot & \cdot \\ A_n & A_1 & \ldots & A_{n-1} \end{vmatrix} = (A + A_1 + \ldots + A_n)(A + i_1A_1 + \ldots + i_1^nA_n) \\ \ldots\ldots\ldots (A + i_nA_1 + \ldots + i_n^nA_n)."$$

No word of proof is added: probably the result was reached by Sylvester's "dialytic" method of elimination. But, however this may be, it should be noted that resolvability into linear factors soon came to be looked on as the fundamental property of the circulant.

It has to be noted that Spottiswoode makes a slip in omitting the sign-factor $(-1)^{\frac{1}{2}n(n-1)}$ from the right-hand member; and that he writes his determinant in such a way as to have it persymmetric with respect to the principal diagonal, whereas Catalan wrote his so as to have it persymmetric with respect to the secondary diagonal. Putting C′ for the functional symbol in the former case we have

$$C(a_1, a_2, \ldots, a_n) = (-1)^{\frac{1}{2}(n-1)(n-2)} \cdot C'(a_1, a_2, \ldots, a_n).$$

If therefore Spottiswoode had followed Catalan's mode of writing, his result would have been strictly accurate.

SYLVESTER, J. J. (1855, April).

[On the change of systems of independent variables. *Quart. Journ. of Math.*, i. pp. 42–56; or *Collected Math. Papers*, ii. pp. 65–85.]

Having reached in the course of his investigation (p. 55) a determinant of the form

$$\begin{vmatrix} a_1+a_2+a_3 & -b_3 & -c_2 \\ -a_2 & b_1+b_2+b_3 & -c_3 \\ -a_3 & -b_2 & c_1+c_2+c_3 \end{vmatrix},$$

the final expansion of which, he says, contains only positive terms with the coefficient unity, Sylvester naturally notes that the number of such terms must be

$$\begin{vmatrix} 3 & -1 & -1 \\ -1 & 3 & -1 \\ -1 & -1 & 3 \end{vmatrix}.$$

He is thus led to the consideration of the n-line circulant

$$\begin{vmatrix} a & -1 & -1 & \ldots & -1 \\ -1 & a & -1 & \ldots & -1 \\ -1 & -1 & a & \ldots & -1 \\ \cdot & \cdot & \cdot & \cdot & \cdot \\ -1 & -1 & -1 & \ldots & a \end{vmatrix},$$

to which he assigns the value

$$(a-n+1)(a+1)^{n-1}.$$

CREMONA, L. (1856).

[Intorno ad un teorema di Abel. *Annali di Sci. mat. e fis.*, vii. pp. 99–105.]

To prove the theorem of Abel referred to in the title, Cremona starts by establishing three lemmas, the first of which is Spottiswoode's theorem regarding circulants. Taking any n quantities

$$a_0,\ a_1,\ a_2,\ \ldots,\ a_{n-1}$$

and denoting

$$a_0+a_1\alpha_r^1+a_2\alpha_r^2+\ldots+a_{n-1}\alpha_r^{n-1} \quad \text{by} \quad \theta_r$$

where α_r stands for α^r and α for a primitive root of the equation $x^n-1=0$, he multiplies the determinant

$$\begin{vmatrix} a_0 & a_1 & a_2 & \ldots & a_{n-1} \\ a_1 & a_2 & a_3 & \ldots & a_0 \\ a_2 & a_3 & a_4 & \ldots & a_1 \\ \cdot & \cdot & \cdot & \cdot & \cdot \\ a_{n-1} & a_0 & a_1 & \ldots & a_{n-2} \end{vmatrix}, \quad \text{or D say,}$$

by the determinant

$$\begin{vmatrix} 1 & 1 & 1 & \ldots & 1 \\ 1 & \alpha_1 & \alpha_2 & \ldots & \alpha_{n-1} \\ 1 & \alpha_1^2 & \alpha_2^2 & \ldots & \alpha_{n-1}^2 \\ \cdot & \cdot & \cdot & \cdot & \cdot \\ 1 & \alpha_1^{n-1} & \alpha_2^{n-1} & \ldots & \alpha_{n-1}^{n-1} \end{vmatrix}, \quad \text{or } \Delta \text{ say,}$$

and obtains a product-determinant from whose columns, he says, the factors $\theta_1, \theta_2, \ldots, \theta_n$ may be removed in order, so that there results

$$\mathrm{D}\Delta = \theta_1\theta_2 \ldots \theta_n \begin{vmatrix} 1 & 1 & 1 & \ldots & 1 \\ 1 & a_{n-1} & a_{n-1}^2 & \ldots & a_{n-1}^{n-1} \\ 1 & a_{n-2} & a_{n-2}^2 & \ldots & a_{n-2}^{n-1} \\ \cdot & \cdot & \cdot & \cdot & \cdot \\ 1 & a_1 & a_1^2 & \ldots & a_1^{n-1} \end{vmatrix}$$

$$= \theta_1\theta_2 \ldots \theta_n \cdot (-1)^{\frac{1}{2}n(n-1)}\Delta,$$

and $\therefore$ $$\mathrm{D} = (-1)^{\frac{1}{2}n(n-1)} \cdot \theta_1\theta_2 \ldots \theta_n.$$

The proof, which is said to be due to Brioschi, is not improved in neatness by introducing the conception of a primitive root, nor by writing the root 1 in a different form from the other roots.

The second lemma concerns the differential-quotient of D with respect to any variable of which the a's are functions. Denoting this differential-quotient by D′, and by D_r the determinant got from D by substituting for each element in the rth column the differential-quotient of that element, Cremona of course has at once

$$\mathrm{D}' = \mathrm{D}_1 + \mathrm{D}_2 + \ldots + \mathrm{D}_n.$$

As, however, D_1 here can be shown by translation of a number of rows and the same number of columns to be equal to any one of the D's following it, there results

$$\mathrm{D}' = n\mathrm{D}_1 = n\mathrm{D}_2 = \ldots.$$

The third lemma is to the effect that the quotient of the determinant

$$\begin{vmatrix} m_0 & q_0 & q_1 d & \ldots\ldots & q_{n-2}d^{n-2} \\ m_1 d & q_1 d & q_2 d^2 & \ldots\ldots & q_{n-1}d^{n-1} \\ \cdot & \cdot & \cdot & \cdot & \cdot \\ m_{n-1}d^{n-1} & q_{n-1}d^{n-1} & q_0 & \ldots\ldots & q_{n-3}d^{n-3} \end{vmatrix}$$

by d is a rational function of d^n. By multiplying the 2nd, 3rd, 4th, ... columns by $d^n, d^{n-1}, d^{n-2}, \ldots$ respectively, and then dividing the corresponding rows by d, d^2, d^3, ... respectively, there is obtained

$$\begin{vmatrix} m_0 & q_0 d^n & q_1 d^n & \dots & q_{n-2}d^n \\ m_1 & q_1 d^n & q_2 d^n & \dots & q_{n-1}d^n \\ m_2 & q_2 d^n & q_3 d^n & \dots & q_0 \\ \cdot & \cdot & \cdot & \cdot & \cdot \\ m_{n-1} & q_{n-1}d^n & q_0 & \dots & q_{n-3} \end{vmatrix},$$

where no power of d occurs except the nth. But, if the original determinant be H, the latter is

$$\frac{\mathrm{H}\cdot d^n d^{n-1} d^{n-2} \dots d^2}{d d^2 d^3 \dots d^{n-1}}, \quad i.e. \quad \frac{\mathrm{H}}{d}\cdot d^n;$$

consequently H/d is of the form asserted.

In connection with this last lemma it is curious to find no note taken of the closely related and more attractive fact that

$$\mathrm{C}(a_1, a_2 d, a_3 d^2, \dots, a_n d^{n-1})$$

is a rational function of d^n.

BELLAVITIS, G. (1857).

[Sposizione elementare della teoria dei determinanti. *Memorie ... Istituto Veneto* .... viii. pp. 67–143.]

Circulants are practically unconsidered by Bellavitis in his exposition, all that appears (§ 85) being two of Laplace's expansions for $\mathrm{C}(a, b, c, d)$ obtained by means of Cauchy's "chiavi algebriche," namely,

$$(a^2-bd)^2-(b^2-ac)^2+(c^2-bd)^2-(d^2-ac)^2-2(ab-cd)(ad-bc)$$

and $$(a^2-c^2)^2-(b^2-d^2)^2-4(ab-cd)(ad-bc).$$

PAINVIN, L. (1858); ROBERTS, M. (1859).

[Questions 432, 465. *Nouv. Annales de Math.*, xvii. p. 185; xviii. p. 117; xix. pp. 151–153, 170–174.]

Here it is special circulants that are set for consideration, namely, by Painvin the circulant whose elements are the first n

integers, and by Michael Roberts the circulant whose elements are $a,\ a+d,\ a+2d,\ \ldots .$, the result in regard to the former circulant being

$$C'(1, 2, \ldots, n) = (-1)^{\frac{1}{2}n(n-1)} \cdot \tfrac{1}{2}n^{n-1}(n+1),$$

and in regard to the latter

$$C'(a, a+d, \ldots, a+\overline{n-1}\cdot d) = (-1)^{\frac{1}{2}n(n-1)} \cdot (nd)^{n-1} \cdot \left(a+\frac{n-1}{2}d\right).$$

The first to offer a proof was Cremona, who, after repeating (xix. pp. 151–153) Brioschi's demonstration regarding the resolvability of a circulant, says that in Roberts' case θ_r being

$$\begin{aligned} &\equiv a\frac{1-a_r^n}{1-a_r} + d\left\{a_r\frac{1-a_r^{n-1}}{(1-a_r)^2} - \frac{na_r^n}{1-a_r}\right\}, \\ &= \frac{nd}{a_r-1} \quad \text{for } r = 1, 2, \ldots, n-1, \end{aligned}$$

and $\qquad \theta_n = na + \tfrac{1}{2}n(n-1)d,$

and that consequently

$$\theta_1\theta_2 \ldots \theta_n = \frac{(nd)^{n-1}}{(a_1-1)(a_2-1)\ldots(a_{n-1}-1)}\left\{na+\tfrac{1}{2}n(n-1)d\right\};$$

whence the desired result readily follows, because the denominator is equal to

$$(-1)^{n-1}(1-\Sigma a_1+\Sigma a_1a_2-\Sigma a_1a_2a_3+\ldots),$$

where $\Sigma a_1 = -1,\ \Sigma a_1a_2 = 1,\ \Sigma a_1a_2a_3 = -1,\ \ldots .$

A proof was also given by G. F. Baehr of Groningen (xix. pp. 170–173), who changes

$$C'(a_1, a_2, \ldots, a_n) \quad \text{into} \quad (-1)^{\frac{1}{2}n(n-1)}C(a_n, a_{n-1}, \ldots, a_1),$$

performs on the latter determinant the operations

$\text{col}_1 - \text{col}_2,\ \ \text{col}_2 - \text{col}_3,\ \ \ldots .$

$\text{row}_1 + \text{row}_2 + \ \ldots\ + \text{row}_n,$

removal of factors d^n and $(-1)^{n-1}\cdot\frac{1}{2}n\left\{2a+(n-1)d\right\}$, leaving as cofactor a determinant of the $(n-1)^{\text{th}}$ order whose diagonal elements are all $1-n$ and non-diagonal elements all 1. On this new determinant he then performs the operations

$$\text{row}_1 + \text{row}_2 + \ldots + \text{row}_{n-1},$$
$$\text{row}_2 + \text{row}_1, \quad \text{row}_3 + \text{row}_1, \quad \ldots\ldots$$

and so finds its value to be

$$(-1)^{n-1} \cdot n^{n-2},$$

which gives for the circulant with which he started the value

$$(-1)^{\frac{1}{2}n(n-1)} \cdot (nd)^{n-1} \cdot \left(a + \frac{n-1}{2}d\right).$$

SOUILLART, C. (1858, May).

[Solution de la question 405. *Nouv. Annales de Math.*, xvii. pp. 192–194; xix. pp. 320–321.]

Michael Roberts having in November 1857 set the problem of finding X, Y, Z as functions of x, y, z, x', y', z', so as to have

$$(x^3+y^3+z^3-3xyz)(x'^3+y'^3+z'^3-3x'y'z') = \text{X}^3+\text{Y}^3+\text{Z}^3-3\text{XYZ},$$

—in other words, having resuscitated Bertrand's exercise of the year 1850,—Souillart showed that the solution

$$\left.\begin{aligned} \text{X} &= (x, y, z \between x', y', z') \\ \text{Y} &= (x, y, z \between y', z', x') \\ \text{Z} &= (x, y, z \between z', x', y') \end{aligned}\right\}$$

is comprised in a general theorem, namely, the theorem which at a later date would have been expressed by saying that the product of two circulants is a circulant. In 1860 he returned to the subject in order to point out that a second suitable set of values is

$$\begin{aligned} \text{X} &= (x, y, z \between z', y', x'), \\ \text{Y} &= (x, y, z \between x', z', y'), \\ \text{Z} &= (x, y, z \between y', x', z'). \end{aligned}$$

In illustrating he uses the fourth order, that is to say, where the initial expression is

$$\begin{vmatrix} x & y & z & u \\ y & z & u & x \\ z & u & x & y \\ u & x & y & z \end{vmatrix} \quad \text{or} \quad - \begin{vmatrix} x & u & z & y \\ y & x & u & z \\ z & y & x & u \\ u & z & y & x \end{vmatrix}$$

or

$$-x^4+y^4-z^4+u^4-4y^2xz-4xu^2z+4x^2yu+4yuz^2+2x^2z^2-2u^2y^2.$$

BAEHR, G. F. (1860).

[Solution de la question 432. *Nouv. Annales de Math.*, xix. pp. 170–174.]

After dealing as we have seen with the circulant whose elements are in equidifferent progression, Baehr proceeds to the circulant whose elements are in equirational progression, namely,

$$C'(a, ar, ar^2, \ldots, ar^{n-1}).$$

This he first changes into

$$a^n \cdot C'(1, r, r^2, \ldots, r^{n-1})$$

and then into

$$(-)^{\frac{1}{2}n(n-1)} \cdot a^n \cdot C(r^{n-1}, r^{n-2}, \ldots, r, 1).$$

On the determinant thus reached the operations

$$r\ \text{row}_1 - \text{row}_2, \quad r\ \text{row}_2 - \text{row}_3, \quad \ldots .$$

are performed, with the result that its value is found to be

$$(-1)^{n-1} \cdot (1-r^n)^{n-1},$$

and thence the value of the original circulant to be

$$(-1)^{\frac{1}{2}n(n-1)} \cdot a^n(r^n-1)^{n-1}.$$

It is worth noting that instead of the last set of operations we might substitute with advantage the set

$$\text{row}_n - r\ \text{row}_{n-1}, \quad \text{row}_{n-1} - r\ \text{row}_{n-2}, \quad \ldots . ;$$

also, that Baehr's circulant is a special case of that referred to under Cremona's third lemma.

CHAPTER XV.

CONTINUANTS, UP TO 1870.

THE more or less disguised use of continued fractions has been traced back to the publication of Bombelli's *Algebra* in 1572, eighty-four years, that is to say, before the publication of Wallis' *Arithmetica Infinitorum*, in which Brouncker's discovery was announced and the fractions explicitly expressed.* The study of the numerators and denominators of the convergents viewed as functions of the partial denominators was first seriously undertaken by Euler in his *Specimen Algorithmi Singularis* of the year 1764, in which denoting by

$$(a),\quad \frac{(a,\, b)}{(b)},\quad \frac{(a,\, b,\, c)}{(b,\, c)},\quad \ldots .$$

the convergents to

$$a + \frac{1}{b} + \frac{1}{c} + \ldots\ldots$$

he established a long series of identities, such as

$$
\begin{aligned}
(a, b, c, d, \ldots) &= a(b, c, d, \ldots) + (c, d, \ldots)\\
(a, b, c, \ldots l) &= (l, \ldots, c, b, a),\\
(a, b)(b, c) - (b)(a, b, c) &= 1,\\
(a, b, c)(d, e, f) - (a, b, c, d, e, f) &= -(a, b)(e, f),\\
\ldots\ldots\ldots
\end{aligned}
$$

* For the early history see Favaro's *Notizie storiche sulle frazioni continue dal secolo decimoterzo al decimosettimo* published in vol. vii. of Boncompagni's *Bollettino*; and as regards Bombelli see a paper by G. Wertheim in the *Abhandl. zur Gesch. d. Math.*, viii. pp. 147–160.

The study was pursued by Hindenburg and his followers during the last twenty years of the eighteenth century, but not with any great profit; and, although in the first half of the nineteenth century considerable attention was given to the theory of continued fractions as a whole, little advance was made in elucidating the properties of the functions referred to.* Their connection with determinants, after the awakening of interest in the latter about 1841, was sure sooner or later to be detected: there is no evidence, however, of the discovery having been made before the year 1853.

SYLVESTER, J. J. (1853, May 13).

[On a remarkable modification of Sturm's theorem. *Philos. Magazine* (4), v. pp. 446–457; or *Collected Math. Papers*, i. pp. 609–619.]

The mention of Sturm's theorem in the title of a paper renders not improbable the occurrence therein of matter connected with continued fractions. Especially likely is this in the case of a writer like Sylvester when in a characteristic mood; and, assuredly, the present communication is in structure, style, and originality redolent of its author. It must have been written in the white heat of discovery. The main part of it consists of six pages: this is followed by a "Remark" a page and a quarter long; then comes a "Postscript" of three and a half pages; and finally a small-page footnote as long as the "Remark."

It is the postscript which particularly concerns us. It begins thus:—

"Suppose that we have any series of terms, $u_1, u_2, u_3, \ldots, u_n$, where

$$u_1 = A_1, \quad u_2 = A_1A_2 - 1, \quad u_3 = A_1A_2A_3 - A_1 - A_3, \quad \ldots$$

and in general

$$u_i = A_i u_{i-1} - u_{i-2},$$

then $u_1, u_2, u_3, \ldots, u_n$ will be the successive principal coaxal determi-

* The state of the theory in 1833 can best be gathered from Stern's monograph, published in vol. x. of *Crelle's Journal*.

nants of a symmetrical matrix. Thus suppose $n=5$; if we write down the matrix

$$\begin{matrix} A_1 & 1 & 0 & 0 & 0 \\ 1 & A_2 & 1 & 0 & 0 \\ 0 & 1 & A_3 & 1 & 0 \\ 0 & 0 & 1 & A_4 & 1 \\ 0 & 0 & 0 & 1 & A_5 \end{matrix}$$

(the mode of formation of which is self-apparent), these successive coaxal determinants will be

$$1,\quad A_1,\quad \begin{vmatrix} A_1 & 1 \\ 1 & A_2 \end{vmatrix},\quad \begin{vmatrix} A_1 & 1 & 0 \\ 1 & A_2 & 1 \\ 0 & 1 & A_3 \end{vmatrix},\quad \begin{vmatrix} A_1 & 1 & 0 & 0 \\ 1 & A_2 & 1 & 0 \\ 0 & 1 & A_3 & 1 \\ 0 & 0 & 1 & A_4 \end{vmatrix},\ \text{etc.},$$

i.e.

$$\begin{aligned} &1,\quad A_1,\quad A_1A_2-1,\quad A_1A_2A_3-A_1-A_3, \\ &A_1A_2A_3A_4-A_1A_2-A_1A_4-A_3A_4+1, \\ &A_1A_2A_3A_4A_5-A_1A_2A_5-A_1A_4A_5-A_3A_4A_5-A_1A_2A_3 \\ &\qquad\qquad\qquad\qquad\qquad\qquad +A_5+A_3+A_1. \end{aligned}$$

It is proper to introduce the unit because it is, in fact, the value of a determinant of zero places, as I have observed elsewhere."

After using this as an aid to prove his proposition regarding Sturm's theorem, he returns to his new determinant in the following words:—

"I may conclude with noticing that the determinative [determinantal?] form of exhibiting the successive convergents to an improper continued fraction affords an instantaneous demonstration of the equation which connects any two consecutive such convergents as

$$\frac{N_{i-1}}{D_{i-1}} \quad \text{and} \quad \frac{N_i}{D_i},$$

namely,

$$N_i \cdot D_{i-1} - N_{i-1}D_i = 1.$$

For if we construct the matrix which for greater simplicity I limit to five lines and columns,

$$\begin{vmatrix} A & 1 & 0 & 0 & 0 \\ 1 & B & 1 & 0 & 0 \\ 0 & 1 & C & 1 & 0 \\ 0 & 0 & 1 & D & 1 \\ 0 & 0 & 0 & 1 & E \end{vmatrix}$$

and represent umbrally as

$$\begin{matrix} a_1 & a_2 & a_3 & a_4 & a_5 \\ b_1 & b_2 & b_3 & b_4 & b_5 ; \end{matrix}$$

and if, by way of example, we take the fourth and fifth convergents, these will be in the umbral notation represented by

$$\frac{\begin{matrix} a_2 & a_3 & a_4 \\ b_2 & b_3 & b_4 \end{matrix}}{\begin{matrix} a_1 & a_2 & a_3 & a_4 \\ b_1 & b_2 & b_3 & b_4 \end{matrix}} \quad \text{and} \quad \frac{\begin{matrix} a_2 & a_3 & a_4 & a_5 \\ b_2 & b_3 & b_4 & b_5 \end{matrix}}{\begin{matrix} a_1 & a_2 & a_3 & a_4 & a_5 \\ b_1 & b_2 & b_3 & b_4 & b_5 \end{matrix}}$$

respectively. Hence

$$N_5D_4 - N_4D_5$$

$$= \begin{matrix} a_2 & a_3 & a_4 & a_5 \\ b_2 & b_3 & b_4 & b_5 \end{matrix} \times \begin{matrix} a_2 & a_3 & a_4 & a_1 \\ b_2 & b_3 & b_4 & b_1 \end{matrix} - \begin{matrix} a_2 & a_3 & a_4 \\ b_2 & b_3 & b_4 \end{matrix} \times \begin{matrix} a_2 & a_3 & a_4 & a_5 & a_1 \\ b_2 & b_3 & b_4 & b_5 & b_1 , \end{matrix}$$

$$= \begin{matrix} a_2 & a_3 & a_4 & a_5 \\ b_2 & b_3 & b_4 & b_5 \end{matrix} \times \begin{matrix} a_2 & a_3 & a_4 & a_1 \\ b_2 & b_3 & b_4 & b_1 \end{matrix} - \overbrace{\begin{matrix} a_2 & a_3 & a_4 & a_5 \\ b_2 & b_3 & b_4 & b_5 \end{matrix} \; \begin{matrix} a_2 & a_3 & a_4 & a_1 \\ b_2 & b_3 & b_4 & b_1 \end{matrix}}$$

$$= \begin{matrix} a_2 & a_3 & a_4 & a_5 \\ b_2 & b_3 & b_4 & b_1 \end{matrix} \times \begin{matrix} a_2 & a_3 & a_4 & a_1 \\ b_2 & b_3 & b_4 & b_5 , \end{matrix}$$

$$= \begin{matrix} a_2 & a_3 & a_4 & a_5 \\ b_1 & b_2 & b_3 & b_4 \end{matrix} \times \begin{matrix} a_1 & a_2 & a_3 & a_4 \\ b_2 & b_3 & b_4 & b_5 , \end{matrix}$$

$$= \begin{matrix} 1 & B & 1 & 0 \\ 0 & 1 & C & 1 \\ 0 & 0 & 1 & D \\ 0 & 0 & 0 & 1 \end{matrix} \times \begin{matrix} 1 & 0 & 0 & 0 \\ B & 1 & 0 & 0 \\ 1 & C & 1 & 0 \\ 0 & 1 & D & 1 , \end{matrix}$$

$$= 1 \times 1 = 1,$$

as was to be proved. And the demonstration is evidently general in its nature."

In regard to this there has to be noted, first the use of

$$\begin{matrix} a_2 & a_3 & a_4 \\ b_2 & b_3 & b_4 \end{matrix}$$

when it would have been equally effective to use

$$\begin{matrix} 2 & 3 & 4 \\ 2 & 3 & 4 ; \end{matrix}$$

and, second, the use of a theorem for expressing the product of a five-line determinant and one of its secondary minors as an aggregate of products of pairs of four-line determinants.

Following on this comes the assertion that

"We may treat a proper continued fraction [*i.e.* with positive unit numerators] in precisely the same manner, substituting throughout $\sqrt{-1}$ in place of 1 in the generating matrix, and we shall thus, by the same process as has been applied to improper continued fractions, obtain

$$\begin{aligned} N_{i+1}D_i - N_iD_{i+1} &= (\sqrt{-1})^i \times (\sqrt{-1})^i \\ &= (-1)^i." \end{aligned}$$

This would seem to imply that as yet Sylvester had not observed that an alternative mode of representation was obtainable by merely changing the sign of the units on one side of the diagonal.

The footnote contains two additional observations, the first being to the effect that the new mode of representation

"gives an immediate and visible proof of the simple and elegant rule for forming any such numerators or denominators by means of the principal terms [term?] in each; the rule, I mean, according to which the ith denominator may be formed from

$$q_1q_2q_3q_4 \cdot \cdot \cdot q_i$$

($q_1, q_2, \ldots, q_i$ being the successive quotients) and the ith numerator from

$$q_2q_3q_4 \cdot \cdot \cdot q_i$$

by leaving out from the above products respectively any pair or any number of pairs of consecutive quotients as $q_\rho q_{\rho+1}$. For instance, from $q_1q_2q_3q_4q_5$ by leaving out q_1q_2, q_2q_3, q_3q_4 and q_4q_5 we obtain

$$q_3q_4q_5 + q_1q_4q_5 + q_1q_2q_5 + q_1q_2q_3 :$$

and by leaving out $q_1q_2 \cdot q_3q_4$, $q_1q_2 \cdot q_4q_5$, $q_2q_3 \cdot q_4q_5$ we obtain

$$q_5 + q_3 + q_1 ;$$

so that the total denominator becomes

$$q_1q_2q_3q_4q_5 + q_3q_4q_5 + q_1q_4q_5 + q_1q_2q_5 + q_1q_2q_3 + q_5 + q_3 + q_1 ;$$

and in like manner the numerator of the same convergent is

$$q_2q_3q_4q_5 \left\{ 1 + \frac{1}{q_2q_3} + \frac{1}{q_3q_4} + \frac{1}{q_4q_5} + \frac{1}{q_2q_3q_4q_5} \right\},$$

i.e.

$$q_2q_3q_4q_5 + q_4q_5 + q_2q_5 + q_2q_3 + 1."$$

The "rule" here spoken of is that enunciated for the more general case of

$$a_1 + \frac{b_1}{a_2 +} \frac{b_2}{a_3 +} \cdots\cdots$$

in Stern's *Theorie der Kettenbrüche*, the fourth section of which is given up to the consideration of such rules (*Crelle's Journ.*, x. pp. 4–7).

The other observation is to the effect that

"every progression of terms constructed in conformity with the equation

$$u_n = a_n u_{n-1} - b_n u_{n-2} + c_n u_{n-3} - \dots$$

may be represented as an ascending series of principal coaxal determinants to a common matrix. Thus if each term in such progression is to be made a linear function of the three preceding terms, it will be representable by means of the matrix

$$\begin{matrix} A & B' & C'' & 0 & 0 \\ 1 & A' & B'' & C''' & 0 \\ 0 & 1 & A'' & B''' & C'''' \\ 0 & 0 & 1 & A''' & B'''' \\ 0 & 0 & 0 & 1 & A'''' \end{matrix}$$

indefinitely continued, which gives the terms

$$1, \quad A, \quad AA' - B', \quad AA'A'' - B'A'' - AB'' + C'', \quad \dots \text{"}$$

This exhausts the paper so far as determinants are concerned: the results announced in it, one can readily own, were such as fairly to entitle the enthusiastic author to express his belief that "the introduction of the method of determinants into the algorithm of continued fractions cannot fail to have an important bearing upon the future treatment and development of the theory of numbers."

SPOTTISWOODE, W. (1853, August).*

[Elementary theorems relating to determinants. Second edition, rewritten and much enlarged by the author. *Crelle's Journ.*, li. (1856), pp. 209–271, 328–381.]

Save the utilisation of the fact that the denominator of any convergent of the continued fraction

$$a_1 + \frac{b_1}{a_2} + \frac{b_2}{a_3} + \dots$$

* This is the author's date at the end of the paper (p. 381). The first two parts of the volume, however, are dated 1855, and the remaining two 1856.

is the differential-quotient of the numerator, Spottiswoode did nothing but report the fundamental result reached by Sylvester. The full passage (p. 374) is as follows:—

"The improper continued fraction

$$\frac{1}{A} - \frac{1}{B} - \frac{1}{C} - \cdots \qquad = \frac{d}{dA}\log_e \nabla$$

where

$$\nabla = \begin{vmatrix} A & 1 & 0 & \dots & 0 & 0 \\ 1 & B & 1 & \dots & 0 & 0 \\ 0 & 1 & C & \dots & 0 & 0 \\ \cdot & \cdot & \cdot & \cdot & \cdot & \cdot \\ 0 & 0 & 0 & \dots & M & 1 \\ 0 & 0 & 0 & \dots & 1 & N \end{vmatrix},$$

in which any number of rows may be taken at pleasure, and the formula will give the corresponding convergent fraction.

The same holds good for the continued fraction

$$\frac{1}{A} + \frac{1}{B} + \cdots$$

if we write

$$\nabla = \begin{vmatrix} A & 1 & 0 & \dots \\ -1 & B & 1 & \dots \\ 0 & -1 & C & \dots \\ \cdot & \cdot & \cdot & \cdot \end{vmatrix}."$$

SYLVESTER, J. J. (1853, Sept.).

[On a fundamental rule in the algorithm of continued fractions. *Philos. Magazine* (4), vi. pp. 297–299; or *Collected Math. Papers*, i. pp. 641–644.]

Without any reference to his previous paper on the subject Sylvester here announces that if

$$(a_1, a_2, \dots, a_i)$$

be the denominator of the ith convergent to

$$\frac{1}{a_1} + \frac{1}{a_2} + \frac{1}{a_3} + \cdots$$

then

$$(a_1, \ldots, a_m, a_{m+1}, \ldots, a_{m+n}) = (a_1, \ldots, a_m)(a_{m+1}, \ldots, a_{m+n}) \\ + (a_1, \ldots, a_{m-1})(a_{m+2}, \ldots, a_{m+n}),$$

—a possibly new result which he considers "the fundamental theorem in the theory of continued fractions." This, he says, is an immediate consequence of the fact that $(a_1, \ldots, a_{m+n})$ can be expressed as a determinant, all that is further necessary being the application of the "well-known simple rule for the decomposition of determinants. Thus, *e.g.*, the determinant

$$\begin{array}{cccccc} a & 1 & & & & \\ -1 & b & 1 & & & \\ & -1 & c & 1 & & \\ & & -1 & d & 1 & \\ & & & -1 & e & 1 \\ & & & & -1 & f \end{array}$$

is obviously decomposable into

$$\begin{array}{ccc} a & 1 & \\ -1 & b & 1 \\ & -1 & c \end{array} \times \begin{array}{ccc} d & 1 & \\ -1 & e & 1 \\ & -1 & f \end{array} + \begin{array}{cc} a & 1 \\ -1 & b \end{array} \times \begin{array}{cc} e & 1 \\ -1 & f, \end{array}$$

or into

$$\begin{array}{cc} a & 1 \\ -1 & b \end{array} \times \begin{array}{cccc} c & 1 & & \\ -1 & d & 1 & \\ & -1 & e & 1 \\ & & -1 & f \end{array} + a \times \begin{array}{ccc} d & 1 & \\ -1 & e & 1 \\ & -1 & f, \end{array}$$

or into

$$a \times \begin{array}{ccccc} b & 1 & & & \\ -1 & c & 1 & & \\ & -1 & d & 1 & \\ & & -1 & e & 1 \\ & & & -1 & f \end{array} + \begin{array}{cccc} c & 1 & & \\ -1 & d & 1 & \\ & -1 & e & 1 \\ & & -1 & f. \end{array}$$

Following this is what is called "Corollary I.," namely,

$$a_1, a_2, \ldots, a_m)\cdot(a_2, a_3, \ldots, a_{m+i}) - (a_2, a_3, \ldots, a_m)\cdot(a_1, a_2, \ldots, a_{m+i}) \\ = (-)^m(a_{m+i}a_{m+i-1} \ldots \text{to } i-1 \text{ factors}),$$

its connection with the expression for the difference of two convergents being illustrated by the instances $i=1, 2, 3, 4, \ldots$

The next "corollary," namely,

$$
\begin{aligned}
&(a_1, \ldots, a_\rho, a_{\rho+1}, \ldots, a_{\rho+f})(a_1, \ldots, a_\rho, a_{\rho+1}, \ldots, a_{\rho+k}) \\
&- (a_1, \ldots, a_\rho, a_{\rho+1}, \ldots, a_{\rho+g})(a_1, \ldots, a_\rho, a_{\rho+1}, \ldots, a_{\rho+h}) \\
&= (-)^\rho\left\{(a_{\rho+1}, \ldots, a_{\rho+f})(a_{\rho+1}, \ldots, a_{\rho+k}) - (a_{\rho+1}, \ldots, a_{\rho+g})(a_{\rho+1}, \ldots, a_{\rho+h})\right\}
\end{aligned}
$$

is clearly incorrect, it being impossible for the value of the left-hand side to be independent of the elements $a_1, a_2, \ldots, a_\rho$. Further, as the author gives no accompanying word of comment, the difficulty of suggesting the true theorem is increased. A "sub-corollary" is appended dealing with the case where all the a's are equal, and leading up, not without some misprints or inaccuracies, to a theorem of Euler's quoted from the *Nouvelles Annales de Math.*, v. (Sept. 1851), pp. 357–358, to the effect that if $T_{n+1} = aT_n - bT_{n-1}$ be the generating equation of a recurrent series, then

$$\frac{T_{n+1}^2 - aT_nT_{n+1} + bT_n^2}{b^n}$$

is a constant with respect to n. Of course the more natural form of this expression is

$$\frac{T_{n+1}^2 - T_nT_{n+2}}{b^n},$$

the numerator of which being

$$\begin{vmatrix} T_{n+1} & T_{n+2} \\ T_n & T_{n+1} \end{vmatrix}$$

is successively transformable by means of the recursion-formula into

$$b\begin{vmatrix} T_n & T_{n+1} \\ T_{n-1} & T_n \end{vmatrix}, \quad b^2\begin{vmatrix} T_{n-1} & T_n \\ T_{n-2} & T_{n-1} \end{vmatrix}, \quad b^3\begin{vmatrix} T_{n-2} & T_{n-1} \\ T_{n-3} & T_{n-2} \end{vmatrix}, \quad \ldots,$$

so that the constant in question is

$$\begin{vmatrix} T_1 & T_2 \\ T_0 & T_1 \end{vmatrix}.$$

This, however, Sylvester does not show.*

* An interesting extension of this is given by Brioschi in the *Nouv. Annales de Math.*, xiv. (Jan. 1854), p. 20.

Finally, and to more purpose, it is noted that if we pass from $(a_1, a_2, \ldots, a_i)$ to the readily-suggested extension

$$\begin{matrix} m_1 & l_1 & & & & & \\ n_1 & m_2 & l_2 & & & & \\ & n_2 & m_3 & l_3 & & & \\ & & \cdot & \cdot & \cdot & \cdot & \\ & & & n_{i-1} & m_i & l_i & \\ & & & & n_i & m_{i+1}, & \end{matrix}$$

the corresponding fundamental theorem is

$$\begin{aligned} \begin{pmatrix} l_1 \ldots\ldots l_{i+j} \\ m_1, m_2, \ldots, m_{i+j+1} \\ n_1 \ldots\ldots n_{i+j} \end{pmatrix} &= \begin{pmatrix} l_1 \ldots\ldots l_{i-1} \\ m_1, m_2, \ldots, m_i \\ n_1 \ldots\ldots n_{i-1} \end{pmatrix} \begin{pmatrix} l_{i+1} \ldots\ldots l_{i+j} \\ m_{i+1}, m_{i+2}, \ldots, m_{i+j+1} \\ n_{i+1} \ldots\ldots n_{i+j} \end{pmatrix} \\ &\quad - l_i n_i \begin{pmatrix} l_1 \ldots\ldots l_{i-2} \\ m_1, m_2, \ldots, m_{i-1} \\ n_1 \ldots\ldots n_{i-2} \end{pmatrix} \begin{pmatrix} l_{i+2} \ldots\ldots l_{i+j} \\ m_{i+2}, m_{i+3}, \ldots, m_{i+j+1} \\ n_{i+2} \ldots\ldots n_{i+j} \end{pmatrix}. \end{aligned}$$

SYLVESTER, J. J. (1853, Oct., Nov.).

[On a theory of the syzygetic relations of two rational integral functions, comprising an application to the theory of Sturm's functions, and that of the greatest algebraical common measure. *Philos. Transac. R. Soc.* (London), cxliii. pp. 407–548; or *Collected Math. Papers,* i. pp. 429–586.]

Although this lengthy memoir in its original form bears date "16th June, 1853," it is the equally lengthy "supplements" added later while passing through the press that claim attention in the present connection. In the first of these (§ i., p. 474) the denominator of the fraction

$$\cfrac{1}{q_1 - \cfrac{1}{q_2 - \ddots - \cfrac{1}{q_n}}}$$

is denoted by $[q_1, q_2, \ldots, q_n]$, and termed a "cumulant," and throughout the later portion of the paper this name constantly recurs. It is not, however, until we come to the second "supplement" that anything apparently new in substance is met with. There, in § α (p. 497), the following lemma occurs:—

"The roots of the cumulant $[q_1, q_2, \ldots, q_i]$, in which each element is a linear function of x, and wherein the coefficient of x for each element has the like sign, are all real: and between every two of such roots is contained a root of the cumulant $[q_1, q_2, \ldots, q_{i-1}]$ and *ex converso* a root of the cumulant $[q_2, q_3, \ldots, q_i]$: and (as an evident corollary) for all values of ζ and ζ' intermediate between 1 and i the greatest root of $[q_1, q_2, \ldots, q_i]$ will be greater, and the least root of the same will be less than the greatest and least roots respectively of

$$[q_\rho, q_{\rho+1}, \ldots, q_{\rho'-1}, q_{\rho'}]."$$

Even this, however, may be placed under the well-known theorem regarding the roots of the equation

$$\begin{vmatrix} a_{11}-x & a_{12} & a_{13} & \ldots \\ a_{12} & a_{22}-x & a_{23} & \ldots \\ a_{13} & a_{23} & a_{33}-x & \ldots \\ \cdot & \cdot & \cdot & \cdot \end{vmatrix} = 0$$

which had been enunciated by Cauchy in 1829.

The next noteworthy result occupies § i. (p. 502). As a preparation for it the theorem

$$\begin{aligned}[a_1, a_2, \ldots, a_m, b_1, b_2, \ldots, b_n] &= [a_1, a_2, \ldots, a_m][b_1, b_2, \ldots, b_n] \\ &\quad - [a_1, a_2, \ldots, a_{m-1}][b_2, b_3, \ldots, b_n]\end{aligned}$$

may be recalled, the group of elements on the left being now viewed as consisting of two sub-groups. This theorem Sylvester writes in the form

$$[\Omega_1\Omega_2] = [\Omega_1][\Omega_2] - [\Omega_1'][{}'\Omega_2],$$

and he succeeds in including it in a general theorem, not explicitly formulated, in which the number of groups is i, the next two cases being

$$\begin{aligned}[\Omega_1\Omega_2\Omega_3] &= [\Omega_1][\Omega_2][\Omega_3] \\ &\quad - [\Omega_1'][{}'\Omega_2][\Omega_3] - [\Omega_1][\Omega_2'][{}'\Omega_3] + [\Omega_1'][{}'\Omega'][{}'\Omega_3],\end{aligned}$$

and

$$\begin{aligned}[\Omega_1\Omega_2\Omega_3\Omega_4] &= [\Omega_1][\Omega_2][\Omega_3][\Omega_4] \\ &\quad - [\Omega_1'][{}'\Omega_2][\Omega_3][\Omega_4] - [\Omega_1][\Omega_2'][{}'\Omega_3][\Omega_4] - [\Omega_1][\Omega_2][\Omega_3'][{}'\Omega_4] \\ &\quad + [\Omega_1'][{}'\Omega_2'][\Omega_3][\Omega_4] + [\Omega_1'][{}'\Omega_2][\Omega_3'][{}'\Omega_4] + [\Omega_1][\Omega_2'][{}'\Omega'][{}'\Omega_4] \\ &\quad - [\Omega_1'][{}'\Omega_2'][{}'\Omega_3'][{}'\Omega_4].\end{aligned}$$

The general theorem is described as giving an expression for $[\Omega_1\Omega_2 \ldots \Omega_i]$ in terms of

$$\begin{array}{l} [\Omega_1],\ [\Omega_2],\ \ldots,\ [\Omega_{i-1}],\ [\Omega_i] \\ [\Omega_1'],\ [\Omega_2'],\ \ldots,\ [\Omega_{i-1}'] \\ \qquad ['\Omega_2],\ \ldots,\ ['\Omega_{i-1}],\ ['\Omega_i] \\ \qquad ['\Omega_2'],\ \ldots,\ ['\Omega_{i-1}']; \end{array}$$

that is to say, in terms of all the unaltered Ω's, all the curtailed Ω's except the last, all the beheaded Ω's except the first, and all the "doubly-apocopated" Ω's except the first and the last; and it is pointed out that the number of products (or terms) in the expansion is 2^{i-1} "separable into i alternately positive and negative groups containing respectively

$$1,\quad (i-1),\quad \tfrac{1}{2}(i-1)(i-2),\quad \ldots,\quad i-1,\quad 1$$

products." Further, it is noted that "in every one of the above groups forming a product the accents enter in pairs and between contiguous factors, it being a condition that if any Ω have an accent on the right the next Ω must have one on the left, and if it have one on the left the preceding Ω must have an accent on the right, and the number of pairs of accents goes on increasing in each group from 0 to $i-1$." *

In a footnote the case where each Ω has only one element, and where, therefore, each singly-accented Ω becomes 1, and each doubly-accented Ω vanishes, is stated to be identical with the "rule"

$$[a_1,\ a_2,\ \ldots,\ a_i] = a_1a_2\ldots a_i - \sum\frac{1}{a_ea_{e+1}}\cdot a_1a_2\ldots a_i + \sum\frac{1}{a_ea_{e+i}a_fa_{f+1}}\cdot a_1a_2\ldots a_i - \ldots$$

formerly given by him in words.

* It is to be regretted that Sylvester did not give the recurrent law of formation

$$[\Omega_1\Omega_2\ldots\Omega_{r-1}\Omega_r] = [\Omega_1\Omega_2\ldots\Omega_{r-1}][\Omega_r] - [\Omega_1\Omega_2\ldots\Omega_{r-1}']['\Omega_r],$$

as this would have made all his statements clear and a number of them unnecessary.

SMITH, H. J. [S.] (1854, May).

[De compositione numerorum primorum formae $4\lambda+1$ ex duobus quadratis. *Crelle's Journ.*, l. pp. 91–92; or *Collected Math. Papers*, i. pp. 33–34.]

Pointing out, as Sylvester had already done, that the writing of Euler's algorithm $(q_1, q_2, \ldots, q_n)$ as a determinant leads easily to the properties

$$\begin{aligned}(q_1, q_2, \ldots, q_i) &= (q_i, q_{i-1}, \ldots, q_1),\\ (q_1, q_2, \ldots, q_n) &= (q_1, q_2, \ldots, q_i)(q_{i+1}, \ldots q_n)\\ &\quad +(q_1, q_2, \ldots, q_{i-1})(q_{i+2}, \ldots, q_n),\end{aligned}$$

Smith therefrom deduces that for centro-symmetric series of elements

$$(q_1, q_2, \ldots, q_i, q_i, \ldots, q_2, q_1) = q_1, q_2, \ldots, q_i)^2+(q_1, q_2, \ldots, q_{i-1})^2,$$

and

$$(q_1, q_2, \ldots, q_{i-1}, q_i, q_{i-1}, \ldots, q_1) = (q_1, q_2, \ldots, q_{i-1})\{q_1, \ldots, q_i)+(q_1, \ldots, q_{i-2}\},$$

noting in regard to the former that the two numbers squared on the right are mutually prime. He then makes application to the theorem referred to in the title of his paper.

SYLVESTER, J. J. (1854, August).

[Théorème sur les déterminants de M. Sylvester. *Nouv. Annales de Math.*, xiii. p. 305; or *Collected Math. Papers.*, ii. p. 28.]

This communication in its entirety is as follows:—

"Soient les déterminants

$$\lambda, \qquad \begin{matrix}\lambda & 1\\ 1 & \lambda,\end{matrix} \qquad \begin{matrix}\lambda & 1 & 0\\ 2 & \lambda & 2\\ 0 & 1 & \lambda,\end{matrix} \qquad \begin{matrix}\lambda & 1 & 0 & 0\\ 3 & \lambda & 2 & 0\\ 0 & 2 & \lambda & 3\\ 0 & 0 & 1 & \lambda,\end{matrix}$$

$$\begin{matrix}\lambda & 1 & 0 & 0 & 0\\ 4 & \lambda & 2 & 0 & 0\\ 0 & 3 & \lambda & 3 & 0\\ 0 & 0 & 2 & \lambda & 4\\ 0 & 0 & 0 & 1 & \lambda,\end{matrix} \quad \ldots\ldots$$

la loi de formation est évidente ; effectuant, on trouve

$$\lambda,\ \lambda^2-1,\quad \lambda(\lambda^2-2^2),\quad (\lambda^2-1^2)(\lambda^2-3^2),\quad \lambda(\lambda^2-2^2)(\lambda^2-4^2),$$
$$(\lambda^2-1^2)(\lambda^2-3^2)(\lambda^2-5^2),\quad \lambda(\lambda^2-2^2)(\lambda^2-4^2)(\lambda^2-6^2),$$

et ainsi de suite."

That Sylvester was the author of the implied theorem may be considered proved by an entry in the index of the volume (See p. 478), and by a statement of Cayley's in the *Quarterly Journal of Mathematics*, ii., p. 163. Probably the title of the communication was prefixed by the editors, who, knowing of Sylvester's papers in the *Philosophical Magazine*, felt themselves justified in applying the name "Sylvester's determinants."

SCHLÄFLI, L. (Nov. 1855).

[Réduction d'une intégrale multiple qui comprend l'arc de cercle et l'aire du triangle sphérique comme cas particuliers. *Journ.* (*de Liouville*) *de Math.*, xx. pp. 359–394.]

Here there appears the equation

$$\frac{\Delta(\alpha,\beta,\ldots,\zeta,\eta)}{\Delta(\beta,\ldots,\zeta,\eta)} = 1 - \frac{\cos^2\alpha}{1} - \frac{\cos^2\beta}{1} - \ddots - \frac{\cos^2\zeta}{1-\cos^2\eta},$$

where, in view of the contents of a subsequent paper (see under year 1858), it would seem that $\Delta(\alpha,\beta,\ldots,\zeta,\eta)$ was used for

$$\begin{vmatrix} 1 & \cos\alpha & & & & & \\ -\cos\alpha & 1 & \cos\beta & & & & \\ & -\cos\beta & 1 & & & & \\ & & & \cdot\ \cdot\ \cdot\ \cdot\ \cdot & & & \\ & & & & -\cos\zeta & 1 & \cos\eta \\ & & & & & -\cos\eta & 1 \end{vmatrix}.$$

No properties, however, of this determinant are given.

RAMUS, C. (1856, March).

[Determinanternes Anvendelse til at bestemme hoven for de convergerende Bröker. *Oversigt . . . danske Vidensk. Selsk. Forhandl.* . . . (Kjøbenhavn), pp. 106–119.]

Ramus' introduction consists in recalling the result of the application of determinants to the solution of a set of linear equations, his mode of stating the result being that given by Jacobi in the *De formatione* . . . of the year 1841,—that is to say, he takes for his set of equations

$$\left.\begin{array}{l} a_0{}^0y_0 + a_1{}^0y_1 + a_2{}^0y_2 + \ldots + a_n{}^0y_n = u_0 \\ a_0{}^1y_0 + a_1{}^1y_1 + a_2{}^1y_2 + \ldots + a_n{}^1y_n = u_1 \\ \cdot \quad \cdot \quad \cdot \quad \cdot \quad \cdot \quad \cdot \quad \cdot \quad \cdot \quad \cdot \quad \cdot \quad \cdot \quad \cdot \\ a_0{}^ny_0 + a_1{}^ny_1 + a_2{}^ny_2 + \ldots + a_n{}^ny_n = u_n \end{array}\right\},$$

and puts the solution in the form

$$\mathrm{R}_n y_r = \mathrm{A}_r{}^0 u_0 + \mathrm{A}_r{}^1 u_1 + \mathrm{A}_r{}^2 u_2 + \ldots + \mathrm{A}_r{}^n u_n, \qquad (\omega)$$

where

$$\begin{aligned} \mathrm{R}_n &= \sum \pm a_0{}^0 a_1{}^1 a_2{}^2 \ldots a_n^n, \\ \mathrm{A}_i^i &= \sum \pm a_0{}^0 a_1{}^1 \ldots a_{i-1}^{i-1} a_{i+1}^{i+1} \ldots a_n^n, \\ \mathrm{A}_\kappa^i &= -\sum \pm a_0{}^0 a_1{}^1 \ldots a_{i-1}^{i-1} a_i^\kappa a_{i+1}^{i+1} \ldots a_{\kappa-1}^{\kappa-1} a_{\kappa+1}^{\kappa+1} \ldots a_n^n.* \end{aligned}$$

He then recalls the further fact that if $y_0, y_1, y_2, \ldots, y_n$ be the numerators of the convergents of the continued fraction

$$a_0 + \frac{b_1}{a_1 +}\,\frac{b_2}{a_2 +}\ldots+\frac{b_n}{a_n}$$

---

* It is in this mode of writing A_κ^i, namely, with the negative sign, that Jacobi's peculiarity consists. Not content with removing from R_n the row and column in which a_κ^i occurs and prefixing to the minor thus obtained the sign-factor $(-1)^{i+\kappa}$, he takes the further step of moving the row with the index κ over $\kappa - i + 1$ rows, thus arriving at

$$\mathrm{A}_\kappa^i = -\sum \pm a_0^0 a_1^1 \ldots a_{i-1}^{i-1} a_{i+1}^{i+1} \ldots a_{\kappa-1}^{\kappa-1} a_i^\kappa a_{\kappa+1}^{\kappa+1} \ldots a_n^n.$$

Of course there is at this second step the option of moving the *column* with the index i over $\kappa - i + 1$ columns, and this Ramus does.

there exists the set of equations

$$\left.\begin{array}{llr} y_0 & & = a_0 \\ -a_1y_0 + \quad y_1 & & = b_1 \\ -b_2y_0 - a_2y_1 + \quad y_2 & & = 0 \\ \quad - b_3y_1 - a_3y_2 + y_3 & & = 0 \\ \cdots\cdots\cdots\cdots\cdots & & \\ & -b_ny_{n-2} - a_ny_{n-1} + y_n & = 0 \end{array}\right\},$$

and he thereupon draws the natural conclusion that the previous result can be applied to the determination of $y_0, y_1, y_2, \ldots, y_n$.

Making the necessary substitution for the u's and for R_n he of course obtains

$$y_n = a_0A_n^{\,0} + b_1A_n^{\,1},$$

$A_n^{\,0}$, $A_n^{\,1}$ being now determinants which for want of Cayley's notation he cannot accurately specify, but which he persists in writing in the form

$$-\sum \pm a_0^{\,n}a_1^{\,1}a_2^{\,2} \ldots a_{n-1}^{n-1}, \quad -\sum \pm a_0^{\,0}a_1^{\,n}a_2^{\,2} \ldots a_{n-1}^{n-1}.$$

From this result he calculates in succession the values of y_1, y_2, y_3, y_4; but it will readily be understood that the process is neither elegant nor short.

In the remainder of the paper (§§ 4–9) no further use of the properties of determinants is made, the contents of the last ten pages being such as might appear in any ordinary exposition of continued fractions. First there is established the old "rule" for writing out the value of y_n, above referred to as being found in Stern's monograph. This is followed by the results

$$\begin{aligned}(y_n)_{\substack{a_0=a_1=\ldots=a \\ b_0=b_1=\ldots=b}} &= \frac{1}{\sqrt{a^2+4b}}\left\{\left(\frac{a+\sqrt{a^2+4b}}{2}\right)^{n+2} - \left(\frac{a-\sqrt{a^2+4b}}{2}\right)^{n+2}\right\}, \\ &= a^{n+1} + C_{n,1}a^{n-1}b + C_{n-1,2}a^{n-3}b^2 + \ldots\ldots,\end{aligned}$$

which by putting $a=1=b$ give the number of terms in y_n, a number also obtained in the form

$$\frac{1}{2^{n+1}}\left\{C_{n+2,1} + C_{n+2,3}\cdot 5 + C_{n+2,5}\cdot 5^2 + \ldots\right\}.$$

Anything else is of small moment.

BRIOSCHI, F. (1856).

[THÉORIE DES DÉTERMINANTS, et leurs principales applications; par le Dr. F. Brioschi: traduit de l'italien par M. Edouard Combescure. xii+216 pp. Paris.]

In the French edition of his text-book Brioschi added an expository note of two pages (pp. 142–144) on the subject, beginning at once with the general form referred to at the end of Sylvester's paper of 1853, Sept., namely,

$$\begin{vmatrix} a_1 & m_1 & . & \cdots & . & . \\ n_1 & a_2 & m_2 & \cdots & . & . \\ . & n_2 & a_3 & \cdots & . & . \\ . & . & . & \cdots & . & . \\ . & . & . & \cdots & a_{r-1} & m_{r-1} \\ . & . & . & \cdots & n_{r-1} & a_r \end{vmatrix}, \quad \text{or } D_r \text{ say.}$$

His first result, obtained rather clumsily, is

$$D_r = a_r D_{r-1} - m_{r-1} n_{r-1} D_{r-2};$$

his second is

$$N_r D_{r-1} - D_r N_{r-1} = m_1 m_2 \ldots m_{r-1} \cdot n_1 n_2 \ldots n_{r-1},$$

where N_r is the cofactor of a_1 in D_r, this being got by consideration of the two-line minor whose elements are the corner elements of the adjugate of D_r; and his third is that given by Sylvester at the place just mentioned.

CAYLEY, A. (1857, April).

[On the determination of the value of a certain determinant. *Quart. Journ. of Math.*, ii. pp. 163–166; or *Collected Math. Papers*, iii. pp. 120–123.]

The determinant in question is rather more general than Sylvester's of the year 1854, being

$$\begin{vmatrix} \theta & 1 & . & . & \cdots & . & . \\ x & \theta & 2 & . & \cdots & . & . \\ . & x-1 & \theta & 3 & \cdots & . & . \\ . & . & x-2 & \theta & \cdots & . & . \\ . & . & . & . & \cdots & . & . \\ . & . & . & . & \cdots & \theta & n-1 \\ . & . & . & . & \cdots & x-n+2 & \theta \end{vmatrix},$$

while the other is obtained from this by putting $x=n-1$. Denoting his own form by U_n, Cayley, with Sylvester's results before him, found

$$\begin{aligned} U_2 &= (\theta^2-1) - (x-1), \\ U_3 &= \theta(\theta^2-4) - 3(x-2)\theta, \\ U_4 &= (\theta^2-1)(\theta^2-9) - 6(x-3)(\theta^2-1) + 3(x-3)(x-1); \end{aligned}$$

so that, if he put H_n for the value of U_n in Sylvester's case (viz., when $x=n-1$), he could write

$$\begin{aligned} U_2 &= H_2 - (x-1)H_0 \\ U_3 &= H_3 - 3(x-2)H_1 \\ U_4 &= H_4 - 6(x-3)H_2 + 3(x-3)(x-1)H_0, \\ &\cdots\cdots\cdots\cdots \end{aligned}$$

and thence, doubtless, divined the generalisation

$$U_n = H_n - B_{n,1}\cdot(x-n+1)\cdot H_{n-2} + B_{n,2}\cdot(x-n+1)(x-n+3)\cdot H_{n-4} - \ldots\ldots,$$

where

$$H_n = (\theta+n-1)(\theta+n-3)(\theta+n-5)\ldots\ldots \text{ to } n \text{ factors}$$

and

$$B_{n,s} = \frac{n(n-1)(n-2)\ldots\ldots(n-2s+1)}{2^s\cdot 1\cdot 2\cdot 3\ldots\ldots s}.$$

The establishment of the truth of this is all that the paper is occupied with, the procedure being to expand U_n in terms of the elements of its last row and their complementary minors, thus obtaining

$$U_n = \theta U_{n-1} - (n-1)(x-n+2)U_{n-2},$$

and thence

$$\begin{aligned} U_n + \Big\{(n-1)(x-n+2) + (n-2)(x-n+3) - \theta^2\Big\}\, U_{n-2} \\ + (n-2)(n-3)(x-n+3)(x-n+4)U_{n-4} = 0, \end{aligned}$$

and showing that the above conjectural expression for U_n satisfies the latter equation. The process of verification is troublesome, and was not viewed with satisfaction by Cayley himself.

As a preliminary the coefficients of the H's in the value of U_n are for shortness' sake denoted by $A_{n,0}$, $-A_{n,1}$, $\ldots$, and for the same and an additional reason the coefficient of U_{n-2} in the difference-equation is denoted by

$$M_{n,s} - \left\{ \theta^2 - (n-2s-1)^2 \right\},$$

which is equivalent to putting

$$M_{n,s} \equiv (n-1)(x-n+2) + (n-2)(x-n+3) - (n-2s-1)^2.$$

The operation to be performed being thus the substitution of

$$A_{n,0}H_n - A_{n,1}H_{n-2} + \ldots . + (-)^s A_{n,s}H_{n-2s} + \ldots .$$

for U_n in the expression

$$U_n + \left[M_{n,s} - \left\{\theta^2 - (n-2s-1)^2\right\}\right]U_{n-2} + (n-2)(n-3)(x-n+3)(x-n+4)U_{n-4},$$

it is readily seen that the result will be an aggregate of expressions like

$$A_{n,s}H_{n-2s} + \left[M_{n,s} - \left\{\theta^2 - (n-2s-1)^2\right\}\right]A_{n-2,s}H_{n-2-2s}$$
$$+ (n-2)(n-3)(x-n+3)(x-n+4)A_{n-4,s}H_{n-4-2s}.$$

Now if we bear in mind that by definition

$$\left\{\theta^2 - (n-2s-1)^2\right\}H_{n-2-2s} = H_{n-2s},$$

the second of the three terms of this

$$= M_{n,s}A_{n-2,s}H_{n-2-2s} - A_{n-2,s}H_{n-2s},$$

or, if we write $s-1$ for s in one case,

$$= - M_{n,s-1}A_{n-2,s-1}H_{n-2s} - A_{n-2,s}H_{n-2s}$$
$$= - H_{n-2s}\left\{M_{n,s-1}A_{n-2,s-1} + A_{n-2,s}\right\};$$

and the third, by writing $s-2$ for s,

$$= (n-2)(n-3)(x-n+3)(x-n+4)A_{n-4,s-2}H_{n-2s}.$$

Consequently the sum of the three will vanish if

$$A_{n,s} - (M_{n,s-1}A_{n-2,s-1} + A_{n-2,s}) + (n-2)(n-3)(x-n+3)(x-n+4)A_{n-4,s-2} = 0,$$

and therefore if

$$B_{n,s}(x-n+1) - B_{n-2,s}(x-n+2s+1)$$
$$- B_{n-2,s-1}M_{n,s-1} + B_{n-4,s-2}(n-2)(n-3)(x-n+4) = 0,$$

that is, if

$$\begin{aligned}(x-n)\Big[\mathrm{B}_{n,s} - \mathrm{B}_{n-2,s} - (2n-3)\mathrm{B}_{n-2,s-1} + (n-2)(n-3)\mathrm{B}_{n-4,s-2}\Big] \\ + \Big[\mathrm{B}_{n,s} - (2s+1)\mathrm{B}_{n-2,s} - \big\{5n - 8 - (n-2s+1)^2\big\}\mathrm{B}_{n-2,s-1} \\ + 4(n-2)(n-3)\mathrm{B}_{n-4,s-2}\Big] = 0.\end{aligned}$$

But this is the case; for, as Cayley shows, both the cofactor of $x-n$ and the other similar expression following it vanish identically. The verification aimed at is thus attained.

PAINVIN, L. (1858, February).

[Sur un certain système d'équations linéaires. *Journ.* (*de Liouville*) *de Math.* (2), iii. pp. 41–46.]

The system of equations referred to in the title of Painvin's paper had presented themselves to Liouville in the course of the research which led to his "Mémoire sur les transcendantes elliptiques . . ." (*Journ. de Liouville* (1), v. pp. 441–464). Painvin's reason for taking up the subject was his belief that one of Liouville's results could be more simply arrived at by the use of determinants; and in a few lines of introduction he succeeds in showing that the result in question can be viewed as merely the resolution of the determinant

$$\begin{vmatrix} r & a & \cdot & \cdot & \ldots & \cdot & \cdot \\ n(a-1) & r-1 & 2a & \cdot & \ldots & \cdot & \cdot \\ \cdot & (n-1)(a-1) & r-2 & 3a & \ldots & \cdot & \cdot \\ \cdot & \cdot & (n-2)(a-1) & r-3 & \ldots & \cdot & \cdot \\ \cdot & \cdot & \cdot & \cdot & \cdot & \cdot & \cdot \\ \cdot & \cdot & \cdot & \cdot & \ldots & r-n+1 & na \\ \cdot & \cdot & \cdot & \cdot & \ldots & a-1 & r-n \end{vmatrix}$$

into factors.

In explanation of the process followed the case of the fourth order

$$\begin{vmatrix} r & a & \cdot & \cdot \\ 3(a-1) & r-1 & 2a & \cdot \\ \cdot & 2(a-1) & r-2 & 3a \\ \cdot & \cdot & a-1 & r-3 \end{vmatrix}$$

will suffice. Increasing each element of the first row by the corresponding elements of the other rows,—an operation which as before we may symbolise by

$$\text{row}_1 + \text{row}_2 + \text{row}_3 + \ldots\ldots,$$

—he removes the factor $r+3a-3$ and finds left the cofactor

$$\begin{vmatrix} 1 & 1 & 1 & 1 \\ 3(a-1) & r-1 & 2a & \cdot \\ \cdot & 2(a-1) & r-2 & 3a \\ \cdot & \cdot & a-1 & r-3 \end{vmatrix}.$$

On this are performed the operations

$$\text{col}_1-\text{col}_2,\quad \text{col}_2-\text{col}_3,\quad \text{col}_3-\text{col}_4,\quad \ldots\ldots$$

the result being a determinant of the next lower order

$$-\begin{vmatrix} 3a-r-2 & r-2a-1 & 2a \\ 2-2a & 2a-r & r-3a-2 \\ \cdot & 1-a & a-r+2 \end{vmatrix}.$$

Finally, after changing the signs of all the elements here, the operations

$$\begin{aligned} &\text{row}_1+\text{row}_2+\text{row}_3+\ldots, \\ &\text{row}_2+\text{row}_3+\ldots, \\ &\text{row}_3+\ldots,\quad \ldots \end{aligned}$$

are performed, the result

$$\begin{vmatrix} r-a & a & \cdot \\ 2(a-1) & r-a-1 & 2a \\ \cdot & a-1 & r-a-2 \end{vmatrix}$$

being a determinant exactly similar in form to the original, but with $r-a$ instead of r. This, therefore, in turn may be transformed into

$$(r+a-2)\begin{vmatrix} r-2a & a \\ a-1 & r-2a-1 \end{vmatrix},$$

and so on.

The value thus obtained for the above-written determinant of the $(n+1)^{\text{th}}$ order is

$$(r+na-n)(r+na-n-2a+1)(r+na-n-4a+2)\ldots(r-na),$$

each factor being less than the preceding by $2a-1$, and the whole a function of $a(a-1)$.

The special case is noted where $a=\frac{1}{2}$, and where therefore all the $n+1$ resulting factors are alike. This Painvin writes in the form

$$\begin{vmatrix} r & \frac{1}{2} & . & . & \cdots & . & . \\ -\frac{n}{2} & r-1 & \frac{2}{2} & . & \cdots & . & . \\ . & -\frac{n-1}{2} & r-2 & \frac{3}{2} & \cdots & . & . \\ . & . & -\frac{n-2}{2} & r-3 & \cdots & . & . \\ \cdot & \cdot & \cdot & \cdot & \cdot & \cdot & \cdot \\ . & . & . & . & \cdots & r-n+1 & \frac{n}{2} \\ . & . & . & . & \cdots & -\frac{1}{2} & r-n \end{vmatrix} = \left(r-\frac{n}{2}\right)^{n+1};$$

but a preferable form is, evidently,

$$\begin{vmatrix} \rho & 1 & . & . & \cdots & . & . \\ -n & \rho-2 & 2 & . & \cdots & . & . \\ . & -n+1 & \rho-4 & 3 & \cdots & . & . \\ . & . & -n+2 & \rho-6 & \cdots & . & . \\ \cdot & \cdot & \cdot & \cdot & \cdot & \cdot & \cdot \\ . & . & . & . & \cdots & \rho-2n+2 & n \\ . & . & . & . & \cdots & -1 & \rho-2n \end{vmatrix} = (\rho-n)^{n+1}.$$

HEINE, E. (1858, Sept.).

[Auszug eines Schreibens über die *Lamé*schen Functionen an den Herausgeber. Einige Eigenschaften der *Lamé*schen Functionen. *Crelle's Journ.*, lvi. pp. 79–86, 87–99.]

In the case of Heine the functions afterwards known as "continuants" made their appearance under totally different circumstances, namely, while he was engaged in transforming a special homogeneous function of the second degree by means of an orthogonal transformation. It will be remembered that if the quadric

$$\begin{array}{l} a_{11}x_1^2 + 2a_{12}x_1x_2 + 2a_{13}x_1x_3 + \dots \\ \qquad + \quad a_{22}x_2^2 + 2a_{23}x_2x_3 + \dots \\ \qquad\qquad + \quad a_{33}x_3^2 \quad + \dots \\ \qquad\qquad\qquad \cdot \quad \cdot \quad \cdot \quad \cdot \quad \cdot \end{array}$$

be transformed by an orthogonal transformation into

$$A_{11}\xi_1^2 + A_{22}\xi_2^2 + A_{33}\xi_3^2 + \dots$$

the coefficients of the latter expression are the roots of the equation

$$\begin{vmatrix} a_{11}-A & a_{12} & a_{13} & \dots \\ a_{12} & a_{22}-A & a_{23} & \dots \\ a_{13} & a_{23} & a_{33}-A & \dots \\ \cdot & \cdot & \cdot & \cdot \end{vmatrix} = 0,$$

Now Heine's peculiar quadric was

$$\begin{array}{l} c_0^2x_0^2 - 2\kappa c_0c_1x_0x_1 \\ \qquad + (c_1^2+c_2^2)x_1^2 - 2\kappa c_2c_3x_1x_2 \\ \qquad\qquad + (c_3^2+c_4^2)x_2^2 - \\ \qquad\qquad\qquad \cdot \quad \cdot \quad \cdot \quad \cdot \quad \cdot \\ \qquad\qquad\qquad\qquad + (c_{2\sigma-1}^2+c_{2\sigma}^2)x_\sigma^2, \end{array}$$

where in every case the coefficient of the product of two x's vanishes if their suffixes differ by more than 1, and where

$$\begin{aligned} c_0^2 &= \tfrac{1}{2}(n)(n+1), \\ c_1^2 &= \tfrac{1}{4}(n-1)(n+2), \\ &\cdot \quad \cdot \quad \cdot \quad \cdot \quad \cdot \\ c_r^2 &= \tfrac{1}{4}(n-r)(n+r+1), \quad (r>0) \\ &\cdot \quad \cdot \quad \cdot \quad \cdot \quad \cdot \\ c^2_{n-1} &= \tfrac{1}{2}n, \\ \text{and } \kappa &= \frac{c^2-b^2}{c^2+b^2}. \end{aligned}$$

He was thus naturally led to the equation in z

$$\begin{vmatrix} z-c_0^2 & \kappa c_0c_1 & \cdot & \dots & \cdot \\ \kappa c_0c_1 & z-c_1^2-c_2^2 & \kappa c_2c_3 & \dots & \cdot \\ \cdot & \kappa c_2c_3 & z-c_3^2-c_4^2 & \dots & \cdot \\ \cdot & \cdot & \cdot & \cdot & \cdot \\ \cdot & \cdot & \cdot & \dots & \kappa c_{2\sigma-2}c_{2\sigma-1} \\ \cdot & \cdot & \cdot & \dots & z-c^2_{2\sigma-1}-c^2_{2\sigma} \end{vmatrix} = 0,$$

where either $c_{2\sigma}^2$ is c_{n-1}^2, or $c_{2\sigma-1}^2$ is c_{n-1}^2 and, if the latter, $c_{2\sigma}^2 = 0$. From a knowledge of Painvin's paper he recognised the left-hand side of the equation as being the numerator of the continued fraction

$$z - c_0{}^2 - \frac{\kappa^2 c_0{}^2 c_1{}^2}{z - c_1{}^2 - c_2{}^2 -} \frac{\kappa^2 c_2{}^2 c_3{}^2}{z - c_3{}^2 - c_4{}^2 -} \ddots$$

but he ventured nothing in elucidation of it. Even the special case where $b = 0$ and where therefore $\kappa = 1$ appears to have proved at the time too troublesome, although he knew otherwise that in this case the continued fraction

$$= \frac{z(z-2^2)(z-4^2) \ldots . (z-n^2)}{(z-1^2)(z-3^2) \ldots . (z-\overline{n-1}^2)} \text{ if } n \text{ be even,}$$

and

$$= \frac{(z-1^2)(z-3^3)(z-5^2) \ldots . (z-n^2)}{(z-2^2)(z-4^2) \ldots . (z-\overline{n-1}^2)} \text{ if } n \text{ be odd;}$$

for his words are—" Einen directen Beweis für diese Summirung des Kettenbruchs habe ich noch nicht aufgefunden."

SCHLÄFLI, L. (1858).

[On the multiple integral $\int^n dx\,dy \ldots dz$ whose limits are $p_1 = a_1x + b_1y + \ldots + h_1z > 0$, $p_2 > 0, \ldots, p_n > 0$, and $x^2 + y^2 + \ldots + z^2 > 1$. *Quart. Journ. of Math.*, ii. pp. 269–301; iii. pp. 54–68, 97–108.]

The determinant which makes its appearance in the course of Schläfli's research is

$$\begin{vmatrix} 1 & -\cos\alpha & \cdot & \cdots & \cdot & \cdot & \cdot \\ -\cos\alpha & 1 & -\cos\beta & \cdots & \cdot & \cdot & \cdot \\ \cdot & -\cos\beta & 1 & \cdots & \cdot & \cdot & \cdot \\ \cdot & \cdot & \cdot & \cdots & \cdot & \cdot & \cdot \\ \cdot & \cdot & \cdot & \cdots & 1 & -\cos\eta & \cdot \\ \cdot & \cdot & \cdot & \cdots & -\cos\eta & 1 & -\cos\theta \\ \cdot & \cdot & \cdot & \cdots & \cdot & -\cos\theta & 1 \end{vmatrix},$$

which for shortness' sake he denotes by

$$\Delta(\alpha, \beta, \gamma, \ldots, \eta, \theta)$$

and whose connection with continued fractions he therefore specifies by the equation

$$\frac{\Delta(\alpha, \beta, \gamma, \ldots, \eta, \theta)}{\Delta(\beta, \gamma, \ldots, \eta, \theta)} = 1 - \frac{\cos^2\alpha}{1} - \frac{\cos^2\beta}{1} - \cdots - \frac{\cos^2\eta}{1-\cos^2\theta}.$$

The first property noticed is, naturally,

$$\Delta(\alpha, \beta, \gamma, \ldots, \theta) = \Delta(\beta, \gamma, \ldots, \theta) - \cos^2\alpha \cdot \Delta(\gamma, \ldots, \theta).$$

Later there is given what may be viewed as an extension of this, viz.,

$$\begin{aligned}\Delta(\alpha, \ldots, \delta, \epsilon, \zeta, \eta, \theta, \ldots, \lambda) = \Delta(\alpha, \ldots, \delta, \epsilon) \cdot \Delta(\eta, \theta, \ldots, \lambda)\\ -\cos^2\zeta \,.\, \Delta(\alpha, \ldots, \delta) \cdot \Delta(\theta, \ldots, \lambda),\end{aligned}$$

the proof being said to present no difficulty. The third is a little more complicated, and is logically led up to by taking four instances of the first property, namely,

$$\begin{aligned}\Delta(\alpha, \beta, \gamma, \ldots, \zeta) &= \Delta(\beta, \gamma, \ldots, \zeta) - \cos^2\alpha \cdot \Delta(\gamma, \delta, \ldots, \zeta),\\ \Delta(\beta, \gamma, \delta, \ldots, \zeta, \eta) &= \Delta(\gamma, \delta, \ldots, \eta) - \cos^2\beta \cdot \Delta(\delta, \ldots, \zeta, \eta),\\ \Delta(\gamma, \delta, \ldots, \zeta, \eta, \theta) &= \Delta(\gamma, \delta, \ldots, \eta) - \cos^2\theta \cdot \Delta(\gamma, \delta, \ldots, \zeta),\\ \Delta(\delta, \ldots, \zeta, \eta, \theta, \alpha) &= \Delta(\delta, \ldots, \zeta, \eta, \theta) - \cos^2\alpha \cdot \Delta(\delta, \ldots, \zeta, \eta),\end{aligned}$$

using in connection with these the multipliers

$$\Delta(\delta, \ldots, \zeta, \eta), \quad -\Delta(\delta, \ldots, \zeta), \quad \Delta(\delta, \ldots, \zeta), \quad -\Delta(\gamma, \delta, \ldots, \zeta),$$

respectively, performing addition, and then showing that the right-hand sum vanishes, the result thus being

$$\begin{aligned}\Delta(\alpha, \beta, \gamma, \delta, \ldots, \zeta) \cdot \Delta(\delta, \ldots, \zeta, \eta) - \Delta(\delta, \ldots, \zeta, \eta, \theta, \alpha) \cdot \Delta(\gamma, \delta, \ldots, \zeta)\\ = \left\{\Delta(\beta, \gamma, \delta, \ldots, \zeta, \eta) - \Delta(\gamma, \delta, \ldots, \zeta, \eta, \theta)\right\} \cdot \Delta(\delta, \ldots, \zeta).\end{aligned}$$

The fourth property concerns the determinant

$$\begin{vmatrix} \Delta(\beta, \gamma, \ldots, \eta, \theta) & \Delta(\alpha, \beta, \gamma, \ldots, \eta, \theta) \\ \Delta(\beta, \gamma, \ldots, \eta) & \Delta(\alpha, \beta, \gamma, \ldots, \eta) \end{vmatrix},$$

which by reason of the first property can be shown equal to

$$\begin{vmatrix} \Delta(\beta,\gamma,\ldots,\eta,\theta) & -\Delta(\gamma,\ldots,\eta,\theta) \\ \Delta(\beta,\gamma,\ldots,\eta) & -\Delta(\gamma,\ldots,\eta) \end{vmatrix} \cos^2\alpha,$$

or

$$\begin{vmatrix} \Delta(\gamma,\ldots,\eta,\theta) & \Delta(\beta,\gamma,\ldots,\eta,\theta) \\ \Delta(\gamma,\ldots,\eta) & \Delta(\beta,\gamma,\ldots,\eta) \end{vmatrix} \cos^2\alpha,$$

and ultimately, "by repeating this sort of transformation," equal to

$$\cos^2\alpha \, \cos^2\beta \, \cos^2\gamma \, \ldots \, \cos^2\theta.$$

If we use for a moment the present-day notation for continuants, viz., where

$$a_1 + \frac{b_1}{a_2 +} \frac{b_2}{a_3 +} \ddots \quad = \quad \frac{K\begin{pmatrix} & b_1 & & b_2 & \\ a_1 & & a_2 & & a_3 \ldots \end{pmatrix}}{K\begin{pmatrix} & b_2 & \\ a_2 & & a_3 \ldots \end{pmatrix}}$$

Schläfli's results are seen to be

$$K\begin{pmatrix} \beta_1 & \beta_2 & \beta_3 & \cdot\cdot \\ 1 & 1 & 1 & \ldots \end{pmatrix} = K\begin{pmatrix} \beta_2 & \beta_3 & \cdot\cdot \\ 1 & 1 & \ldots \end{pmatrix} + \beta_1 K\begin{pmatrix} \beta_3 & \beta_4 & \cdot\cdot \\ 1 & 1 & \ldots \end{pmatrix},$$

$$\left.\begin{aligned} K\begin{pmatrix} \beta_1 & \beta_2 \ldots & \beta_\kappa \ldots & \beta_n \\ 1 & 1 \;\ldots 1 & 1 \ldots & 1 \end{pmatrix} &= K\begin{pmatrix} \beta_2 & \beta_3 \cdot\cdot & \beta_{\kappa-1} \\ 1 & 1 \;\ldots & 1 \end{pmatrix} \cdot K\begin{pmatrix} \beta_{\kappa+1} \ldots & \beta_n \\ 1 & 1 \ldots \; 1 \end{pmatrix}, \\ &\quad - \beta_\kappa K\begin{pmatrix} \beta_1 \ldots & \beta_{\kappa-2} \\ 1 & 1 \ldots \; 1 \end{pmatrix} \cdot K\begin{pmatrix} \beta_{\kappa+2} \ldots & \beta_n \\ 1 & 1 \ldots \; 1 \end{pmatrix}, \end{aligned}\right\}$$

$$\left.\begin{aligned} &K\begin{pmatrix} \beta_1 \ldots & \beta_{n-2} \\ 1 \; 1 \ldots & 1 \end{pmatrix} \cdot K\begin{pmatrix} \beta_4 \ldots & \beta_{n+1} \\ 1 \; 1 \ldots & 1 \end{pmatrix} \\ -&K\begin{pmatrix} \beta_4 \ldots \beta_n & \beta_1 \\ 1 \; 1 \ldots 1 & 1 \end{pmatrix} \cdot K\begin{pmatrix} \beta_3 \ldots & \beta_{n-2} \\ 1 \; 1 \ldots & 1 \end{pmatrix} \end{aligned}\right\} = \left\{\begin{aligned} &K\begin{pmatrix} \beta_2 \ldots & \beta_{n-1} \\ 1 \; 1 \ldots & 1 \end{pmatrix} \\ -&K\begin{pmatrix} \beta_3 \ldots & \beta_n \\ 1 \; 1 \ldots & 1 \end{pmatrix} \end{aligned}\right\} K\begin{pmatrix} \beta_4 \ldots & \beta_{n-2} \\ 1 \; 1 \ldots & 1 \end{pmatrix},$$

$$\begin{vmatrix} K\begin{pmatrix} \beta \ldots & \beta_n \\ 1 \; 1 \ldots & 1 \end{pmatrix} & K\begin{pmatrix} \beta_1 \ldots & \beta_n \\ 1 \; 1 \ldots & 1 \end{pmatrix} \\ K\begin{pmatrix} \beta_2 \ldots & \beta_{n-1} \\ 1 \; 1 \ldots & 1 \end{pmatrix} & K\begin{pmatrix} \beta_1 \ldots & \beta_{n-1} \\ 1 \; 1 \ldots & 1 \end{pmatrix} \end{vmatrix} = (-1)^n \beta_1 \beta_2 \beta_3 \ldots \beta_n,$$

the only change being the writing of β_1, β_2, ... for $-\cos^2\alpha$, $-\cos^2\beta$, ...

WORPITZKY, [J. D. T.] (1865, April).

[Untersuchungen über die Entwickelung der monodromen und monogenen Functionen durch Kettenbrüche. (Sch. Progr.) 39 pp., Berlin.]

Of the six sections into which the paper giving the results of Worpitzky's painstaking investigation is divided it is only the first headed "Fundamentalrelationen" which concerns us, these relations being nothing else than what we should now call "properties of continuants."

He takes his continued fraction in the same form as Schläfli, viz.,

$$1+\frac{a_1}{1}+\frac{a_2}{1}+\cdots+\frac{a_n}{1},$$

showing of course that it equals

$$\frac{N_{1,n}}{N_{2,n}},$$

where

$$N_{\kappa,n}=\begin{vmatrix} 1 & 1 & . & \cdots & . & . & . \\ -a_\kappa & 1 & 1 & \cdots & . & . & . \\ . & -a_{\kappa-1} & 1 & \cdots & . & . & . \\ . & . & . & \cdots & . & . & . \\ . & . & . & \cdots & -a_{n-1} & 1 & 1 \\ . & . & . & \cdots & . & -a_n & 1 \end{vmatrix}.$$

The first matter of interest is the expansion of $N_{\kappa,n}$ as a sum of products of $a_\kappa, a_{\kappa-1}, \ldots, a_n$, *e.g.*,

$$N_{1,3}=1+(a_1+a_2+a_3)+a_1a_3.$$

This is written in the form

$$1+\overset{1}{a}_{\kappa,n}+\overset{2}{a}_{\kappa n}=\ldots\ldots,$$

where, he says, "$\overset{r}{a}_{\kappa,n}$ die Summe aller möglichen (als Producte aufgefassten) Combinationscomplexionen ohne Wiederholung bedeutet, welche sich aus $a_\kappa, a_{\kappa+1}, \ldots, a_n$ so zu je r Elementen bilden lassen, dass nicht zwei neben einander stehende Elemente

a_s, a_{s+1} dieser Reihe in den einzelnen Producten zugleich vorkommen." By way of proof it is pointed out (1) that the term independent of all the a's is

$$\begin{vmatrix} 1 & 1 & & & & \\ 0 & 1 & 1 & & & \\ & \cdot & \cdot & \cdot & \cdot & \cdot \\ & & & 0 & 1 & 1 \\ & & & & 0 & 1 \end{vmatrix} \quad i.e. \ +1;$$

(2) that the cofactor* of $(-a_r)(-a_s)(-a_t)\ldots\ldots$ when two of the a's are consecutive is

$$\begin{vmatrix} 1 & 1 & & & & & & & & \\ 0 & 1 & 1 & & & & & & & \\ & \cdot & \cdot & \cdot & \cdot & & & & & \\ & & 0 & 1 & 1 & & & & & \\ & & & 1 & 0 & 0 & & & & \\ & & & & 1 & 0 & 0 & & & \\ & & & & & 0 & 1 & 1 & & \\ & & & & & & \cdot & \cdot & \cdot & \cdot \\ & & & & & & & 0 & 1 & 1 \\ & & & & & & & & 0 & 1 \end{vmatrix} \quad i.e. \ 0;$$

and (3) that the cofactor of $(-a_r)(-a_s)(-a_t)\ldots\ldots$ when no two of the a's are consecutive and their number is p, is another long-drawn-out continuant found equal to

$$\begin{vmatrix} 1 & 1 \\ 1 & 0 \end{vmatrix}^p \quad i.e. \ (-1)^p,$$

and that, therefore, the cofactor of $a_r a_s a_t \ldots$ in this case is $+1$.

In exactly similar fashion by partitioning $N_{\kappa,n}$ into terms which contain $-a_s$ and terms which do not, he finds

$$N_{k,n} = D_0 - a_s D_s,$$

where D_0 and D_s, determinants equally greedy of page space, are

* To obtain the cofactor of the product of a number of a set of elements in a determinant Worpitzky puts a 1 in the determinant in place of each element occurring in the said product, 0's in all the other places of the rows to which these elements belong, and 0's for all the other elements of the set.

shown to be equal to $N_{k,s-1} \cdot N_{s+1,n}$ and $-N_{k,s-2}N_{s+2,n}$ respectively, and he thus reaches the result

$$N_{k,n} = N_{k,s-1}N_{s+1,n} + a_s N_{k,s-2}N_{s+2,n}$$

already obtained in a different way by Schläfli.

Lastly, taking a determinant of the same form as $N_{k,n}$, but having

$$-a_s, \ -a_{s-1}, \ldots, \ -a_{k+1}, \ -a_k, \ -a_k, \ -a_{k+1}, \ldots, \ -a_{n-1}, \ -a_n$$

for its minor diagonal of a's, he obtains for it by isolating the first a_k the expression

$$N_{s,k+1}N_{k,n} + a_k N_{s,k+2}N_{k+1,n},$$

and by isolating the second a_k

$$N_{s,k}N_{k+1,n} + a_k N_{s,k+1}N_{k+2,n};$$

and thus deduces

$$N_{k,n}N_{k+1,s} - N_{k,s}N_{k+1,n} = -a_k(N_{k+1,n}N_{k+2,s} - N_{k+1,s}N_{k+2,n}).$$

It is then noted that the bracketed expression on the right differs from the expression on the left merely in having $k+1$ in place of k; so that there results

$$\begin{aligned} N_{k,n}N_{k+1,s} - N_{k,s}N_{k+1,n} &= (-1)^2 a_k a_{k+1}(N_{k+2,n}N_{k+3,s} - N_{k+2,s}N_{k+3,n}) \\ &= \cdots\cdots\cdots\cdots \\ &= (-1)^{s-k+1} a_k a_{k+1} \ldots\ldots a_{s+1} N_{s+3,n}. \end{aligned}$$

This also, it will be seen, is connected with a result of Schläfli's; for putting $s = n-1$ we have *

$$\begin{vmatrix} N_{k+1,n-1} & N_{k,n-1} \\ N_{k+1,n} & N_{k,n} \end{vmatrix} = (-1)^{n-k} a_k a_{k+1} \ldots a_n,$$

which becomes identical with Schläfli's last proposition on transposing the two rows of the determinant and (what is equally immaterial) putting $k=1$.

* In giving to $N_{s+1,s}$, $N_{s+2,s}$, $N_{s+3,s}$ the values 1, 1, 0 which are necessitated by assuming the generality of the recursion-formula

$$N_{k,n} = N_{k+1,n} + a_k N_{k+2,n},$$

Worpitzky forgets to note that in these cases the proposition $N_{k,n} = N_{n,k}$, used by him in the demonstration, does not hold.

MAZZA, F. (1866).

[ELEMENTI DI ALGEBRA, par R. Rubini. Terza edizione, accrescinta e migliorata. iv+295 pp. Napoli.]

In his chapter on determinants (Cap. x. pp. 249–292) Rubini gives (p. 270) the result

$$\begin{vmatrix} \lambda & a_1 & \cdot & \cdot & \cdots & \cdot & \cdot \\ a_n & \lambda & a_2 & \cdot & \cdots & \cdot & \cdot \\ \cdot & a_{n-1} & \lambda & a_3 & \cdots & \cdot & \cdot \\ \cdot & \cdot & \cdot & \cdot & \cdots & \cdot & \cdot \\ \cdot & \cdot & \cdot & \cdot & \cdots & a_{n-1} & \cdot \\ \cdot & \cdot & \cdot & \cdot & \cdots & \lambda & a_n \\ \cdot & \cdot & \cdot & \cdot & \cdots & a_1 & \lambda \end{vmatrix} = \begin{vmatrix} \lambda & a_n \\ a_n & \lambda \end{vmatrix} \begin{vmatrix} \lambda & a_1 & \cdot & \cdot & \cdots \\ a_{n-1} & \lambda & a_2 & \cdot & \cdots \\ \cdot & a_{n-2} & \lambda & a_3 & \cdots \\ \cdot & \cdot & \cdot & \cdot & \cdots \end{vmatrix}$$

where $a_1, a_2, \ldots, a_n$ are elements increasing by the common difference a_1. This is established by performing the operations which would at a later date have been denoted by

$$\begin{array}{l} \text{row}_1 + \text{row}_3 + \text{row}_5 + \cdots \\ \text{row}_2 + \text{row}_4 + \text{row}_6 + \cdots \\ \text{row}_3 + \text{row}_5 + \text{row}_7 + \cdots \\ \cdot \quad \cdot \quad \cdot \quad \cdot \quad \cdot \quad \cdot \quad \cdot \\ \text{col}_{n+1} - \text{col}_{n-1}, \quad \text{col}_n - \text{col}_{n-2}, \quad \text{col}_{n-1} - \text{col}_{n-3}, \quad \cdots \end{array}$$

The process is said to be due to Francesco Mazza, and the result to degenerate into Sylvester's of the year 1854 on putting $a_1 = 1$. It is not noticed, however, that on the other hand the result may be viewed as a special case of Sylvester's, namely, where λ/a_1 is put for λ.

THIELE, T. N. (1869, 1870).

[Bemærkninger om Kjædebrøker. *Tidsskrift for Math.* (2), v. pp. 144–146.

Den endelige Kjædebrøksfunktions Theori. *Tidsskrift for Math.* (2), vi. pp. 145–170.]

The first of the two notes comprising Thiele's first paper contains only one result, namely,

$$a_1 + \frac{b_1}{a_2 + \frac{b_2}{a_3 + \ddots + \frac{b_n}{a_n}}} = \frac{(a_1, a_2, \ldots, a_n)}{(a_2, \ldots, a_n)},$$

where $(a_1, a_2, \ldots, a_n)$ is used to stand for

$$\begin{vmatrix} a_1 & b_1 & \cdot & \cdots & \cdot & \cdot \\ -1 & a_2 & b_2 & \cdots & \cdot & \cdot \\ \cdot & \cdot & \cdot & \cdot & \cdot & \cdot \\ \cdot & \cdot & \cdot & \cdots & a_{n-1} & b_{n-1} \\ \cdot & \cdot & \cdot & \cdots & -1 & a_n \end{vmatrix}$$

There is nothing to indicate that this is not viewed as a fresh discovery, notwithstanding the fact that Ramus' paper of 1856 containing virtually the same identity had been published in the same city.

The other paper may be described as a careful study of finite continued fractions with the help of determinants. Instead of $b_1, b_2, \ldots$ are used $a_{12}, a_{23}, \ldots$; and

$$\begin{vmatrix} a_p & a_{p,p+1} & \cdot & \cdots & \cdot & \cdot \\ 1 & a_{p+1} & a_{p+1,p+2} & & \cdot & \cdot \\ \cdot & \cdot & \cdot & \cdot & \cdot & \cdot \\ \cdot & \cdot & \cdot & \cdots & a_{q-1} & a_{q-1,q} \\ \cdot & \cdot & \cdot & \cdots & 1 & a_q \end{vmatrix}$$

is denoted by

$$\mathrm{K}(p,q).$$

Further, this determinant is spoken of as a "Kjædebrøksdeterminant," or, shortly, a "K-Determinant"; and a section (§ 3, pp. 149–152) is devoted to a statement of its properties.

There is no need to rehearse all of these, the last portion (D) of the section being alone that which contains fresh matter. Opening with the double use of a previous property, viz.,

$$\mathrm{K}(h,m) = \mathrm{K}(h,k-1)\cdot\mathrm{K}(k,m) - a_{k-1,k}\mathrm{K}(h,k-2)\cdot\mathrm{K}(k+1,m),$$
$$\mathrm{K}(h,n) = \mathrm{K}(h,k-1)\cdot\mathrm{K}(k,n) - a_{k-1,k}\mathrm{K}(h,k-2)\cdot\mathrm{K}(k+1,n),$$

where h, k, m, n are in ascending order of magnitude, the author eliminates $\mathrm{K}(h,k-1)$ and obtains

$$\begin{vmatrix} \mathrm{K}(h,m) & \mathrm{K}(k,m) \\ \mathrm{K}(h,n) & \mathrm{K}(k,n) \end{vmatrix} = a_{k-1,k}\cdot\mathrm{K}(h,k-2)\cdot\begin{vmatrix} \mathrm{K}(k,m) & \mathrm{K}(k+1,m) \\ \mathrm{K}(k,n) & \mathrm{K}(k+1,n) \end{vmatrix}. \qquad (\alpha)$$

Then by taking the particular case of this where k appears in place of h and $k+1$ in place of k there results

$$\begin{vmatrix} \mathrm{K}(k,m) & \mathrm{K}(k+1,m) \\ \mathrm{K}(k,n) & \mathrm{K}(k+1,n) \end{vmatrix} = a_{k,k+1}\begin{vmatrix} \mathrm{K}(k+1,m) & \mathrm{K}(k+2,m) \\ \mathrm{K}(k+1,n) & \mathrm{K}(k+2,n) \end{vmatrix},$$

which when applied to one of the determinants occurring in itself gives

$$\begin{vmatrix} \mathrm{K}(k,m) & \mathrm{K}(k+1,m) \\ \mathrm{K}(k,n) & \mathrm{K}(k+1,n) \end{vmatrix} = a_{k,k+1}a_{k+1,k+2}\begin{vmatrix} \mathrm{K}(k+2,m) & \mathrm{K}(k+3,m) \\ \mathrm{K}(k+2,n) & \mathrm{K}(k+3,n) \end{vmatrix},$$

and finally

$$= a_{k,k+1}a_{k+1,k+2}\ \ldots\ a_{m,m+1}\cdot\begin{vmatrix} \mathrm{K}(m+1,m) & \mathrm{K}(m+2,m) \\ \mathrm{K}(m+1,n) & \mathrm{K}(m+2,n) \end{vmatrix},$$

$$= a_{k,k+1}a_{k+1,k+2}\ \ldots\ a_{m,m+1}\cdot\ \mathrm{K}(m+2,n). \qquad (\beta)$$

Further, by using this to make a substitution in the previous result (α) there is obtained

$$\begin{vmatrix} \mathrm{K}(h,m) & \mathrm{K}(k,m) \\ \mathrm{K}(h,n) & \mathrm{K}(k,n) \end{vmatrix} = a_{k-1,k}a_{k,k+1}\ \ldots\ a_{m,m+1}\cdot\mathrm{K}(h,k-2)\mathrm{K}(m+2,n), \quad (\gamma)$$

which on putting $k=h+1$ and $m=n-1$ becomes

$$\begin{vmatrix} \mathrm{K}(h,n-1) & \mathrm{K}(h+1,n-1) \\ \mathrm{K}(h,n) & \mathrm{K}(h+1,n) \end{vmatrix} = a_{h,h+1}a_{h+1,h+2}\ \ldots\ a_{n-1,n},$$

—a result which may be compared with one of Schläfli's and Worpitzky's, but which is more general in that the main diagonal of each "K-Determinant" does not consist of units.

CHAPTER XVI.

THE LESS COMMON SPECIAL FORMS, UP TO 1860.

THERE now only remain for consideration those special forms which, prior to 1860, had not received any noteworthy attention. These will be found to include: (α) permanents, which are touched on by three authors; (β) determinants with the typical element $a_{rs}+b_{rs}i$, which are referred to in four memoirs; (γ) two other forms, which are each dealt with in two papers; and (δ) nine others, which make their appearance only once. It will also be appropriate to collect in a note (ϵ) the facts ascertained up to 1860 regarding the census of terms in special forms of determinants.

(α) PERMANENTS.

As we have already seen, Cauchy, in his memoir of 1812, widened the ordinary meaning of the term "symmetric function," and was consequently led to call such expressions as

$$a_1b_2 + a_2b_1, \quad a_1b_2 + a_2b_3 + a_3b_1 + a_1b_3 + a_2b_1 + a_3b_2, \quad \ldots$$

"fonctions symétriques permanentes," denoting them by $S^2(a_1b_2)$, $S^3(a_1b_2)$, ...

In the same year, as we have also noted, Binet gave the identities

$$\begin{aligned} \Sigma ab' &= \Sigma a \Sigma b - \Sigma ab, \\ \Sigma ab'c'' &= \Sigma a \Sigma b \Sigma c + 2\Sigma abc - \Sigma a \Sigma bc - \Sigma b \Sigma ca - \Sigma c \Sigma ab, \\ &\ldots\ldots\ldots\ldots\ldots\ldots \end{aligned}$$

which in Cauchy's notation would have been written

$$\mathrm{S}^n(a_1b_2) = \mathrm{S}^n(a_1)\mathrm{S}^n(b_1) - \mathrm{S}^n(a_1b_1),$$
$$\mathrm{S}^n(a_1b_2c_3) = \mathrm{S}^n(a_1)\mathrm{S}^n(b_1)\mathrm{S}^n(c_1) + 2\mathrm{S}^n(a_1b_1c_1) - \;\ldots\ldots,$$

but which, in reality, are due to Waring, who, denoting the sum of the pth powers of $\alpha, \beta, \gamma, \ldots$ by s_p asserted in his *Miscellanea Analytica* of the year 1762 that

$$\sum \alpha^p\beta^q \;\;= s_p \cdot s_q - s_{p+q},$$
$$\sum \alpha^p\beta^q\gamma^r = s_p \cdot s_q \cdot s_r + 2s_{p+q+r} - \ldots$$
$$\ldots\ldots\ldots\ldots\ldots$$

Proofs of Waring's identities were given by Paoli in his *Supplemento agli Elementi di Algebra*, published in 1804 (See *Op.*, ii. § 28), and by Meier Hirsch in his *Sammlung von Aufgaben aus der Theorie der algebraischen Gleichungen*, published in 1809 (See pp. 34–41).

It is only symmetric functions like $\mathrm{S}^2(a_1b_2)$, $\mathrm{S}^3(a_1b_2c_3)$, $\mathrm{S}^4(a_1b_2c_3d_4)$, ..., whose every term involves the full number of letters, that at the present day are spoken of as *permanents*.

BORCHARDT, C. W. (1855).

[Bestimmung der symmetrischen Verbindungen vermittelst ihrer erzeugenden Function. *Monatsb.... Akad. d. Wiss.* (Berlin), 1855, pp. 165–171; or *Crelle's Journ.*, liii. pp. 193–198; or *Gesammelte Werke*, pp. 97–105.]

Having already fully dealt with this paper under the heading *Alternants*, it suffices merely to recall the identity therein given, namely,

$$\sum\left(\frac{1}{t-a}\cdot\frac{1}{t_1-a_1}\cdots\frac{1}{t_n-a_n}\right)\times\sum\left(\pm\frac{1}{t-a}\cdot\frac{1}{t_1-a_1}\cdots\frac{1}{t_n-a_n}\right)$$
$$=\sum\left(\pm\frac{1}{(t-a)^2}\cdot\frac{1}{(t_1-a_1)^2}\cdots\frac{1}{(t_n-a_n)^2}\right),$$

where the first factor on the left differs from the determinant which is its cofactor merely in having the signs of all its terms positive.

JOACHIMSTHAL, F. (1856, September).

[De æquationibus quarti et sexti gradus quæ in theoria linearum et superficierum secundi gradus occurrunt. *Crelle's Journ.*, liii. pp. 149–172.]

Joachimsthal, requiring the use of the so-called "Binet's" identities, devotes section iii. of his paper to them, combining them in one proposition, and showing more or less satisfactorily, after the manner of Meier Hirsch, how the proof of each case can be made dependent on the previous case. His proposition is—*There being* m *rows each of* z *quantities*

$$\begin{array}{cccc} \alpha_1 & \alpha_2 & \dots & \alpha_z \\ \beta_1 & \beta_2 & \dots & \beta_z \\ \cdot & \cdot & \cdot & \cdot \\ \lambda_1 & \lambda_2 & \dots & \lambda_z \\ \mu_1 & \mu_2 & \dots & \mu_z \end{array}$$

and z *being not less than* m, *the sum*

$$\sum \alpha_1\beta_2 \dots \mu_m,$$

consisting of $z(z-1)(z-2)\dots(z-m+1)$ *terms, can be expressed as an integral function of the sums arranged in the following rows:*

$$\begin{array}{llll} \Sigma\alpha_1 & \Sigma\beta_1 & \dots & \Sigma\mu_1 \\ \Sigma\alpha_1\beta_1 & \Sigma\alpha_1\gamma_1 & \dots\dots & \Sigma\lambda_1\mu_1 \\ \Sigma\alpha_1\beta_1\gamma_1 & \Sigma\alpha_1\beta_1\delta_1 & \dots\dots\dots & \Sigma\kappa_1\lambda_1\mu_1 \\ \cdot & \cdot & \cdot & \cdot \\ \Sigma\alpha_1\beta_1\gamma_1\dots\mu_1, & & & \end{array}$$

each sum consisting of z *terms: further, the said function when* $z < m$ *vanishes identically.*

To prove the proposition when $m=3$ he takes the previous case

$$\Sigma\alpha_1\beta_2 = \Sigma\alpha_1\Sigma\beta_1 - \Sigma\alpha_1\beta_1 \qquad (x_2)$$

and multiplies both sides by $\Sigma\gamma_1$, thus obtaining

$$\Sigma\alpha_1\beta_2\gamma_3 + \Sigma\alpha_1\gamma_1\beta_2 + \Sigma\beta_1\gamma_1\alpha_2 = \Sigma\alpha_1\Sigma\beta_1\Sigma\gamma_1 - \Sigma\alpha_1\beta_1\cdot\Sigma\gamma_1,$$

in which the previous case enables him to replace

$$\Sigma a_1\gamma_1\beta_2 \quad \text{by} \quad \Sigma a_1\gamma_1\cdot\Sigma\beta_1 - \Sigma a_1\beta_1\gamma_1,$$

and

$$\Sigma\beta_1\gamma_1 a_2 \quad \text{by} \quad \Sigma\beta_1\gamma_1\cdot\Sigma a_1 - \Sigma a_1\beta_1\gamma_1,$$

with the result that

$$\Sigma a_1\beta_2\gamma_3 = \Sigma a_1\Sigma\beta_1\Sigma\gamma_1 - \Sigma a_1\beta_1\cdot\Sigma\gamma_1 - \Sigma a_1\gamma_1\cdot\Sigma\beta_1 - \Sigma\beta_1\gamma_1\cdot\Sigma a_1 + 2\Sigma a_1\beta_1\gamma_1 \quad (x_3)$$

as desired. Similarly, on multiplying both sides of this by $\Sigma\delta_1$ there is obtained on the left

$$\Sigma a_1\beta_2\gamma_3\delta_4 + \Sigma a_1\delta_1\beta_2\gamma_3 + \Sigma\beta_1\delta_1 a_2\gamma_3 + \Sigma\gamma_1\delta_1 a_2\beta_3,$$

the last three terms of which have only to be replaced by expressions warranted from (x_3) in order to give the desired equivalent * for $\Sigma a_1\beta_2\gamma_3\delta_4$.

It is then pointed out that when $z=m$ the sum $\Sigma a_1\beta_2 \ldots \mu_m$ "tantum a determinante differt, quod omnes ejus termini sunt positivi," and that therefore when $z < m$ the sum vanishes.

The three sections following (iv., v., vi.) are occupied, as has been noted elsewhere, with the generalisation of Borchardt's theorem of the previous year.

CAYLEY, A. (1857).

[Note sur les normals d'une conique. *Crelle's Journ.*, lvi. pp. 182–185; or *Collected Math. Papers*, iv. pp. 74–77.]

In dealing with essentially the same geometrical subject as Joachimsthal, Cayley gives, in support of part of his demonstration, the identity

$$\left\{\begin{matrix} x_1 & y_1 & z_1 \\ x_2 & y_2 & z_2 \\ x_3 & y_3 & z_3 \end{matrix}\right\} \times \begin{vmatrix} x_1 & y_1 & z_1 \\ x_2 & y_2 & z_2 \\ x_3 & y_3 & z_3 \end{vmatrix} = \begin{vmatrix} x_1^2 & y_1^2 & z_1^2 \\ x_2^2 & y_2^2 & z_2^2 \\ x_3^2 & y_3^2 & z_3^2 \end{vmatrix} + \begin{vmatrix} y_1z_1 & z_1x_1 & x_1y_1 \\ y_2z_2 & z_2x_2 & x_2y_2 \\ y_3z_3 & z_3x_3 & x_3y_3 \end{vmatrix}$$

where the first factor on the left is what Cauchy denoted by $S^3(x_1y_2z_3)$.

* By an oversight three terms of this are left out by Joachimsthal.

(β) DETERMINANTS WITH COMPLEX ELEMENTS.

HERMITE, C. (1854).

[Extrait d'une lettre .... sur le nombre des racines d'une équation algébrique comprises entre des limites données. *Crelle's Journ.*, lii. pp. 39–51; or *Œuvres*, i. pp. 397–414.]

On p. 40 it is pointed out that any determinant whose conjugate elements are of the form $a_{rs}+b_{rs}\sqrt{-1}$, $a_{rs}-b_{rs}\sqrt{-1}$, and whose diagonal elements are therefore of the form a_{rr}, must be real, for the reason that it is not altered in value by changing $\sqrt{-1}$ into $-\sqrt{-1}$.

HERMITE, C. (1855, August).

[Remarque sur un théorème de M. Cauchy. *Comptes rendus .... Acad. des Sci.* (Paris), xli. pp. 181–183; or *Œuvres*, i. pp. 479–481.]

The remark concerns the determinant just referred to, and is to the effect that the equation

$$\begin{vmatrix} a_{11}-x & a_{12}+b_{12}i & \dots & a_{1n}+b_{1n}i \\ a_{21}+b_{21}i & a_{22}-x & \dots & a_{2n}+b_{2n}i \\ \cdot & \cdot & \cdot & \cdot \\ a_{n1}+b_{n1}i & a_{n2}+b_{n2}i & \dots & a_{nn}-x \end{vmatrix} = 0,$$

where $a_{rs}=a_{sr}$, $b_{rs}=-b_{sr}$, $i=\sqrt{-1}$, has all its roots real if the a's and b's be real,—a result which degenerates into one previously known (Lagrange, 1773; Cauchy, 1829) when all the b's vanish. No proof is given, but it is stated that one is obtainable by transforming "le déterminant en un autre à éléments réels, d'un nombre double de colonnes et symétrique par rapport à la diagonale." A rule is formulated for determining the number of roots of the equation which lie between two limits. Lastly, it is remarked that the equation arises in connection with the study of forms of the type

$$\begin{array}{cc|c} x+x'i & y+y'i & \\ \hline a_{11} & a_{12}-\beta_{12}i & x-x'i \\ a_{12}+\beta_{12}i & a_{22} & y-y'i, \end{array}$$

that is to say,

$$\left.\begin{array}{l} a_{11}x^2 + 2a_{12}xy + a_{22}y^2 \\ + a_{11}x'^2 + 2a_{12}x'y' + a_{22}y'^2 \end{array}\right\} + 2\beta_{12}|xy'|.$$

RUBINI, R. (1857, May).

[Applicazione della teorica dei determinanti. *Annali di Sci. mat. e fis.*, viii. pp. 179–200.]

In treating of determinants with binomial elements Rubini's most interesting example is that in which the element in the $(r, s)^{\text{th}}$ place is $a_{rs}+b_{rs}\sqrt{-1}$. By substitution in his general result he readily obtains the expansion of the determinant in the form $C+D\sqrt{-1}$, which is seen to alter into $C-D\sqrt{-1}$ on changing the signs of the b's. The product of $|a_{1n}+b_{1n}\sqrt{-1}|$ and $|a_{1n}-b_{1n}\sqrt{-1}|$ is consequently expressible as the sum of two squares. His next point is that on using the ordinary multiplication-theorem the same product is got in the form

$$\begin{vmatrix} a_{11} & a_{12}-\beta_{12}\sqrt{-1} & \dots & a_{1n}-\beta_{1n}\sqrt{-1} \\ a_{12}+\beta_{12}\sqrt{-1} & a_{22} & \dots & a_{2n}-\beta_{2n}\sqrt{-1} \\ \cdot & \cdot & \cdot & \cdot \\ a_{1n}+\beta_{1n}\sqrt{-1} & a_{2n}+\beta_{2n}\sqrt{-1} & \dots & a_{nn} \end{vmatrix},$$

and that a comparison of the two forms may be fruitful of results. When $n=2$, the identity resulting from such comparison is

$$(ad-bc-\alpha\delta+\beta\gamma)^2 + (a\delta-b\gamma+\alpha d-\beta c)^2$$
$$=(a^2+\alpha^2+b^2+\beta^2)(c^2+\gamma^2+d^2+\delta^2)-(ac+\alpha\gamma+bd+\beta\delta)^2-(a\gamma-\alpha c+b\delta-\beta d)^2,$$

a result which gives the product of two sums of four squares as a like sum.

In connection with this special example, however, note should be taken that Hermite in a letter to Jacobi published in 1850 (see *Crelle's Journ.*, xl. p. 297), had pointed out that it followed from the row-by-row multiplication of

$$\begin{vmatrix} a+\alpha\sqrt{-1} & b+\beta\sqrt{-1} \\ -b+\beta\sqrt{-1} & a-\alpha\sqrt{-1} \end{vmatrix} \text{ by } \begin{vmatrix} -c+\gamma\sqrt{-1} & -d+\delta\sqrt{-1} \\ d+\delta\sqrt{-1} & -c-\gamma\sqrt{-1} \end{vmatrix}.$$

CLEBSCH, A. (1859).

[Theorie der circularpolarisirenden Medien. *Crelle's Journ.*, lvii. pp. 319–358.]

In § 3 (pp. 324–330) Clebsch is led to consider the nature of the roots of the equation dealt with by Hermite in 1855, not knowing, apparently, what the latter had done. Unfortunately the proof given of the reality of the roots is not effected without the use of a set of unessential equations of which the determinant is the eliminant.

The interesting fact is noted that when $n=3$ the equation can be changed into

$$\begin{vmatrix} a_{11}-x & a_{12} & a_{13} \\ a_{12} & a_{22}-x & a_{23} \\ a_{13} & a_{23} & a_{33}-x \end{vmatrix} - \begin{vmatrix} & b_{23} & -b_{13} & b_{12} \\ b_{23} & a_{11}-x & a_{12} & a_{13} \\ -b_{13} & a_{12} & a_{22}-x & a_{23} \\ b_{12} & a_{13} & a_{23} & a_{33}-x \end{vmatrix} = 0.$$

(γ_1) DETERMINANTS CONNECTED WITH ANHARMONIC RATIOS.

CAYLEY, A. (1854, February).

[On some integral transformations. *Quart. Journ. of Math.*, i. pp. 4–6; or *Collected Math. Papers*, iii. pp. 1–4.]

This paper opens with two statements in reference to the determinant

$$\begin{vmatrix} 1 & \alpha & \alpha' & \alpha\alpha' \\ 1 & \beta & \beta' & \beta\beta' \\ 1 & \gamma & \gamma' & \gamma\gamma' \\ 1 & \delta & \delta' & \delta\delta' \end{vmatrix}, \quad \text{or } \Psi \text{ say.}$$

The first is to the effect that the equation

$$\Psi = 0$$

asserts the equality of the anharmonic ratios of α, β, γ, δ and

$\alpha', \beta', \gamma', \delta'$: and the second that the said equation may also be expressed in the forms*

$$\begin{aligned} \mathrm{K}\alpha &= -\{\gamma\delta(\gamma'-\delta')(\alpha'-\beta') + \delta\beta(\delta'-\beta')(\alpha'-\gamma') + \beta\gamma(\beta'-\gamma')(\alpha'-\delta')\}, \\ \mathrm{K}(\alpha-\beta) &= (\delta-\beta)(\beta-\gamma)(\gamma'-\delta')(\alpha'-\beta'), \\ \mathrm{K}(\alpha-\gamma) &= (\beta-\gamma)(\gamma-\delta)(\delta'-\beta')(\alpha'-\gamma'), \\ \mathrm{K}(\alpha-\delta) &= (\gamma-\delta)(\delta-\beta)(\beta'-\gamma')(\alpha'-\delta'), \end{aligned}$$

if we use K to stand for

$$\beta(\gamma'-\delta')(\alpha'-\beta') + \gamma(\delta'-\beta')(\alpha'-\gamma') + \delta(\beta'-\gamma')(\alpha'-\delta').$$

Accepting the first statement, and knowing that the equality referred to is

$$\frac{(\gamma-\alpha)(\beta-\delta)}{(\alpha-\beta)(\gamma-\delta)} = \frac{(\gamma'-\alpha')(\beta'-\delta')}{(\alpha'-\beta')(\gamma'-\delta')},$$

we readily make the deduction that

$$\Psi = (\gamma-\alpha)(\beta-\delta)(\alpha'-\beta')(\gamma'-\delta') - (\alpha-\beta)(\gamma-\delta)(\gamma'-\alpha')(\beta'-\delta').$$

By accepting the second statement, like conclusions may be drawn; for then the elimination of K from any two of the

---

* These may be established as follows. By separating the terms of K which involve α' from those which do not, we see that

$$\mathrm{K} = -\begin{vmatrix} . & 1 & . & \alpha' \\ 1 & \beta & \beta' & \beta\beta' \\ 1 & \gamma & \gamma' & \gamma\gamma' \\ 1 & \delta & \delta' & \delta\delta' \end{vmatrix},$$

a determinant differing from Ψ in the first row only, and consequently on multiplying by α and adding we obtain

$$\Psi + \mathrm{K}\alpha = \begin{vmatrix} 1 & . & \alpha' & . \\ 1 & \beta & \beta' & \beta\beta' \\ 1 & \gamma & \gamma' & \gamma\gamma' \\ 1 & \delta & \delta' & \delta\delta' \end{vmatrix} = \begin{vmatrix} \beta & \beta'-\alpha' & \beta\beta' \\ \gamma & \gamma'-\alpha' & \gamma\gamma' \\ \delta & \delta'-\alpha' & \delta\delta' \end{vmatrix} = \ldots.$$

Similarly,

$$\Psi + \mathrm{K}(\alpha-\beta) = \begin{vmatrix} 1 & \beta & \alpha' & \beta\alpha' \\ 1 & \beta & \beta' & \beta\beta' \\ 1 & \gamma & \gamma' & \gamma\gamma' \\ 1 & \delta & \delta' & \delta\delta' \end{vmatrix} = \begin{vmatrix} 1 & . & \alpha' & . \\ 1 & . & \beta' & . \\ 1 & \gamma-\beta & \gamma' & (\gamma-\beta)\gamma' \\ 1 & \delta-\beta & \delta' & (\delta-\beta)\delta' \end{vmatrix} = \ldots.$$

and so of the others.

In doing this we learn, too, that

$$\Psi+\mathrm{K}\alpha = \Psi_{\alpha=0}, \quad \Psi+\mathrm{K}(\alpha-\beta) = \Psi_{\alpha=\beta}, \quad \ldots.$$

equations involving it must of course lead us back to some form or other of the equation with which we started. Thus, multiplying K by α and using the first of the four derived equations we obtain by subtraction

$$0 = (\alpha\beta+\gamma\delta)(\gamma'-\delta')(\alpha'-\beta') \\ - (\alpha\gamma+\beta\delta)(\beta'-\delta')(\alpha'-\gamma') + (\alpha\delta+\beta\gamma)(\beta'-\gamma')(\alpha'-\delta'),$$

whence we deduce in the same manner as before that the expression * on the right when changed in sign is equal to Ψ; and using any pair of the remaining equations we reach either the form of Ψ previously obtained or one of the two forms derivable from it by means of the simultaneous circular substitutions

$$\beta, \gamma, \delta = \gamma, \delta, \beta,$$
$$\beta', \gamma', \delta' = \gamma', \delta', \beta'.$$

CAYLEY, A. (1858, February).

[A fifth memoir on quantics. *Philos. Transac. R. Soc.* (London) cxlviii. pp. 429–460; or *Collected Math. Papers*, ii. pp. 527–557.]

The second part (§§ 96–114) of the memoir deals with two or more quadrics, and forming part of it is a digression (§§ 105–114) on involution and the anharmonic relation. The determinant Ψ thus again makes its appearance, and associated with it is the determinant

$$\begin{vmatrix} 1 & \alpha+\alpha' & \alpha\alpha' \\ 1 & \beta+\beta' & \beta\beta' \\ 1 & \gamma+\gamma' & \gamma\gamma' \end{vmatrix}, \quad \text{or } \Upsilon \text{ say,}$$

for the reason that, when $\delta=\alpha'$ and $\delta'=\alpha$, Ψ is readily shown to be equal to

$$(\alpha'-\alpha)\Upsilon.$$

* This second form of Ψ may be got directly from the determinant by expanding in terms of the two-line minors formable from the first and third columns, and the minors complementary to these. Of course we also have

$$\Psi = (\alpha'\beta'+\gamma'\delta')(\alpha-\beta)(\gamma-\delta) - (\alpha'\gamma'+\beta'\delta')(\alpha-\gamma)(\beta-\delta) + (\alpha'\delta'+\beta'\gamma')(\alpha-\delta)(\beta-\gamma).$$

To obtain the required non-determinant forms of the two the multiplication-theorem is used with pleasing effect. In the first place Υ is multiplied row-wise by

$$\begin{vmatrix} u^2 & -u & 1 \\ v^2 & -v & 1 \\ w^2 & -w & 1 \end{vmatrix},$$

the result being, of course,

$$\Upsilon\cdot(w-v)(w-u)(v-u) = \begin{vmatrix} (u-\alpha)(u-\alpha') & (v-\alpha)(v-\alpha') & (w-\alpha)(w-\alpha') \\ (u-\beta)(u-\beta') & (v-\beta)(v-\beta') & (w-\beta)(w-\beta') \\ (u-\gamma)(u-\gamma') & (v-\gamma)(v-\gamma') & (w-\gamma)(w-\gamma') \end{vmatrix}.$$

In this Cayley then puts $u=\alpha$, $v=\alpha'$, obtaining

$$\Upsilon\cdot(\alpha'-\alpha) = (\alpha-\beta)(\alpha-\beta')(\alpha'-\gamma)(\alpha'-\gamma') - (\alpha'-\beta)(\alpha'-\beta')(\alpha-\gamma)(\alpha-\gamma'),$$

and putting $u, v, w=\alpha, \beta, \gamma$ obtains

$$\Upsilon = (\alpha-\beta')(\beta-\gamma')(\gamma-\alpha') - (\alpha-\gamma')(\beta-\alpha')(\gamma-\beta'),$$

a result known to Hesse in 1849 (see *Crelle's Journ.* l. p. 265).

In the next place (§ 114) Ψ is multiplied by the similar determinant

$$\begin{vmatrix} ss' & -s' & -s & 1 \\ tt' & -t' & -t & 1 \\ uu' & -u' & -u & 1 \\ vv' & -v' & -v & 1 \end{vmatrix}, \quad \text{or } \Psi' \text{ say,}$$

the result being

$$\Psi\Psi' = \begin{vmatrix} (s-\alpha)(s'-\alpha') & (t-\alpha)(t'-\alpha') & (u-\alpha)(u'-\alpha') & (v-\alpha)(v'-\alpha') \\ (s-\beta)(s'-\beta') & (t-\beta)(t'-\beta') & (u-\beta)(u'-\beta') & (v-\beta)(v'-\beta') \\ (s-\gamma)(s'-\gamma') & (t-\gamma)(t'-\gamma') & (u-\gamma)(u'-\gamma') & (v-\gamma)(v'-\gamma') \\ (s-\delta)(s'-\delta') & (t-\delta)(t'-\delta') & (u-\delta)(u'-\delta') & (v-\delta)(v'-\delta') \end{vmatrix},$$

so that on putting

$$\left.\begin{matrix} s, & t, & u, & v, \\ s', & t', & u', & v', \end{matrix}\right\} = \left\{\begin{matrix} \alpha, & \beta, & \gamma, & \delta \\ \beta', & \alpha', & \delta', & \gamma' \end{matrix}\right.$$

the product becomes

$$\begin{vmatrix} \cdot & \cdot & (\gamma-\alpha)(\delta'-\alpha') & (\delta-\alpha)(\gamma'-\alpha') \\ \cdot & \cdot & (\gamma-\beta)(\delta'-\beta') & (\delta-\beta)(\gamma'-\beta') \\ (\alpha-\gamma)(\beta'-\gamma') & (\beta-\gamma)(\alpha'-\gamma') & \cdot & \cdot \\ (\alpha-\delta)(\beta'-\delta') & (\beta-\delta)(\alpha'-\delta') & \cdot & \cdot \end{vmatrix},$$

and there is obtained

$$\begin{vmatrix} 1 & \alpha & \alpha' & \alpha\alpha' \\ 1 & \beta & \beta' & \beta\beta' \\ 1 & \gamma & \gamma' & \gamma\gamma' \\ 1 & \delta & \delta' & \delta\delta' \end{vmatrix} \cdot \begin{vmatrix} 1 & \alpha & \beta' & \alpha\beta' \\ 1 & \beta & \alpha' & \beta\alpha' \\ 1 & \gamma & \delta' & \gamma\delta' \\ 1 & \delta & \gamma' & \delta\gamma' \end{vmatrix} = \left\{ \begin{array}{l} \quad (\alpha-\gamma)(\beta-\delta)(\alpha'-\delta')(\beta'-\gamma') \\ -(\alpha-\delta)(\beta-\gamma)(\alpha'-\gamma')(\beta'-\delta') \end{array} \right\}^2 .$$

From this Cayley concludes (1) that Ψ is not changed by the transposition

$$\begin{pmatrix} \alpha' & \gamma' \\ \beta' & \delta' \end{pmatrix},$$

and (2) that either form equals

$$(\alpha-\gamma)(\beta-\delta)(\alpha'-\delta')(\beta'-\gamma') - (\alpha-\delta)(\beta-\gamma)(\alpha'-\gamma')(\beta'-\delta').$$

SARDI, C. (1864).

[Quistione 39. *Giornale di Mat.*, p. 256, pp. 315–316.]

On the determinant Ψ Sardi performs the operation which we may indicate by

$$\mathrm{col}_4 - \beta\,\mathrm{col}_3 - \delta'\,\mathrm{col}_2 + \beta\delta'\,\mathrm{col}_1,$$

thus obtaining

$$\begin{vmatrix} 1 & \alpha & \alpha' & (\alpha-\beta)(\alpha'-\delta') \\ 1 & \beta & \beta' & \cdot \\ 1 & \gamma & \gamma' & (\gamma-\beta)(\gamma'-\delta') \\ 1 & \delta & \delta' & \cdot \end{vmatrix},$$

in which the cofactor of $(\alpha-\beta)(\alpha'-\delta')$ is

$$\begin{vmatrix} \beta-\gamma & \beta'-\gamma' \\ \delta-\gamma & \delta'-\gamma' \end{vmatrix}$$

and the cofactor of $(\gamma-\beta)(\gamma'-\delta')$ is

$$\begin{vmatrix} \beta-\alpha & \beta'-\alpha' \\ \delta-\alpha & \delta'-\alpha' \end{vmatrix},$$

where, be it observed, it is the rows 1, β, β' and 1, δ, δ' that are diminished on both occasions. There is thus obtained

$$\begin{aligned}\Psi = \quad & (\alpha-\beta)(\alpha'-\delta')\{(\beta-\gamma)(\delta'-\gamma') - (\delta-\gamma)(\beta'-\gamma')\} \\ & - (\gamma-\beta)(\gamma'-\delta')\{(\beta-\alpha)(\delta'-\alpha') - (\beta'-\alpha')(\delta-\alpha)\}, \\ = \; & (\alpha-\beta)(\alpha'-\delta')(\gamma-\delta)(\beta'-\gamma') + (\gamma-\beta)(\gamma'-\delta')(\beta'-\alpha')(\delta-\alpha), \\ = \; & -(\alpha-\beta)(\gamma-\delta)(\alpha'-\delta')(\gamma'-\beta') + (\alpha'-\beta')(\gamma'-\delta')(\alpha-\delta)(\gamma-\beta).^*\end{aligned}$$

(γ_2) SYLVESTER'S UNISIGNANT.

SYLVESTER, J. J. (1855, April).

[On the change of systems of independent variables. *Quart. Journ. of Math.*, i. pp. 42–56; or *Collected Math. Papers*, ii. pp. 65–85.]

In the course of Sylvester's investigations a peculiar three-line determinant turns up, which he considers deserving of attention on its own account, namely, the determinant

$$\begin{vmatrix} a_1+a_2+a_3 & -a_2 & -a_3 \\ -b_1 & b_1+b_2+b_3 & -b_3 \\ -c_1 & -c_2 & c_1+c_2+c_3 \end{vmatrix},$$

the final expansion of which consists of 16 terms, all positive. To obtain this expansion a "simple rule" is laid down, namely, to substitute

$$\left.\begin{matrix} a & a_b & a_c \\ b_a & b & b_c \\ c_a & c_b & c \end{matrix}\right\} \quad \text{for} \quad \left\{\begin{matrix} a_1 & a_2 & a_3 \\ b_1 & b_2 & b_3 \\ c_1 & c_2 & c_3 \end{matrix}\right.$$

* It will be seen that merely by accident the three ways in which Ψ can be expressed as the difference of two products have turned up in succession, and that they may be written

$$|PQ'|, |QR'|, |PR'|$$

if we put

$$\begin{aligned}(\alpha-\beta)(\gamma-\delta) &= P, \\ (\alpha-\gamma)(\beta-\delta) &= Q, \\ (\alpha-\delta)(\beta-\gamma) &= R.\end{aligned}$$

and then multiply together the elements of the diagonal, rejecting every term such as $a_b b_a$, $a_b b_c c_a$, ... in which the letters form a cycle. Two examples are given, but no justification of the "rule" is vouchsafed. The examples are—

$$\begin{vmatrix} a+a_b+a_c & -a_b & -a_c \\ -b_a & b+b_c+b_a & -b_c \\ -c_a & -c_b & c+c_a+c_b \end{vmatrix} = \begin{aligned} abc+(c_a+c_b)ab+(a_b+a_c)bc+(b_c+b_a)ca \\ +a(b_a c_a+b_a c_b+c_a b_c) \\ +b(c_b a_b+c_b a_c+a_b c_a) \\ +c(a_c b_c+a_c b_a+b_c a_b), \end{aligned}$$

$$\begin{vmatrix} a+a_b+a_c+a_d & -b_a & -c_a & -d_a \\ -a_b & b+b_c+b_d+b_a & -c_b & -d_b \\ -a_c & -b_c & c+c_d+c_a+c_b & -d_c \\ -a_d & -b_d & -c_d & d+d_a+d_b+d_c \end{vmatrix}$$

$$= abcd + \sum abc(d_a+d_b+d_c) + \sum ab(c_d d_a+ \;....) + \sum a(b_c c_d d_a+\;....).$$

The arrangement of the two developments almost raises doubts as to whether the "rule" had been utilised, suggesting indeed that in the latter instance, for example, the cofactor of ab was first obtained in the form

$$\begin{vmatrix} c_d+c_a+c_b & -d_c \\ -c_d & d_a+d_b+d_c \end{vmatrix},$$

and the cofactor of a in the form of a similar determinant of the third order. The "rule," however, is noted by Cayley in *Crelle's Journal*, lii. (1855), p. 279.

The number of terms is $(n+1)^{n-1}$, n being the order-number of the determinant. This Sylvester obtains by putting $a, a_b, a_c, \ldots$ all equal to 1. It will be observed that from the form of the development we thus have

$$1 + 3\cdot2 + 3\cdot3 = 4^2$$

$$1 + 4\cdot3 + 6\cdot8 + 4\cdot16 = 5^3$$

$$1 + 5\cdot4 + 10\cdot15 + 10\cdot50 + 5\cdot125 = 6^4$$

. . . .

BORCHARDT, C. W. (1859, May).

[Ueber eine der Interpolation entsprechende Darstellung der Eliminations-Resultante. *Crelle's Journ.*, lvii. pp. 111–121; or *Monatsb. d. Akad. d. Wiss.* (Berlin), pp. 376–388; also abstract in *Annali di Mat.*, ii. pp. 262–264.]

The representation in question is in terms of the values which the two functions $\phi(x)$ and $\psi(x)$, both of the nth degree, assume for the values $a_0, a_1, a_2, \ldots, a_n$ of x. It emerges as a special determinant of the form

$$\begin{vmatrix} \sigma_1-(11) & -(12) & \ldots\ldots & -(1n) \\ -(21) & \sigma_2-(22) & \ldots\ldots & -(2n) \\ \cdot & \cdot & \cdot & \cdot \\ -(n1) & -(n2) & \ldots\ldots & \sigma_n-(nn) \end{vmatrix},$$

where

$$\sigma_r=(r0)+(r1)+\ldots\ldots+(rn) \quad \text{and} \quad (rs)=(sr),$$

a form which we readily recognise to be the axisymmetric case of Sylvester's determinant of the year 1855. To the consideration of it Borchardt, probably supposing it to be new, devotes the last six pages of his paper.

Denoting it by $(0, 1, 2, \ldots, n)$, since it is a function of the $\frac{1}{2}n(n+1)$ quantities,

$$\begin{array}{cccc} (01) & (02) & \ldots\ldots & (0n) \\ & (12) & \ldots\ldots & (1n) \\ & & \cdot\ \cdot\ \cdot\ \cdot\ \cdot\ \cdot & \\ & & & (n-1,\, n), \end{array}$$

he first shows with some prolixity that the cofactor of (01) in it is $(\overline{0+1}, 2, 3, \ldots, n)$, next that the cofactor of $(01)(02)\ldots(0i)$ is

$$(\overline{0+1+\ldots\ldots+i}, i+1, i+2, \ldots\ldots, n),$$

and finally that

$$\begin{aligned} (0, 1, 2, \ldots, n) = & \sum(01)(1, 2, \ldots, n) \\ & +\sum(01)(02)(\overline{1+2}, 3, \ldots, n) \\ & +\ \cdot\ \cdot\ \cdot\ \cdot\ \cdot\ \cdot\ \cdot\ \cdot\ \cdot\ \cdot\ \cdot \\ & +\sum(01)(02)\ldots(0k)(\overline{1+2+\ldots+k}, k+1, \ldots, n) \\ & +\ \cdot\ \cdot\ \cdot\ \cdot\ \cdot\ \cdot\ \cdot\ \cdot\ \cdot\ \cdot\ \cdot \\ & +(01)(02)\ldots(0n). \end{aligned}$$

Resuming consideration, but proceeding on a different tack, he arrives at Sylvester's "rule," namely, that $(0, 1, 2, \ldots, n)$ is "gleich der Summe aller nicht-cyclischen Producte, die aus je n jener $\frac{1}{2}n(n+1)$ Elemente $(i\,k)$ gebildet werden können." Unlike Sylvester, however, he is careful to give a justification of it based on four observed facts, namely, (1) that $(0, 1, 2, \ldots, n)$ is unaltered by interchanging any two of the umbræ; (2) that the coefficient of the term $(01)(02)\ldots(0n)$ is 1; (3) that none of the terms is free of the umbra 0; (4) that, as already mentioned, the cofactor of (01) is $(\overline{0+1},\ 2, \ldots, n)$.* As the proof, which extends to two pages (pp. 119–120), applies only to the case of axisymmetry, it need not be given.

Lastly, the number of terms in the development of $(0, 1, 2, \ldots, n)$ is investigated, the result obtained agreeing with Sylvester's.

We may note for ourselves in passing that the first three of the basic facts of the proof are, like the last, most readily appreciated by observing the determinant form, the case where $n=3$, namely,

$$\begin{vmatrix} 10+12+13 & -12 & -13 \\ -21 & 20+21+23 & -23 \\ -31 & -32 & 30+31+32 \end{vmatrix}$$

being amply sufficient. Thus, increasing any column by all the others, and thereafter increasing the corresponding row by all the other rows, we obtain the first result, learning at the same time that it only holds when axisymmetry exists; the second is self-evident; and the third follows from the fact that the aggregate of the terms which are free of 0, being got by deleting 10, 20, 30, is expressible as a vanishing determinant.

(δ) MISCELLANEOUS SPECIAL FORMS.

CAYLEY, A. (1845).

[On certain results relating to quaternions. *Philos. Magazine*, xxvi. pp. 141–145; or *Collected Math. Papers*, i. pp. 123–126.]

Assuming that in each term of the development of a deter-

* As (01) occurs only in the element $\sigma_1 - (11)$, its cofactor is the primary minor obtained by deleting the first row and the first column, and this is seen to be $(\overline{0+1}, 2, \ldots, n)$ by definition.

minant the elements are arranged in the order of the columns from which they are taken, Cayley points out that if the elements be quaternions

$$\begin{vmatrix} \pi & \pi' \\ \pi & \pi' \end{vmatrix} = \pi\pi' - \pi\pi' = 0,$$

but

$$\begin{vmatrix} \pi & \pi \\ \pi' & \pi' \end{vmatrix} = \pi\pi' - \pi'\pi \neq 0.$$

He is thus led to inquire what the non-zero value is in this latter case and in other similar cases. Taking

$$\begin{aligned} \pi &= x + iy + jz + kw, \\ \pi' &= x' + iy' + jz' + kw', \\ \pi'' &= x'' + iy'' + jz'' + kw'', \end{aligned}$$

he says it is easy to show * that

$$\begin{vmatrix} \pi & \pi \\ \pi' & \pi' \end{vmatrix} = -2 \begin{vmatrix} i & j & k \\ y & z & w \\ y' & z' & w' \end{vmatrix},$$

$$\begin{vmatrix} \pi & \pi & \pi \\ \pi' & \pi' & \pi' \\ \pi'' & \pi'' & \pi'' \end{vmatrix} = -2 \begin{vmatrix} 3 & i & j & k \\ x & y & z & w \\ x' & y' & z' & w' \\ x'' & y'' & z'' & w'' \end{vmatrix},$$

---

*Probably the easiest way is to express the determinant as a sum of determinants with monomial elements. In the case of the third order the number of such determinants is 64, of which 40 vanish, the sum remaining being

$$\begin{aligned} &123+132+213+231+312+321 \\ &+124+142+\ldots\ldots \\ &+134+143+\ldots\ldots \\ &+234+243+\ldots\ldots \end{aligned}$$

where rst stands for the determinant whose columns are in order the rth, s^{th}, t^{th} columns of the array

$$\begin{matrix} x & iy & jz & kw \\ x' & iy' & jz' & kw' \\ x'' & iy'' & jz'' & kw'' \end{matrix}$$

and where therefore

$$123+132+\ldots = |\, xy'z'' \,|\cdot(ij+ij+ij-ij-ij+ij) = 2k\,|\, xy'z'' \,|,$$

and so on. The multiplication table of i, j, k, it may be recalled, is

$$\begin{pmatrix} ii & ij & ik \\ ji & jj & jk \\ ki & kj & kk \end{pmatrix} = \begin{pmatrix} -l & k & -j \\ -k & -l & i \\ j & -i & -l \end{pmatrix}.$$

but that for higher orders the result is 0. He next notes the identity

$$\begin{vmatrix} \phi & \chi \\ \phi' & \chi' \end{vmatrix} + \begin{vmatrix} \chi & \phi \\ \chi' & \phi' \end{vmatrix} = \begin{vmatrix} \phi & \phi \\ \chi' & \chi' \end{vmatrix} - \begin{vmatrix} \phi' & \phi' \\ \chi & \chi \end{vmatrix},$$

adding "etc. for determinants of any order";* and then from this set of identities and the previous set he concludes that *if any four adjacent columns of a quaternion determinant be transposed in every possible manner, the sum of the determinants thus obtained vanishes*—a property which, he says, is much less simple than the analogous one for the rows, this last being the same that holds in the case of determinants with ordinary elements. Lastly, he gives the important warning that the eliminant of

$$\left.\begin{aligned} \pi\Pi + \phi\Phi &= 0 \\ \pi'\Pi + \phi'\Phi &= 0 \end{aligned}\right\}$$

is neither $\pi\phi' - \pi'\phi$ nor $\pi\phi' - \phi\pi'$, but

$$\pi^{-1}\phi - \pi'^{-1}\phi'.$$

TISSOT, A. (1852, May).

[Sur un déterminant d'intégrales définies. *Journ.* (*de Liouville*) *de Math.*, xvii. pp. 177–185.]

The subject here is the evaluation of the determinant of the $(n+1)^{\text{th}}$ order whose $(r, s)^{\text{th}}$ element is

$$\int_{a_{s-1}}^{a_s} e^{-x} \frac{x^r dx}{(\phi)_{s-1}(x)},$$

---

* Very probably the next case is the identity

$$\begin{vmatrix} \phi & \chi & \psi \\ \phi' & \chi' & \psi' \\ \phi'' & \chi'' & \psi'' \end{vmatrix} + \begin{vmatrix} \phi & \psi & \chi \\ \phi' & \psi' & \chi' \\ \phi'' & \psi'' & \chi'' \end{vmatrix} + \begin{vmatrix} \chi & \phi & \psi \\ \chi' & \phi' & \psi' \\ \chi'' & \phi'' & \psi'' \end{vmatrix} + \dots + \begin{vmatrix} \psi & \chi & \phi \\ \psi' & \chi' & \phi' \\ \psi'' & \chi'' & \phi'' \end{vmatrix}$$

$$= \begin{vmatrix} \phi & \phi & \phi \\ \chi' & \chi' & \chi' \\ \psi'' & \psi'' & \psi'' \end{vmatrix} - \begin{vmatrix} \phi & \phi & \phi \\ \chi'' & \chi'' & \chi'' \\ \psi' & \psi' & \psi' \end{vmatrix} - \begin{vmatrix} \phi' & \phi' & \phi' \\ \chi & \chi & \chi \\ \psi'' & \psi'' & \psi'' \end{vmatrix} + \dots - \begin{vmatrix} \phi'' & \phi'' & \phi'' \\ \chi' & \chi' & \chi' \\ \psi & \psi & \psi \end{vmatrix},$$

where, as in the other cases, the rth determinant on the left is equal to the aggregate of the rth terms of all the determinants on the right.

where

$$\phi_i(x) = (x-a_0)^{m_0}(x-a_1)^{m} \ldots . (x-a_i)^{m_i}(a_{i+1}-x)^{m_{i+1}} \ldots . (a_n-x)^{m_n},$$

the m's are all less than 1, and $a_{n+1}=\infty$. The simplest example is

$$\begin{vmatrix} \int_{a_0}^{a_1} e^{-x}\frac{dx}{\phi_0(x)} & \int_{a_1}^{\infty} e^{-x}\frac{dx}{\phi_1(x)} \\ \int_{a_0}^{a_1} e^{-x}\frac{x\,dx}{\phi_0(x)} & \int_{a_1}^{\infty} e^{-x}\frac{x\,dx}{\phi_1(x)} \end{vmatrix} = \Gamma(1-m_0)\cdot\Gamma(1-m_1)\cdot(a_1-a_0)^{1-m_0-m_1}\cdot e^{-a_0-a_1}.$$

In establishing the result, use is made of the fact that the determinant is expressible also as a multiple integral: for example, the two-line determinant just written is equal to

$$\int_{a_0}^{a_1}\int_{a_1}^{\infty}\frac{e^{-x-x_1}(x_1-x)\,dx\,dx_1}{(a-x_0)^{m_0}(a_1-x)^{m_1}(x_1-a_0)^{m_0}(x_1-a_1)^{m_1}}.$$

BAZIN, [H.] (1854, July).

[Démonstration d'un théorème sur les déterminants. *Journ. (de Liouville) de Math.*, xix. pp. 209–214.]

The theorem in question is to the effect that if there be two n-by-m arrays R, R′ with integral elements, and such that the ratio of any n-line minor of R to the corresponding minor of R′ is constant and integral, and if the n-line minors of R have 1 for their highest common factor, then it is possible to find a determinant S of the nth order with integral elements so that the product of S by any n-line minor of R′ shall equal the corresponding minor of R. For example, it being given that

$$k\begin{Vmatrix} b_1 & b_2 & b_3 \\ c_1 & c_2 & c_3 \end{Vmatrix} = \begin{Vmatrix} x_1 & x_2 & x_3 \\ y_1 & y_2 & y_3 \end{Vmatrix}$$

where all the letters denote integers, and that the highest common factor of $|b_1c_2|$, $|b_1c_3|$, $|b_2c_3|$ is 1, four integers α, β, γ, δ can be found such that

$$\begin{vmatrix} \alpha & \beta \\ \gamma & \delta \end{vmatrix}\cdot\begin{Vmatrix} b_1 & b_2 & b_3 \\ c_1 & c_2 & c_3 \end{Vmatrix} = \begin{Vmatrix} x_1 & x_2 & x_3 \\ y_1 & y_2 & y_3 \end{Vmatrix}.$$

BRIOSCHI, F. (1855).

[Additions à l'article No. 15, page 239 de ce tome. *Crelle's Journ.*, L. pp. 318–321; or *Opere mat.*, v. pp. 271–276.]

The determinant here (pp. 320–321) dealt with is, for shortness' sake, taken to be of the 4th order, namely, $|m_1\ \delta_2\ b_3\ c_4|$, in which $\delta_1, \delta_2, \delta_3, \delta_4$ stand for

$$\begin{gathered} ax_1 + ex_2 + fx_3 + gx_4\,, \\ ex_1 + bx_2 + hx_3 + kx_4\,, \\ fx_1 + hx_2 + cx_3 + lx_4\,, \\ gx_1 + kx_2 + lx_3 + dx_4\,, \end{gathered}$$

and where the δ's, b's, c's are such that

$$\begin{aligned} \delta_1x_1 + \delta_2x_2 + \delta_3x_3 + \delta_4x_4 &= 0\,, \\ b_1x_1 + b_2x_2 + b_3x_3 + b_4x_4 &= 0\,, \\ c_1x_1 + c_2x_2 + c_3x_3 + c_4x_4 &= 0\,. \end{aligned}$$

The cofactor of m_r in $|m_1\ \delta_2\ b_3\ c_4|$ being denoted by M_r, we see that, as an example,

$$M_4^2 = -\begin{vmatrix} a & e & f & \delta_1 & b_1 & c_1 \\ e & b & h & \delta_2 & b_2 & c_2 \\ f & h & c & \delta_3 & b_3 & c_3 \\ \delta_1 & \delta_2 & \delta_3 & \cdot & \cdot & \cdot \\ b_1 & b_2 & b_3 & \cdot & \cdot & \cdot \\ c_1 & c_2 & c_3 & \cdot & \cdot & \cdot \end{vmatrix}$$

which after performance of the operations

$$\begin{gathered} \mathrm{col}_4 - x_1\mathrm{col}_1 - x_2\mathrm{col}_2 - x_3\mathrm{col}_3\,, \\ \mathrm{row}_4 - x_1\mathrm{row}_1 - x_2\mathrm{row}_2 - x_3\mathrm{row}_3\,, \end{gathered}$$

becomes

$$M_4^2 = -\begin{vmatrix} a & e & f & g & b_1 & c_1 \\ e & b & h & k & b_2 & c_2 \\ f & h & c & l & b_3 & c_3 \\ g & k & l & d & b_4 & c_4 \\ b_1 & b_2 & b_3 & b_4 & \cdot & \cdot \\ c_1 & c_2 & c_3 & c_4 & \cdot & \cdot \end{vmatrix} x_4^2, \quad \text{or say } -x_4^2\Delta\,.$$

As it can be shown similarly that

$$M_3^2 = -x_3^2\Delta, \quad M_2^2 = -x_2^2\Delta, \quad M_1^2 = -x_1^2\Delta,$$

Brioschi obtains*

$$|m_1\delta_2b_3c_4| = (m_1x_1+m_2x_2+m_3x_3+m_4x_4)\sqrt{-\Delta},$$

nothing being said as to the sign to be taken in extracting the square root of x_r^2.

We have only to add for ourselves that the first of the conditioning equations is the vanishing of the quaternary quadric

$$\begin{array}{cccc|c} x_1 & x_2 & x_3 & x_4 & \\ \hline a & e & f & g & x_1 \\ e & b & h & k & x_2 \\ f & h & c & l & x_3 \\ g & k & l & d & x_4 \end{array},$$

and that the δ's are the halved differential quotients of this with respect to x_1, x_2, x_3, x_4.

HERMITE, C. (1855, January).

[Sur la théorie de la transformation des fonctions abéliennes. *Comptes rendus .... Acad. des Sci.* (Paris), xl. pp. 249–254: or *Œuvres*, i. pp. 444–478.]

The special determinant here considered, as being auxiliary to Hermite's main purpose, is $|a_1b_2c_3d_4|$ with its elements subject to the conditions

$$\left.\begin{array}{l} |a_1d_2| + |b_1c_2| = 0 = |a_1d_3| + |b_1c_3| \\ |a_1d_4| + |b_1c_4| = k = |a_2d_3| + |b_2c_3| \\ |a_2d_4| + |b_2c_4| = 0 = |a_3d_4| + |b_3c_4| \end{array}\right\},$$

and the results in regard to it are:—(1) that it is equal to k^2; (2) that the row-by-row product of two such determinants is a determinant of the same type. No proof is given, but from the way in which Hermite writes the conditions, it would appear

* The minus sign is omitted by him throughout. If the number of x's had been odd, the sign would have been +.

that the first was obtained by multiplying the given determinant columnwise by itself in the form

$$\begin{vmatrix} d_1 & d_2 & d_3 & d_4 \\ c_1 & c_2 & c_3 & c_4 \\ -b_1 & -b_2 & -b_3 & -b_4 \\ -a_1 & -a_2 & -a_3 & -a_4 \end{vmatrix}.$$

A generalisation by Brioschi (1855) has already been dealt with under Skew Determinants.

ZEHFUSS, G. (1858).

[Uebungsaufgaben für Schüler. *Archiv d. Math. u. Phys.*, xxxi. p. 246; or *Nouv. Annales de Math.*, xviii. p. 171; (2) ii. pp. 60–61.]

The proposition offered for proof by Zehfuss is in modern phraseology to the effect that the determinant of the difference of the two square matrices

$$\begin{matrix} a_1 & a_1 & \dots & a_1 \\ a_2 & a_2 & \dots & a_2 \\ \cdot & \cdot & \cdots & \cdot \\ a_n & a_n & \dots & a_n, \end{matrix} \qquad \begin{matrix} b_1 & b_2 & \dots & b_n \\ b_1 & b_2 & \dots & b_n \\ \cdot & \cdot & \cdots & \cdot \\ b_1 & b_2 & \dots & b_n \end{matrix}$$

vanishes for all orders higher than the second. The proof given by Gustave Harang in the *Nouvelles Annales* rests on the operations

$$\mathrm{col}_1 - \mathrm{col}_2, \quad \mathrm{col}_2 - \mathrm{col}_3, \quad \dots\dots$$

When $n=2$ we have

$$\begin{vmatrix} a_1-b_1 & a_1-b_2 \\ a_2-b_1 & a_2-b_2 \end{vmatrix} = (a_1-a_2)(b_1-b_2).$$

CAYLEY, A. (1859, March).

[On the double tangents of a plane curve. *Philos. Transac. R. Soc.* (London), cxlix. pp. 193–212; or *Collected Math. Papers*, iv. pp. 186–206.]

The theorem on which an important part of this investigation rests is enunciated by its author as follows: *If the* 2n − 1 *columns of the special three-row matrix*

$$\begin{matrix} a_0 & a_1 & a_2 & \dots & a_{n-1}; & a'_0 & a'_1 & \dots & a'_{n-2} \\ a_1 & a_2 & a_3 & \dots & a_n\ ; & a'_1 & a'_2 & \dots & a'_{n-1} \\ a'_0 & a'_1 & a'_2 & \dots & a'_{n-1}; & a''_0 & a''_1 & \dots & a''_{n-2} \end{matrix}$$

be represented by

$$1 \quad 2 \quad 3 \quad \dots \quad n \ ; \qquad (1)\ (2)\ \dots\ (n-1)$$

respectively: the determinant whose columns are those thus represented by r, s, (t) *be denoted by* {r, s, (t)}; *and the determinant aggregates*

$$\begin{aligned} &\{n, n-1, (2)\} + \{n, n-2, (3)\} + \dots + \{n, 2, (n-1)\}, \\ -&\{n, n-1, (1)\} - \{n, n-2, (2)\} - \dots - \{n, 2, (n-2)\} - \{n, 1, (n-1)\}, \\ -&\{1, 2, (n-1)\} - \{1, 3, (n-2)\} - \dots - \{1, n-1, (2)\} - \{1, n, (1)\}, \\ &\{1, 2, (n-2)\} + \{1, 3, (n-3)\} + \dots + \{1, n-1, (1)\}, \end{aligned}$$

by I, II, III, IV; *then*

$$a_0\text{I} + a_1\text{II} + a_{n-1}\text{III} + a_n\text{IV} = 0.$$

The mode of verification suggested consists in showing that there exist six quantities (12), (13), (14), (23), (24), (34), say, such that

$$\begin{aligned} \text{I} &= \quad a_0\cdot 0 \quad + a_1(12) + a_{n-1}(13) + a_n(14), \\ \text{II} &= -a_0(12) + a_1\cdot 0 \quad + a_{n-1}(23) + a_n(24), \\ \text{III} &= -a_0(13) - a_1(23) + a_{n-1}\cdot 0 \quad + a_n(34), \\ \text{IV} &= -a_0(14) - a_1(24) - a_{n-1}(34) + a_n\cdot 0; \end{aligned}$$

and then taking the sum of the requisite multiples. The six quantities in question are actually found for the cases where $n=3, 4, 6$. In the last case, the matrix being

$$\begin{matrix} a & b & c & d & e & f & a' & b' & c' & d' & e' \\ b & c & d & e & f & g & b' & c' & d' & e' & f' \\ a' & b' & c' & d' & e' & f' & a'' & b'' & c'' & d'' & e'', \end{matrix}$$

their values are written by Cayley in the form

$$
\begin{aligned}
(12) &= -e''g && +f'f', \\
(13) &= b''f+c''e+d''d+e''c && -b'f'-c'e'-d'd'-e'c'-f'b', \\
(14) &= -b''e-c''d-d''c && +b'e'+c'd'+d'c'+e'b', \\
(23) &= -a''f-b''e-c''d-d''c-e''b && +a'f'+b'e'+c'd'+d'c'+e'b'+f'a' \\
(24) &= a''e+b''d+c''c+d''b && -a'e'-b'd'-c'c'-d'b'-e'a', \\
(34) &= -a''a && +a'a'.
\end{aligned}
$$

The final lemma used in the verification may be formulated thus:

If from the n *quantities* $x_1, x_2, \ldots, x_n$ *and the* $\frac{1}{2}n(n-1)$ *others*

$$
\begin{array}{cccc}
12, & 13, & \ldots, & 1n \\
 & 23, & \ldots, & 2n \\
 & & \cdot\ \cdot & \cdot\ \cdot \\
 & & & nn
\end{array}
$$

there be formed n *lineo-linear functions of the two sets, namely,*

$$
\left.
\begin{aligned}
f_1 &= \quad x_1\cdot 0 + x_2(12) + x_3(13) + \ldots + x_n(1n) \\
f_2 &= -x_1\cdot(12) + x_2\cdot 0 + x_3(23) + \ldots + x_n(2n) \\
f_3 &= -x_1(13) - x_2(23) + x_3\cdot 0 + \ldots + x_n(3n) \\
&\cdot\ \cdot\ \cdot\ \cdot\ \cdot\ \cdot\ \cdot\ \cdot\ \cdot\ \cdot\ \cdot\ \cdot \\
f_n &= -x_1(1n) - x_2(2n) - x_3(3n) - \ldots + x_n\cdot 0
\end{aligned}
\right\},
$$

then

$$x_1f_1+x_2f_2+\ldots+x_nf_n=0.^*$$

It may be viewed as included in the identity

$$
\begin{array}{ccccc|c}
x_1 & x_2 & x_3 & \ldots & x_n & \\
\hline
 & 12 & 13 & \ldots & 1n & x_1 \\
-12 & \cdot & 23 & \ldots & 2n & x_2 \\
-13 & -23 & \cdot & \ldots & 3n & x_3 \\
\cdot\ \cdot & \cdot\ \cdot & \cdot\ \cdot & \cdot\ \cdot & \cdot & \ldots \\
-1n & -2n & -3n & \ldots & \cdot & x_n
\end{array}
= 0;
$$

or in the statement that *Any quadric whose discriminant is a zero-axial skew determinant vanishes identically.*

---

* When the coefficient of x_r in f_r is not 0 but (rr), the result of course is

$$x_1f_1+x_2f_2+\ldots+x_nf_n = x_1^2(11)+x_2^2(22)+\ldots x_n^2(nn);$$

and in this connection it may be well to recall a step in Hermite's mode of effecting the automorphic transformation of a quadric (See under *Orthogonants*).

HIRST, T. A. (1859)

[Question 489. (A determinant which vanishes for every order higher than the fourth.) *Nouv. Annales de Math.*, xviii. p. 358; (2) ix. pp. 561–563.]

Hirst's theorem is that *if*

$$a_{rs} = (a_r + \beta_r s)\cos s\phi + (\gamma_r + \delta_r s)\sin s\phi$$

then the determinant

$$\begin{vmatrix} a_{1,s} & a_{1,s+1} & \cdots & a_{1,s+n-1} \\ a_{2,s} & a_{2,s+1} & \cdots & a_{2,s+n-1} \\ \cdot & \cdot & \cdot & \cdot \\ a_{n,s} & a_{n,s+1} & \cdots & a_{n,s+n-1} \end{vmatrix}$$

vanishes when $n > 4$, *and has a non-zero value independent of* s *when* n=4; and the real significance of it is best grasped by noting—as is not done in the *Annales*—that the determinant is the product

$$\left\|\begin{matrix} a_1 & \beta_1 & \gamma_1 & \delta_1 \\ a_2 & \beta_2 & \gamma_2 & \delta_2 \\ \cdot & \cdot & \cdot & \cdot \\ a_n & \beta_n & \gamma_n & \delta_n \end{matrix}\right\|$$

$$\cdot\left\|\begin{matrix} \cos s\phi & \cos(s+1)\phi & \cdots & \cos(s+n-1)\phi \\ s\cos s\phi & (s+1)\cos(s+1)\phi & \cdots & (s+n-1)\cos(s+n-1)\phi \\ \sin s\phi & \sin(s+1)\phi & \cdots & \sin(s+n-1)\phi \\ s\sin s\phi & (s+1)\sin(s+1)\phi & \cdots & (s+n-1)\sin(s+n-1)\phi \end{matrix}\right\|.$$

The vanishing of it when $n > 4$ is then self-evident, and its value when $n = 4$ being

$$|a_1\beta_2\gamma_3\delta_4| \cdot \begin{vmatrix} \cos s\phi & s\cos s\phi & \sin s\phi & s\sin s\phi \\ \cos(s+1)\phi & (s+1)\cos(s+1)\phi & \sin(s+1)\phi & (s+1)\sin(s+1)\phi \\ \cos(s+2)\phi & (s+2)\cos(s+2)\phi & \sin(s+2)\phi & (s+2)\sin(s+2)\phi \\ \cos(s+3)\phi & (s+3)\cos(s+3)\phi & \sin(s+3)\phi & (s+3)\sin(s+3)\phi \end{vmatrix},$$

we have only to show that the second determinant here is independent of s. The solver (Lucien Bignon) does this by multiplying the determinant by itself in the form

$$\begin{vmatrix} s\cos s\phi & -\cos s\phi & s\sin s\phi & -\sin s\phi \\ (s+1)\cos(s+1)\phi & -\cos(s+1)\phi & (s+1)\sin(s+1)\phi & -\sin(s+1)\phi \\ \cdot & \cdot & \cdot & \cdot \end{vmatrix},$$

and so finding for its square a determinant whose every element is independent of s, the element in the place i,j being in fact

$$(j-i)\cos(j-i)\phi.$$

He does not note, however, that such a determinant is zero-axial and skew, and that its value is thus readily seen, by a theorem of Cayley's, to be

$$(\cos^2\phi - 4\cos^2 2\phi + 3\cos\phi\cos 3\phi)^2,$$

i.e.

$$(-4\sin^4\phi)^2.$$

CAYLEY, A. (1859).

[Note on the value of certain determinants, the terms of which are the squared distances of points in a plane or in space. *Quart. Journ. of Math.*, iii. pp. 275–277; or *Collected Math. Papers*, iv. pp. 460–462.]

The five results given in the paper are more important than the title would imply, being true when instead of Cayley's elements $\overline{12}^2$, $\overline{13}^2$, ... we write any elements whatever, namely, 12, 13, ... This change being made, the fourth and fifth are

$$\begin{vmatrix} \cdot & 12 & 13 & 14 \\ 21 & \cdot & 23 & 24 \\ 31 & 32 & \cdot & 33 \\ 41 & 42 & 43 & \cdot \end{vmatrix} = \sum 12\,21\,.\,34\,43 - \sum 12\,23\,24\,41,$$

$$\begin{vmatrix} \cdot & 12 & 13 & 14 & 15 \\ 21 & \cdot & 23 & 24 & 25 \\ 31 & 32 & \cdot & 34 & 35 \\ 41 & 42 & 43 & \cdot & 45 \\ 51 & 52 & 53 & 54 & \cdot \end{vmatrix} = \begin{cases} \sum 12\,23\,34\,45\,51 \\ -\sum 12\,23\,31\,.\,45\,54, \end{cases}$$

where the Σ's cover 3, 6, 24, 20 terms respectively. No commentary is added, nor any indication of a law including both results. The three of the other set are less general, namely,

$$\begin{vmatrix} . & 12 & 12 & 1 \\ 21 & . & 23 & 1 \\ 31 & 32 & . & 1 \\ 1 & 1 & 1 & . \end{vmatrix} = \sum 12\,21 - \sum 12\,23,$$

where the Σ's cover 3 and 6 terms respectively;

$$\begin{vmatrix} . & 12 & 13 & 14 & 1 \\ 21 & . & 23 & 24 & 1 \\ 31 & 32 & . & 34 & 1 \\ 41 & 42 & 43 & . & 1 \\ 1 & 1 & 1 & 1 & . \end{vmatrix} = \sum 12\,23\,34 - \sum 12\,34\,43 - \sum 12\,23\,31,$$

where the Σ's cover 24, 12, 8, terms respectively;

$$\begin{vmatrix} . & 12 & 13 & 14 & 15 & 1 \\ 21 & . & 23 & 24 & 25 & 1 \\ 31 & 32 & . & 34 & 35 & 1 \\ 41 & 42 & 43 & . & 45 & 1 \\ 51 & 52 & 53 & 54 & . & 1 \\ 1 & 1 & 1 & 1 & 1 & . \end{vmatrix} = \begin{cases} -\sum 12\,23\,34\,45 \quad -\sum 12\,21\,.\,34\,43 \\ +\sum 12\,23\,.\,45\,54 + \sum 12\,23\,34\,41 \\ +\sum 12\,23\,31\,.\,45, \end{cases}$$

where the Σ's cover 120, 15, 60, 30, 40 terms respectively. Here again no generalisation is attempted.

(ϵ) CENSUS OF TERMS IN SPECIAL DETERMINANTS.

The first instance of the finding of the number of terms in the final development of a determinant of special form has already been drawn attention to, the investigator being Scherk, and the date 1825.

During the period now occupying us, the earliest suggestion on the subject occurs in 1844 in *Crelle's Journ.*, xxviii. pp. 191–192, the determinant being of the 8th order, and the specialisation consisting in having a zero in the places

16, 17, 18, 27, 28, 31, 38, 41, 42
83, 82, 81, 72, 71, 68, 61, 58, 57.

The proposer of the problem, so far as it appears, received no satisfaction. The next instance occurred to Sylvester, who in 1855 having hit upon a peculiar determinant whose terms were all positive, ascertained the number of them by evaluating a special circulant. (See *Quart. Journ. of Math.*, i. pp. 42–56, or our notice of it given above, pp. 406–407.) The third instance, like the first, arose as a problem and remained long unsolved. It appeared in 1858 in the *Nouv. Annales de Math.*, xvii. p. 262, under the heading "Question 445," the requirement being to find the number of terms remaining in the case of a determinant of the nth order when all those terms have been deleted which contain two or more diagonal elements. The fourth instance —which is more closely connected with the third than might at first appear—was the actual but incidental determination by Cayley in 1859 of the number of terms in a zero-axial determinant whose order is not greater than the 7th. The numbers found were 9, 44, 265, for the 4th, 5th, 6th orders respectively. (See *Quart. Journ. of Math.*, iii. pp. 275–277, or our notice of it given above, pp. 469–470.)

LIST OF AUTHORS WHOSE WRITINGS ARE REPORTED ON.

GLASGOW: PRINTED AT THE UNIVERSITY PRESS BY ROBERT MACLEHOSE AND CO. LTD.

www.ingramcontent.com/pod-product-compliance
Lightning Source LLC
LaVergne TN
LVHW011300110826
845149LV00001B/198

* 9 7 8 1 4 1 8 1 8 4 7 1 1 *